THE FORESTS HANDBOOK
Volume 2

The Forests Handbook

VOLUME 2

APPLYING FOREST SCIENCE
FOR SUSTAINABLE MANAGEMENT

EDITED BY

JULIAN EVANS

OBE BSc, PhD, DSc, FICFor

T.H. Huxley School of Environment, Earth Sciences and Engineering,
Imperial College of Science, Technology and Medicine, University of London, United Kingdom

**Blackwell
Science**

© 2001 by
Blackwell Science Ltd
Editorial Offices:
Osney Mead, Oxford OX2 0EL
25 John Street, London WC1N 2BL
23 Ainslie Place, Edinburgh EH3 6AJ
350 Main Street, Malden
  MA 02148 5018, USA
54 University Street, Carlton
  Victoria 3053, Australia
10, rue Casimir Delavigne
  75006 Paris, France

Other Editorial Offices:
Blackwell Wissenschafts-Verlag
  GmbH
Kurfürstendamm 57
10707 Berlin, Germany

Blackwell Science KK
MG Kodenmacho Building
7–10 Kodenmacho Nihombashi
Chuo-ku, Tokyo 104, Japan

First published 2001

Set by Best-set Typesetter Ltd.,
Hong Kong
Printed and bound in Great Britain
at MPG Books Ltd, Bodmin,
Cornwall

The Blackwell Science logo is a
trade mark of Blackwell Science Ltd,
registered at the United Kingdom
Trade Marks Registry

A catalogue record for this title
is available from the British Library

ISBN 0-632-04821-2 (vol. 1)
    0-632-04823-9 (vol. 2)
    0-632-04818-2 (set)

Library of Congress
Cataloging-in-Publication Data

The forests handbook / edited by
  Julian Evans.
      p.   cm.
   Includes bibliographical
references.
   Contents: v. 1. An overview of
forest science—v. 2. Applying
forest science for sustainable
management.
   ISBN 0-632-04821-2 (v. 1)—
    ISBN 0-632-04823-9 (v. 2)
   1. Forests and forestry.
I. Julian Evans.
SD373 .F65   2000
333.75—dc21   00-021516

DISTRIBUTORS
Marston Book Services Ltd
PO Box 269
Abingdon, Oxon OX14 4YN
(Orders: Tel: 01235 465500
         Fax: 01235 465555)

USA
Blackwell Science, Inc.
Commerce Place
350 Main Street
Malden, MA 02148 5018
(Orders: Tel: 800 759 6102
                781 388 8250
          Fax: 781 388 8255)

Canada
Login Brothers Book Company
324 Saulteaux Crescent
Winnipeg, Manitoba R3J 3T2
(Orders: Tel: 204 837-2987)

Australia
Blackwell Science Pty Ltd
54 University Street
Carlton, Victoria 3053
(Orders: Tel: 3 9347 0300
          Fax: 3 9347 5001)

For further information on
Blackwell Science, visit our website:
www.blackwell-science.com

Contents

List of Contributors

H. LEE ALLEN *Department of Forestry, North Carolina State University, Raleigh, NC 27695-8008, USA* *Lee_Allen@ncsu.edu*

SIMMATHIRI APPANAH *FORSPA, FAO Regional Office for Asia and the Pacific 39 Phra Atit Road, Bangkok 10200, Thailand* *Simmathiri.appanah@fao.org*

PETER M. ATTIWILL *School of Botany, University of Melbourne, Victoria 3010, Australia* *attiwill@unimelb.edu.au*

STEPHEN BASS *International Institute for Environment and Development, 3 Endsleigh Street, London WC1H 0DD, UK* *steve.bass@iied.org*

TIMOTHY J.B. BOYLE *UNDP, 1, UN Plaza, New York, USA* *tim.boyle@undp.org*

PETER BRANG *Swiss Federal Institute for Forest Snow and Landscape Research, Zuercherstrasse 111, CH-8903 Birmensdorf, Switzerland* *brang@wsl.ch*

HUGH F. EVANS *Forest Research, Alice Holt Lodge, Wrecclesham, Farnham, Surrey GU10 4LH, UK* *h.evans@forestry.gov.uk*

JULIAN EVANS *T.H. Huxley School of Environment, Earth Sciences and Engineering, Imperial College of Science, Technology and Medicine, Silwood Park, Ascot, Berkshire SL5 7PY, UK* *julian.evans@ic.ac.uk*

JANE M. FEWINGS *School of Botany, University of Melbourne, Victoria 3010 Australia* *j.fewings@botany.unimelb.edu.au*

BARRY GARDNER *Forest Research, Northern Research Station, Roslin, Midlothian EH25 9SU, UK* *b.gardner@forestry.gov.uk*

JOHN N. GIBBS *Forest Research, Alice Holt Lodge, Wrecclesham, Farnham, Surrey GU10 4LH, UK* *j.gibbs@forestry.gov.uk*

GRAHAM GILL *Forest Enterprise, Kielder Forest District, Eals Burn, Bellingham, Hexham, Northumberland NE48 2AJ, UK* *graham.gill@forestry.gov.uk*

ROGER GOOD *NSW National Parks and Wildlife Service, PO Box 2115, Queanbeyan 2620, Australia* *roger.good@npws.nsw.gov.au*

BOB McINTOSH *Forest Enterprise, 231 Corstorphine Road, Edinburgh EH12 7AT, UK* *r.mcintosh@forestry.gov.uk*

PETA MASSON *SAPPI-Usutu, Private Bag, Mbabane, Swaziland*
peta@iafrica.sz

LAWRENCE A. MORRIS *Daniel B. Warnell School of Forest Resources, University of Georgia, Athens, GA 30602-2152, USA*
morris@smokey.forestry.uga.edu

KJELL NILSSON *Danish Forest and Landscape Research Institute, Horsholm Kongevej 11, 2970 Horsholm, Denmark*
kjn@fsl.dk

ERNST F. OTT *Im Später 20, CH-8906 Borstetten, Switzerland*
ott@wsl.ch

ROBERT F. POWERS *US Forest Service Pacific Southwest Research Station, 2400 Washington, Avenue, Redding, CA 96001, USA*
rpowers@c-zone.net

CINDY PRESCOTT *Faculty of Forestry, University of British Columbia, 3041-2424 Main Mall, Vancouver, British Columbia V6T 1Z4, Canada*
cprescott@interchg.ubc.ca

THOMAS B. RANDRUP *Danish Forest and Landscape Research Institute, Horsholm Kongevej 11, 2970 Horsholm, Denmark*
tr@fsl.dk

SIMON RIETBERGEN *IUCN—World Conservation Union, 28 rue Mauverney, CH-1196, Gland, Switzerland*
SPR@hq.iucn.org

DAVID ROSE *Forest Research, Alice Holt Lodge, Wrecclesham, Farnham, Surrey GU10 4LH, UK*
d.rose@forestry.gov.uk

NARESH C. SAXENA *Department of Rural Development, Ministry of Rural Areas and Employment, 197 Krishi Bhavan, New Delhi–110001, India*
secyrd@parlis.nic.in

WALTER SCHÖNENBERGER *Swiss Federal Institute for Forest Snow and Landscape Research, Zuercherstrasse 111, CH-8903 Birmensdorf, Switzerland*
schönenberger@wsl.ch

BARBARA L.M. WANDALL *Danish Forest and Landscape Research Institute, Horsholm Kongevej 11, 2970 Horsholm, Denmark*
blmw@fsl.dk

GORDON WEETMAN *Department of Forest Sciences, University of British Columbia, 3041-2424 Main Mall, Vancouver, British Columbia V6T 1Z4, Canada*
gweetman@interchg.ubc.ca

Preface to Volume 1

The Forests Handbook joins an eminent series of Blackwell's 'Handbooks'. However, the connotation of 'Handbook', for those brought up in the English tradition, is that of a practical manual telling the reader how to do things. This 'handbook' is not like that at all. What we have done in these two volumes is to assemble a unique compilation, and I employ that overused adjective advisedly. It is unique in the sense of bringing together eminent foresters, biologists, ecologists, scientists, academics and managers from many countries to tell a story that embraces much of the span of what we call 'forest science' though not all of operational forestry.

Volume 1 seeks to present an overview of the world's forests from what we know about where they occur and what they are like, about the way they function as complex ecosystems, as 'organisms' interacting with their environment, and about their interface with people—at least in part. We have tried to present the state-of-the-art science, and to present it in a way accessible to the general reader with an interest in trees, forests and forestry. It is, after all, sound science that underpins sustainable forest management.

Volume 2 seeks to apply this science to good practice. It is focused on operations and their impacts, on principles governing how to protect forests and on how to harness in sustainable ways the enormous benefits forests confer in addition to supply of wood. Volume 2 also contains valuable and highly informative case studies drawn from several countries to illustrate key points and show good management in practice.

In only two volumes one cannot cover everything, nor everything in detail and in depth. What is presented is a series of overviews of interrelated topics written by highly competent authorities. To achieve this authors were given considerable latitude to interpret their topic. In effect, although chapters are linked logically they do not necessarily trace a coherent flowing story, but are more a suite of essays that move through their subjects as the authors have chosen to address them. The value of this approach is to allow development of topics and themes by those best placed to write about them in ways that are largely unfettered.

While it is impossible to claim worldwide coverage, there is a deliberate attempt to make the text relevant across the globe and to draw on examples, cite practices and relate experience from a great many countries. Omissions are largely what might be termed 'forestry' such as forest economics, forest logging and harvesting operations, the whole matter of utilization and trade in forest products, and there is only limited coverage of forest fires and the pressing problem of tropical deforestation *per se*. There is insufficient treatment of social and policy-related factors in forest development. And, of course, not all the riches in the world's forest types can be mentioned. Indeed, apart from Chapter 2 in Volume 1, authors do not treat their subjects by taxonomic groupings but take a more functional approach. In contrast to perceived omissions, several chapters deal with aspects of ecophysiology and forest–environment interaction and some three chapters address the important subject of forest soils and/or their management. Foresters and forest scientists are notorious for neglecting this aspect of the ecosystem and the emphasis on soil is deliberate and welcome even if in a few places there is some repetition as the different authors provide their own perspectives. We recognize all these imbalances and ask for readers' forbearance. While coverage may not be comprehensive, we hope anyone interested in the world's forests will find much here to welcome. We hope, too, that it helps forward the goal of sustainable forest management

But why have a Forests Handbook? We believe that nowhere else will such a broad sweep of modern forest science and its application be found in two books, and in two books that bring together 45 authors from 12 countries to present their own perspectives. It was a monumental task, but worth it.

Julian Evans
April 2000

Preface to Volume 2

Volume 2 seeks to apply the science presented in Volume 1 to good practice as the foundation for sustainable forest management. It focuses on recognizing the enormous benefits forests confer in addition to the supply of wood, how to sustain these forest influences, on operations and their impacts on sustainable wood production, and on principles governing how to protect forests. But these environmental and biological imperatives cannot be viewed apart from the people who enjoy the benefits forests confer, and three chapters explore this tree/forest–people interface. Volume 2 also contains the valuable feature of six case studies drawn from several different countries to illustrate key points and show genuine progress towards the goal of sustainable forest management in practice.

However, Volume 2 begins with a survey of the history and impact of forest management. Too easily we forget all that previous generations have learnt, too easily we erect myths and hold to half-truths. The opening chapter addresses just such issues.

Julian Evans
June 2000

Acknowledgements

I am greatly indebted to Douglas Malcolm and Larry Morris for taking on the role of editorial advisers. Not only did they guide development of the book itself but in many instances undertook the arduous task of refereeing chapters. Thanks expressed here cannot do justice to the significance of their contribution to *The Forests Handbook*. That said, I must accept responsibilities for any mistakes, inaccuracies or omissions in the text.

A book of this type is primarily judged by the quality and diligence of its many authors, not of the editor. I am full of admiration for what the authors collectively have achieved and I record my thanks without reservation for all their hard work and dedication. We even kept close to the set deadlines!

I must also thank Claire Holmes of Forest Research (UK) and Delia Sandford of Blackwell Science who successively fulfilled that crucial role of editorial assistant. Their persistence in chasing up authors and keeping a check on this sizable undertaking was enormously appreciated. Added to this, Blackwell Science's editors were supportive throughout: Susan Sternberg and subsequently Delia herself.

I would also like thank Gus Hellier, who helped with chapter formatting and undertook the tedious task of checking references for me.

Finally, I am grateful to Imperial College of Science, Technology and Medicine for enabling me to continue this project during the two years since joining their staff.

1: The History and Impact of Forest Management

SIMON RIETBERGEN

1.1 INTRODUCTION

Why should *The Forests Handbook* have a chapter dedicated to the history of forest management? Even if we admit that history is an intellectually stimulating subject, a worthy pass-time for academics and interested lay people, the question remains: what practical value has it for the management of present-day forests? There are no doubt many answers to this question, but possibly most pertinent is the one given by Oliver Rackham (1990): 'Forestry is an art in which, because of the long time-scale, failures tend to be forgotten, not learnt from, and thus later repeated.' So, knowledge of the history of forests and their management may help human societies to avoid repeating such failures—rather than having to live with the consequences of poor decisions yet again. In addition, improved knowledge and interpretation of the history of forests and their management can help forest managers to increase visitor interest and enjoyment.

Educating the general public about forests and their history is essential in an age in which rural pursuits such as forest management and farming are increasingly poorly understood by largely urban electorates. As Rackham has highlighted, these electorates have been fed mainly pseudo-history, made up of 'factoids' and having no connection with the real world. As a consequence, they can be easily misled by decision-makers, for example where timber harvesting decisions, environmental grants and forestry development aid are concerned.

1.1.1 Environmentalist pseudo-histories

There are various kinds of pseudo-histories about forests and their management. Among the more prominent are the environmentalist and forestry professional pseudo-histories. The environmentalist versions are histories of wood use that are passed off as a history of woodlands and their non-management. Such wood-use histories are based on two assumptions: one is that most wood products were derived from unmanaged forests and the other is that harvesting wood from forests leads to deforestation as a rule. Both of these assumptions may hold true for certain areas and periods, but are patently untrue when generalized to the overall history of humanity worldwide.

The well-known environmentalist and author John Perlin (1989), in his otherwise delightful book *A Forest Journey: The Role of Wood in the Development of Civilization*, clearly falls into this trap. Yet even the illustrations in his own work tell a different story. His original caption to the fifteenth-century woodcut depicted in Fig. 1.1 refers to people carrying pieces of wood home from the forest. But closer inspection reveals that the head-loads involved consist of neatly stacked, uniform rods, such as one would harvest from a managed coppice woodland or from trimming a hedgerow.

Outsized timber used for prestigious construction projects and fleet expansion often derived from further afield, no doubt in many cases from overharvesting old-growth forests. If such timber-harvesting operations were carried out in hilly terrain, they led to forest and soil degradation, and maybe in some cases even outright deforestation. But how common were such occurrences really?

Fig. 1.1 A misinterpreted woodcut? This fifteenth-century woodcut depicts three people carrying home from the forest pieces of wood with which to cook and heat their homes. (From Perlin 1989 with permission.)

And how important were the forest areas affected, given the fact that wood is a high-bulk low-value product that could not be transported very far over land economically? Detailed research carried out in many developing countries indicates that most of the wood used on a daily basis by people derives not from forests, but from trees in the agricultural landscape: live fences and hedgerows, scattered trees on farms, small coppice woodlots, etc. Needless to say, such resources are extremely valuable and therefore under secure private or communal ownership, and without exception intensively managed. There are strong indications that this was the same in regions that were densely populated early on in history, such as parts of China and Europe. In my own country, the Netherlands, the nobility sold the right to plant trees alongside roads and canals to tenant farmers as early as the sixteenth century, and much of the wood used locally was produced in this way.

1.1.2 Forestry profession pseudo-history

Foresters have what they like to call their professional history, although as with many professions, it is really more a myth of origin, comprising both true and invented elements. In outline, this pseudo-history goes as follows. First there were forests almost everywhere. Then shortsighted and ignorant people, more often than not poor peasants, destroyed almost all of these forests. Subsequently, foresters fought a heroic battle to protect the remaining forests against these people and re-green the wastelands—thus saving human societies from almost certain doom. As we shall see below, reality is much more complex than this.

1.1.3 Definition of forest management

Many publications purporting to chart the history of forest management are really histories of silviculture, and, narrowing down even further, silviculture as practised by recognized forestry professionals since the eighteenth or nineteenth century. Although this is a fascinating field that has given rise to some real classics (e.g. Dawkins & Philip 1998 on the tropics), it only deals with a small part of the purposeful activities people have undertaken with regards to forests over the centuries in many different parts of the world. Therefore, we would like to define forest management for the purposes of the present chapter in a much wider sense.

With Menzies (1995) this chapter defines a forest management system as a 'set of rules and techniques that people devise to maintain forested land in a desired condition, including the processes through which the rules and techniques are adapted to deal with changing circumstances'. At a minimum, management activities would include enforcing boundaries, setting yield levels for forest product harvesting and controlling them, and allocating the benefits and costs derived from the resource. Such a forest-management system is said to be sustainable when it is able to adapt to changes in the pressures acting on it, and to maintain the desired forest condition over time. It is important to note that this definition does not imply that full-time forestry professionals or written management plans are involved. Indeed, many local management systems are based on unwritten rules applied by farmers and herders, rather than on plans implemented by forestry professionals.

1.1.4 Forest management typology

Forest management systems can be typified by three attributes: market orientation, scale, and locus of decision-making. Each of these three attributes can be located on a continuum: from solely subsistence or no use to commercial market production; from smallholder woodlots to large forest management units, for example industrial timber concessions or large forest protected areas; and from local to distant decision-makers (Menzies 1995). Over time, there has been a shift away from small-scale, subsistence-orientated, locally controlled forest management towards regimes that are larger-scale, market (and, more recently, preservation) orientated, and with control vested in remote authorities. In the latter case, management is often caried out by professional foresters, whereas smaller-scale, locally controlled systems that mix market and subsistence values are routinely managed by non-professionals.

1.2 STUDYING FOREST MANAGEMENT HISTORY

There is increasing evidence that human societies all over the world managed forests long before they started to write. This has two major implications. An obvious one is that there exists a prehistory of forest management. In other words, the study of written sources from the past does not suffice to describe the history of forest management and the interpretation of the archaeological record is crucial to shed light on the subject.

Archaeological finds often surprise us through their sophistication. Exceptional ones frequently push the dates for technologies that were thought to be more recent way back into the past. Woodworking and woodsmanship are no exceptions to this. Well-known examples of such finds are Otzi the iceman (see below) and the Fenland trackways.

1.2.1 Evidence for woodland management in prehistoric Britain

In order to cross soft peat in the Somerset Levels during the Neolithic period, people built many trackways (wooden walkways), some of which were entombed and preserved by the subsequent growth of the peat. The earliest example that has been found, the Sweet Track, thought to date from 3900 BC, is also the most elaborate and sophisticated. It contains poles of oak, ash, lime, hazel, alder and holly, of different sizes and selected for particular functions in the structure. Rods, poles and even most timber are likely to have been harvested from mixed coppice, i.e. managed woodland, rather than from wildwood. (The latter would have yielded occasional giant trees used for making dugout boats—but for most purposes such huge timbers were impractical given the effort needed to cut them up and transport them.)

The Sweet Track and other similar finds have tremendous implications for how we think about the way in which Neolithic people related to forests. With the spread of agriculture, parts of the original wildwood would certainly have been grubbed out and converted into arable land or pasture. But others were turned into managed woodland, coppiced at frequent intervals to yield the variety of underwood products that constituted the bulk of the material culture of that age: wattle for hurdles used in fencing and house and road building; fuel for cooking, heating, metalworking and pottery. The occasional standard would be left to grow for several coppice rotations to fulfil the more occasional need for structural timber (Rackham 1990).

The second implication is perhaps less obvious. If forest management originated early, before communications were made more frequent and reliable by written records and long-distance transport facilities, it must have been 'invented' independently many times over—as was agriculture (Fagan 1995). This implies there is not a single history of forest management, but rather many different histories—conditioned by different biophysical and sociocultural and economic contexts in different localities.

1.2.2 Recent advances in forest history

Until about two decades ago, the history of forests and their management has largely been written by the forestry profession—which has thereby sat in judgement of a case in which it is also a party. Thankfully, this situation has changed over the past few decades, due to a number of trends both in the history and forestry professions.

In the forestry profession, initially unsuccessful forays into aid projects in developing countries have led to a reassessment of the relationship between rural people and forest resources worldwide. In-depth work by foresters, anthropologists and ecologists on 'traditional' farming and forest management systems in many areas of the world has demonstrated the validity and adaptability of these systems—and has cast renewed doubt on the presumed ignorance of the peasants whom the forestry professionals replaced as forest managers in developed countries.

Among academic historians, there has been a move from writings concentrating on the lives of important people to those focusing on the lives of common people: peasants, tradesmen, etc. (Le Roy Ladurie 1979). This has allowed a critical reassessment of histories reflecting mainly the ideology of the dominant social classes. It is not surprising that this shift in perspective has been particularly fertile in the case of forests, given the contested nature of forest resources and their management through the ages. This trend has been accompanied by an increasing emphasis on quantification. New or improved research techniques such as the analysis of pollen diagrams, deep-sea sediment cores and carbon dating have allowed historians to address questions they had been unable to tackle (Box 1.1). For example until recently, pollen analysis gave only a broad-scale picture of forest changes over centuries. Recent fine-scale resolution techniques applied to small hollow sediments have allowed researchers to distinguish shorter-term trends within individual stands (Rackham 1990).

1.2.3 Problems in finding and interpreting the evidence

In some regions, there is a multitude of written evidence regarding forest history and past management practices. Records of land transactions, woodland boundaries, court cases, time series of market prices for forest products going back more than a thousand years are available, for example for parts of Europe and in China. But in other regions, and for earlier times, written sources on the history of forests and their management are absent and archaeological research has to provide the bulk of the evidence.

As with most historical evidence, written or unwritten, the results generated by the archaeologists' new toolbox remain open to interpretation. Two common problems in interpreting archaeological finds are that organic matter is only rarely preserved, and that where it is, it is often difficult to interpret the finds in terms of the contemporary landscape.

What archaeologists usually find is only a very small part of the material culture of the people involved, for example stone and bone tools are usually preserved, but wood and bark (rope) rarely are. Parts of the diet can be easily identified, especially bones of animals, whereas vegetable materials such as fruits are often decomposed beyond recognition. Archaeological research has therefore often concentrated on wet environments such as lakeside settlements, where conditions for the preservation of organic material are more favourable. But the resulting gain in tangible evidence may give rise to other sources of bias: for example population densities may have been much higher in such areas than elsewhere, and people without easy access to lake or river fisheries may have had quite different hunting and gathering, and farming and forest management practices.

The next level of complexity is due to the problems of interpreting what is found in terms of what the surrounding landscape looked like and how it was used or managed. For example, a relative decrease of shade-tolerant tree species in the pollen record is routinely attributed to the spread of farming, although the purposeful use of fire to facilitate hunting would similarly favour light-demanding over shade-tolerant trees. The incessant crop-breeding efforts of Neolithic farmers have enabled prehistorians to distinguish the seeds and pollen of cereals from their wild ances-

Box 1.1 Wildfire or intentional combustion? Prehistoric humanity's fingerprints in the ocean

Over the past one million years, forest cover receded during the ice ages, known to scientists as glacial periods, while regaining lost ground during the so-called interglacial periods in between, when the climate became both warmer and wetter. During the ascendancy of the human species over roughly the same era, there was a period when the overlay of natural variation with human intervention created a grey area: which forest cover changes were due to nature and which to culture?

The potency of fire as an agent of environmental change has led researchers to speculate that prehistoric humans may have been actively engaged in modifying vegetation patterns through their burning practices. It has been suggested that in Africa, prehistoric human use of fire, in combination with the domestication of livestock, has led to the expansion of the Sahel in the past few thousand years. But other observers favour natural mechanisms to explain the advance of the desert.

Marine sediments 'downwind' of continents accumulate wind-blown debris derived from woodland and steppe fires. Sediments from the deep sea are particularly suitable for establishing long-term trends in fire incidence, as they provide relatively undisturbed, continuous records and are readily dated. Sediment samples need to be taken sufficiently far from the coast to ensure the 'signal' is integrated over a large continental area and therefore not sensitive to latitudinal changes in vegetation types in response to climate changes/fluctuations. In order to establish a million-year record of fire incidence in sub-Saharan Africa, a core of 21 m of deep-sea sediment taken from the Sierra Leone rise was analysed. The sediment is a muddy ooze consisting of the remains of foraminifers, a marine invertebrate species, amorphous silica and wind dust—the latter containing elemental carbon from the burning of terrestrial biomass.

Analysis of the organic carbon content of this sediment gives rise to two major conclusions. One is that over the past 400 000 years, the highest intensity of vegetation fires occurred during periods when the global climate was changing from interglacial to glacial mode. This is to be expected, as during such transitions the climate would have become increasingly variable, leading to the build-up of fuel loads during wet periods followed by intense biomass burning in subsequent dry periods. Neither interglacials, thought to have been characterized by warm, wet conditions and the northward expansion of forest over savannah regions, nor glacials, characterized by comparatively cool, dry conditions and the southward expansion of desert, would have been periods of high fire incidence. There is one telling exception to this pattern: the occurrence of a peak in elemental carbon abundance in sediments from the present interglacial period. This seems to suggest that humans in sub-Saharan Africa have actively shaped fire regimes at least since the last 10 000 years (Bird and Cali 1998).

The second conclusion of the analysis is that fire was not a common phenomenon in sub-Saharan Africa before 400 000 years ago. Is the increased fire incidence after this date a reflection of the taming of fire by populations of *Homo erectus* and *H. sapiens*? We may never know the answer to this question.

tors. But as most trees used by humanity for a variety of purposes were never truly domesticated, archaeologists interested in forest management cannot draw a similar line between wildwood and managed woodland. Often, triangulation of different types of evidence is needed to 'read' the prehistoric landscape (Box 1.2).

Problems of interpretation of historical evidence are not restricted to the archaeological record. Written sources need to be interpreted with similar care (Table 1.1).

1.3 THE HISTORY OF FOREST COVER

The history of forest management is a subset of the history of forest cover. The latter would seem to be less equivocal, but is nevertheless a matter of great controversy. The pendulum has swung all

---

Box 1.2 Reading the prehistoric landscape: the changing nature of forest cover in the Lake Constance region during the Stone and Bronze Ages

Manfred Roesch (1996) gathered botanical on- and off-site data in the western Lake Constance region in south-western Germany. Off-site data consisted of high-resolution pollen diagrams derived from naturally accumulating deposits such as lake sediment and peat, while on-site data were based on pollen gathered from cultural layers, usually revealed in the course of archaeological excavation, at prehistoric lake-shore settlements. The data show large differences between the Neolithic and Bronze Age as regards forest composition, crops and crop weeds, and charcoal inputs. These differences are explained in terms of changes in farming systems and hence cultural landscapes.

In the late Neolithic, from 4300 BC onwards, there was a significant decline in tall, long-lived shade-tolerant trees such as beech, elm and lime. Because according to on-site wood analyses these species seem to have been little used for wood, the practice of shifting cultivation involving slash-and-burn is the most likely explanation for this decline. When the arable fields were abandoned, vigorous redevelopment of coppice of species such as hazel and birch prevented the development of both shade-tolerant trees and typical arable weed communities, resulting in a landscape largely dominated by tall shrubs and other plants of forest margins and clearings.

In the Bronze Age, similar to the early medieval period, forests were substituted to a certain degree by more or less permanent pastures and arable fields with only short fallow phases, and remains of crop weeds and fallow plants predominate. The forests that remained showed little change in composition, except for a rise in oak, suggesting forest pasture and/or a forest economy based on coppicing.

---

the way from a few centuries ago when even large-scale deforestation was no great cause for concern to most people, to the present day when the loss of any forest anywhere is deemed to be just that. It seems that the level of public concern has not been matched—or, as some observers would say, has been hindered—by a similar level of scientific accuracy.

There are no unequivocal estimates of past forest cover and even those of the present are controversial (see Fairhead & Leach 1998 for a well-documented example from Guinea). Problems include those of definition (what constitutes a forest?), measurement (especially of small-scale mosaics of farm fields and forests) and choice of baseline (human impact on forests dating tens and possibly hundreds of thousands of years back in some continents).

Recent research has shown that human impact on the extent and nature of forest cover dates back much further than was previously assumed, mainly through burning and hunting by pre-agricultural societies. One example of this is 'fire-stick farming', a shorthand for the hunting and gathering practices of Australian Aborigines, which goes back tens of thousands of years

(Diamond 1998). Similar practices in East Africa may be more than 100 000 years old (Burgess and Clark 2000).

The advent of agriculture was a watershed in humanity's impact on forest cover. But even if over the longer term human societies have converted large areas of forests and woodlands to farms and pasture, this conversion was rarely a gradual, unidirectional process. Deforestation has proceeded with stops and starts, and with important reversals both in recent and more ancient times (Box 1.3). Large population influxes, sudden economic booms and new technologies gave rise to periods of rapid deforestation in some places, whereas outmigration caused by socioeconomic marginalization, and more recently the generation of non-land-based wealth, led to prolonged periods of spontaneous forest regeneration in others. The remaining two sections in this part of the chapter discuss some of the lesser-known ways in which people have influenced forest cover.

1.3.1 Early human influence on forest cover: fire and hunting

More than 130 million years ago, broad-leaved

Table 1.1 Examples of bias and errors in written evidence for forest management history.

Source of information	Nature of evidence	Reasons for bias or errors	Reference
Myths, legends and other literature	Names of trees used for construction and other purposes	Variable use of names in different areas and over time	Meiggs 1982
Toll books and other (export) trade records	Volume and nature of forest products traded	Overemphasis on export trade versus local use Misinterpretation of changes in trade flows	Tossavainen 1996 Tossavainen 1996
Court books and other legal documentation	Forest-use rules and violations	Overemphasis on forests subject to higher jurisdictions (e.g. state and formal common property) Overestimate of extent to which rules were violated	Menzies 1995
Accounts written by conquerors and travellers	Nature and importance of forests in landscape	Misinterpretation by authors or later interpreters	Leach & Fairhead 1996 Baasen 1940
Forest histories	Quantitative estimates of tree product use	Equating tree use with loss of forest cover	Leach & Mearns 1988
Forest history written by professional foresters	Nature of non-professional management Superiority of professional management	Misinterpretation of management rules and objectives Self-justification by forestry profession	Veer 1980
Local case studies	Details on forest use and management	Overgeneralization of localized/time-bound studies	Perlin 1989

trees started to compete for space with the existing forest cover consisting of conifers, palms and tree ferns. From that time onwards until humans started exercising a decisive influence on the Earth's vegetation a few thousand (or tens of thousands of) years ago, the history of forest cover was mainly determined by cyclical changes in climate.

The invention of agriculture is often taken as a starting point to chart the influence of human societies on the earth's forest. However, prior to the spread of agriculture, humans already exerted significant impact on forest cover, through the use of fire and through hunting of large herbivore species.

The earliest way in which humans influenced forest cover was no doubt through the domestication of fire. Combustion has underpinned the technological ascent of the genus *Homo* (Bird & Cali 1998). Although the earliest traces of what may be humanly tamed fire encountered in South Africa and the Kenya Rift Valley date to about 1.6 million years ago, it is not until between 1 million and 700 000 years ago that fire is well documented in the archaeological record — in the remains of hearths found in temperate China.

Fire confers tremendous competitive advantage on human populations able to master it. It offers protection against predators and an easy way of hunting game, from large mammals down to rodents and even insects, fleeing from a line of flames. Where fire was used to make otherwise toxic vegetable foods edible, through roasting or parching in hot ashes, it allowed people to use a wider range of plants in their diet — and possibly conquer new habitats. The earliest human settlements of the colder latitudes of Europe and Asia

---

Box 1.3 More forested today than in the historical past?

North-eastern USA

At the turn of the century, north-eastern states such as Vermont and New Hampshire had about 10% forest cover, most of the original forest cover having been transformed into arable land and pasture by New England colonists. The most recent statistics show that these states now have over 80% forest cover, mainly due to spontaneous forest regeneration on abandoned farmland. Indeed, in the whole of the United States, 239 000 km^2 of cleared land is estimated to have reverted to forest in the period 1910–79 (Tuiner *et al.* 1990).

Congo Basin

Not so long ago, mainstream historians assumed that most of the migration routes of the so-called Western Bantu expansion, which populated much of the centre, east and south of the African continent, circumvented the Congo Basin forests. But recent research has cast doubt on the status of the Congo Basin as 'the last pristine rainforest area in Africa'. Fay (1997) carbon dated oil palm kernels

he sampled in 16 streams in the Nouabale-Ndoki National Park, a presently virtually uninhabited forest area in Northern Congo, to between 2340 (±90) and 990 (±80) BP. The enormous quantities of palm kernels he encountered demonstrate convincingly that oil palm, a tree originating from the West African coastal savannah, was extensively grown and utilized by farmers in Northern Congo. This, in combination with historical records, and ancient pottery found over large areas in the Congo Basin, indicates that many of the forest we like to call 'pristine' today are more likely to have regenerated on abandoned farm land over the past millennium and a half.

Many foresters I have worked with in Central Africa have remarked on how some of the major timber species such as sipo and sapelli (*Entandophragma utile* and *E. cylindricum* respectively) regenerate better in logged-over than in unlogged forests. This indicates that their present widespread distribution in the Congo Basin may well be due to earlier human forest clearing.

---

may have been made possible by the prior taming of fire by *Homo erectus* in Africa (Fagan 1995). Hunters would have intentionally fired vegetation in the dry season, to encourage new grass growth likely to attract the large herbivores that constituted their favoured prey animals. But even unintentional use would have had major impact during droughts (Goudsblom 1992).

Much prejudice exists about the alleged negative impact of fire in the forest environment. But fire has long been an indispensable forest management tool, both for non-professional and professional forest managers. Indian cultures in much of the North American continent used fire as a technology to change the environment for their own ends, and that inevitably entailed the destruction of certain plants and trees and the disruption of some animal habitats. But this technology was used, and used carefully, with a largely positive effect. Regular and controlled ground-burnings such as the Indians practised, whether in the prairie grasslands or the hardwood forests, increased the number and diversity of

species, the level of their populations, the amount of nutrients in the soils, the quantities of forbs and grasses, and the quality of available forage for all herbivores. In the forests regular annual fires averted the danger of hot-burning and damaging wildfires in the accumulated underbrush and stimulated the growth of such desired and fire-resistant hardwoods as chestnuts and oaks; in the prairies they held back the growth of forest cover altogether to promote the populations of bison and small-game animals and birds (Sale 1991).

Many large animals became extinct throughout the world at the end of the Pleistocene, but nowhere as drastically as in the Americas, where three-quarters of the genera of large mammals vanished abruptly at this time. The catastrophic extinction of ice age big-game animals in the Americas after 10 000 BC is often linked to the simultaneous arrival of sophisticated hunters, the so-called Clovis people, whom most archaeologists believe to have been the first humans to populate the New World. Many scientists believe climate change also played a role, especially the more pronounced seasonal contrasts in the temperate zone, which would have had a harder impact on the young of species that are born in small litters, after long gestation periods and at fixed times of the year—traits characteristic of larger mammals—many of which became extinct during the late Pleistocene.

1.3.2 Before farming: producing food by keeping forests in check?

Archaeologists increasingly believe that the transition from hunting and gathering to food production by farming and livestock raising was gradual in most cases. An interesting hypothesis about the origin of food production in south-eastern Europe stresses the role of postglacial forest regrowth in 'forcing' people to farm. As the climate started to get wetter and warmer after the last ice age, grasses like einkorn and barley became more abundant, growing in dense stands that supported sizeable Mesolithic foraging populations. When trees started to spread subsequently, grass stands were scattered, decreasing the amount of food available to humans. The latter reacted by ring-barking tree trunks or burning the forest, to clear space for the wild grasses to grow. Harvested plots would gradually revert to woodland, providing cover and browsing for deer, wild sheep and other animals hunted by human populations—until the next cycle of firing and grass growth. As sheep were domesticated, they could be used to keep woodland open for cereal growing. Thus, productive mosaics of different types of vegetation cover were formed (Fagan 1995).

Over time farming came to transform the forest landscape more profoundly than any human activity before or since. Many more forests have been converted to farming and other land uses than are now under management. But in other areas, farming was only a temporary affair, and forests regained lost ground. Famous examples are the ancient temples in Cambodia (Angkor Wat), Mexico (Maya) and Peru (Machu Picchu), which were overgrown by jungle at the time they were rediscovered. But this pattern is much more common than is generally known (Box 1.3).

1.4 FROM FOREST USE TO FOREST MANAGEMENT

1.4.1 Forest and tree use in prehistory

In 1991, the body of a man was found protruding from the ice in the Italian part of Tyrol, near the border with Austria. After some initial speculation about the identity of the person, involving a rambler who disappeared earlier this century and a sixteenth-century soldier among others, the body was carbon dated to 3300–3200 BC! The valid scientific designation of the find is 'Late Neolithic glacier corpse from the Hauslajoch, Municipality Schnalls (Senales), Autonomous Province Bolzano/South Tyrol, Italy', so I am sure the reader will forgive me for referring to him as the iceman (otzi) henceforth (Spindler 1995).

The find of this member of a late Neolithic farming community was unique not only because soft body tissues and clothes had been preserved in the ice but also because it was not a burial. Therefore, what was found on the body is likely to represent the man's day-to-day clothes and equipment rather than what his kinsmen thought he would need in the hereafter. Why should anyone interested in forest history wax lyrical over a corpse more than 5000 years old? The eloquence of the iceman's silence lies in his well-preserved equipment, which tells us much about forest and tree use in the Neolithic.

As the iceman's kit contained materials derived from 17 different tree species (Table 1.2), it can be safely assumed that Neolithic farmers had a sophisticated understanding of the uses of a wide range of tree species. This is hardly surprising, considering how closely their livelihoods depended on the use of a large variety of natural resources. Among the tree materials were found non-wood products such as birch tar, 'the all-purpose glue of prehistory'. But the iceman knew a thing or two about wood as well. First of all, he seems to insist that small is beautiful: most of the wood in his kit is less than 3 cm in diameter. There is a certain logic to this: thicker cords of hazel wood would have made his backpack frame too heavy and making arrow shafts out of thin cornel wood is much less time consuming than

Table 1.2 Tree species used by the iceman more than 5000 years ago.

Tree	Scientific name	Part used	Type of use
Alder	*Alnus viridis*	Wood	Fuel
Amelanchier	*Amelanchier ovalis*	Wood	Fuel
Ash	*Fraxinus excelsior*	Wood	Dagger handle
Birch	*Betula* sp.	Bark	Container
		Sap	Tar (glue)
Blackthorn	*Prunus spinosa*	Fruit	Food
Cornel/dogwood tree	*Cornus* sp.	Wood (shoot)	Arrow shaft
Elm	*Ulmus* sp.	Wood	Fuel
Hazel	*Corylus avellana*	Wood (stem)	U-frame of backpack, quiver bracing
Juniper	*Juniperus* sp.	Needles	?
Larch	*Larix decidua*	Wood	Boards of backpack, fuel
Lime	*Tilia* sp.	Wood (branch)	Retouching tool
		Bast	Cord, binding material
Norway maple	*Acer platanoides*	Leaves	Insulating material
Norway spruce	*Picea abies*	Wood	Fuel
		Needles	?
Pine	*Pinus* sp. (not *cembra*)	Wood	Fuel
Reticulate willow	*Salix reticulata* T.	Wood	Fuel
Wayfaring tree	*Viburnum lantana*	Wood (shoot)	Arrow shaft
Yew	*Taxus baccata*	Wood	Bow, axe-helve

trying to carve them out of big logs. Furthermore, unlike today's professional foresters, the iceman seems to have had a need for wood that was not straight. For example, his axe handle was made by the crafty use of a natural fork in a yew tree. Of course the trees that delivered such products don't look much like those in today's textbook production forests, with straight, thick trunks, that are free of branches at least halfway up the tree.

1.4.2 Why manage forests? Why not?

The iceman knew a lot about trees and what useful products could be made from them, but did he also know about forest and tree management? Before we can answer such questions, we have to go back to basics first: why and how does a forest come under management? First of all, the wooded component of the landscape must be recognized as a resource and, second, actions must be taken to control the way in which the resource is used, in order to maintain its usefulness to the users. The concept of scarcity implied by the use of the word 'resource' should be conceived not just in its economic sense, but also in terms of political and ethical needs.

Scarcity of forest or forest products might be a necessary condition for forest management, but it is not by itself a sufficient one. There are three main strategies other than forest management to adapt to scarcity, including importing forest products from elsewhere, substituting them with non-forest products or integrating trees in farming systems.

Mediterranean civilizations started importing timber as long as 4000 years ago; indeed timber scarcity seems to have been a motive not only in establishing trade relations but even in territorial conquest (Perlin 1989). The earliest recorded timber imports in Britain were from Scandinavia in 1230. Many countries the world over are net timber importers. Some of these countries have large populations and little forest, such as the Netherlands. Others need to import timber because most of their forests are not available for timber harvesting. Japan, for example, is the most forested OECD country after Finland with 68% forest cover, but most of this forest is on steep, inaccessible slopes.

Box 1.4 The trees before the forest: the importance of chestnuts as food in southern Europe

In the late Middle Ages in Europe, with growing populations and stagnating cereal yields, famine was never far away. Trees played an important role in food security. Chestnuts yielded flour that was used to make a biscuit known as 'tree bread' in the Cévennes and Corsica. In Italy, if grain was scarce the people also had to eat *pane di castagni e legume* (bread made of chestnuts and pulses). In Aquitaine (where they were called *ballotes*) and elsewhere, they often filled the role taken over by potatoes in the nineteenth century. People in southern countries relied on chestnuts to a larger degree than is usually thought. Charles V's major-domo, living with his master at Jarandilla near Yuste in the Castilian Estremadura, noted in 1556: 'It is the chestnuts that are good here, not the wheat, and what wheat there is is horribly expensive'. However, consumption of acorns and roots as in Dauphiné during the winters of 1674–76 was quite exceptional, and a symptom of terrible famine (Braudel 1991).

Were these chestnut trees grown in forests? Probably most of them were not. Chestnut trees in woodlands were more often than not coppiced for

Average consumption per capita in Piedmont (Italy) in about 1750 (Braudel 1991)

Crop	Amount in hectolitres	Percentage of total
wheat	0.94	35
rye	0.91	34
other grains	0.41	15
chestnuts	0.45	17
total	2.71	101 (not 100 due to rounding)

producing fence posts, and would not have yielded large quantities of nuts. But many trees scattered in the agricultural landscape (e.g. in wood pasture) would be left to grow or be pollarded and would have yielded copiously, given their favourable exposure and shape. Two miles from where I sit, in the foothills of the Swiss Jura, I encountered large numbers of chestnut trees during a recent forest hike. Closer inspection revealed that almost all of them had regrown from large stumps. Some of these may well date from the days when most of the hills were covered in pasture with scattered trees, rather than in vineyards and forests as is the case now.

Another common response to scarcity is to substitute forest products with non-forest products, for example coal for woodfuel, and fertilizer instead of green mulch (Gilmour in Arnold & Dewees 1997). Finally, trees can be integrated into farming systems in a variety of ways, for example scattered in fields and pastures, as hedgerows or live fences bordering fields, or as line plantings along roads and canals. This adaptation is especially common where most land is suitable for agriculture and population pressure is high (Box 1.4).

1.4.3 When does use become management?

If the seasons of husbandry be not interfered with, the grain will be more than can be eaten. If close nets are not allowed to enter the pools and ponds, the fishes and turtles will be more than can be consumed. If the axes and bills enter hills and forest (in the plains) only at the proper time, the wood will be more than can be used. When the grain and fish and turtles are more than can be eaten and there is more wood than can be used this enables the people to nourish their living and mourn for their dead, without any feeling against any. This condition, in which the people nourish their living and bury their dead without any feeling against any, is the first step of royal government.

Mencius 372–289 BC (in Legge 1894)

As for so many other subjects, China is a case apart where forest history is concerned. As can be seen from the above quotation (from Mencius 1970), written texts referring to controlled use of forests and other natural resources in China go back at least 2300 years. Did such controlled use—limiting the time allowed for timber harvesting—add up to forest management? According to this chapter's definition of forest management, it did. But whatever the definition one adopts, there is always a grey area where use shades into management, and where it may be hard to tell the two apart.

1.5 FOREST MANAGEMENT BY NON-PROFESSIONALS

1.5.1 Introduction

Over most of the history of civilization, forests were managed by non-professionals, if at all, and large-sized timber was not the most important product derived from woodmanship. Woodlands and trees were an integral part of farming systems, both as a source of leaf litter for maintaining permanent field farming systems and as a source of livestock feed, and of people's spiritual culture (Table 1.3).

The following case descriptions of forest management in the more or less remote past by non-professionals in different parts of the world give an idea of the tremendous range of practices, and of the *savoir-faire* of the managers involved.

1.5.2 Neolithic forest management in Central Europe

About 7500 years ago, farmers originating from the western Hungarian plain first migrated to the north and west, marking the start of the Neolithic period in Central Europe. They colonized dry and warm areas with less than 600 mm annual precipitation and about 8°C annual average temperature. Their culture is called Bandkeramik, because of the striking ornaments of band-like grooves on their pottery. The earliest Band-

keramik culture extended over a quite considerable area between the River Rhine and the western Ukraine, the northern Harz mountains and the foothills of the Alps. The find of a well clad with oak boards, carbon dated to around 5300 BC, shows that these farmers were competent wood workers but were they also able woodland managers?

The earliest Central European farmers left behind pits, ditches and postholes. An interdisciplinary research team analysed these features and their contents. The identification of charred plant remains (seeds, fruit and wood) from 10 settlements located in western Germany and Austria was carried out based on wet sieving more than 1000 samples with a total volume of almost 20 000 litres. This plant material was then interpreted to attempt a reconstruction of the sites' rural economy and environment.

The tree species assemblages derived from the charcoal analysis were remarkably similar for all 10 sites, and very much at odds with the natural species distribution in the surrounding woodlands, as derived from pollen analysis.

Most of the charcoal would have been burned intentionally as firewood. Neolithic long-houses of 30×7 m needed to be heated about half the year and required significant amounts of fuelwood. How did Neolithic people acquire firewood? The possibility that they used mainly wood refuse or dead wood collected from the nearby forest is unlikely, given the fact that the charcoal analysis referred to above points to much more selective use of a smaller range of tree species. In addition, the fungal hyphae that are expected to occur in wind-fallen wood are absent from the charcoal

Woody vegetation type	Dominant use/management activity
Sacred groves in high forest	Integral protection (ritual use only) and replanting
High forest	Extractive use (hunting, gathering, pig mast, leaf litter)
Secondary forest	Slash-and-burn shifting cultivation
Coppice stands	Coppicing for fuel, fodder and occasional timber
Hedgerows and fence rows	Coppicing and pollarding for fuel and fodder

Table 1.3 Woody biomass used and/or managed since prehistory.

samples. This would imply that the most likely source of firewood was the surrounding wildwood, maybe at times as a by-product of land clearing for farming.

The absence from the pollen analysis of any sign of big forest clearances, in combination with the fact that most species used for firewood occur at the edges of woods, in brush and in hedges, points to a rather different scenario. Firewood is likely to have been derived from hedges growing alongside fields. How did these hedges come into being? Land clearing by farmers causes a specific woodland community of light-demanding species to arise at the edge of the field. Subsequent land clearing would be carried out, leaving the woodland edge community intact for reasons discussed below. This would form the beginning of a hedge, which would need to be managed in order to prevent it from being overgrown by thorny species. Hedges are essential as live fences for farmers with free-roaming livestock, which can do tremendous damage to crops. But the practice of leaving some forest vegetation between fields is not only a feature of farming areas where livestock can do major damage to crops. Even in tropical moist forest areas, where there is often little livestock, the practice is frequent. Given the frequency of boundary disputes from the earliest days for which we have legal records until the present, one possible explanation for this practice is that it would reduce conflict between neighbouring farmers.

There is no doubt that early Neolithic woodmanship, such as pollarding, coppicing, shredding, etc., is very difficult to prove directly, if there is no preservation of wood or wooden structures under moist conditions. But all the indirect evidence seems to indicate that Neolithic wildwoods in Central Europe were converted to managed hedges and coppice woodlands, as well as to arable land.

1.5.3 Sacred groves in classical Eurasia

Up to the fourth century AD, pagan shrines from Britain to Syria were surrounded by carefully tended sacred groves, in towns as well as in the countryside. These woods, consisting of oaks in northern Lydia, cypresses in Cos and pines in the cults of Attis, attained a great age and maturity. They housed deer and dogs, cattle and snakes, and unusual birds—and on occasion escaped prisoners. At Lagina in south-western Asia, eunuchs and public slaves tended the noble trees of Hecate's great shrine, working under strict orders from the priests to replant any specimens which died. A text cut into the rock of the acropolis of Lindos, in Rhodes, in the third century, honoured Aglochartos, a priest of Athena, for feats of silviculture that were alleged to rival those of the gods of myth. To that bare, dry site, he had transplanted the goddess's grey green olives, 'decorating it with the scented branch', at his own expense.

The persistence of civic decrees concerned with preservation and replanting, and rules against illicit 'wooding' and gathering in these groves demonstrate how important they were in everyday life. Local magistrates were to extract fines from freeborn offenders and whip the slaves, one or two strokes being equivalent to every drachme of a free man's fine, reminding us that the offence must have been tempting, especially for someone short of fuel during an unusually harsh Mediterranean winter (Lane Fox 1988).

1.5.4 Beyond mere subsistence: the value of forests in early medieval Britain and late medieval Normandy

In medieval times, the place of woodland in the British countryside was well established. Woods producing underwood and timber were differentiated, as in earlier centuries, from the various categories of wood pasture. Although some woods had small common rights, they were usually pieces of private property. Most woods had well-defined boundaries, consisting of a bank and ditch combined with either a fence or a hedge. More often than not, woods were part of the farming estate of a lord of the manor, with tenants holding customary rights to limited quantities of forest produce for specific purposes, such as fuel, fencing and building materials.

Woodlands had great economic value, but timber was rarely the most important forest product. Underwood for a variety of materials and grazing and browsing for livestock were often more important in terms of cash income than agriculture. A mere hazel stool in medieval South Wales was worth 3.75 sheep. The high value of woodlands was also apparent from the fact that they became the subject of complicated lawsuits as early as 825 (Rackham 1990).

A detailed diary kept by Gilles de Gouberville, a sixteenth-century nobleman who managed a large estate in the 'bocage' of Normandy, provides an interesting insight into the relative income derived from arable and woodland. He had devoted much of his land to extensive grain growing, practising a triennial rotation of wheat–mixed fodder–improved fallow sowed with peas, followed by a long interval during which the land was rested for several years. But pork and everything connected with its production was top of the list of sources of cash income. Gouberville sold pannage, that is the right to graze acorns, in his forests to the villagers at extremely high rates: up to 50 livres a year, more than he got for all the grain he sold in the market. In addition, his own well-fed pigs were killed in winter, salted and sent as far away as Paris, fetching another 60–80 livres a year (LeRoy Ladurie 1979). (Ham of pigs fattened on acorn mast still fetch premium prices in present-day Spain: 5–10 times the price of normal ham.)

1.5.5 Intensive forest management by farmers in late nineteenth century China

Many of the forest management systems practised in imperial China in the seventeenth to nineteenth century are still current today. They vary from closely spaced, fast-growing trees within predominantly agricultural systems, to successional systems similar to taungya (intercropping trees and food in young plantations), to the complete enclosure and preservation of natural forest in imperial hunting reserves (Menzies 1995).

Parts of China became very densely populated early on, and farmers managed the forest areas they did not convert to agriculture intensively, both for wood and non-timber forest products. Extensive areas of forest in late nineteenth century upland Manchuria were carefully managed. Tree growth on private landholdings in the hills consisted of scattering pine seldom higher than 25 feet (8 m) with an understorey of

oak, standing at a height of 2–4 feet (60–120 cm) and stretching over the slopes much like a regular crop. Narrow strips of tree growth stretching directly up the slope were clear-cut every 3–5 or perhaps 10 years for sale as fuel and poles, although on some strips the forest growth had been allowed to stand undisturbed for 20 or more years. Even stumps and the large roots were dug and used for fuel. This extensive digging caused new trees to spring up quickly from scattered seed and from the roots, so that planting was not generally required. Little pine nurseries were maintained in suitable places throughout the woods, to beat up in places where the volunteer growth had not been sufficiently dense.

These upland forests yielded not only wood, but also large quantities of non-timber forest products. Under favourable conditions, mushrooms cultivated in such stands yielded up to 2.5 times the value of the wood produced. The oaks were repeatedly coppiced at short intervals (2–3 years) mainly for use as green manure, but also for rearing wild silkworms, which thrive on succulent young oak leaves.

Domestic silkworm rearing requires large amounts of mulberry tree leaves, ranging from 123 to 164 pounds (56–75 kg) of leaves for 1 pound (0.45 kg) of silk. In southern China, mulberry trees were commonly grown from low cuttings rooted by layering. Mulberry stands were intensively managed and fertilized, for example with canal mud. Up to half the arable land in some districts was planted with these trees and yields, according to an early report that has not been verified, could be as high as 13 tons of green leaves per acre. In more southerly latitudes, the first crop of leaves was harvested by pruning the previous year's shoots, and the second and third by picking the leaves. (In northerly latitudes, leaf harvesting was by picking only.) Leaves were sold to silk producers and the limbs, once stripped of their leaves, bundled and sold as fuel (King 1911).

1.5.6 Beating the foresters at their own game: *Shorea javanica* forests planted by farmers in Sumatra

On the west coast of South Sumatra, in the foothills of the Bukit Barisan mountain range, lie the damar gardens of Krui. To casual observers these gardens look like natural forests: uneven aged, with a large variety of tree species and lianas growing up to 45 m high. But many of the trees have been planted by local farmers, who have harvested and exported the damar resin produced by a dipterocarp tree (*Shorea javanica*) since the third century AD. Initially they probably tapped naturally occurring trees in the forest, but as early as 1782, British explorers had remarked on the large extent of planted damar forests.

The gardens are interesting for a number of reasons. One of these is technical. Professional foresters have never managed to successfully grow dipterocarp species in plantations, although recently there has been some success on an experimental scale. This is a serious setback as the dipterocarps are the major commercial timber species of South-East Asia, and are being rapidly depleted by unsustainable logging operations. And yet villagers in this part of Sumatra have planted as much as 54 000 ha of damar gardens. One explanation for their success is that they collect wildlings for planting in their gardens from the forest with the soil, thus retaining the mycorrhiza crucial for tree growth—but this may not be the whole story. In the meantime, it works!

From an environmental point of view, the gardens are also extremely valuable. A recent comparative study of vegetation sample plots in the region found that rainforest plots had 230 species, and damar 120 species, whereas rubber estates only had 10. The gardens are also providing critical habitat for threatened animals such as the Sumatran rhinoceros and tiger (de Foresta & Michon in Halladay & Gilmour 1995).

Economically, the gardens are high fliers as well: production of damar resin is combined with that of commercial fruit trees such as durian (*Durio zibethinus*), rambutan (*Nephelium lappaceum*), nangka (*Artocarpus heterophyllus*), menteng (*Baccaurea racemosa*), duku (*Aglaia dookkoo*), manggis (*Garcinia mangostana*), mango (*Mangifera indica*) and petai (*Parkia speciosa*). Farming households prepared to invest 127 person-days in a mixed damar garden of 1 ha can expect to earn $1200–1800 per year. As the people in Krui have successfully maintained many of the traditional social safeguards such as limits on the sale of land, the distribution of this income is fairly equitable.

In the early 1990s however, the Indonesian government leased the 29 000 ha of damar gardens which are formally located within the State Forestry Zone to a timber concessionaire. (Indonesia's laws do not recognize pre-existing customary law and rights.) The company threatened to log the three million or so valuable trees planted by the villagers. At the same time, oil palm companies, with the support of local government, began encroaching on Krui's agroforests. In 1996 one company clear cut dozens of hectares of community-planted damar on the southern border of Krui. More recently, a combination of non-governmental organizations and researchers have obtained a stay on the destruction of damar agroforests by large-scale corporate interests in the Krui region, and the government has agreed to accord the region a special legal status as an experimental community forestry area (Poffenberger 1999).

1.6 FOREST MANAGEMENT BY PROFESSIONALS

The previous section showed how initially no doubt primitive ways of manipulating the wild vegetation evolved over time into highly sophisticated adaptive management systems run by non-professionals. However many rulers throughout history did not hesitate to conjure up accusations of mismanagement when they felt it necessary to promote their own interests in timber or wild game at the expense of their subjects (Kuechli 1998). Thus, the forestry profession was founded to finally bring some rationality into what was perceived as a history of waste and destruction of forests by ignorant peasants. Common use or even property rights were then extinguished, often causing considerable hardship for poor rural communities.

The professional foresters serving these rulers understandably found many ways to justify such depredations. Von Carlowitz, who in 1713 wrote the first treatise solely devoted to forestry in Germany (Anonymous 1989), refers to local forest users as follows: 'Basketmakers and coopers, mushroom collectors, hop-pole cutters, fowlers and shepherds and other damagers of trees'!

Over time, increasing numbers of foresters managed to find more collaborative ways to work with rural communities. The following case descriptions give an idea of the various ways in which the forestry profession has been involved in forest management, together with or against the local people.

1.6.1 Repressive roots: protecting the king's forests against the people in medieval England

To secure the enjoyment of the chase for themselves and their associates, the kings of medieval England took measures to preserve for their hunting the wild beasts of the field, and more especially 'those the flesh of which was delicate to the taste'. In many instances, the kings not only claimed all beasts and birds which were wild by nature—so that it was unlawful for any man to kill, take or hurt any wild beast or bird even within his own grounds—but also forbade any cutting of forests that might shelter wild beasts.

Whereas the Anglo-Saxon king Canute had in a code of laws of 1016 confirmed to his subjects the full power to hunt in their own lands, provided they abstained from the royal forests, William the Conqueror and his son William Rufus established a more tyrannical forest regime from 1066 onwards. Not content with the forests that existed, 'William the Conqueror destroyed 60 parishes, and drove out their inhabitants, in order that he might turn their lands into a forest, to be used as a hunting-ground for himself and his posterity; and he punished with death the killing of a deer, wild-boar, or even a hare'. William Rufus met his death in 'a new forest which he had caused to be made of 18 parishes, which he had destroyed', an occurrence that was spoken of by the people as 'judgements of God passed upon them for their oppressive selfish appropriation of land which under culture had yielded food for man and beast!'

It was only under Richard I, more than a hundred years later, that the then existing penalties of loss of eyes and of cutting off hands and feet for transgressions committed in hunting outside royal forests were repealed. Penalties were still stiff, however, consisting of considerable fines or, if unable to pay that, imprisonment for a year and a day. Anyone convicted of unauthorized hunting inside royal forests would still lose his eyes and testicles. These draconian measures were finally repealed in the Magna Charta (1215) and Charta Forestae (1225), which the English nobility negotiated with Kings John and Henry III.

Much later, Charles I revived the latent powers of forest laws that had gradually fallen into disuse. He summoned the forest courts, and called forth the full extent of their powers to his assistance, not so much because of any hunting passion, but to extort revenue independent of the grant of Parliament (Brown 1883).

1.6.2 The deep roots of modern silviculture

Silvicultura Oeconomica, written in 1713 by Hans Carl von Carlowitz, was the first publication in continental Europe to focus exclusively on forestry. Von Carlowitz, however, was not himself a forester but the director of metallurgy in the electorate of Saxony. His book, which is full of exhortations to take better care of forests and establish new ones, should be seen against the background of the perceived threat of a great shortage of wood in a Europe that was being rebuilt after the Thirty Years War, and of von Carlowitz's industry as a major user of fuelwood.

According to von Carlowitz, the forestry knowledge he compiled in his book was not new, and most of it had existed since ancient times. Silvicultural measures known to von Carlowitz included:
- Sowing: seed collection and storage, soil preparation, including ploughing, hoeing and furrowing, seed preparation for sowing (including viability testing), sowing (preferably in autumn, and taking account of the weather and phase of the waxing moon).
- Planting: growing seedlings from seed, cuttings and root suckers in nurseries (both exotic and native spp.), root pruning, conditioning of planting pits, planting and manuring.
- Silviculture: tending of newly established stands (including the removal of harmful moss!), pruning, selecting seed trees and leaving individuals or groups of them standing in cutting areas, coppicing, understorey and overstorey silviculture.

1.6.3 Oberwolfach: from Femelwald to chessboard forestry and back

Rural communities in the Black Forest have long had a forest-based economy. Seven centuries ago, they already produced commercial quantities of spruce resin for caulking and many other wood and non-wood products. These days, many farmers depend on sustainable timber production for a regular income, both on private and communal land. The cultivation of annual crops in this mountainous region is a risky activity that only contributes marginally to farm revenue.

Forest management centres around the production of large trees that fetch premium prices, through single tree selection felling, a system called Femelwald in the Black Forest but better known internationally under its Swiss synonym Plenterwald. Concentrated felling only takes place when the community has to invest in roads and other infrastructure, or when a farm needs to be saved from failure. Such temporary over-harvesting would always be compensated for later by a period of restricted felling. As a forest management system, Femelwald has been practised by farmers in the German Black Forest for centuries and is highly sustainable. Of course, it is not without impact on the forest ecosystem. Farmers systematically favoured fir and spruce over beech, which became progressively rarer in the species mix as a consequence.

In the early nineteenth century, there were enormous profits to be made from the timber trade with Holland and other wood-deficit countries. Powerful private interests teamed up with public sector foresters to rapidly introduce scientific 'chessboard' forestry, based on clearfelling and replanting of large squarish areas, in the northern part of the Black Forest. By 1833 scientific forestry was acknowledged in the forestry law of Baden, which summarily forbade the selective felling of trees—a practice described in the law as an inefficient and uneconomical method that violated all the rules of forestry. But replacement of single-tree selection with large-scale clearfelling left many areas denuded. Farmers in the region reacted strongly against these authoritarian attempts to direct their activities. Numerous communes repeatedly petitioned the government and forestry officials became the targets of violence.

In the latter part of the nineteenth century, the ravages of storms, snow and pests started to cast doubts in the minds of many foresters on the merits of clearfelling and replanting. Populations of bark beetles, for example, cause widespread deterioration in spruce monocultures but not in Plenterwald. Furthermore, Plenterwald is highly wind resistant as the young trees develop deep roots that anchor and stabilize them while they remain suppressed for decades beneath the crown cover of mature trees.

Joseph Schatzle, the son of a Black Forest carpenter, who had become the district forester in Wolfach in the 1870s, refused to acknowledge the legal ban on the Plenterwald system. The stiff opposition from farmers, combined with the more passive resistance of its own officials, forced the forestry administration of Baden to refrain from enforcing the ban on single tree selection felling—although it remained formally in force until 1976!

But in the meantime, much damage has been done. Whereas Plenterwald systems largely preserved existing diversity, biodiversity and genetic variation in chessboard forests have been drastically reduced, as the result of clearcutting and the planting of monocultures, with a corresponding reduction in the resilience of the forests with respect to changing environmental conditions (Kuechli 1998).

1.6.4 Operation successful, patient died: the Malayan Uniform System

One of the potentially most effective tropical silvicultural systems, the Malayan Uniform System (MUS), was developed in the 1960s to manage the very productive lowland dipterocarp forests of Peninsular Malaysia. The system appeared effective not only in terms of sustaining timber producion, but also in maintaining considerable forest biodiversity (Poore *et al.* 1995). MUS was designed in response to the introduction of mechanized extraction and the shift in market demand from slow-growing, shade-tolerant, heavy and durable hardwoods to lighter timber species (Dawkins & Philip 1998).

Under the MUS, the felling cycle corresponded to the length of rotation (approximately 60–70 years) and the next crop depended upon tending the seedling regeneration on the ground at the time of felling. As with any silvicultural system, there were hitches; for example, sometimes dense undergrowth after felling led to insufficient recruitment of new seedlings, but early results with the system were most promising. Unfortunately, the forests to which it was applied were situated on soils which were eminently suitable for agricultural cash crops and, as a matter of government policy, appropriate stands were alienated after the first cutting cycle for conversion to plantations of rubber and oil palm. Furthermore, uniform methods such as MUS are ill adapted to the conditions prevailing in the remaining hill dipterocarp forests (Poore *et al.* 1995).

1.6.5 Seeing the broadleaved forest for the conifer trees: participatory forest restoration in Nepal

The Nepal Australia Community Forestry Project (NACFP) is a bilateral aid project between the Australian International Development Assistance Bureau and Her Majesty's Government of Nepal, operating in two districts in the Middle Hills Region: Sindhu Palchok and Kabhre Palanchok. Prior to the start of the project in 1978, forest cover in the area had remained stable for 14 years, but tree density had decreased substantially—indicating that a process of forest degradation was underway.

From 1978 to 1995, about 18 000 ha of new community plantations, mainly of *Pinus* spp., were established in the project area on grasslands, degraded shrublands and abandoned agricultural land. This amounts to 14% of the total forest area measured in 1978. When broadleaf species started to re-establish naturally in the older pine stands, the project's foresters noted the enthusiasm of the local people about the return of these species. In response, the project helped user groups to develop silvicultural systems for gradually favouring preferred indigenous broadleaf trees over pines. As a consequence, most of the plantations are likely to be converted to natural forests as local managers take advantage of preferred species returning to previously degraded sites

over time. In addition, the project responded by helping local people to bring 200 patches of residual natural forest under improved management, covering about 5500 ha between 1988 and 1995 (Ingles in Halladay & Gilmour 1995).

1.7 THE IMPACT OF FOREST MANAGEMENT ON FORESTS

It is impossible in the scope of this chapter to do justice to this subject. Slight differences in management systems or silvicultural practices can have considerable impact on forests and their biological diversity.

The impact of forest management, especially for timber production, on biodiversity has had much public attention. One factor which makes it extremely hard to judge the impact of forest management practices is that most forest ecosystems were fundamentally altered by hunting, gathering and firing of vegetation long before any kind of forest management was practised. In addition, considerable areas of forests existing today were converted to farming at some point during their existence. So one cannot deduce the impact of forest management by a direct comparison between the biodiversity of wildwoods (to the extent they still exist) with that of managed forest.

A recent assessment of tree species threatened with extinction worldwide sheds some light on the impact of forest management on biodiversity (Anon. 1998). Out of an estimated global tree and woody shrub flora of about 100 000 species and varieties, the conservation status of just over 10 000 was assessed. Some 8753 were found to be globally threatened.

As is clear from Table 1.4, most of these 'tree' species are threatened by habitat loss (due to conversion for farming, settlements, grazing; and wildfires and invasive species) and overharvesting of timber and non-timber forest products, rather than by forest management systems or practices. Not included in the latter category is the practice of clear-felling natural forest and woodlands and replanting them with exotic species, which most observers would characterize as forest conversion not management. Some examples of tree species threatened by forest management practices are given in Box 1.5.

Table 1.4 Most fequently recorded threat to globally threatened tree* species (Source: Anonymous 1998).

Threat	Critically endangered	Endangered	Vulnerable	Total
Felling	168	360	762	1290
Agriculture	127	232	560	919
Settlements	119	209	423	751
Grazing	97	122	198	417
Burning	50	77	158	285
Invasive plants	88	78	79	245
Forest management	12	61	141	220
Local use	13	55	105	173
Mining	19	31	101	151
Tourism	23	51	60	134

*Tree is defined as a woody plant growing on a single stem usually to a height of over 2 m.

Box 1.5 Examples of trees threatened by forest management practices. Source: Anonymous (1998)

Shorea bentongensis; Dipterocarpaceae Malaysia (Peninsular Malaysia); endangered
Endemic to Peninsular Malaysia, this tree is locally common in deep valleys. The wood is used as white meranti. However, the tree is slow growing and cannot withstand logging because the reproductive cycle exceeds cutting cycles.

Gonystylus bancanus;Thymelaeaceae Brunei, Indonesia (Kalimantan, Sumatra), Malaysia (Peninsular Malaysia, Sabah, Sarawak); vulnerable
A gregarious, often dominant tree of lowland freshwater swamp and peat-swamp forest. Populations have been heavily depleted as the most important source of ramin timber. The species is also threatened in parts of its range by habitat loss. A recent investigation by Dutch and Malaysian experts, following CITES debates on the species, concluded that *G. bancanus* is not threatened with extinction in Malaysian swamp forests although regeneration in overexploited forests may be a cause for concern.

Nothofagus alessandri; Fagaceae Chile (Maule); endangered
Endemic to Maule region, the species was once more widespread but is now restricted to eight scattered localities in a small area of deciduous forest in the Coastal Cordillera. These stands all represent secondary growth from stump sprouts, and between 1983 and 1991 their extent of occurrence was reduced by almost 60% as a result of the establishment of plantations of *Pinus radiata*. About 13% of the species range is covered by protected areas. It is recognized as a very primitive member of the genus.

1.7.1 Biodiversity loss through discontinuing forest management practices

So-called natural forests often have a long history of human intervention, some of it deliberate and intensive enough for it to be referred to as management and other interventions being more unintentional or extensive. Whatever the intentions of the users and managers, biologists are increasingly aware that many species depend on continued human disturbance in order to thrive. A very common threat to biodiversity in Western Europe is the discontinuation of ancient forest management practices.

This issue will be explored in more detail using the case of Bernwood forest and its rare butterflies.

The medieval landscape around Bernwood Forest was mainly wooded until about 1600, when large areas of wood and wood pasture were cleared to make way for permanent pasture. Before 1600, the habitat mosaic was extensive, with its different elements—woodland, coppice, woodpasture, clearfell, wooded meadows, scrub, pasture, arable—occurring over the whole forest and grading into one another. After 1600, woodland habitats were restricted to woods located within a less diversified agricultural landscape.

Table 1.5 The impact of woodland management practices on Lepidoptera in Bernwood forest. (Based on R.C. Thomas in Kirby & Watkins 1998.)

Species	First present	Last signalled	Reason for decline/increase
Catocala promissa Light crimson underwing	Before 1700?	1947	Caterpillar hides by day in bark crevices of mature oak, which disappeared when oak coppice with standards was progressively converted between 1900 and 1952
Dicycla oo Heart moth	Before 1700?	1940	Similar dependence on mature oak as species above
Hemaris tityus Bee hawkmoth	Before 1400?	1952	Occurs in areas of damp rough meadows (where larvae feed on devil's bit scabious) occurring near woodlands (where adults seek nectar from bugle, common on rides). Disappearance of this species is due as much to agricultural intensification as it is to changes in forest management practices
Hyloicus pinastri Pine hawkmoth	1900s	NA	Scots pine has been present since it was first planted in the forest in the 1930s, but the pine hawkmoth prefers mature trees
Pseudopanthera macularia Speckled yellow	Before 1600?	1920s	Moth feeds on plants of open woodlands (wood sage, hedge woundwort, deadnettle), which declined after coppicing was abandoned
Strymonidia pruni Black hairstreak	Before 1400?	NA	The species prefers mature blackthorn at wood edges in glades and in mature hedges. Its persistence, albeit in reduced numbers, is due no doubt to the fact that much of Bernwood's scrub vegetation is specifically managed for it

NA, not applicable.

Structural variation within the woods increased, however, as a more structured system of coppice management was applied across the forest, creating a network of permanent and temporary open space and increasing the extent of internal wood edge, especially around coppice coupes. With the loss of systematic coppicing in the first half of the twentieth century this habitat largely disappeared, until glade and wood-edge scrub management for insects was developed as a part of modern conservation forestry. The latter is all the more important since agricultural intensification has made the surrounding farming landscape such an inhospitable area for most insect species and their plant hosts. Drainage, fertilization and early mowing have put paid to species-rich meadows, and arable and pasture occur right up to the wood edge, so that external wood-edge scrub is virtually non-existent (R.C. Thomas in Kirby & Watkins 1998).

1.8 CONCLUSIONS: LEAVING THE MARGINS OF HISTORY

In many parts of the world, farmers, herders and other non-professionals have managed forests for centuries if not millennia, and they continue to do so today. In their eyes, underwood and non-timber forest products are often more important than timber. Wood and other forest products are derived from trees on farms as well as from locally managed woodlands. Management has generally been based on simple unwritten rules rather than on complex management plans. As a result, management systems and methods often differ considerably from best practice as defined by the forestry profession. Nevertheless, local managers have shown tremendous capacity for adaptation and innovation.

Much has been made in the literature of the breakdown of 'traditional' management systems due to demographic factors and on the inherent tensions present in collective management, the 'tragedy of the commons'. The reality is often more complex. Population growth and migration do play a role in forest degradation and deforestation, but more often than not the main problem is the inability of village groups to defend their forests against powerful outsiders seeking to exploit them. Once locally managed forests are 'legally' disowned and destructive exploitation commences, village communities often join the fray in order to secure at least some benefit, 'stealing' from forests they used to own.

1.8.1 The forestry profession: getting better with age?

Most professional foresters to date have worked for public sector agencies, although this is rapidly changing. The paradox of many of these public sector forest agencies is that while they have been created to protect the public interest they have rarely been able to do so. More often than not, professional foresters have been grudging instruments, or even willing collaborators, of powerful vested interests aiming to privatize forest benefits and socialize the costs of their destruction. The result has been that local rights were extinguished and rural communities disenfranchised, while the environmental objectives public forest agencies set out to achieve have only been partly achieved (Westoby 1987).

Even if professional foresters have routinely prided themselves on their farsightedness, in reality the time factor has often proven a major problem for them. Once the stand they had established had come to maturity, or the silvicultural technique they had experimented with had proven its usefulness, important shifts in socioeconomic circumstances ensured that their results (however impressive technically) came to nothing. The dense plantings of *Pinus nigra* for the production of mine props (and sometimes to fix shifting sands) in the Netherlands come to

mind: the coal mines were closed before the timber was ready to be harvested. Forest managers are now experimenting with ways to open up these stands to increase diversity without destabilizing them. The Malayan Uniform System was a successful silvicultural system, but the lowland dipterocarp forest for which it was designed has now been converted to rubber and oil palm, and uniform methods are ill adapted to the remaining hill dipterocarp forests.

There are also signs of positive change in the forestry profession. Some of this is born of need: much of traditional professional practice in industrialized countries with proud forestry traditions such as Germany and Japan has become unaffordable due to the heavy burden imposed by the high costs of labour and the limits to further mechanization. In Germany and some other Western European countries, the problem of high labour costs has become an opportunity for less interventionist and more environmentally friendly practices (e.g. the uneven-aged, mixed-species silviculture promoted by the organization PROSYLVA).

1.8.2 Devising forest policies that work

One lesson that is particularly relevant to forest policy-makers and corporate decision-makers is that well-documented forest management systems extending over large areas and long periods of time are the exception rather than the rule. The same holds true for reforestation, even in many of the so-called developed countries with proud forestry traditions, such as the USA. In Vermont and New Hampshire, for example, forest cover increased from 10 to 80% over this century, mainly through spontaneous regeneration of abandoned farmland rather than through the efforts of foresters. As large-scale, government-sponsored forest management and reforestation are often stated to be an indispensable part of any sustainable future for humankind, it is important to consider the reasons why they should be so exceptional. On the one hand, this is mainly due to the dynamic socioeconomic and political context of human land use, to which large-scale, bureaucratic systems of professional forest management are often ill adapted. On the other hand, there are some woodlands that have been managed sustainably for very long periods of time, some of them in all likelihood for more than 5500 years. This was mainly as coppice and most of the time by people without a 'sound, theoretical basis in forestry', as the father of silviculture on the European continent, Hans von Carlowitz, would have it. Clearly, forest policy-makers need to focus much more on enabling a variety of stakeholders to manage forests in a variety of ways, rather than to continue to think in terms of government-led and managed blueprints.

1.8.3 Forestry development aid: learning to overcome Euro-professionalism

The particular direction the forestry profession took in most of Europe, i.e. separating forestry totally from agriculture and focusing on tree planting, arguably left foresters working for colonial governments and later aid agencies ill prepared for assisting rural people in developing countries. The realization that agroforestry is not some backward concept but rather a logical step in agricultural intensification in many countries (Scherr 1997) is now finally sinking in, but it took a lot of time. Forestry projects faced with abject failure have been forced to experiment, some with encouraging results, such as in Nepal. Learning from what farmers and herders do with trees and forests, and experimenting with mechanisms to improve their management is the way forward for forestry aid—if forestry aid has a future.

1.8.4 Let a thousand forests flourish

It would have been preferable to end this chapter on an unequivocally positive note, but unfortunately the jury is still out. Yes there is an amazing diversity of forest management systems, whether developed by farmers and herders or professional foresters, and many of these systems strike an appropriate balance between environmental, social and economic imperatives. Furthermore, most natural and seminatural forest ecosystems have proven to be remarkably resilient, and are able to regenerate quite well once outright conversion and other destructive practices are discontinued. However, many more forests have been converted to farming and other land uses than are now under any form of management. Deforestation has accelerated once again in the tropics, where fires are finding easy fuel in fragmented forests perforated by logging and other incursions with every recurring El Niño event. Extractive use still routinely leads to overexploitation, although simple ways to limit damage, control yields and encourage regeneration are known at least in outline, both for timber species and non-timber forest products. If business as usual continues, forest management as an activity will be relegated to the margins of history.

External demands on forest resources are still increasing, and there is no particular reason why rural forest-dependent communities should do better out of the 'new forestry' (environmental 'concessions' for carbon sequestration, watershed protection and biodiversity conservation) than they did out of industrial timber concessions in the recent past. Significantly, the first ever debt-for-nature swap was concluded without any participation from the indigenous communities involved. Already, there is evidence of considerable inequity in forestry carbon sequestration deals concluded under the Framework Convention on Climate Change, with prices paid in developing countries ranging from $3 to $18 per tonne CO_2.

Experience has shown that durable forest management arrangements are unlikely to result from such lop-sided arrangements. If rural people are disenfranchised from their rights to use and manage the forests surrounding them, these forests will become symbols of oppression and flashpoints for destructive action during the next political upheaval. Or they will suffer from neglect by politicians reading the minds of a disaffected electorate. The challenge for government forest agencies will be to prevent further disempowerment of local stakeholders that are managing their forests sustainably and also to define much more precisely what is the public interest in each and every forest that they are responsible for, and find ways of rewarding local people for the global public goods they produce, and compensating them where their interests are in conflict with national or global priorities. Only then will a thousand forests flourish.

REFERENCES

Anonymous (1989) *An Account of Silvicultura Oeconomica by Hans Carlowitz, 1713.* Johan Friedrich Braun, Leipzig.

Anonymous (1998) *Threatened Trees of the World.* World Conservation Monitoring Centre & IUCN Species Survival Commission, Cambridge.

Baasen, C. (1940) *Wald und Bauerntum* [in German].

Bird, M.I. & Cali, J.A. (1998) A million-year record of fire in sub-Saharan Africa. *Nature* **394**, 767–9.

Braudel, F. (1991) *The Structures of Everyday Life. The Limits of the Possible. Civilization and Capitalism 15th–18th Century*, Vol. I. University of California Press, Berkeley.

Brown, J.C. (1883) *The Forests of England and the Management of them in Byegone Times.* Oliver & Boyd and Marshall, Edinburgh.

Burgess, N. & Clark, G.P. eds (2000) *Coastal Forests of Eastern Africa.* IUCN—the World Conservation Union, Gland, in press.

Dawkins, H.C. & Philip, M.S. (1998) *Tropical Moist Forest Silviculture and Management. A History of Success and Failure.* CAB International, Oxford.

de Foresta, H. & Michon, G (1995) The Indonesian agroforest model. In: Halladay, P. & Gilmour, D.A., eds. *Conserving Biodiversity Outside Protected Areas: The Role of Traditional Agro-ecosystems.* IUCN—The World Conservation Union, Gland.

Diamond, J. (1998) *Guns, Germs and Steel: A Short History of Everybody for the Last 13000 Years.* Vintage, London.

Fagan, B.M. (1995) *People of the Earth: An Introduction to World Prehistory.* Harper Collins, New York.

Fairhead, J. & Leach, M. (1996) *Misreading the African landscape: society and ecology in a forest-savanna mosaic.* Cambridge University Press, Cambridge.

Fay, J.M. (1997) *The Ecology, Social Organization, Populations, Habitat and History of the Western Lowland Gorilla (Gorilla gorilla gorilla Savage and Wyman 1847).* Unpublished PhD Thesis, Washington University, Saint Louis, Missouri.

Goudsblom, J. (1992) *Fire and Civilization.* Allen Lane, Penguin Press.

Halladay, P. & Gilmour, D.A., eds (1995) *Conserving Biodiversity Outside Protected Areas: The Role of Traditional Agro-ecosystems.* IUCN—The World Conservation Union, Gland.

Ingles, A. (1995) Community forestry in Nepal. In: Halladay, P. & Gilmour, D.A., eds. *Conserving Biodiversity Outside Protected Areas: The Role of Traditional Agro-ecosystems.* IUCN—The World Conservation Union, Gland.

King, F.H. (1911) *Farmers of Forty Centuries, or Permanent Agriculture in China, Korea and Japan.* Rodale Press (reprint), Emmaus, Pennsylvania.

Kirby & Watkins (1998) *Ecological History of European Woodlands.* CAB International, Oxford.

Kreuz, A. (1992) Charcoal from ten early neolithic settlements in Central Europe and its interpretation in terms of woodland management and wildwood resources. *Bulletin de la Societe Botanique de France* **139**, Actual. bot. (2/3/4), 383–94.

Kuechli, C. (1998) *Forests of Hope. Stories of Regeneration.* Earthscan Publications, London.

Lane Fox, R. (1988) *Pagans and Christians.* Harper Collins, New York.

Leach, G. & Mearns, R. (1988) *Beyond the Woodfuel Crisis: People, Land and Trees in Africa.* Earthscan, London.

LeRoy Ladurie, E. (1979) In Normandy's woods and fields. In: *The Territory of the Historian*, pp. 133–71. University of Chicago Press.

Meiggs, R. (1982) *Trees and Timber in the Ancient Mediterranean World.* Clarendon Press, Oxford.

Mencius (1970) *The Works of Mencius, Translated and Introduced by James Legge, Dover (1894).*

Menzies, N. (1995) *Forest and Land Management in Imperial China.* St Martin's Press, London.

Perlin, J. (1989) *A Forest Journey: The Role of Wood in the Development of Civilization.* W.W. Norton, New York.

Poffenberger, M., ed. (1999) *Communities and Forest Management in South East Asia: A Regional Profile of the Working Group on Community Involvement in Forest Management.* IUCN, Gland.

Poore, D., Burgess, P., Palmer, J., Rietbergen, S. & Synnott, J. (1995) *No Timber Without Trees: Sustainability in the Tropical Forest.* Earthscan, London.

Rackham, O. (1990) *Trees and Woodland in the British Landscape*, revised edition. J.M. Dent & Sons, London.

Roesch, M. (1996) New approaches to prehistoric land-use reconstruction in south-western Germany. *Vegetation History and Archaeobotany* **5**, 65–79.

Sale, K. (1991) *The Conquest of Paradise. Christopher Columbus and the Columbian Legacy.* Plume, New York.

Scherr, S.J. (1997) Meeting household needs: farmer tree-growing strategies in Western Kanga. In: Arnold, M. & Dewees, P. eds. *Farms, Trees and Farmers: Responses to Agricultural Intersification.* Earthscan Publications, London.

Spindler, K. (1995) *The Man in the Ice.* Phoenix, London.

Tossavainen, J. (1996) The collapse of Danzig's timber export in the first half of the 17th century: a case of ill-considered forest exploitation? In: *Stages and Trends in the Interaction between Economic Development, Forestry and Environmental Protection from the Past to Present Times*. Proceedings of the meeting of Subject Group S6.07 'Forest History' during the XX IUFRO World Congress (Tampere, Finland, August 1995), IUFRO, Vienna.

Turner II, B.L., Clark, W.C., Kates, R.W., Richards, J.F., Mathews, J.T. & Meyer, W.B. (1990) *The Earth as Transformed by Human Action: Global and Regional Changes in the Biosphere over the Past 300 years*. Cambridge University Press, Cambridge.

Veer, C. (1980) *The Sociology of Forestry*. Unpublished lecture notes. Agricultural University, Wageningen (NL).

Westoby, J. (1987) *The Purpose of Forests*. Basil Blackwell, Oxford.

Part 1
Sustaining Forest Influences

Forests and forest land still provide some of the world's great wildernesses and incredibly rich ecosystems, and hence need conserving in their own right as unique entities. They confer many benefits beyond simply the supply of wood products. And, with appropriate interventions in their management or not as the case may be, forest biodiversity can sometimes be enriched and wildlife value enhanced. Of course, forests offer many more non-timber benefits as well, but in Part 1 we focus on these three crucial aspects.

Roger Good (Chapter 2) sketches the uncertain status of parks and reserves worldwide, and the place of forest in them, and then analyses the history of wilderness in Australia and how that nation has grasped fundamental issues to ensure their appropriate conservation. The chapter provides a good example of developing thinking around the issue and how simple neglect is no solution at all. While the main 'drivers' in Australia were recognition of wildlife richness and value, and the importance of having large extant areas of pristine forest and forest land, in Switzerland conservation of forest cover was driven by the overwhelming need to minimize avalanches in the Alps. This is the basis of Peter Brang and colleagues' approach (Chapter 3) to reviewing

forests as a beneficial and protective influence. Protection forest for avalanche control on snow-covered slopes is, of course, only one of several beneficial roles forests may play, such as in catchment hydrology, soil erosion control or dune stabilization. The detailed treatment of the role in Chapter 3 illustrates the essential holistic approach – ecology, silviculture, management – to harness forest influences to the full and is welcome here as providing an accessible account in English of perhaps a less well-known role for forest cover. Also welcome is Barry Gardner's short contribution concerning the provision of shelter from wind by forest shelterbelts.

Tim Boyle (Chapter 4) provides a thorough overview of how to conserve the richness of forest ecosystems by logically taking the reader from what biodiversity is and why it is important, through the tangled web of trying to assess 'it', to the impact people have for better or for worse. So often biodiversity is held out as some great good or desirable goal but with remarkably few aids to help achieve it beyond empirically based recommendations or 'leave it alone'. Boyle's review provides the framework for a practical understanding, recognizing, as he does, that economic dictats are part of the equation.

2: Forest Reserves, Parks and Wilderness: An Australian Perspective

ROGER GOOD

2.1 INTRODUCTION

Forestry in Australia has changed dramatically during the past two decades, with new management approaches being implemented to meet the increasing demands for ecologically sustainable production strategies and practices. Public demand for the setting aside of forested land as national parks, nature reserves and other protected natural areas has also resulted in considerable changes in forestry management practices. Considerable commercial and social readjustment has been forced on the forest industries as a response to the changes in government policies and the shift to ecologically sustainable forestry from a reduced timber resource base as additional national parks and other conservation reserves including wilderness areas are established

Comprehensive forest assessments have recently been undertaken to determine an equitable allocation of forests to commercial timber production and to the establishment of a comprehensive, adequate and representative conservation reserve system of much relevance internationally.

2.2 WORLDWIDE STATUS OF PARKS AND RESERVES

2.2.1 A global perspective

The past 5 years has seen a great increase in the area and number of protected areas around the world, but the accelerated expansion of the reserve network commenced in the 1960s. The area of protected land has increased threefold since 1968 and presently some 13 236 330 km^2 or 8.85% of the world land area now falls within protected areas, in categories I–VI, as defined by the World Commission on Protected Areas (WCPA) and the World Conservation Union (International Union for Conservation of Nature and Natural Resources—IUCN) (Davey & Phillips 1998). These categories are:

I   Strict protection:
(a) Strict Nature Reserve;
(b) Wilderness Area.
II Ecosystem conservation and recreation (National Park).
III Conservation of natural features (National Monument).
IV Conservation through active management (Habitat/Species Management Area).
V Landscape/seascape conservation and recreation (Protected Landscape/Seascape).
VI Sustainable use of natural ecosystems (Managed Resource Protected Area).

The increase in protected areas reflects the worldwide concern for the conservation status of natural areas and the declining quality of life in many countries as a result of development and its environmental consequences, including greenhouse gas emissions and global warming. As a response many countries have become signatories to several significant international agreements to preserve the world's natural biological resources, one being the Convention on Biological Diversity. This Convention was signed in 1992 and identified specific requirements of signatories relating to protected areas. These being to:
• establish a system of protected areas or areas where special measures need to be taken to conserve biological diversity; and
• develop, where necessary, guidelines for the selection, establishment and management of

protected areas where special measures need to be taken to conserve biological diversity.

The Convention (Article 8) calls for signatory States to develop a protected area system plan in which protected areas are recognized as having a major role in biodiversity conservation. The baseline for the conservation of biodiversity is identified as the world biomes. Fourteen major biomes are recognized (Green & Paine 1997) within which biodiversity conservation is a priority. Six of these are forest biomes, two are mixed mountain and island systems which include some forests, two desert systems, two grasslands systems, tundra and lake systems. (Table 2.1).

Approximately 30500 protected areas in 243 countries (and dependencies) have been established, with 12 countries each having protected areas in excess of 20000000 ha, these being Australia (98 661 141 ha including 38 908 358 ha in marine protected areas), Brazil (32 189 837 ha), Canada (82 545 492 ha), China (58 066 563 ha), Russia (65 536 759 ha), USA (104 172 478 ha) and Venezuela (26 322 306 ha). Eight countries have

protected areas exceeding 10 000 000 ha, with a further seven countries exceeding 5 000 000 ha.

Although the largest areas of protected lands occur in the larger countries, it is important to recognize that many small countries have a large proportion of their land area within protected areas and hence make a valuable pro-rata contribution to nature conservation and the maintenance and conservation of biodiversity. Davey and Phillips (1998) give Loa PDR as an example, where 20 National Biological Conservation Areas cover about 30 000 km² or 12.5% of the country. The protected areas of countries such as Loa PDR contribute greatly to the conservation of the regionally significant forests and a number of threatened species (Berkmuller *et al.* 1995; Chape 1996, in Davey & Phillips 1998).

2.2.2 Australian National Parks and Conservation Reserves

The development of national parks, conservation reserves and wilderness in Australia has been

Table 2.1 Area of the world's major biomes and area protected. (After Green & Paine 1997.)

Biome	Protected area			
	Area (km²)	Number	Extent (km²)	% biome protected
Tropical humid forests	10 513 210	1 030	922 543	8.77
Subtropical/temperate rainforests/woodlands	3 930 979	977	404 497	10.29
Temperate needle-leaf forests	15 682 817	1 492	897 375	5.72
Tropical dry forests/ woodlands	17 312 538	1 290	1 224 566	7.07
Tropical broadleaf forests	11 216 659	3 905	403 298	3.60
Evergreen sclerophyllous forests	3 757 144	1 469	164 883	4.39
Warm deserts/semi-deserts	24 279 843	605	1 173 025	4.83
Cold-winter deserts	9 250 252	290	546 168	5.90
Tundra communities	22 017 390	171	1 845 188	8.38
Tropical grasslands/ savannahs	4 264 832	100	316 465	7.42
Temperate grasslands	8 976 591	495	88 127	0.98
Mixed mountain systems	10 633 145	2 766	967 130	9.10
Mixed island systems	3 252 563	1 980	530 676	16.32
Lake systems	517 695	66	5 814	1.12
Total	145 605 658	16 636	9 489 665	6.52

based on similar philosophies and concepts to that evident in several other countries, particularly that of the USA. The 'national park' concept was first promoted soon after European settlement, and parks for the preservation of flora and for recreation were established as early as 1872. the first formally recognized national park was the Royal National Park, on the southern outskirts of Sydney, gazetted in 1879. This was the genesis of the current protected area network in Australia (Cresswell & Thomas 1997) and the start of the many years of debate over the use of forests for commercial timber production, and reservation for nature conservation. The need to maintain large tracts of native forests in an undisturbed state, as primitive areas, was also recognized in the late 1800s, although the first primitive area reserve was not gazetted until 1934. Primitive areas were considered in the same context as wilderness and several areas were formally gazetted as wilderness some years later in the 1960s (Fig. 2.1).

Since European settlement the area of forested land has been continually and greatly reduced by competing land uses; the area now accessible for commercial forest production is approximately 25% of the original forest area

At the time of settlement forests covered approximately 690 000 km^2 and woodlands 1 570 000 km^2 of the Australian continent (Fig. 2.2). Land clearing for agriculture, particularly during the early to mid 1900s, reduced this area to approximately 390 000 km^2 and 1 070 000 km^2, respectively, while open woodland has increased in area from about 1 650 000 to 2 000 000 km^2 following regeneration of previously cleared or logged forest areas. These eucalypt hardwood forests occur primarily in Australia and hence the greater part of this evergreen sclerophyllous forest biome of 3 757 144 km^2 occurs in Australia.

Fig. 2.1 Kosciuszko Primitive Area. (Photo: C. Totterdell.)

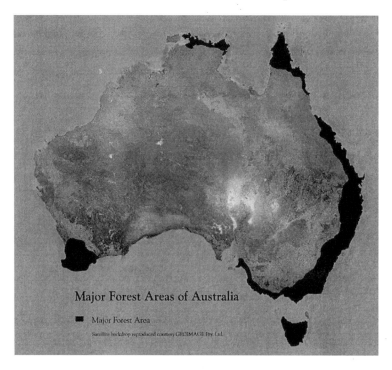

Major Forest Areas of Australia

Major Forest Area

Satellite backdrop reproduced courtesy GEOIMAGE Pty. Ltd.

Fig. 2.2 Major areas of eucalypt forest in Australia. (After State of the Environment, Australia, 1996.)

Unfortunately only $164\,883\,km^2$ occur in protected areas. This is only about 4.4% of this significant world biome and, as such, Australia has a major international obligation and role to play in ensuring its adequate conservation and management. Similarly, the tropical humid forests and subtropical/temperate rainforests/woodlands which are represented in the Australian landscape, albeit in relatively small areas on a global scale, are also poorly protected, with 8.7% and 10.3%, respectively, occurring in secure protected areas around the world (Davey & Phillips 1998). At the continental level, approximately $936\,000\,km^2$ or 12% of the land area is within some 6600 conservation reserves, but the majority of this land is non-forest grassland or shrub communities. This is most evident in the State of Western Australia where some 2500 conservation reserves cover an area of $305\,000\,km^2$, the major area being shrubland (Graetz *et al.* 1992). Forest-dominated conservation reserves in all States cover less than 15% of the total area within all reserves

The area of sclerophyllous forests managed by the State Government forestry agencies is less than $150\,000\,km^2$, approximately two-thirds of the area being managed for hardwood timber production and one-third for forest protection and other purposes, including small forest conservation reserves (Dargavel 1987).

In the late 1800s State and Federal Governments recognized the already extensive area of cleared forest lands and the continuing and increasing rate of clearing of natural areas for expanding agricultural industries. Large areas of remaining forests were, as a consequence, protected by legislation to ensure continued supplies of timber for forest industries. In many other areas where forest logging was permitted, local timber mills were supplied by a once-only logging operation before the lands were fully cleared for agriculture. At the same time (1880s) the first professional forest agencies were established to manage forests and to provide an adequate and orderly supply of sawlogs to the increasing number of private sector sawmills being established. These State Government forestry agencies also managed large areas of non-commercial forests, much of which, fortuitously, was set aside for forest conservation purposes or forest reserves. In so doing these agencies established themselves as the first nature conservation agencies and continued that role until the 1950s when national parks and wildlife management agencies were established in each State by the respective State Governments.

With the establishment of the various State nature conservation agencies and the demand for the creation of new conservation reserves and protected areas, competition between government agencies for use of forested lands increased. In the 1970s, this competition reached a peak and many conflicts arose between conservation activists, supporting the reservation of lands as national parks, and the timber industry opposed to any further removal of any areas of forest from commercial forest management. The intensity of these conflicts forced the State and Commonwealth Governments to find a new agenda within which natural area land-use conflicts could be resolved.

Environmental awareness in the general public had been engendered by these conflicts and demands were made on forestry agencies for implementation of sound ecological and sustainable forest production practices. A number of significant forestry, nature conservation, biological diversity and sustainable development policies, strategies and agreements were subsequently developed between the State and Federal Governments. These provided the policy framework and established the new approach to sustainable forest use and the allocation of forested lands to both commercial timber production and a comprehensive conservation reserve system in the forest environments. This approach was the regional forest assessment process and forests agreements, between the State and Commonwealth Governments. An objective of the forest agreements was an ecologically sustainable timber resource supply. An outcome of the regional forest assessment process was also to be the identification of areas meeting specific criteria relevant to wilderness values and wilderness quality, and the establishment of forest wilderness areas, within or additional to any existing or proposed comprehensive, adequate and representative conservation reserve system.

Prior to the comprehensive regional forest assessment programmes, the identification and selection of lands for management as wilderness were made very subjectively on size, naturalness and remoteness of an area. As part of an early (1997) comprehensive regional forest assessment programme in New South Wales, a quantitative (numerical) approach to wilderness identification was developed and adopted as the basis for all future wilderness assessments in that State. A similar approach to wilderness assessment is also proposed for adoption in other States.

The establishment of wilderness areas in existing national parks has been widely promoted by conservationists for many years and this objective is now widely accepted by the general public. However, the creation of wilderness in areas managed as production forests remains an issue of considerable debate and conflict between conservation agencies, non-government conservation groups and the timber industry. The impasse between the opposed parties remains to be resolved but sound land-use allocation and planning, providing for both biological diversity conservation and sustainable forestry should provide for this as the various forest agreements are negotiated in the next few years.

2.3 A SHORT HISTORY OF FORESTRY IN AUSTRALIA

Forestry in Australia has contributed much to the shaping of the existing Australian vegetation landscape. In the early years of European settlement the government forestry agencies were also the conservation agencies as only low-impact selective logging took place in native forests. In more recent years the clearfelling of native forests for woodchips and exotic plantation forests has sculptured a different vegetation landscape over large tracts of forest lands. Forestry as such has had a very turbulent history, particularly through the last four or five decades. Carron (1985) provides a detailed account of the development of forestry in Australia in his definitive book on forest history. The following short story is mainly referenced from this book and from Carron (1993).

The forest industry commenced with the extraction of logs from the extensive forests on the eastern seaboard of Australia, only a few years after the establishment of the first permanent colony in 1788. As land was cleared for agriculture the fallen timber was milled for construction timbers, burned as firewood and used in other infant industries such as mining.

Local small-scale forestry operations were well established by the late 1800s but forest operation and management were somewhat *ad hoc* and poorly planned. Sawlog supplies to the sawmills were erratic and often did not meet demand. Establishment of the government forestry agencies was a response to the need for a regular supply of timber to the local sawmills that were operative in many small towns. The professional forest management practices introduced by the forestry agencies established a well-managed sustainable timber industry based on a selective logging (although at this time the total volume of sawlogs taken from the forests was very low). The sawlog industry remained the dominant forest industry for the next five to six decades until the 1920s and 1930s, when mechanization of the industry commenced. This enabled the timber industry to expand rapidly until the 1960s when new timber processing techniques developed overseas; this led to a new pulpwood industry in Japan, supplied by woodchips exported from New South Wales, Victoria, Tasmania and Western Australia. Prior to this time, the processing of eucalypt hardwood had not been possible due to the high tannin and gum contents of the timber, and the fibre structure of the eucalypt hardwoods. Rapid expansion of woodchip exports from Tasmania, New South Wales, Victoria and Western Australia occurred through the 1970s to meet the developing pulpwood and paper industry. Selective logging as the dominant forest industry was quickly replaced in many areas by clearfelling of forests for the woodchip industry.

The extensive clearing of large tracts of high-quality coastal and near-coastal forests for woodchip production, more than any other forest management programme, incited much public antipathy, criticism and anger towards the timber industry and forestry agencies. It also initiated an unprecedented level of active hostility by conservation groups to forest management operations

in all states where woodchip production was planned or was being carried out. Many local environmental groups were concerned as to the direction Australian forestry had taken and the obvious detrimental impacts the woodchip operations were having on the extensive near-pristine native forest landscapes. The regular publication of articles and photographs in the media depicting extensive soil erosion and creek sedimentation in areas where clearfelling had taken place only served to enhance the public's perception of extensive environmental devastation in the forests. An image of an industry having a 'frontier mentality' to forest management was rightly or wrongly created. Forestry was depicted as an industry challenging the untamed forest 'wilderness'; certainly not one given to identifying wilderness or conservation areas, for which the many and various environmental groups had been actively lobbying for years.

At the time of the first woodchip exports the environmental movement was made up of local interest groups but their concerns over the clearfelling of high-quality forests was the stimulus to establish themselves as a coordinated conservation lobby group under the pre-eminent environmental organization, the Australian Conservation Foundation. Together with other lobby groups such as the Wilderness Society, the environmental movement established a very vocal presence that was to influence all State and Commonwealth Governments on many environmental issues, including the forestry conservation debate. Both government forestry agencies and the timber industry have, as a consequence, been under close and continuous scrutiny by both government and non-government environmental groups since the 1970s. Unfortunately, the continuous scrutiny and criticism created, for some years, a 'siege mentality' within the forestry agencies. This situation prevailed against any moves to change from production forestry to multiple-use forest management, encompassing catchment management, recreation, biological diversity conservation and wilderness preservation. The debates and conflict over timber production versus conservation and wilderness preservation, which had prevailed for many years, became an even more bitter conflict. This conflict escalated until very recent times (1996) when the need for compromise and sound land-use planning in forests was finally recognized by governments, government agencies, the timber industry and non-government conservation agencies. In response, the Commonwealth and State Governments enacted legislation, and developed and implemented forestry agreements and protocols to ensure changes in forest industry attitudes such that conservation, heritage, national estate and wilderness values were addressed in forest management programmes and operations. Interestingly, the potential to establish extensive wilderness areas over large tracts of forest land that remained in the 1960s and 1970s had indirectly been provided for by the forestry agencies which had, prior to the development of the woodchip industry, managed forests such that wilderness values had been maintained. Many forestry personnel actually made claims to being the first real conservationists and that only through the efforts of early foresters did the wilderness lobby groups now have the large areas of near-pristine forest ecosystems over which to argue (Legg 1988). This was a legitimate claim but the entrenched conservatism of the forestry agencies only further polarized the environmental movement, and any claims of the forestry industries as the 'fathers of conservation' was lost in the debate over ecologically sustainable forestry, biological diversity and forest conservation strategies.

Wootten (1987), in response to the conservation claims made by the forest agencies, suggested that the problems of the forestry agencies were manifest in their continuous attempts to divert attention from the forest industries by blaming the conservation movement for shortcomings in meeting their responsibilities in terms of forest ecosystem management and conservation. Several very notable, emotional and at times heated conflicts over the conservation of forest ecosystems exemplified the opinion the major non-government conservation groups held over forest management and which they had expressed through the 1970s, 1980s and early 1990s. Many blockades of forests and forest activities were put in place, often requiring police intervention to disperse the 'green' activists; or government intervention to resolve the disputes, or enforce compromises between the conservation movement and forest industries.

The debate over forest ecosystem conservation reached a peak and gained international recognition in the 1970s with the conflicts over the planned exploitation of the high-quality eucalypt forests in Tasmania and the tropical rainforests in Queensland (Figs 2.3 and 2.4). The public expression of concern for the conservation of the Lemonthyme forests in Tasmania, was the catalyst for the Commonwealth Government's first direct involvement in the development of forestry policies and the management of publicly owned forests in the States although it had earlier, in 1967, granted licences for the export of woodchips from the near-coastal forests of south-eastern New South Wales. In the same year, the Commonwealth Government also approved a softwood afforestation programme aimed at making Australia self-sufficient in softwood products over a 10–15-year period (Softwood Forestry Agreement Act of 1967).

The emphasis on increased softwood plantations by the forestry agencies was widely accepted by the conservation movement until it became evident that native hardwood forests in several States would be cleared to establish these softwood plantations. Again, the political influence of the conservation movement was to come to the fore and further forestry legislation was enacted (Softwood Forestry Bill of 1972), which included the requirement to consider environmental matters when softwood plantations were being established. As noted by Carron (1985, 1993), about this time the Commonwealth Government also became a signatory to several international agreements, which gave support to the conservation movement and its considerable influence on the future of Australian forestry. Arguably, the most significant of the international agreements was the Convention Concerning the Protection of World Cultural and Natural Heritage, to which Australia became a signatory in 1974.

Earlier, in 1973, the Commonwealth Government had set up a Committee of Inquiry into the National Estate which prepared a number of reports on nature conservation and the forestry industry, particularly that of the woodchip industry. The creation of a Heritage Commission to preserve and enhance properties and places nominated for listing on the National Estate list, including some native forest areas, was also

Fig. 2.3 Queensland Tropical Rainforest World Heritage Area. (Photo: C. Totterdell.)

planned. The Heritage Commission was established in 1976 following the enactment of the Australian Heritage Commission Act of 1975. This provided the Commonwealth Government with an indirect but considerable influence over the States with respect to forest management, even though State Governments still retained rights under the Constitution for the management of all lands within the respective States.

Other legislation enacted during the 1970s included the National Parks and Wildlife Conservation Act of 1975 and the Environment Protection (Impact of Proposals) Act of 1974. These two Acts, together with several conservation strategies developed in the same period, particularly the National Conservation Strategy, National Biological Diversity Strategy and the Strategy on Ecologically Sustainable Development, provided the Commonwealth with additional support to influence the direction and management of forests and forestry industries within the States.

In the early 1980s the forestry debate was focused on Tasmania where the planned construction of several dams for hydroelectricity threatened inundation of large areas of pristine forest with the loss of wilderness values and a number of significant rare plant species and communities. The Commonwealth Government sought to prevent the construction of the dams but the Tasmanian Government challenged the Commonwealth's role in the dams issue and its right of intervention in the destruction of forest resources and wilderness areas. The Commonwealth Government responded by passing the World Heritage Conservation Act in 1983, in an endeavour to prevent the flooding of any wilderness or old-growth forest areas. Following further challenges by the State, the High Court ruled that the application of the relevant World Heritage Act regulations to the Tasmanian State forests was valid and constitutional. As Carron (1993) noted:

From the forestry viewpoint, its influence was enormous; the most important aspect of

Fig. 2.4 Eucalypt forest in Tasmania. (Photo: C. Totterdell.)

the judgement was the reliance of the validity of the World Heritage Act on the external affairs powers (Sec.51(29)) of the Constitution and it was abundantly clear this could be applied by the Commonwealth in any State forestry environment vs. development matter from then on.

The Commonwealth Government's powers through the judgement were soon applied to the Tasmanian and New South Wales woodchip industries which were exporting over one million tonnes of woodchips annually. Environmental impact legislation was used by the Commonwealth Government to subsequently not renew the majority of woodchip export licences, although several licences were renewed for shorter 15-year terms, subject to conditions relating to the preservation and protection of places on the National Estate listings. Once again, the environmental movement was angrily aroused by the renewal of any licences, since any logging operations in the Lemonthyme and Southern forests of Tasmania, were considered a threat to the World Heritage values, including significant wilderness areas. Very active lobbying by the conservation movement to the International Union for the Conservation of Nature, World Heritage Committee was pursued for many months. The Commonwealth Government responded by passing the Lemonthyme and Southern Forests Commission of Inquiry Act. This Act established a Commission of Inquiry into many aspects of forestry in Tasmania. This inquiry, widely recognized as the Helsham Inquiry, is considered a cornerstone in Australian forestry. Following submission to the Government of its report, many challenges were made to the legality of the Lemonthyme and Southern Forests Inquiry Act. Again, the High Court ruled that this Act and the World Heritage Properties Convention Act of 1983 were valid, based on the constitutional external powers of the Commonwealth to implement international conventions. Eventually, an agreement between the Commonwealth and Tasmanian State Governments, with support from the Green Independents in the Tasmanian Government, was reached which provided for nearly 1 400 000 ha of World Heritage area through the addition of 600 000 ha of forest to the existing West Tasmania Wilderness National Parks.

Similar issues and conflicts were evident in forest operations and proposals in the tropical wet forests of Queensland, the temperate rainforests of northern New South Wales and the hardwood eucalypt forests of south-eastern New South Wales. The rainforest issue, particularly that of the wet tropical forests, was another cornerstone issue in Australian forestry. The conservation movement had made a long and vocal argument through the 1960s and 1970s for the reservation and preservation of all remaining wet tropical and temperate rainforest areas. As noted by Just (1987) in Carron (1993), 'Possibly no other issue in recent forestry history has captured the attention and polarized opinion in the community and within the forestry profession itself, as the issue of rainforests'.

The wet tropical forest issue again brought the Commonwealth Government into conflict with a State Government, in this case Queensland, who opposed the moves to have all the remaining tropical rainforest nominated for World Heritage listing. Bitter public debate ensued between the two governments to the extent that the Queensland Government sent a delegation to the 1988 meeting of the World Heritage Bureau Committee, to lobby against the Commonwealth Government's nomination of the wet tropics for World Heritage listing. Despite this lobbying by the Queensland delegation, a recommendation to the World Heritage Committee was made to include much of the wet tropical forest area on the World Heritage list. The Commonwealth Government had again been successful in using its powers to make international treaties and agreements as a means for enforcing forest conservation in the States.

A similar situation existed in the 1960s and 1970s with respect to the subtropical rainforests of northern New South Wales, except that the New South Wales Government was not opposed to the establishment of large areas of rainforest national park. Several large rainforest parks were actually gazetted in the mid 1970s, and in the mid 1980s the New South Wales Government agreed to a nomination by the Commonwealth Government of the subtropical rainforest parks as World Heritage areas. The listing of the areas subsequently occurred in 1986.

Another significant conflict over forest resource use and management occurred in south-eastern New South Wales, a geographical area of some 800 000 ha, of which approximately 425 000 ha are forested. Approximately 106 000 ha exist in State forests and 136 000 ha in national parks. Additional forest areas have been transferred to National Parks (June 2000) following the south-east Comprehensive Regional Forest Assessment Program (2.4). Woodchip production from these forests commenced in the mid 1960s. In 1985 renewal of the export licence was sought and logging in several biologically significant areas was planned. The environmental movement once again took action with a well-planned and bitter response. It sought to have old-growth forests and other ecologically significant forests listed on the National Estate. The forest industries argued against this as it would restrict access to forests deemed necessary for woodchip production. The latter was primarily an excuse, as the early predictions on the return cycles for harvesting indicated they could not be met and access to much larger areas of forest would be necessary to supply the export woodchip demands.

To broker a compromise between the forest industries and the conservation movement, the Commonwealth Government established a Joint Commonwealth–New South Wales Scientific Committee to 'assess, review and undertake studies to determine the biological values of the region'. This committee produced a report in 1990, which recommended the creation of additional conservation reserves and diversification of the forest industry and logging in some National Estate areas. Clearly, this was an advance in resolving the conflict between the forest industry and the conservation movement but the issue of timber resource security (guaranteed wood supplies) prevailed. This remained largely unresolved until 1998 when intensive biological studies of forests to address the issue were commenced as a basis of Commonwealth–State forest conservation and resource security agreements (Comprehensive Regional Forest Assessments and Commonwealth State Forest Agreements).

2.4 COMPREHENSIVE REGIONAL FOREST ASSESSMENTS

The long history of controversy, conflict and very polarized debate over the use and conservation of forests failed to satisfy the majority of stakeholders even with Commonwealth Government intervention. Neither the timber industry nor the conservation movement has readily accepted the decisions, which have been made since the 1960s to establish and maintain viable and sustainable forest industries.

As a way forward, the Commonwealth, State and Territory Governments developed and in 1992 signed a National Forest Policy Statement Commonwealth of Australia (1992). This statement established broad environmental and economic goals for the sustainable use of forests:

• To maintain an extensive and permanent native forest estate in Australia.

• To manage that estate in an ecologically sustainable manner so as to conserve the suite of values that forests can provide for future generations.

• To develop internationally competitive and ecologically sustainable forest-based industries that maximize value-adding opportunities and efficient use of resources.

The Statement further outlined a basis and approach for a joint Commonwealth–States' programme for comprehensive regional assessments of Australia's forests covering heritage, environmental, economic and social values. Following the completion of these detailed assessments, regional forest agreements between the Commonwealth and individual State or Territory Governments were planned. Scoping agreements were also signed committing the Commonwealth and individual State Governments to establish the procedures and processes for developing a regional forest agreement and a timetable for completion. The agreements also committed the Governments to negotiate regional forest agreements consistent with a range of Commonwealth and State policies and legislation, principally:

• the specific criteria developed by a Joint ANZECC–MCFFA National Forest Policy Statement Implementation Sub-committee (referred to as the JANIS criteria); the guiding principles of the National Strategy for Ecologically Sustainable Development, and the Intergovernmental Agreement on the Environment (Commonwealth of Australia 1997a, b);

• Commonwealth legislative requirements of the Environmental Protection (Impact of Proposals) Act of 1974, the Endangered Species Protection Act of 1992, the World Heritage Properties Act of 1983 (Commonwealth of Australia 1998) and the Australian Heritage Act of 1975;

• relevant forestry, conservation, heritage, threatened species, cultural, mining, land-use and wilderness legislation in the individual States and Territories.

The core issues of the JANIS criteria establish regional conservation objectives and strategies; the basis for conservation planning; biological diversity guidelines; old-growth forest, and wilderness components of a comprehensive, adequate and representative reserve system conserving samples of all forest ecosystems. For biological diversity conservation a minimum reservation of 15% of the original occurrence (pre 1750) of forest ecosystems on public lands was set as a basis. Other objectives include the reservation of ecosystems across their geographical range; reserve size to maintain rare and threatened species viability, quality and integrity of populations; minimum areas for reservation of vulnerable forest ecosystems (60% of extent of existing forest); conservation of old-growth forest (60% of identified old-growth areas); rare or very depleted forests (less than 10% of extant distribution), protection of all viable remnants; and for wilderness values 90% or more of the area of high-quality wilderness that meets minimum area requirements to be protected in reserves.

2.4.1 Regional forest assessments and agreements

Regional forest assessments have been carried out in a number of areas in the southern States and Western Australia, and several Common-

JANIS, Joint ANZECC–MCFFA National Forestry Policy Statement Implementation Subcommittee;
ANZECC, Australian and New Zealand Environment and Conservation Council;
MCFFA, Ministerial Council on Forestry, Fisheries and Agriculture.

wealth–State Government Regional Forest Agreements have been signed to date. Notable among these is the agreement covering the East Gippsland forest region in the State of Victoria and the agreement covering all of the State of Tasmania.

The Tasmanian forest agreement, being a whole-State agreement, serves as a good example of the comprehensive regional forest assessment and agreement process. It is also a good example of the benefits that accrue to the forestry industry, through security of resource supply, and to conservation through the reservation of additional forests as national parks and wilderness areas. The Tasmanian–Commonwealth Regional Forest Agreement is detailed in *Options for the Tasmanian–Commonwealth Regional Forest Agreement: A Strategic Approach* (Commonwealth of Australia 1997c), from which the following summary is drawn.

Tasmania depends on its forest resources and industries more than any other State, in terms of employment and contribution to the State's economy. Timber production and valued added timber products contribute in excess of 7% of the State's gross product. Of the total Tasmanian land area of 6.8 million ha, 3.4 million ha are covered by native forests and plantations. Approximately 70% of the forests occur on public lands, 40% as commercial forests, and 30% on protected public lands. The remainder of the forests are held in private ownership. Commercial forests cover approximately 1.6 million ha and dedicated reserves cover approximately 1.65 million ha. A further 650 000 ha of forest occur in informal reserves, but are legislatively insecure. They are available for mineral exploration and other uses. Prior to the commencement of the Tasmanian comprehensive regional forest assessment programme, approximately 320 000 ha were set aside as deferred forests or forest areas not available to the timber industry until the comprehensive assessment process was completed and the conservation values of the State fully examined and determined.

The comprehensive regional forest assessment had an objective of finding a compromise between the demands of the forest industries on forested lands and those of the conservation movement for the reservation of a comprehensive, adequate and representative sample of all forest ecosystems. This was achieved in Tasmania through the setting aside of additional forest lands as conservation reserves, while still providing for a sustainable forest industry based on the supply of up to 300 000 m^3 of high-quality hardwood sawlogs annually, as well as providing forests for a sustainable hardwood pulp log industry. Tasmania produces approximately three million tonnes of pulpwood products a year and, as in other States, the pulpwood forest industry has been the most controversial.

Additional forest areas were identified in the assessment process and set aside as multiple-use forests, where a range of values and a variety of uses could be effectively managed without compromising the overall forest values. These were identified as having a high potential for nature-based tourism and could provide an alternative to the increasing use and visitor pressure on national parks for ecotourism, adventure tourism, vehicle-based recreation and environmental education.

2.4.2 Outcomes of the Tasmanian Forest Assessment Process

Fifty forest communities were recognized in the forest assessment process as representing the biological diversity of Tasmanian forests. Only 18 of the communities were identified as existing in conservation reserves at levels that met, or exceeded, the reservation criteria (JANIS criteria) agreed to prior to the forest assessment being undertaken. Approximately 90% of the original (pre-1750) area of the 18 forests communities still exist, providing an opportunity to ensure comprehensive and adequate representation of the 18 forest ecosystems in conservation reserves. Less than 53% of the pre-1750 area of the other 32 forest communities currently exists, these communities being very poorly represented in existing conservation reserves. For 17 of the communities, there is not sufficient public land available to meet the conservation criteria. The overall shortfall in forest area required to meet the conservation criteria is about 215 000 ha, much of which would need to be accessed from privately owned forest lands.

Forty-three forest communities were assessed

for old-growth attributes and 11 communities were identified as rare or depleted. Fourteen of the 43 communities met the JANIS criteria in terms of reservation while 10 of the 14 also met the JANIS forest biological diversity criteria. As with the JANIS reservation criteria for pre-1750 forest communities, there is not enough public land to meet reservation requirements for old-growth forest for 16 of the 29 communities that do not meet the criteria. The overall shortfall between existing old-growth forest and the area needed to be reserved to meet the JANIS criteria is approximately 17 300 ha.

In terms of meeting other native flora reservation criteria, 124 threatened species were identified for which the percentage of core populations to be protected ranged from 30% for 24 category 4 species, to 100% for 30 category 1 species. (Categories for formal listing of rare and threatened native plant species.)

Tasmania has vast areas of undisturbed forest lands but a further 16 areas of high-quality wilderness exceeding 8000 ha in area (JANIS criteria) were identified, of which 86% (approximately 1.65 million ha) occurred within existing dedicated and informal reserves. An additional 275 000 ha was identified as high-quality wilderness on other public lands, with a very small area on private lands. Forty-two per cent of the high-quality wilderness (approximately 824 000 ha) is forested land. The additional area on public lands will enable the JANIS criteria for wilderness to be readily met. A large proportion of the identified wilderness area occurs in the Tasmanian World Heritage Area, an area of 1.38 million ha.

2.5 WILDERNESS AND FOREST MANAGEMENT

While the JANIS criteria now requires consideration of wilderness values and wilderness conservation in regional forest assessment programmes, the evolution of the wilderness ethic in Australia predates the regional forest assessment programme. The early European settlers in Australia were confronted by a generally harsh and forbidding landscape and most viewed the country as one large wilderness to be tamed and opened up to agriculture to feed the young and growing colony.

The conservation concept of wilderness emerged relatively quickly, as notable botanists and senior government officers recognized the increasing rate of destruction of forests as settlers moved inland from the coastal colony settlements. Only 100 years after the first settlement in 1788, over 25% (95 million ha) of the forests in the colony of New South Wales had been cleared. Concerns were expressed in the nineteenth century by a number of noted botanists including Charles Moore, Director of the Sydney Botanic Gardens, W.B. Clarke, and Baron Ferdinand Von Mueller, the first Victorian Government Botanist. The concern for the rate and extent of forest destruction in Australia by the early settlers was even recognized in New Zealand in the late nineteenth century. As Hall (1992) records, a John Matson of Christchurch, New Zealand, wrote a letter in 1892 to a Sydney newspaper appealing for the creation of Australian indigenous parks for the preservation of wildlife and its habitat. Matson drew parallels between Australia and New Zealand in nature conservation and wilderness establishment (or the lack of it) and while he singularly had little influence, the New Zealand experience had begun to influence Australian policy-makers and several committed conservationists.

The wilderness movement in the USA also had a considerable influence on the development of the Australian wilderness ethos. In the early 1920s, the United States Forest Service and the National Parks Service, who between them managed vast tracts of near-pristine natural land, agreed to control access and resource development in key primitive areas under their control in order to protect wilderness recreation values (Robertson *et al.* 1992). The American wilderness movement was closely followed by Dunphy (often called the John Muir of Australian conservation) and colleagues who increasingly agitated for the establishment of national parks and wilderness areas in Australia, particularly near to Sydney and in the south-east of the country (Mosley 1978). As Mosley notes: 'the Australian wilderness reserve movement developed spontaneously to meet a local need, but owes something to the awareness of the success of this method in the USA'.

Hendee *et al.* (1990) note that the common

English colonial background and historical frontier experience of the USA, Canada, New Zealand, South Africa and Zimbabwe has given rise to very similar concepts of wilderness and wilderness ethos in all of these countries. Interestingly, it was only in 1964 that the United States National Wilderness Preservation System was established through legislation. The legislation established 54 wilderness areas encompassing 3.65 million ha of national forests in 13 States. Ten States now have wilderness areas exceeding 600 000 ha. The total number of wilderness areas now exceeds 600 covering some 41 million ha, or 4.5% of the total US land area. The majority of areas are in the western States, principally California, but more than half of the total wilderness area occurs in the State of Alaska (23.2 million ha).

Mosley (1978), Thompson (1986), Robertson *et al.* (1992) and Hall (1992) provide detailed histories of the evolution of the wilderness movement in Australia and all note 'the change in Euro–Australian attitudes from the pejorative to the protective', and the role played in influencing this change by Myles Dunphy; the dominant person in the early push for wilderness conservation in Australia. Dunphy and colleagues campaigned for several decades from the 1920s for the protection of significant natural areas as national parks and wilderness areas, although little action was taken in terms of wilderness protection until the 1960s. In 1920 Dunphy expressed his concern for the lack of wilderness and appealed to State and Commonwealth Governments 'to set aside (primitive areas) in time before alienation can take place, or the wilderness qualities be leased away forever'. Dunphy formed the Mountain Trails Club in 1914 and it was this Club which was central to the development of an Australian wilderness ethic (Hall 1992).

The subsequent establishment of many other bushwalking clubs was instrumental in establishment of the National Parks and Primitive Areas Council in 1932. This was described by Thompson (1986) as the first wilderness society. It maintained an anthropocentric view of wilderness and the American concept of wilderness: 'that one may be able to travel on foot in any direction for at least a full day without meeting a road or highway'.

The National Parks and Primitive Areas Council, while promoting the establishment of both national parks and wilderness, recognized that dedicating natural areas as national parks did not necessarily provide for the conservation of wilderness values. The Council as a response campaigned for primitive areas both inside and outside parks to be gazetted as wilderness areas. Unfortunately, this had little success, but the basis for wilderness had at least been well established for later years when public appreciation and expectation of natural area land managers demanded the identification and gazetting of wilderness areas, particularly in forest lands under threat from commercial exploitation.

Frawley (1989) and Ramson (1991) note that as wilderness has become less available, greater appreciation of the need to respect and protect the Australian environment has developed, to the extent that wilderness protection has actually become a justification for the establishment of additional conservation reserves. This is now reflected in the requirement to undertake assessment of wilderness values in any regional forest assessment.

2.6 WILDERNESS ASSESSMENT AND IDENTIFICATION

McCloskey and Spalding (1987) published the first world-wide inventory of wilderness in which they identified 1050 wilderness areas around the world which exceeded 4 000 000 ha in area. They identified 82 areas in Australia meeting their criteria, these potential wilderness areas totalling in excess of 250 million ha or approximately 4.9% of the total global wilderness. Based on their surveys only the Soviet Union (now Russian Federation), Antarctica and Canada had larger areas of wilderness (Robertson *et al.* 1992).

Earlier in the 1970s and 1980s a number of systematic surveys of wilderness values and wilderness potential were undertaken in Australia by Helman *et al.* (1976), Stanton and Morgan (1977), Feller *et al.* (1979), Russell *et al.* (1979), Kirkpatrick (1979), Hawes and Heatley (1985) and

Prineas *et al.* (1986). Prineas *et al.* (1986) identified 157 areas 'throughout Australia and its Territories having wilderness values and worthy of consideration for protection and management as wilderness areas'. These authors also noted (at the time) that the identified areas having wilderness values were not reserved in conservation reserves and that much work still needed to be done in terms of wilderness identification.

The need for an Australian-wide wilderness survey and evaluation was recognized by the Australian Heritage Commission in 1986 when the National Wilderness Inventory (NWI) was commenced (Lesslie *et al.* 1993). As these authors note, the NWI was not undertaken to list or catalogue wilderness areas but only established a database and set of techniques designed to enable evaluation and identification of wilderness quality across the Australian landscape. It is a computer-based mapping system which conceives wilderness as being part of a spectrum of remote and natural conditions which vary in intensity from undisturbed to urban (Fig. 2.5) (Lesslie & Taylor 1985). As such, the NWI assessment process is not matched to any biocentric or anthropogenic concept of wilderness, as was the case with the wilderness groups.

The NWI measures wilderness quality across the landscape using four wilderness indicators that represent the two essential attributes of wilderness, these being remoteness and naturalness. The indicators are derived from the definition of wilderness quality, as to the extent to which a location is remote from and undisturbed by the influence of modern technological society. These indicators are distance from settlements and access, apparent naturalness (the degree to which the landscape is free from the presence of permanent structures) and biophysical naturalness (the degree to which the natural environment is free from biophysical disturbance caused by modern technological society) (Lesslie *et al.* 1993; Lesslie & Maslen 1995). The areas of high-quality wilderness values, as determined by the application of NWI criteria and process, are indicated in Fig. 2.6. It is to be noted that a number of smaller wilderness areas have been established along the eastern seaboard in the areas of high-quality forest, even though the areas do not rank highly in the NWI assessment process. By contrast, the desert wilderness areas are the largest areas having high NWI values but these have experienced little conflict in their establishment compared with that of the wilderness areas in the forest environments. This has been due mainly to the extreme remoteness of the desert areas, the low level of past pastoral use, and a very limited demand for recreation in these areas. The desert wilderness areas are arguably the closest to the general public's sense of true wilderness.

2.6.1 Wilderness assessment in the comprehensive forest: assessment process

Several approaches and techniques have been applied to wilderness identification and assessment in the comprehensive regional forest assessment programmes in the various States. All of these approaches are partly or fully subjective in setting or applying remoteness, size, disturbance and other criteria. Legislative requirements established in each State, while having a common objective, vary considerably. Hence the JANIS wilderness criteria have been applied as the base criteria and objective for the identification of wilderness in the comprehensive regional assessment process. The JANIS Criteria require that at least 90% of all forested areas in excess of 8000 ha having identified wilderness values, to be placed in conservation reserves.

The New South Wales State Wilderness Act of 1987 was the first wilderness legislation in Aus-

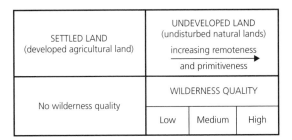

Fig. 2.5 The wilderness continuum concept. (After Lesslie & Taylor 1985.)

Fig. 2.6 Wilderness areas in Australia. (After Prineas *et al.* 1986.)

tralia and the assessment, identification, declaration and management of wilderness is principally guided by it. New South Wales serves as an example of the relationship between wilderness identification at the State level and that of the comprehensive regional forest assessments. In the comprehensive regional forest assessments being undertaken in New South Wales the JANIS criteria and the requirements under the NSW Wilderness Act are addressed concurrently, through a recently developed process which aims to quantify criteria in numerical terms, particularly that of 'naturalness'. In general if the requirements of the NSW Wilderness Act are met, the JANIS criteria are more than adequately met as the NSW Wilderness Act provides for larger areas of wilderness than would be the case when only applying the JANIS criteria. This occurs because the NSW Wilderness Act acknowledges that not all areas within a nominated wilderness area at the time of assessment required to have high-quality wilderness values, if they could in reasonable time regain wilderness values through the

input of appropriate restoration measures. It also provides for some areas which do not meet the wilderness criteria to be considered as wilderness when the lands are required to protect the integrity of the larger part of a wilderness area, or are required to assist wilderness management. The Act states that a wilderness is 'an area of land that:

(a) together with its plant and animal communities is in a state that has not been substantially modified by humans and their works or is capable of being restored to such a state;

(b) is of a sufficient size to make its maintenance in such a state feasible; and

(c) is capable of providing opportunities for solitude and appropriate self-reliant recreation.'

It further states that 'a wilderness shall be managed so as:

(a) to restore (if applicable) and to protect the unmodified state of the area;

(b) to preserve the capacity of the area to evolve in the absence of significant human interference; and

(c) to permit opportunities for solitude and appropriate self-reliant recreation.'

2.6.2 Assessment criteria

Wilderness areas are identified and assessed through the application of criteria pertaining to naturalness, size, solitude and remoteness and capability of restoration.

Naturalness

The naturalness of a potential wilderness area is considered to be its persistence or capacity to remain in a state substantially unmodified by existing and future human activities. Quantitative measurement of naturalness is difficult, but a measure of disturbance against the pristine state can be determined which reflects the naturalness of an area. Any measure of naturalness is therefore not a measure of presence or absence of modification but an index of the degree of modification. (See also discussion by Peterken (Volume 1, Chapter 3))

The definition of wilderness as derived from the Wilderness Act places these areas at the 'least modified' end of the spectrum of land uses and human impacts on the landscape but not necessarily in an untouched and pristine condition. The Act's provisions as to the capacity of an area to be restored to an essentially unchanged natural condition also indicates that some degree of modification can be tolerated within the bounds of potential for restoration within a reasonable timeframe. The modifications or disturbances that are evident in an area are considered in terms of their effects on, or changes to, the key components of the ecosystems that determine an area's naturalness. Examples of these key components are: functional components of the ecosystems; spatial arrangement or distribution of species; species diversity, abundance, age-classes; and successional patterns or changes over time. Changes in structure and composition are readily measured and generally have a direct relationship with the anthropocentric view of recreation and experiential values of wilderness. Two methods have previously been used to categorize the degree of naturalness in wilderness assessments (Laut *et al.*

1977; Lesslie *et al.* 1987). They developed broad descriptive categories of native vegetation disturbance. Laut *et al.* used four descriptors: undisturbed natural, disturbed natural, degraded natural and cultural. Lesslie *et al.* developed descriptors in a hierarchy of degrees of biophysical alteration, using five classes of naturalness, unused by Europeans: low-intensity use now ceased, high-intensity use now ceased, low-intensity use continuing and high-intensity use continuing.

The two systems are qualitative and are highly subjective and have been used in one form or another in a number of wilderness assessments. Being subjective they are not entirely appropriate for the assessments of wilderness areas as they do not provide for consistent replication when used by different personnel with a range of land assessment skills.

The degree or extent of modification of the natural ecosystems and the capacity of an area to be restored to a substantially unmodified state is linked to the above categories but is assessed on:
• the extent of substantially unmodified vegetation cover;
• the extent, location and distribution of modified areas;
• the past or continuing disturbances;
• the degree of modification evident;
• the potential for restoration of all or the majority of modified areas; and
• the probable timeframe for restoration to a substantially unmodified state.

The implication here is that an area does not have to have the potential to be restored to the pristine state to qualify as wilderness. Recognition is also given to the variation in time that various forest ecosystems will take to return to the substantially unmodified state.

Size of a potential wilderness area

The size of a parcel of land has in the past been a core attribute in wilderness assessment, based on an area being large enough to sustain its natural systems and large enough for users to feel satisfied they have established contact with the wilderness (Helman *et al.* 1976). The International Union for the Conservation of Nature (IUCN) considers that the minimum size for a wilderness is determined by the total area required to ensure the integrity of the ecosystems and to accomplish the management objective which provides for its protection (IUCN 1990).

The general perception is that the larger the area the better, although Ledec and Goodland (1988) note that the area required to support a viable population varies with the species and ecosystems, and this is scientifically highly uncertain. Slatyer (1975), for example, calculated that in arid areas of Australia, to sustain a minimum viable population of 5000 Red Kangaroos, a reserve or wilderness area would have to be in the order of 500 000 ha. Alternatively, in the higher rainfall areas towards the coast an area of only 50 000 ha would be sufficient to support a minimum viable population of Grey Kangaroos.

Large areas also generally contain a greater species diversity and a greater genetic diversity within species, providing for a greater range of ecosystem processes to occur without human influence. Management in terms of maintaining viability, self-sustainability and biological integrity is also increasingly less demanding as areas get larger. A minimum area of 25 000 ha has been for many years the baseline for the establishment of wilderness areas in Australia, but following the work of Lesslie *et al.* (1993) the Commonwealth Government adopted 8000 ha of high-quality (category 12 or above) wilderness from the National Wilderness Inventory (NWI) as the minimum area to be considered in the identification of wilderness in the Comprehensive Regional Forest Assessment process (as stated in the JANIS criteria).

Recreation and solitude

Recreation and solitude are generally recognized by the public as being central to the concept of wilderness and numerous attempts have been made to define recreational and experiential indicators. Lesslie *et al.* (1987) used remoteness from access as an indicator of solitude. This can be readily measured (quantified) using geographical information system (GIS) techniques, but it is a very artificial measure as even small but remote areas can provide for solitude and wilderness-like experiences for many people. The concept of what

is appropriate self-reliant recreation is also a very subjective measure, as what is appropriate in one wilderness area may not be in another and will vary from one user to another. The 'appropriateness' of a recreational activity is again difficult to define but may be considered in terms of one that does not influence or diminish the wilderness values, particularly biological integrity.

2.7 A NEW APPROACH TO WILDERNESS ASSESSMENT

An assessment of the wilderness values of the extensive forested lands in the south-east of New South Wales was undertaken as part of the South-east (Eden) Comprehensive Regional Forest Assessment programme carried out in that part of the State between 1996 and 1998. In undertaking the wilderness assessment, it was acknowledged that existing assessment procedures relied very much on the perceptions of those carrying out the programme. To ameliorate the limitations of such subjective approaches, a quantitative (analytical) technique was developed to enable consistent and repeatable assessments, and to provide for unbiased comparisons between potential wilderness areas. The development of this quantitative assessment process ('Brogo' approach) was not the first attempted in Australia, as Kirkpatrick and Haney (1980) proposed a quantitative system in which they assessed wilderness quality against criteria related to recreational use of natural areas in Tasmania. The Kirkpatrick and Haney system assigned numerical values to remoteness and naturalness at points in the landscape which they equated to primitiveness. The points to which they assigned values for wilderness quality were the highest points in each 4-km^2 grid cell of a wilderness assessment area. The highest point was used as the assignment point, as they considered wilderness values were more influenced by the view walkers could get from peaks than other features and hence the greater wilderness experience they would gain.

Kirkpatrick and Haney considered remoteness to be the walking time from the highest point(s) to the closest entry point to the wilderness area from a vehicular access site. They used a continuum approach for assigning values for remoteness, starting at a walking time of 4 hours from the vehicle access sites. For example, 50% of the total wilderness value for remoteness is attributed at the 8-hour point, with 100% wilderness value accruing at a walking time in excess of 48 hours. This concept of wilderness, that wilderness is at least a half-day walk from vehicular access, is very much a North American concept of wilderness and is generally not considered appropriate to Australia. Kirkpatrick and Haney also measured naturalness by assigning a value to the degree-of-disturbance to the viewfield of the natural landscape (Fig. 2.7). Wilderness values increased when the viewfield extended beyond 5 km from a viewing point (the highest point). A disturbance within less than 5 km was considered to negate the wilderness values.

The Kirkpatrick and Haney approach was an advance over earlier subjective wilderness assessment approaches in that it applied numerical values to the two core criteria but the surrogates for remoteness and naturalness were not widely accepted by those involved in wilderness assessment in other States.

2.7.1 The Brogo Wilderness Assessment Programme

The Brogo approach, like most assessment techniques, uses the core criteria of naturalness, size and capacity for solitude and self-reliant recreation but quantifies these against predetermined criteria and limits.

Naturalness assessment

As with other assessment techniques surrogates for naturalness are used in the Brogo system, these being past land-use disturbance and eco-system recoverability. A weighting system was developed and applied to the disturbance and recoverability of any unit of land within the wilderness study area, to provide a classification of an area into one of four 'naturalness' categories

Fig. 2.7 Disturbances (roads) in the wilderness viewfield. (Photo: C. Totterdell.)

as defined by Laut *et al.* (1977). These areas, as noted earlier, are defined as 'undisturbed natural; disturbed natural; degraded natural and cultural'. Naturalness is identified and measured as 'substantially modified; modified but capable of restoration in a reasonable timeframe, and modified and not capable of restoration but needed for management purposes'. Specific criteria and weightings are applied to each category.

The weighting system

Any disturbances, both past and present, that affect the naturalness of a site are identified for the entire area being assessed for wilderness values. Each disturbance factor is assigned an independent disturbance and recoverability weighting using a Delphi approach, which is a systematic process to obtain and integrate expert judgements on a subject (McAllister 1990). An expert panel is used to assign a numerical value to each criteria and a ranking to each identified unit within an assessment area. Weightings are then applied for the range of disturbance levels identified. For the Brogo Wilderness Assessment for which the Brogo approach was developed, a weighting factor between 1 and 30 was applied. This range would vary in the assessment of other wilderness areas, depending on the range of disturbances identified. A weighting factor of 1 is taken as correlating to minimal disturbance and 30 being the worse-case ecosystem disturbance.

The expert panel is also used to assign recoverability weightings to each unit area according to a prediction of the time an area or ecosystem would take to rehabilitate to a near-natural state or recover its 'naturalness'. A scale of 1–10 was used, with a weighting of 1 indicating a minimum timeframe for an ecosystem to recover, and with a weighting of 10 representing a recovery period in excess of 25 years. Again, the upper weighting figure depends on the ecosystems within any assessment area. For example, a rainforest ecosystem may recover considerably faster than a dry sclerophyllous forest, and hence the range of the weighting values would be varied to reflect these different rates of recovery. The disturbance weighting and the corresponding recoverability weightings are multiplied to give an 'impact weighting', which reflects the compound effect of disturbance and recoverability, and the time since occurrence of the disturbance factor (Table 2.2).

For multiple impacts in any one area, the weightings are additive. For example, in an area where selective logging has taken place, planned fire regimes have been imposed for management purposes and recreation is taking place, individual weightings are simply added together to give a cumulative weighting.

The range of analytically derived 'impact weightings' is then correlated with the four disturbance categories (as defined by Laut *et al.* 1977) following identification/determination of acceptable cumulative impact thresholds for each disturbance category. For the Brogo wilderness assessment, the resultant match of derived impact weightings to the disturbance categories were:
• undisturbed natural equated to an impact weighting range of 50–100;
• disturbed natural, 80–220;
• degraded natural, 185–555;
• cultural equated to a figure at or above 500 (generally beyond any wilderness identification threshold).

A value of 250 was determined through a match of legislative requirements and operational factors as the upper threshold value for the Brogo wilderness identification process. Again, the threshold value would vary for other wilderness assessments in different environments and ecosystems, as a response to changes in the range of weighting values applied to the four disturbance categories. The threshold of 250 represents the lower 'least disturbed' end of the entire spectrum of natural lands as considered in the four Laut *et al.* categories. The relationship of the Laut *et al.* (1977) classification, the definitions of disturbed land in the NSW Wilderness Act, and the analytically derived impact weightings are shown in Fig. 2.8.

As noted earlier, the majority of wilderness areas require a buffer or management zone around them to mitigate against incompatible wilderness activities such as agriculture and commercial forestry operations which may be carried out on neighbouring lands. In this situation, bordering areas that have impact weightings exceeding the

Table 2.2 Disturbance and recoverability ratings as applied in the Brogo Wilderness Assessment process.

Disturbance	Source of impact	Damage rating	Recoverability rating
Fires (no. since 1970)	1 fire	1	1
	2 fires	3	3
	3 fires	8	5
	4 fires	15	8
	5 fires	30	10
Logging	Minor	2	1
	Light selective	10	5
	Moderate selective	15	7
	Heavy (pulp and sawlog)	30	10
Grazing (non-cleared)	Open range	1	2
	Seasonal	2	3
	Continuous	6	4
Agriculture	Native pasture	1	2
	Native pasture fenced	2	3
	Chemically improved	10	6
	Physically improved	20	10
	Apiary	2	2
Cultural sites	Domestic (hut ruins)	2	1
	Industry (e.g. mines)	3	2
Recreation	Bushwalking	1	1
	Horseriding	1	1
	Vehicle based		
	4-wheeled driving	3	1
	2-wheeled driving	4	1
Access	Bridletrail/walking	2	1
	Vehicle based		
	4-wheeled driving	6	5
	2-wheeled driving	8	5

threshold of 250 may be included in the identified wilderness if there is a justifiable management purpose. This can be noted in Fig. 2.8, where areas classified under the Wilderness Act, viz. 'modified, not restorable but required for management' are acceptable as wilderness even though they would be classified as 'degraded natural' and unacceptable under the Laut *et al.* (1977) disturbance categories. The acceptance of lands having an impact weighting exceeding 250 only applies to boundary areas where management issues demand that the lands be part of the identified wilderness area.

Other assessment criteria

Size criteria. The early concepts of wilderness were based on very large, undisturbed natural areas where little or no recreational activities took place. The very large area concept also was seen as providing for a 'let nature take its course approach' to management and one where ecosystems could function without any interference by external factors, particularly humans. As noted earlier, what was a very large area was generally not defined but was, by inference, somewhere in the order of 25 000–250 000 ha. The acceptable

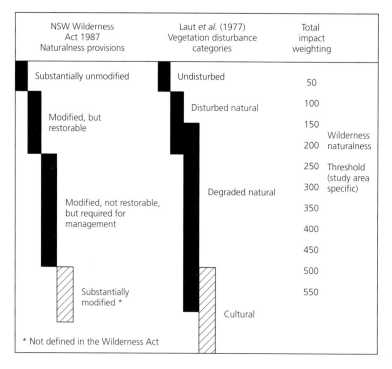

Fig. 2.8 Relationship of New South Wales Wilderness Act definitions, Laut *et al.* (1977) naturalness categories and analytically derived impact weightings on the assessment of naturalness (specific to the Brogo Wilderness Study Area).

size for wilderness in most situations was based on the minimum area to maintain a viable population of the native animals that inhabited an area. While nature conservation is an obvious component of wilderness conservation, the concept of wilderness is one of remoteness, solitude and experiential enjoyment which can be gained from areas much less than 25 000–250 000 ha. The more commonly accepted minimum area is now in the order of 15 000 ha, while the National Wilderness Inventory (Lesslie *et al.* 1994) considers 8000 ha of high-quality wilderness (NWI 12 areas) as the minimum area. Smaller areas can be considered to meet the NWI size criteria if 10% or more of the total boundary length of a wilderness adjoins an existing identified or declared wilderness. An area of 5000 ha or larger is acceptable where it adjoins the coastline or would be topographically protected.

Wilderness areas assessed using the 'Brogo' approach would in most assessments result in areas exceeding the 8000 ha (NWI 12 or above)

minimum and generally the 15 000-ha area considered as the minimum area in earlier New South Wales wilderness assessments.

Solitude and self-reliant recreational experiences. These are individual perceptions and vary from one wilderness user to another, and hence are not given actual values in the Brogo approach. It is accepted that wilderness areas meeting the core assessment criteria and that of minimum size will provide a more than adequate range of self-reliant recreational experiences while still providing for solitude and inspirational experiences for those who seek it. The solitude and self-reliant recreation criteria are therefore only 'modifiers' in the final application of numerical values in the identification of a wilderness area through the Brogo process.

Cultural heritage

Wilderness areas provide many aesthetic, spiritual, psychological and educational benefits for all Australians, including both Aboriginal and non-Aboriginal people. For Aboriginal people the land and wilderness is central to their culture and they hold a very strong affinity to both. Part of this affinity with the land was and is the spirituality which the Aboriginal population attribute to the earth and land.

The very sensitive land management practices carried out by the Aboriginal tribes to ensure their very existence also shaped and influenced the lands which now still retain wilderness values so important to the non-Aboriginal population. The Aboriginal populations were as such an integral part of the evolution and ecology of the land and the Australian natural landscape for thousands of years. Much of this has been lost but wilderness areas now provide the opportunity for the non-Aboriginal population to maintain and preserve this significant cultural entity. As Robertson *et al.* (1992) consider, 'indigenous culture and land management factors were, and in many places remain, vital factors in ongoing ecological processes, the relationship of Aborigines and Islanders to their country is inherently an important wilderness value'.

Wilderness assessments now encompass both Aboriginal and non-Aboriginal cultural considerations, this being particularly important in the forested wilderness areas where so much of past cultural values have been lost through ignorance and poor land-use management practices. As Robertson *et al.* (1980) acknowledge:

> Recognition of Aboriginal and Islander relationships with Australia's wilderness holds the potential for increased non-Aboriginal respect for something that has not ceased to be a primary indigenous concern: caring for country. Wilderness not only holds a key to the continued growth of non-Aboriginal, Aboriginal and Islander identity, but also to the ability of each to respect and learn from the other.

The assessment of cultural values in terms of spiritual, psychological and aesthetic values is almost impossible to quantify but is a significant consideration which in most situations would raise a marginal wilderness proposal to one worthy of establishment. As with size and remoteness criteria, cultural criteria are important modifiers to any outcome from the analytical assessment of wilderness.

2.8 SUMMARY

Forestry and the forest industries have contributed much to the Australian economic and social well-being. In the early years of European settlement forestry was a major primary industry supporting the early development of the country. The early recognition by Governments during the nineteenth century of extensive clearing of forests for agriculture led to the reservation of large tracts of forest land for future commercial forestry. Indirectly, the reservation of these large tracts of forest provided the opportunity, some 100 years later, for permanent reservation and conservation in national parks and other conservation reserves of representative samples of the original forest communities and ecosystems.

In the mid twentieth century the establishment of the various State conservation agencies commenced, although the conservation movement evolved much earlier (early twentieth century). It was at this time that conflicts between forestry and conservation agencies and the conservation movement commenced. Much of the area reserved in the 1800s for future commercial forestry was subsequently, in the 1960s, proposed as national parks, nature reserves, other protected areas and wilderness, resulting in several decades of very bitter conflict over natural area management and resource use.

A resolution to the conflicts was sought in the mid 1990s by State and Commonwealth Governments and a Regional Forest Assessment process was put in place, with an objective of securing timber resource supply and the equitable allocation of forest lands in a comprehensive, adequate and representative conservation reserve system. This has been achieved through detailed studies of the forest ecosystems and their biological diversity, together with regional assessments of the social and economic impacts on local communities of changes in land use of large areas of forest from commercial forests to conservation reserves and wilderness areas.

The process has met many of the objectives set down in the scoping agreements but several issues remain unresolved. These include issues derived from the different State legislation relevant to forest and land management, and the different State environmental policies and directions. Irrespective of the differences between the States (and Commonwealth Government), the *Comprehensive Regional Forest Assessment process has proven to be a very effective way of finding a balance between production forestry and forest ecosystem conservation, including Aboriginal and non-Aboriginal cultural values.* The biological diversity survey techniques, wilderness assessment programme, community involvement and negotiation processes used in determining the allocation of resources are major contributions to the future management of forest ecosystems in the areas retained as commercial forests, national parks and conservation reserves, and in wilderness areas established as an outcome of the regional forest assessments. The quantitative wilderness assessment approach developed as part of the south-east (Eden) regional forest assessment process is an advance in transparent, consistent and repeatable wilderness identification.

The regional forest assessment process is recommended for consideration by forestry and conservation agencies in other countries where forest land use and management conflicts similar to those in Australia are evident. The quantitative wilderness assessment programme is also an advance over earlier subjective assessment programmes; however, this programme approach would only be applicable in countries where the wilderness concept and ethic is similar to that in Australia. In developing countries where the indigenous population remains a very integral part of the natural landscape and draws from it its continuing existence and well-being, other international conservation concepts such as world biosphere areas may be more appropriate than the establishment of wilderness areas. Nevertheless, the recent approach taken to forest management and resource allocation in Australia is a model with much to offer to the future management of forested lands around the world.

ACKNOWLEDGEMENTS

The development of the 'Brogo' approach to wilderness assessment was a collaborative project by staff of the National Parks and Wildlife Service, particularly Allison Treweek, Kelly Mullen, Imogen Fullager, Tony Baxter, Allan Ginns and Ross Constable. The product of their combined efforts is acknowledged and recorded in the above overview of the process.

REFERENCES

Berkmuller, K., Southammakoth, S. & Vongphet, V. (1995) *Protected Area System. Planning and Management in Loa PDR: Status Report to Mid-1995.* IUCN and Loa-Swedish Forestry Co-operation Program, Forest Resource Conservation Project, Vientiane.

Carron, L.T. (1985) *A History of Forestry in Australia.* Pergamon–Australia National University Press, Canberra.

Carron, L.T. (1988) The state of knowledge of the history of public forests and forestry in Australia. In: Frawley, K.J. & Semple, N., eds. *Australia's Ever Changing Forests. Proceedings of the First National Conference on Australian Forest History*, pp. 217–21. Department of Geography and Oceanography, Australian Defence Force Academy, Campbell ACT.

Carron, L.T. (1993) Changing nature of federal-state relations in forestry. In: Dargavel, J. & Feary, S., eds. *Australia's Ever-Changing Forests II. Proceedings of the Second National Conference on Australian Forest History.* Centre for Resource and Environmental Studies, Australian National University, Canberra. pp. 207–39.

Chape, S. (1996) Biodiversity conservation, protected areas and the development imperative in Lao PDR: forging the links. IUCN–Lao PDR Discussion Paper No. 1. Vientianne, Lao PDR.

Commonwealth of Australia (1992) *National Forestry Policy, A New Focus for Australia's Forests*, 2nd edn. Australian Government Printing Service, Canberra.

Commonwealth of Australia (1997a) *National Agreed Criteria for the Establishment of a Comprehensive, Adequate and Representative Reserve System for Forests in Australia.* A Report to the Joint ANECC/MCFFA National Forest Policy Statement Implementation Sub-committee, Canberra.

Commonwealth of Australia (1997b) *National Strategy for the Conservation of Australia's Biological Diversity.* Commonwealth Department of the Environment, Sport and Territories, Canberra.

Commonwealth of Australia (1997c) *Options for the Tasmanian–Commonwealth Regional Forest Agreement: A Strategic Approach.* Tasmanian–Commonwealth Joint Regional Steering Committee. Hobart, Tasmania.

Commonwealth of Australia (1998) *World Heritage Report.* Record of the World Heritage Expert Panel

Meeting: Western Australia, New South Wales and Queensland.

Cresswell, I.D. & Thomas, G.M. (1997) *Terrestrial and Marine Protected Areas in Australia*. Environment Australia, Biodiversity Group, Canberra.

Dargavel, J. (1987). Prospects—present and preferred. In: Dargaval, J. & Sheldon, G., eds. *Prospects for Australian Hardwood Forests*, pp. 1–12. Centre for Resource and Environmental Studies, Australian National University, Canberra.

Davey, A.G. & Phillips, A. (1998) *National Systems Planning for Protected Areas. Best Practice Protected Area Guidelines Series 1*. World Commission on Protected Areas (WCPA), IUCN/Cambridge University, Gland, Switzerland/ Cambridge, UK.

Feller, M., Hooley, D., Drecher, T., East, I. & Jung, R. (1979) *Wilderness in Victoria: An Inventory*. Department of Geography, Monash University, Clayton, Victoria.

Frawley, K. (1989) Cultural landscapes and national parks: philosophical and planning issues. *Australian Parks and Recreation* **25** (3), 16.

Graetz, D., Fisher, R. & Wilson, M. (1992) *Looking Back. The Changing Face of the Australian Continent. 1972–92*. CSIRO Division of Wildlife and Ecology, Canberra.

Green, M.J.B. & Paine, J.R. (1997) *State of the World's protected areas at the end of the twentieth century*. Paper presented at Protected Areas in the 21st Century: From Islands to Networks, Albany, Australia, November 1997.

Hall, C.M. (1992) *Wasteland to World Heritage- Preserving Australia's Wilderness*. Melbourne University Press, Carlton, Victoria.

Hawes, M. & Heatley, D. (1985) *Wilderness Assessment and Management*. The Wilderness Society, Hobart.

Helman, P.M., Jones, A.D., Pigram, J.J. & Smith, J.M. (1976) *Wilderness in Australia: Eastern New South Wales and Southeastern Queensland*. Department of Geography, University of New England, Armidale.

Hendee, J.C., Stankey, G.H. & Lucas, R.C. (1990) *Wilderness Management*, 2nd edn. International Wilderness Leadership Foundation. North American Press, Colorado.

International Union for the Conservation of Nature and Natural Resources (1990) *A framework for the classification of terrestrial and marine protected areas*. Submitted to the Commission on National Parks and Protected Areas Meeting in Perth, Australia.

Just, T.E. (1987) Management of tropical rainforests in North Queensland. In: *Forest Management in Australia. Proceedings of Conference of the Institute of Foresters. Perth 1987*, pp. 299–312. Australian Institute of Foresters.

Kirkpatrick, J.B. (1979) *Hydroelectric Development and Wilderness in Tasmania*. Department of Environment, Hobart.

Kirkpatrick, J.B. & Haney, R.A. (1980) Quantification of developmental wilderness loss: the case of forestry in Tasmania. *Search* **11**, 331–5.

Laut, P., Heyligers, P.C., Keig, P. *et al.* (1977) *Environments of South Australia*. CSIRO Division of Land Use Research, Canberra.

Ledec, G. & Goodland, R. (1988) *Wildlands: Their Protection and Management in Economic Development*. World Bank, Washington DC.

Legg, S.M. (1988) Re-writing the history of forestry? Changing perceptions of forest management in the New World. In: Frawley, K.J. & Semple, N., eds. *Australia's Ever Changing Forests. Proceedings of the First National Conference on Australian Forest History*. Special Publication no. 1. Department of Geography and Oceanography, Australian Defence Force Academy, Campbell ACT.

Lesslie, R., Mackay, B.G. & Preece, K.M. (1987) *National Wilderness Inventory: A Computer Based Methodology for the Survey of Wilderness in Australia*. A report to the Australian Heritage Commission, Canberra.

Lesslie, R. & Maslen, M. (1995) *National Wilderness Inventory—Handbook of Procedures, Content and Usage*, 2nd edn. Australian Heritage Commission, Canberra.

Lesslie, R. & Taylor, S. (1985) The wilderness continuum concept and its implications for wilderness preservation policy. *Biological Conservation* **32**, 309–33.

Lesslie, R., Taylor, D. & Maslen, M. (1993) The National Wilderness Inventory. In: Barton, W., ed. *Wilderness—The Future*, pp. 29–41. Papers from the Fourth National Wilderness Conference. Envirobook, Sydney.

McAllister, D.M. (1990) *Evaluation in Environmental Planning: Assessing Environmental, Social, Economic and Political Trade-offs*. MIT, Cambridge.

McCloskey, J.M. & Spalding, H. (1987) The Wilderness Act of 1964: its background and meaning. *Oregon Law Review* **45**, 288–314.

Mosley, G. (1978) A history of the wilderness reserve idea in Australia. In: Mosley, G., ed. *Australia's Wilderness: Conservation Progress and Plans*. Proceedings of the First National Wilderness Conference, Australian Academy of Science, Canberra. Australian Conservation Foundation, Hawthorn Victoria.

Prineas, P., Lembit, R. & Fisher, N. (1986). *Australia's Wilderness: An Inventory*. A report to the Australian Conservation Foundation. Australian Conservation Foundation, Melbourne.

Ramson, W. (1991) Wasteland to wilderness: changing perceptions of the environment. In: Mulvaney, D.J., ed. *The Humanities and the Australian Environment*. pp. 5–19 Australian Academy of the Humanities, Canberra.

Robertson, R., Helman, P. & Davey, A. eds (1980) *Wilderness Management in Australia.* Proceedings of Symposium, College of Advanced Education, Canberra, July 1978.

Robertson, M., Vang, K. & Brown, A.J. (1992) *Wilderness in Australia—Issues and Options.* A discussion paper. Australian Heritage Commission, Canberra.

Russell, J.A., Matthews, J.H. & Jones, R. (1979) *Wilderness in Tasmania.* Occasional Paper 10. Centre for Environmental Studies, University of Tasmania, Hobart.

Slatyer, R.O. (1975) Ecological reserves: size, structure, and management. In: Fenner, F., ed. *A National System of Ecological Reserves in Australia.* Symposium Proceedings, Australian Academy of Science, Canberra.

Stanton, J.P. & Morgan, M.G. (1977) *The Rapid Selection and Appraisal of Key and Endangered Sites: The Queensland Case Study.* Report no. PR4. School of Natural Resources, University of New England, Armidale.

State of the Environment Advisory Council (1996) *Australia State of the Environment 1996.* Independent Report to the Minister for the Environment. CSIRO Publishing, Collingwood.

Thompson, P., ed. (1986) *Selected Writings of Myles Dunphy.* Ballagirin, Sydney.

Wootten, H. (1987) Conservation objectives. In: Dargavel, J. & Sheldon, G., eds. *Prospects for Australian Hardwood Forests.* CRES Monograph 19. Centre for Resource and Environmental Studies, Australian National University.

3: Forests as Protection from Natural Hazards

PETER BRANG, WALTER SCHÖNENBERGER, ERNST OTT
AND BARRY GARDNER

3.1 WHAT IS A PROTECTION FOREST?

3.1.1 Chapter overview

This chapter will give a short account of the ecology and management of forests that serve as protection against natural hazards or adverse climate. Consistent with our expertise, the focus is on natural hazards in mountain areas, especially in the European Alps. Nevertheless, the chapter illustrates well the importance of non-timber benefits of forest and how accumulation of knowledge has led to sustainable management and beneficial use of forest in this way.

Section 3.1 explains the basic concept of protection forests and necessary terminology. An historical account of protection forests is given in section 3.2, and their distribution is outlined. In section 3.3, some of the ecological foundations that underpin natural forest dynamics in protective mountain forests (section 3.4) are provided. Section 3.5 advocates mimicking nature when managing protection forests, and a general procedure for the management of protection forests is suggested (section 3.6). Section 3.7 gives examples of successful management, and section 3.8 covers shelterbelts as a special case of protection forests. After examining future issues in the management of protection forests (section 3.9), our conclusions are presented in section 3.10.

3.1.2 An early example

In 1397, the people of Andermatt in central Switzerland banned any further cutting of the remaining Norway spruce (*Picea abies* (L.) Karst.) forests on the steep slopes above their houses

(Fig. 3.1). Moreover, it was forbidden to remove trees, branches or even cones (Sablonier 1995). This ban is an early recognition that forests, in addition to their function as a resource of timber and other products, are able to protect people, settlements, traffic routes, dams, other infrastructures and agricultural land against natural hazards. Without protection forests, many villages in mountain regions would be uninhabitable.

3.1.3 A definition of 'protection forest'

The terms used in this chapter are defined in Box 3.1. A protection forest is a forest that has as its primary function the protection of people or assets against the impacts of natural hazards or adverse climate. This definition implies the simultaneous presence of (i) people or assets that may be damaged, (ii) a natural hazard or a potentially adverse climate that may cause damage and (iii) a forest that has the potential to prevent or mitigate this potential damage (Wasser & Frehner 1996). The term 'protection' designates the protection of humans from natural hazards or adverse climate, and not of nature from human impact as in the notion 'wilderness protection' (Chapter 2).

Protection forests may or may not be managed. Whether they fulfil other functions besides their protective function is irrelevant to their status as protection forests. In protection forests, however, the protective function is dominant.

3.1.4 Natural hazards that protection forests protect against

The natural hazards that protection forests protect against include snow avalanches, rock-

Fig. 3.1 The protection forest of Andermatt in central Switzerland.

falls, shallow landslides, debris flows, surface erosion (by precipitation or by wind) and floods (also in flooded lowland areas, mangrove and riparian forests). The different natural hazards are determined by relief, and may therefore occur in combination. Snow avalanches and rockfalls, for example, often occur on the same steep slope. The same applies to surface erosion and shallow landslides. Forests may also protect people and assets by improving (moistening, cooling, equalizing) the local and regional climate.

The effectiveness and reliability of the protection provided by a forest depends on the natural hazards involved, the frequency and the intensity of damaging events and the condition of the protection forest. In many cases, protection forests effectively protect against natural hazards; however, they cannot provide complete protec-

tion from all damage. As forests are usually unable to protect against the effects of earthquakes and volcanic eruptions, these natural disasters are not included in the list of natural hazards even though, on a global scale, these disasters cause the highest death toll and damage (Hewitt 1997; p. 380).

From an ecological viewpoint, natural hazards

Box 3.1 Selected terms in protection forests

Forest function = a role that man attributes to a forest
Protective function = a protective role that man attributes to a forest
Direct protective function = a protective function that is bound to the presence of a forest at a particular location
Indirect protective function = a protective function that is bound to the presence of a certain portion of forest at the landscape level
Protection forest = a forest that has as its primary function the protection of people or assets against the impacts of natural hazards or adverse climate
Natural hazard = a natural factor that has the potential to cause damage to people or assets
Protective effect = an effect of a forest in preventing damage that natural hazards or adverse climate would otherwise cause to people or assets, or in mitigating such damage
Protective potential = a protective effect that a forest is likely to have over the long term if properly managed
Disturbance = the sudden destruction of a single canopy tree, or of several canopy trees. This definition is narrower than the broad one that is sometimes used for disturbance: 'Any relatively discrete event in time that disrupts ecosystem, community or population structure and changes resources, substrate availability or the physical environment' (van der Maarel 1993)
Resilience (of a forest ecosystem) = ability to return to its original state (and dynamics) after disturbance
Resistance (of a forest ecosystem to disturbance) = ability to resist disturbances without significant changes in ecosystem components, structures and processes

are simply factors contributing to the natural disturbance regime. At any given site, the disturbances may occur very frequently or their impact may be substantial. If one or both of these conditions apply, a forest is either unable to become established, or is regularly destroyed in an early successional stage, and is therefore ineffective as a protection forest. This is the case with avalanche tracks (Johnson 1987).

The protective ability of a protection forest is mainly provided by the presence of trees. Tree stems halt falling stones (Cattiau *et al.* 1995, Fig. 3.2), tree crowns prevent, by snow interception and by snow release, the build-up of a homogeneous snow layer that may glide as a compact blanket (In der Gand 1978), tree roots reduce shallow landslide hazard (Hamilton 1992), and the permanent input of litter reduces surface erosion and increases the water-holding capacity of the soil by the build-up of an organic layer (cf. Hamilton 1992). Even dead trees lying on the ground may act as dams to downslope mass transfers (Mössmer *et al.* 1994). The canopy may mitigate adverse climate impacts, reducing radiative

transfers both to the ground and into the atmosphere, and wind speed (Chen *et al.* 1995).

3.1.5 An important distinction: direct and indirect protection

Protection forests may be classified into forests offering direct and indirect protection. This classification is used, for example, in Italy and Switzerland. A given forest has a *direct protective function* if the protective effect depends on the presence of the forest at a particular location. An example is a forest protecting a village against snow avalanches (Fig. 3.1). A forest with a direct protective function is usually restricted to a small area and protects a limited area below and close to the protection forest. *Indirect protection* depends only on the presence of a certain portion of forest at the landscape level, but not on its exact location. Examples include shelterbelts and, in catchment areas, forests that have the potential to reduce soil erosion and peak flows, at least at a local scale (Hamilton 1992). The claim that forests in the headwaters of a river mitigate or even

Fig. 3.2 A rock stopped by a tree.

Fig. 3.3 Temporary avalanche barriers established after storm damage, protecting the village of Curaglia, central Switzerland.

prevent floods in remote lowland regions is being increasingly challenged (Hamilton 1992); several interacting factors seem to override clear effects at large scales (see also Volume 1, Chapter 12).

The landscape or watershed level is important for the management of forests with an indirect protective function, but not of forests with a direct protective function. Such forests are mainly managed at the stand level. Indirect protection is to some extent provided by any forest, as forests generally act against soil erosion and affect the local climate. Forests offering direct protection therefore always offer indirect protection, whereas the converse is not true.

3.1.6 Replacement of protection forests by artificial constructions

Artificial constructions, such as avalanche barriers, terraces, dams and galleries, can provide protection against natural hazards. They are used if the protective effectiveness of a forest is temporarily or permanently impaired, or if complete protection is required. However, an artificial construction can never replace all the functions of a protection forest and is usually more expensive to build and maintain. In Switzerland, avalanche barriers (Fig. 3.3) have been, over the long term, estimated to be about 1000 times more expensive than the financial costs of silvicultural interventions aimed at the maintenance of effective

protection (Altwegg 1991). In different socioeconomic settings, such as in the Himalayas, such interventions may even cover their costs.

3.2 OCCURRENCE AND HISTORY OF PROTECTION FORESTS

3.2.1 Occurrence of protection forests

As many types of natural hazards are gravity-driven and therefore restricted to a minimum slope of the terrain, protection forests are prominent on steep slopes in mountain regions. In mountain protection forests in the temperate zone, the prevalent natural hazards are snow avalanches, torrents and rockfall, whereas in the protection forests of the subtropics and tropics, soil erosion and landslides, caused by high-rainfall events (and forest clearance), are more prominent. However, protection forests occur also in lowland areas where they protect against wave erosion (e.g. mangrove forests) and wind erosion (shelterbelts).

In the Swiss National Forest Inventory, traces of erosion were recorded on 16% of the plots in mountain regions, evidence of moving snow on 37% and evidence of rockfall on 31% (Mahrer *et al.* 1988, p. 85). In publicly owned French mountain forests, the dominant natural hazards were torrent erosion (65% of the area), snow avalanches (14%), rockfall (10.5%) and landslides

(10.5%) (Sonnier 1991). In the Bavarian Alps of south-eastern Germany, 63% of the forests are estimated to provide protection against soil erosion and debris flows, 42% against avalanches and 64% against floods (Plochmann 1985). Despite obvious differences in methodology, these figures clearly reveal the importance of natural hazards in mountain regions.

3.2.2 History of protection forests

The history of protection forests is closely linked to the colonization of the land. The removal of many forests in the Mediterranean region more than 2000 years ago caused extensive soil erosion and land degradation (Thirgood 1981; Brückner 1986). In many valleys of the European Alps, human colonization left only some forest patches with an important protective role, whereas vast forest areas were cleared. The motivation for clearing included demand for arable land as the population increased, and timber demand for heating, construction, financial income for the local communities, mining and industry. In Austria, France and Switzerland, the decrease in forest area caused several serious floods in the nineteenth century, which also affected lowland areas (Schuler 1995). In Switzerland, it took some 50 years to put countermeasures into practice and to reverse the decreasing trend in forest area. Clearcutting was prohibited by law, and an afforestation policy was established. This development was favoured by the railways as new means of transportation that enabled local firewood to be replaced by coal, by emigration after repeated famines, and, in the last few decades, by the decreasing importance of agricultural land and forests as sources of income. Similar developments have occurred in other countries in the European Alps. The early recognition of the problems associated with natural hazards may also be illustrated by the Bavarian Forest Act of 1852, which specifically uses the term 'protection forests' ('Schutzwaldungen').

As a consequence of tough regulations, establishment of a forest service, incentives to afforestation and socioeconomic changes, the Swiss forest area has increased by about 60% since 1860, and by about 4% between 1985 and 1995 (Brassel & Brändli 1999), with the majority of the regained forest area being located in the Alps. Similar developments are taking place elsewhere in the European Alps (Russ 1997; French National Forest Inventory 1998). On a global scale, however, the area of protection forests is decreasing, and many important but often undesignated protection forests in developing countries are today under similar pressures to the central European protection forests of the nineteenth century (Ives 1985; FAO 1997, p. 41).

In the European Alps, prohibiting large clearcuts or even the removal of trees was often insufficient to ensure effective long-term protection. The remaining patches of protection forests were in many cases under heavy and permanent human impact (Ott *et al.* 1997). Extensive grazing and litter raking for instance have the potential to endanger a forest in the long term. The same applies to browsing or fraying ungulates such as red deer (*Cervus elaphus*) or chamois (*Rupicapra rupicapra*), which can impede seedling establishment of all or some tree species. An additional problem caused by human impact is the introduction of tree species or lowland provenances unadapted to mountainous site conditions. And last but not least the structure of many forests was simplified, by large clearcut areas regenerating simultaneously, by regular thinnings adapted from lowland practices, and by large afforestation efforts. The resulting homogeneous structures are less able to resist climate impacts (Fig. 3.4). Even complete bans were usually not effective over the long term since natural stand dynamics often include disturbances that impair effective protection (section 3.4).

Besides their protective role, protection forests in mountains contribute to the socioeconomic development of rural areas: they provide renewable resources and jobs. In many countries, a policy to support the maintenance of protection forests has been established, particularly in relation to subsidies for expensive silvicultural interventions (e.g. Austria, France, Germany, Switzerland) and in relation to prevention of further deforestation.

Over the past decades, protection forests have become increasingly important because our demands for protection have increased; remote mountain areas that were formerly avoided during winter are now expected to be perma-

Fig. 3.4 Homogeneous stand structure in a 90-year-old subalpine afforestation in the Swiss Alps: such stands are prone to storm damage and snow break.

nently accessible for tourists, and settlements have been spreading into areas that were considered unsafe by our ancestors.

3.3 ECOLOGICAL CHARACTERISTICS OF MOUNTAIN FORESTS

This section will provide some of the ecological foundations that are necessary for a sound ecosystem-based management of protection forests in mountain regions.

3.3.1 Environment of subalpine protection forests

Mountain forests with an important protective role are mostly located in the montane and subalpine belts. There is no generally accepted definition of the limit between the montane and subalpine belts. In the Alps, the absence of tree species that do not usually reach the treeline (*Abies alba*, *Fagus sylvatica*), a stand structure

where trees are aggregated to small groups with a common long crown and a low cover of the canopy with many permanent gaps, are often used to designate a forest as subalpine (see Ott *et al.* 1991).

Subalpine forests are close to the alpine treeline, the ecotone between forest and alpine meadow, with its harsh environment. In the Alps, many subalpine forest environments exhibit high precipitation, high and long-lasting snow cover and low temperatures (Reisigl & Keller 1987, p. 18). The rugged topography with steep slopes causes important ecological differences between aspects. While north-facing slopes are often cool and moist, south-facing slopes are considerably warmer and drier (Mettin 1977; Alexander 1984).

A distinctive feature of subalpine forests is the importance of microsites (Schönenberger *et al.* 1995; Ott *et al.* 1997). This is most obvious close to the treeline where trees are restricted to a few suitable microsites, the remaining area being permanently devoid of trees. Such microsites are the

Fig. 3.5 Open, park-like stand structure of a subalpine Norway spruce forest.

warmest spots in moist to wet climates (micro-elevations including nurse logs, Spittlehouse *et al.* 1990), and the moistest spots in dry climates (micro-depressions). In both cases, the microsite differentiation leads to an open forest or even to a park-like appearance (Fig. 3.5). The limiting factors are, in the case of moist sites in the Alps, long-lasting snowpack and moving snow, dense vegetation, pathogenic fungi and the short growing season, all leading to high seedling mortality. As suitable microsites are usually already occupied by canopy trees with long crowns, they are only partly available for seedlings, resulting in a sparse seedling bank under canopy (Ott *et al.* 1997). At lower altitudes, microsite variations are less important, until, in the montane belt, almost the whole ground is suitable for tree seedling establishment (Brang 1997) and usually covered with a seedling bank (Burschel *et al.* 1985). However, the presence of a seedling bank is not only correlated with altitude; it is also more

likely if sufficient soil moisture is available, and if shade-tolerant species are present (especially *Abies* spp., Kohyama 1984; Veblen *et al.* 1989).

3.3.2 Environment of montane protection forests

The climate of montane forests is less harsh than the climate of subalpine forests: air and soil temperatures are higher (average annual air temperature decreases by about 0.6–0.7°C for every 100 m increase in altitude, Reisigl & Keller 1987, p. 18; Mayer & Ott 1991, p. 36), and the growing season is longer (by about 1 week for every 100 m decrease in altitude, Reisigl & Keller 1987, p. 18). Similarly, the length of the frost-free season increases. Tree growth is environmentally limited to a lesser extent (Tranquillini 1979). In many montane forests, a persistent and dense seedling bank ensures rapid regrowth of trees after tree- or stand-replacing disturbances (e.g. Burschel *et al.* 1985).

Chapter 3

3.3.3 Tree species in protection forests

As the most important tree species reaching the canopy are covered in Chapter 2 in Volume 1 of *The Forests Handbook*, this section will be kept very short. With increasing harshness of the climate from the montane to the subalpine belt, the number of tree species that are able to cope with the environment decreases. In the subalpine belt, the resulting stands are sometimes even monospecific. In montane and subalpine protection forests of the Northern Hemisphere, the genera *Abies, Acer, Betula, Cedrus, Juniperus, Larix, Quercus, Picea, Pinus, Sorbus* and *Tsuga* prevail, whereas in the Southern Hemisphere the genera *Nothofagus* and several *Podocarpaceae* are prominent.

3.4 STAND DYNAMICS, STAND STRUCTURE AND STABILITY PROPERTIES IN PROTECTION FORESTS IN THE ALPS

This section will first focus on natural disturbance regimes in protection forests in the Alps and their impact on stand structure, then which stand structures are desirable and, finally, the question of 'ecological stability' in protection forests will be addressed.

3.4.1 Why small-scale disturbances prevail in protection forests in the Alps

In protection forests in the Alps, both large-scale and small-scale disturbance regimes are present, but the latter prevail (Zukrigl 1991), with single disturbance events affecting patches that are usually smaller than 0.1 ha. This is especially true under extreme site conditions, for example in subalpine forests close to the treeline—with the exception of very dry forests that are prone to fire disturbance—and in the montane belt (mixed *Fagus sylvatica–Abies alba–Picea abies* forests). The small-scale disturbance regime leads to the continuous dominance of late-seral species. Real primary successions occur only as exceptions.

The large-scale disturbance agents affecting mountain forests in the Alps are storms, snow load, bark beetles (*Ips typographus*, affecting *P. abies*) and fire. Fire is important in *Pinus sylvestris* forests in continental climates of the inner Alps, where summer drought periodically leads to high fire hazard. Fire may also periodically disturb on a large scale various forest ecosystems in the southern Alps, as a consequence of winter and early spring droughts. The forest ecosystems in the Alps that are susceptible to large-scale disturbances other than fire are mostly forests with a monospecific tree layer of *P. abies* (in the montane and subalpine belts). Monospecific *P. abies* stands tend, with the exception of extreme site conditions in the subalpine belt, to form relatively dense homogeneous stands, with high intraspecific competition, short crowns and similar tree age and height (Korpel' 1995). This natural tendency has often been reinforced by large-scale clearcutting, followed by rapid natural regeneration or by afforestation with a single species.

Large-scale disturbances are uncommon in the Alps for several reasons. First, the forests are often patchy, interrupted by avalanche tracks, mountains reaching the alpine and nival belt, and agricultural land. Second, the terrain is often very rugged, resulting in small-scale variability in site conditions and, consequently, tree species composition and seral stages. This reduces large-scale fire, storm and bark beetle disturbance. Third, the high precipitation with a summer maximum, the low intensity of thunderstorms and the low fuel loads resulting from past intensive forest use mean that large-scale fires are rare events in the Alps. This situation is completely different to that found in many North American mountain forests where vast continuous forest areas, high fuel loads, summer drought and intensive thunderstorms result in frequent large-scale fires (Kimmins 1997, p. 296).

The impact of large-scale disturbances on the protective effect of a mountain forest varies considerably. While fires have the potential to damage uniformly the protective effect on affected areas, snow load usually leaves many surviving trees that partly maintain the protective effect. The impact of storms on the protective effect is highly variable; some storms have only a thinning effect, others destroy entire stands. Bark beetles often complete a partial stand destruction caused by storm or snow load, especially the beetle *Ips typographus* in *P. abies* forests.

3.4.2 Duration of successional cycles

With increasing altitude, tree growth decreases as a consequence of restricted resources (especially energy) and adverse impacts (especially climatic impacts). Successional cycles are therefore longer in high-elevation forests. In the Carpathians, which are similar and nearby to the European Alps and which contain a few 100-ha remaining old-growth forests (Korpel' 1995), the duration of secondary successional cycles varies from 250 to 300 years (*Fagus sylvatica*), over 400 years (*P. abies*), to 800 years (*Pinus cembra*). Cycles shorter than 500 years prevail.

3.4.3 Characteristics favouring and impeding resistance of the tree layer to disturbance

A stand with a direct protective function should be permanently effective. This is only the case if a certain tree layer is permanently in place. Several ecological characteristics of subalpine and montane forests favour the permanence of the tree layer, while others counteract it. In subalpine forests, the resistance of the tree layer to disturbance from agents such as storms and snow load is enhanced by the low coefficient of slenderness of the stems (Petty & Worrell 1981; Rottmann 1985; Strobel 1997), the low centre of gravity of the trees (Strobel 1997) and the clumped stand structure (Fig. 3.5) that is caused by the microsite differentiation and favoured by steep environmental gradients within gaps. 'Clumped' stands contain long internal edges that often act as breaks to natural disturbances, preventing a 'domino' effect where each fallen tree triggers the fall of adjacent trees. However, the high resistance of clumped stand structures is only intuitively understood (see Mlinsek 1975), whereas the low resistance to disturbance of homogeneous stand structures, with slender stems, short crowns with high centres of gravity and no internal edges, is obvious and has been repeatedly experienced (Cerny 1980; Petty & Worrell 1981; Savill 1983; Rottmann 1985). Homogeneous stand structures are common in forest ecosystems with low microsite variability, large-scale disturbance regimes and only one canopy tree species. With increasing altitude, homogeneous stand structures become increasingly rare in natural forest ecosystems, and heterogeneous ones dominate. Homogeneous stand structures are often a consequence of natural regeneration on clearcuts or of planted areas that were subsequently left untended, a common situation in the Alps.

3.4.4 Factors favouring and impeding recovery from disturbance

The disturbance regime of a forest ecosystem always includes disturbances of a magnitude that break the resistance of one to many trees, by uprooting, breaking or killing them as standing trees. This frees resources (growing space, light, water, nutrients), and initiates a recovery process. In subalpine forests, the infrequent years with seed production (Mencuccini *et al.* 1995), the sparse or missing seedling bank and the slow tree growth mean that recovery of the tree layer after major disturbances usually takes several decades. The recovery depends in many environments on nurse logs favourable to coniferous regeneration (Harmon *et al.* 1986; Imbeck & Ott 1987), although the mechanisms involved are only partly understood. The sparse occurrence of nurse logs (Fig. 3.6) in most utilized forests (Bretz Guby & Dobbertin 1996) is therefore a serious impediment to natural seedling establishment (Brang 1996). The commonly occurring long crown of canopy trees in clumps often locally suppresses vegetation development, so that, after removal of a tree clump by a disturbance, there is an additional opportunity for tree seedling establishment. Such microsites, however, are suitable for regeneration only for a limited time, as the released resources are competed for not only by tree seedlings but also by other vegetation. A dense cover of mosses, herbs, grasses and shrubs often completely prevents tree seedling establishment (Imbeck & Ott 1987; Coates *et al.* 1991; Brang 1996). Seedling establishment may also be impaired by pathogenic fungi (e.g. *Herpotrichia*, *Gremmeniella*) and snow gliding. Browsing ungulates are another impediment to tree seedling establishment. Browsing often reduces height growth, prolonging the phase where the seedlings are susceptible to further damage by factors such as pathogenic fungi and snow press. Heavy browsing may change the species composition or even completely prevent tree seedling establishment (Ott *et al.* 1997).

Fig. 3.6 Naturally established seedlingson a nurse log in a subalpine *Abies lasiocarpa–Picea engelmannii* forest in British Columbia, Canada.

The frequency and intensity of some disturbances (e.g. snow load, wind) usually increases from montane to subalpine forests, but this may be balanced by the increasing resistance (for definition, see Box 3.1) of the trees. Hence, there is no general relationship between persistence of the tree layer and altitude. Recovery times, however, become longer with increasing altitude, as the harsh climate in subalpine forests causes tree seedlings to grow slowly. For coniferous seedlings suppressed by canopy trees, annual height growth rates are often lower than 1 cm (Mettin 1977; see also Koppenaal *et al.* 1995). In subalpine *P. abies* forests, it often takes 30–50 years until a seedling has outgrown the average snowpack depth. In montane forests, suppressed *P. abies* seedlings grow considerably faster, reaching minimum annual height growth rates of 3–4 cm (Schütz 1969; Mosandl & El Kateb 1988), similar to height growth rates in *Tsuga mertensiana* and *Abies* sp. in south-western Oregon, USA (Seidel 1985).

3.4.5 What is the ideal stand structure of a protection forest?

There is no general rule as to how a protection stand should be structured. The ideal stand structure depends on the range of possible structures (that are part of natural forest dynamics), the disturbance regime (including natural hazards) and on the importance of the protective function (including the distinction between direct and indirect protective function). Natural or anthropogenic disturbances (e.g. storm gaps, bark beetle damage resulting in snags, cuttings) may temporarily impair the protection provided by a protection forest. Stand renewal is therefore often 'the weak link in the chain of forest dynamics', and must be integrated into a scheme of ideal stand structures.

The natural hazards at a given site determine the minimum requirements for stand structures to ensure permanent protection. If rockfall is prevalent, dense stands, with only small stemless distances (gaps) parallel to the slope, are most effective (Box 3.2, Omura & Marumo 1988; Zinggeler *et al.* 1991; Gsteiger 1993; Cattiau *et al.* 1995); if snow avalanches are prevalent, gap size requirements are similar (Box 3.3), but stand density is less important, and coniferous trees are more effective than deciduous trees in preventing avalanches starting in the forest (Schneebeli & Meyer-Grass 1993). The minimum structural requirements for a given stand should include a margin of safety, because disturbances can easily impair effective protection.

Stands with high stem densities are not the best solution for effective protection against rockfall and snow avalanches over the long term; such stands are usually susceptible to storm damage and snow break, and they do not allow for sufficient regeneration. In many cases, a natural or

Box 3.2 Threshold values for rockfall hazard in the Alps (Cattiau *et al.* 1995; Wasser & Frehner 1996)

Characteristic	*Critical value*
Number of stems with diameter at breast height (d.b.h.) > 40 cm in the transit and run-out zones	< 400 ha^{-1}
Gap width parallel to the slope	> 20 m

Note: If the given number of stems does not allow for continuous regeneration, it has to be reduced in favour of sufficient regeneration

Box 3.3 Threshold values for starting zones of snow avalanches within forests in the Alps (Meyer-Grass & Schneebeli 1992; Berger 1996; Wasser & Frehner 1996)

Characteristic	*Critical value*
Slope	30–55° (if grasses are dominant, 28–55°)
Snow depth	> 50 cm
Canopy cover	< 250 stems ha^{-1} (should be as high as possible, but allow for continuous regeneration), or < 50% cover
Gap size (from stem to stem)	> 30 m parallel to the slope, > 50 m perpendicular to the slope

Note: If the threshold values for slope and snow depth are exceeded, insufficient canopy cover or gap size may cause ineffective protection. Additional unfavourable conditions include south-facing slopes, edaphic gaps (e.g. rock faces where avalanches may start), broadleaved trees, stands accumulating snow by snow drift. Gap size values parallel to the slope are valid for 45° slope; add 10 m for every reduction by 5° in slope. Similar values for five forest types that vary in species composition are given in Schneebeli and Meyer-Grass (1993)

anthropogenic small-scale disturbance regime that creates a mosaic of developmental phases is therefore the best way to achieve permanent protection (Chauvin *et al.* 1994; Ott *et al.* 1997). This will result in a small-scale patchwork of trees of all ages and sizes, with only small treeless areas. On many sites in the European Alps, however, the multistoreyed structure of a classic single-tree selection forest (Schütz 1994) is not achievable. In subalpine forests without abundant precipitation, soil moisture directly under canopy trees is temporarily insufficient for *P. abies* seedlings and saplings (Ott *et al.* 1997; Brang 1998), resulting in an open forest with tree clumps that are spatially separated. This structure is called 'mountain selection forest'. Similarly, in subalpine *P. abies* forests with abundant precipitation, the areas between canopy trees seem to be permanently free of trees (Ott *et al.* 1997), mainly because they remain covered by the winter snowpack for such a long time that pathogenic fungi (*Herpotrichia* spp.) are able to kill seedlings (Bazzigher 1976). Such treeless microsites can only be colonized by input of nurse logs that provide a suitable regeneration niche.

An alternative to mountain selection forests are coppice forests with a high stem density. Such forests are effective in halting falling rocks in run-out zones (Gsteiger 1993).

3.4.6 How natural should the structure of a protection forest be?

Natural stand dynamics often do not result in the patchwork described above as the ideal for a pro-

tection forest. In a protection forest, the first goal is to ensure effective protection, and not to closely imitate natural stand dynamics. However, natural stand structures and dynamics remain very important (see section 3.5). Finding the balance between protective requirements and high naturalness of the desired stand structures is one of the challenges for the management of protection forests. In addition, modifications of the stand structures are not simply feasible in the short term; the current structures determine possible pathways of management (Ott *et al.* 1997). It may take many decades, sometimes even centuries, of silvicultural interventions to reach optimum stand structures for continuous protection.

3.4.7 What is 'stability' in a protection forest?

The term 'ecological stability' has been used inconsistently and is therefore rather confusing. It should be avoided, and instead several stability properties be used that have been clearly defined (Gigon 1983; Gigon & Grimm 1997). In the context of protection forests, the following two stability properties are the most important:

• 'Resistance' is the ability of an ecosystem to keep its characteristics (including dynamics) unchanged, despite the presence of potentially disturbing factors.

• 'Resilience' is the ability of an ecosystem to return to its original state (and dynamics) after disturbances, by restoring altered ecosystem characteristics (Gignon & Grimm 1997).

To use resistance or resilience, an 'original', unchanged, predisturbance state needs to be defined. As this predisturbance state is, in the European Alps, in most cases a state resulting from substantial human disturbance, it is unlikely that, in the absence of such disturbance or under a changed human disturbance regime, a disturbed forest ecosystem will return to the predisturbance state. It can rather be expected that a disturbed forest ecosystem will develop along the natural successional pathways. This leads us to use the natural forest ecosystem that was not subject to direct and substantial human disturbance as the 'original', unchanged state, even if this is difficult in several situations, such as:

• if the forest established under more favourable climatic conditions than the actual ones;

• if species migration after extreme disturbances such as ice ages is not completed.

It should also be noted that resistance and resilience should only be used in the context of forest ecosystems at long time scales of several decades to several centuries, thereby allowing for forest succession. Both terms, resistance and resilience, are also useful for describing the reaction of trees to disturbance.

In a forestry context, 'stability' has often been used with a narrow focus on the resistance of trees to internal or external mechanical stress factors such as storms, snowload or snow gliding (Langenegger 1979), or as 'physical resistance, continuity and persistence' (Zeller 1982). Such definitions focus on the resistance of individual trees or tree clumps to disturbance. However, in his stability assessments, Langenegger (1979) included an assessment of the ability of the forest to recover from disturbance, by recording present regeneration and the conditions for future regeneration. Although the characteristics of the tree layer, such as stand structure, species composition, stem taper and crown length, are very important in forests with a direct protective function, the existence and the protective effect of trees depend on the condition of the forest ecosystem as a whole.

Forests with a direct protective function should continuously provide effective protection, and therefore be as resistant and as resilient as possible. These requirements can be lowered in forests which have a solely indirect protective function. However, even in the case of direct protection, high resistance and resilience is not needed in all ecosystem characteristics. We need them only for ecosystem characteristics (i) that are directly related to effective protection (trees and other vegetation), and (ii) that are essential to the functioning of the ecosystem (e.g. nutrient pools, water-holding capacity of the soil). Changes in species composition may not impair effective protection if several species are functionally redundant.

Generally, mountain forest ecosystems of the temperate zone seem to be rather resilient. In Europe, this is shown by the fast recovery of forests in the nineteenth and twentieth century, following long periods of continuous and heavy human disturbance. This recovery process was only partly a consequence of interventions such as afforestation. Mountain forest ecosystems with low resilience occur on sites that are susceptible to surface and mass erosion. If erosion has taken place on such sites, for example after large-scale disturbances such as storms, landslides, clearcutting and road construction, recovery may take several centuries or may even be impossible due to changes in the regional climate.

3.5 SUGGESTED APPROACH AND PRINCIPLES FOR THE MANAGEMENT OF PROTECTION FORESTS

How should protection forests be managed to ensure their long-term effectiveness? How can the necessary stand structures be achieved and continuously maintained? In the European Alps, similar questions have been asked for many decades, and it seems now that we are close to practical, though preliminary, solutions. The following suggestions are a generalization of a procedure that was developed and is now being applied in Switzerland (section 3.7.1; Wasser & Frehner 1996).

3.5.1 Goal and restrictions for the management of protection forests

The overall management goal in a protection forest is to ensure that the forest will reliably and continuously act as a protection forest. If possible, a protection forest should remain effective even if interventions are suspended for several decades. In many cases, this goal implies that the present forest condition must be improved, increasing the resistance of the protection forest to disturbance, and its resilience. To achieve this management goal, certain restrictions that limit the range of possible solutions need to be considered. One restriction is that ecosystem degradation must be prevented. Expressed positively, this means that components, structures and processes that are essential for the functioning of the forest as an ecosystem must be preserved or, where they are already impaired, restored. Another restriction is that the costs of any interventions that are aimed to maintain or enhance effective protection of a forest must be covered by receipts, or subsidized.

3.5.2 Management approach: mimic natural ecosystem dynamics

For the design of any interventions in protection forests, we recommend, as a general approach, remaining close to natural ecosystem dynamics. With 'natural ecosystem dynamics', we mean the development of an ecosystem without direct human disturbance. By 'naturalness' we mean the degree to which ecosystem characteristics (components, structures, processes, Box 3.4) resemble the characteristics of an ecosystem without direct human disturbance. A close affinity to naturalness is justifiable by two facts. First, deviations from natural ecosystem dynamics introduced by forest management have often impaired effective

---

> **Box 3.4 Changes in characteristics of mountain forest ecosystems that are due to human disturbance, and potential causes. These changes occur frequently, but not in every case**
>
> **1** Reduced tree species diversity (clearcutting, planting, selective browsing)
> **2** Homogenized stand structure (clearcutting, planting, regular thinning)
> **3** Reduced resistance of canopy trees to storms and snow load (clearcutting and planting without thinning)
> **4** Reduced variation in age of the canopy trees (clearcutting)
> **5** Reduction of coarse woody debris, especially nurse logs (timber harvesting)
> **6** Increase in cover of ground vegetation (regular thinning)
> **7** Impeded natural regeneration (as a consequence of points 3, 5 and 6)

---

protection over the long term, especially in the subalpine belt (see end of section 3.2; Ott *et al.* 1997). Second, our knowledge about ecosystem processes is too limited to enable us to 'design' protective forest ecosystems, although we may be able to design protective tree stands. It is therefore advisable to rely on natural ecosystem dynamics that seem to have ensured a permanent and effective forest cover over centuries and millennia.

To implement the approach of high naturalness, a three-level sequence of interventions with decreasing naturalness for the management of protection forests is advocated:

1 *Utilize natural ecosystem characteristics.* Natural characteristics (components, structures and processes) of an ecosystem at a given site that enhance its resistance to disturbance, and its capacity to recover from disturbance (resilience), should be integrated and utilized. The occurrence of stabilizing components, for example broadleaved species intermingled in coniferous stands, and nurse logs suitable as regeneration niches for coniferous seedlings, can be promoted. Resistant structures such as tree clumps, and diverse stand structures with a mosaic of trees of all ages and sizes at short distances can be maintained or promoted. One example of a stabilizing ecosystem process is the variation in germination times within a season that makes it more likely for a given species that the weather conditions are favourable for successful emergence for at least some of the seeds (Brang 1996).

2 *Slightly modify natural ecosystem characteristics.* When required, the natural ecosystem dynamics should be slightly modified, but only by introducing stabilizing ecosystem characteristics (components, structures and processes) that sometimes occur during the natural ecosystem dynamics on the sites concerned. Shifts in the occurrence of existing tree species, planting and sowing are examples of such interventions (Fig. 3.7). Another example is the promotion of diverse stand structures on sites where such structures

occur only temporarily. This is practised now in homogeneous montane *P. abies* afforestations in the Swiss and French Alps, by cutting all trees in strips of 8–10 m width around tree clumps of 8–15 m diameter (Ott *et al.* 1997, p. 86).

3 *Introduce artificial protective components.* As an exception, and only when required to achieve effective protection, artificial stabilizing components should be introduced. Examples used to prevent snow gliding include microterraces, or poles and tripods driven into the ground, or avalanche barriers (Fig. 3.7b). Such measures may be temporary or permanent.

In cases where natural stabilizing ecosystem characteristics (according to the preceding paragraph) have been impaired or destroyed by human

a   b

Fig. 3.7 Two regeneration methods with different degrees of naturalness: (a) seedlings sown using mini-greenhouses; (b) planting in combination with temporary avalanche barriers. Compare with Fig. 3.6 where natural ecosystem dynamics are relied on.

impact, interventions that correspond to the second and third level of the above-mentioned three levels are often required.

3.5.3 Three management principles

In addition to the above, the following three management principles are recommended:

1 Use all accessible sources of relevant information.
2 Standardize management procedures, including the step of decision making.
3 Design interventions as controlled experiments.

The first principle is derived from the frequent experience that very similar investigations and assessments are repeated, wasting considerable resources, simply because the existing information is not used. Sources of information include the often rich experience and tradition of local people (here placed first deliberately), management documentation, papers in scientific and professional journals, treatises on the local history and thematic maps. In many protection forests, much information is available, as the multiple resources that they provide have attracted human interest for centuries (section 3.2). The second principle—standardize management procedures—should be followed because it helps to avoid the omission of relevant considerations, because it makes all procedures, especially the decision making, transparent to others and because it can serve as a first step in monitoring the success of management actions. The third principle—design interventions as controlled experiments, or short 'adaptive management' (Grumbine 1994; Holling 1996)—is important to track the effectiveness of any intervention. Field examples of many promising interventions in forest management remain controversial because suitable control treatments are lacking.

3.6 SUGGESTED PROCEDURE TO MANAGE PROTECTION FORESTS

3.6.1 Overview

In a particular protection stand, or in a watershed dominated by protective forests, a four-step proce-

---

Box 3.5 Suggested procedure for the management of protection forests

1 Collect general information on the management area
  1.1 Get knowledge about the relevant legislation
  1.2 Delineate the management area
  1.3 Identify and involve interest groups
  1.4 Resolve conflicts due to contrasting interests
  1.5 Collect information on natural hazards
  1.6 Establish a site classification
  1.7 Collect information on disturbance regimes and stand dynamics
  1.8 Assess the protective potential of different stand structures
2 Design interventions at the stand level
  2.1 Assess stand conditions
  2.2 Describe target stand conditions
  2.3 Compare actual and target stand conditions
  2.4 Evaluate alternative interventions
  2.5 Take a decision
3 Carry out interventions
4 Monitor forest development
  4.1 Keep track of the effectiveness of the interventions
  4.2 Document the whole procedure (steps 1–4)

---

dure is suggested to put the above ideas into practice (Box 3.5).

In Box 3.5 it has been attempted to provide a detailed procedure that is applicable in most cases. If the protection forest in question has only an indirect protective function, the suggested procedure can be considerably simplified, and there is much more flexibility to integrate other forest uses. The steps are in a logical order, but the procedure should not be used as a stepwise procedure since feedback loops are required, especially in community involvement (points 1.3–1.4).

3.6.2 First step: collect general information on the management area

The first information that is collected concerns the socioeconomic context of the protection forest. Knowledge of the relevant legislation is a prerequisite, because it often restricts the range of possible management solutions, and identifies those that may be subsidized financially. Once the management area is delineated and the ownership is clear, all interest groups should be identified and involved in a transparent decision-finding process (Hamilton *et al.* 1997). Successful public involvement leads to management solutions that are widely accepted, and may reveal sources for additional funding. Potential interests include, aside from protection, timber production, animal husbandry, tourism and recreation, nature conservation, hunting, gathering of wild fruits, fungi and medicinal plants, and use for military training. As, in most cases, protection is not the only forest function, a spatially explicit hierarchy of the functions of the forest should be established, resolving problems due to contrasting interests.

Collect information on natural hazards

Spatially explicit information on the prevalent natural hazards in the management area is especially important in the case of direct protection (section 3.1). This information is usually easy to obtain as damaging events often induce action in protection forests, and may even have been recorded for many decades. Natural hazards are bound to terrain conditions, enabling the establishment of critical values. Field observations provide additional hints to natural hazards as events leave traces.

Fresh stones on the surface and scars on tree stems indicate recent rockfall. Starting points of rockfall (usually unstable rockfaces) and the area affected are often easy to identify. Models to assess rockfall hazard are in use (Cattiau *et al.* 1995), and minimum slopes for the transit and the deposition zones of rockfall have been established (Box 3.2; Zinggeler *et al.* 1991).

Snow avalanche hazard strongly depends on the amount of snowfall, the terrain (exposure, slope, roughness) and the stand structure, and corresponding critical values have been established (Box 3.3; Meyer-Grass & Schneebeli 1992). The main traces left by snow avalanches are broken stems and scars on canopy trees. In avalanche tracks, shrub communities that are tolerant of frequent discharges develop (Johnson 1987). Within protection forests, avalanches often leave no traces that can be used as early warning signals and that would confirm the map-based hazard assessment. In contrast to avalanches, gliding snow may be identified because it leaves branches on the ground that are wrapped up in grass and may serve as 'silent witnesses' (Bayerische Staatsforstverwaltung 1994, p. 150).

Whether a site is susceptible to soil erosion is not easy to establish, because mass transportation is heavily determined by extreme events that are unlikely to be observed (Kirchner *et al.* 1998). The most important catchment features making soils susceptible to erosion are particle size distribution, high slope, occurrence of heavy precipitation, absence of trees or herbaceous vegetation, and soil disturbance by logging, road construction or grazing (Hamilton 1992).

Establish a site classification

A site classification is very helpful because disturbance regimes, plant establishment and growth are site specific. A site type can serve as an ecological 'coordinate' helpful in retrieving relevant published information. Implementing the general attitude of high naturalness is impossible if site types and, accordingly, the site-specific natural stand dynamics are unknown.

The sites should therefore be classified, adopting or adapting an existing classification or developing a new one, and at least coarsely mapped. If the costs for site mapping are justifiable anywhere, then it should be in forests with a direct protective function.

Collect information on disturbance regimes and stand dynamics

Site-specific information on stand dynamics and natural disturbance regimes is usually sparse. If the stands are in a near-natural state, this information may be partly derived from published records (e.g. Mayer *et al.* 1989; Korpel' 1995). However, care is required when extrapolating results from one site to another. In the European Alps, knowledge of natural stand dynamics will probably remain very limited for a long time because the few remaining untouched forests cover only a few site types, and most are heavily affected by excessive browsing.

Because many protection stands have been strongly modified by human impact, disturbance regimes and stand dynamics deviate from natural conditions. The lack of firm information on stand dynamics is a serious handicap to the sustainable management of protection forests. Some preliminary information on disturbance regimes and stand dynamics may be derived from simple field observations (Box 3.6) that are in any case part of step 2.1 of the suggested procedure (Box 3.5).

Ungulates that browse on tree seedlings and damage saplings by fraying are important disturbance agents in many European protection forests. This problem has arisen in the last few decades due to a strong increase in ungulate popu-

lations. In protection forests, it is irrelevant if the actual population densities are at natural or elevated levels; the question is whether ungulates impair effective protection or not. As tree recruitment—be it seedlings or saplings—is a crucial component of ecosystem resilience, information on browsing and fraying damage is important.

Assess the protective potential of different stand structures

In this step, knowledge of natural hazards and forest dynamics is combined. The effectiveness of a forest to protect against natural hazards varies between different stand structures. Stand structures depend on forest dynamics. Each seral stage exhibits a range of possible stand structures. Some of these structures may cause ineffective protection, and these should be identified. They will often include the renewal phase. Failure to reach effective protection during this phase is again a site-specific issue. On sites where stands tend towards homogeneous structures and large-scale disturbances, the renewal phase is a weak link. On other sites with a continuous small-scale disturbance regime, stands may be sufficiently heterogeneous and continually regenerating, and interventions to sustain effective protection may be unnecessary.

3.6.3 Second step: design interventions

In this step, the interventions are designed (Box 3.5). The additional information needed now is stand-specific.

Assess stand conditions

When assessing actual stand conditions, remote sensing imagery is helpful (Kusché *et al.* 1994). At least in the case of direct protection, however, field observations in each stand are required. Useful parameters are listed in Box 3.6. The field assessment should include an estimate of the probable stand development without interventions.

Describe target stand conditions

Target stand conditions should be described as

Box 3.6 Useful parameters for field observations to assess stand conditions with respect to effective protection against natural hazards

1 Tree species composition
2 Developmental stage, variability in stem diameter
3 Age distribution of a stand (e.g. from fresh stumps)
4 Spatial distribution of the trees (aggregation)
5 Vertical distribution of the tree crowns in canopy layers
6 Gap size distribution, gap shape and orientation
7 Physical stability of the dominant trees conferring the stability of the stand (coefficient of slenderness, signs of root and stem rot, crown length)
8 Visible damage on the trees that could impair their resistance to disturbance
9 Occurrence (density), condition (e.g. annual height increment of tree seedlings and saplings on light gradients, browsing damage), height, spatial distribution and microsite dependence of existing regeneration
10 Conditions for future natural regeneration (e.g. dense vegetation cover, nurse log occurrence)

long-term minimum conditions for effective protection at the stand level. This requires the integration of all preceding steps. It is necessary to
know how important the protective function is,
which natural hazards the forest is expected to
protect against, how the stands will develop on
the relevant sites and which stand structures and
seral stages may provide ineffective protection
(step 1 in Box 3.5).

A description of the target stand conditions
should include information on ecosystem characteristics (especially components and structures)
that are important to ensure effective protection.
Possible parameters are those mentioned in Box
3.6. The description should be detailed in cases
where the minimum standard for effective protection is high or not reached, but can be very rough
where the minimum standards are low or clearly
exceeded.

Compare actual and target stand conditions

To design interventions, actual and target stand
conditions are compared. This includes a comparison of target conditions to the actual stand conditions, and to the conditions anticipated in a
mid-term perspective, without interventions. As
a mid-term perspective, we recommend 10 and 50
years. If stand dynamics are very slow (e.g. near
the treeline, or in cold and dry climates), longer
periods may be appropriate. If in a mid-term perspective a stand is unlikely to meet the target
stand conditions, there is a need for intervention.

*Evaluate alternative interventions and
take a decision*

Interventions must obviously be site-specific,
otherwise the general approach of naturalness
cannot be implemented.

Any intervention must meet three criteria to
be realized: it must be effective, justifiable (cost-
effective) and feasible. If one of these three criteria
is not met, no intervention should be taken, or
alternative interventions should be examined. In
forests with a direct protective function, effective
and feasible measures are almost always justifiable. In the case of interventions that do not cover
their costs, their necessity is an additional fourth

criterion: Only interventions that are required to
achieve or maintain effective protection are done.
The costs of interventions are in most cases an
important issue, as cable or helicopter logging
are often the only alternatives available for timber
harvest due to the rugged terrain. The interventions may therefore not cover their costs. It is
important to keep in mind that in some cases no
intervention is the most appropriate solution:
'The wisdom of a forester does not become evident
in what he does, but in what he leaves' (Zeller,
cited in Ott *et al.* 1997, p. 82). The slow development of many mountain forests requires a patient
approach to management. If we utilize natural
processes to achieve desired changes in stand composition and structure—and this should be our
general approach—these changes may take
several decades. Any action should be taken at the
time when it has the highest effect. This may not
be now, but only in 10, 30 or 50 years.

Stem extraction (timber harvest) is often the
most effective intervention to direct a stand
towards target stand conditions, especially as it
changes the stand structure. Often it is also justifiable, because it may cover its costs. However, the
removal of trees often temporarily reduces the
resistance of a stand to storms and snow load,
leading to the destruction of the tree layer (Ott *et
al.* 1997). Such a destabilization can be mitigated
or prevented if the dominant trees with a compara-
bly low coefficient of slenderness that confer stability to a stand are left on site, and if tree clumps
are either left or removed entirely (Ott *et al.* 1997).

Stem extraction is not the only intervention
that should be considered. Other measures may be
cheaper and more effective. Alternative interventions include felling without extraction, creation
of microsites that are favourable for natural
seedling establishment such as mounds (Bassman
1989) and mineral soil (Prévost 1992), sowing,
planting, vegetation control, game control and
tending of young stands (e.g. artificial creation of
clumps).

The decision-making process should be documented (principle 2, section 3.5.3), and interventions should be designed as simple controlled
experiments (principle 3, section 3.5.3). However,
any treatments in forests with a direct protective
function must be very unlikely to impair effective

protection. To test a wide range of experimental treatments, forests with only an indirect protective function are more appropriate.

3.6.4 Third step: carry out interventions

The realization of interventions usually requires technical and organizational skills. The technicians that will plan and survey the interventions should be integrated into the decision-finding process, especially at the stand level (step 2 in Box 3.5), because they are able to judge the feasibility of the operation.

3.6.5 Fourth step: monitor forest development

Monitoring will determine to what extent stand development is approaching target conditions, and how interventions contribute to stand development. Monitoring helps to demonstrate successes to investors (usually, the state), and to learn from failures. Monitoring can provide firm evidence for unexpected outcomes, which may lead to revising traditional assumptions.

Monitoring parameters include those that are used to describe target stand conditions. In addition, any realized interventions should be described, and an event history of major disturbances and natural hazards established. Parameters that are measurable and not only observed are preferable. Even if observations are correct, they are subjective and may not be repeatable by other experts. Terrestrial (Magill & Twiss 1965) and aerial photography are helpful and objective means for monitoring. Terrestrial monitoring should be carried out on permanently marked plots. In forests with a direct protective function, monitoring should focus on the crucial spots for effective protection. In such cases, we recommend establishing representative indicator plots (0.5–2.0 ha) instead of sampling in systematic grids.

Monitoring intensity depends on the importance of the protective function. In forests with a direct protective function, intensity will be higher than in forests with only an indirect protective function.

All monitoring activities should be properly documented. The documentation for forest management should thus include information on the whole management procedure, from the collection of information over the field assessments, the decision-finding process and the realized interventions to the monitoring activities.

When interpreting monitoring results, it is important to recognize that short-term management successes may be long-term failures. This is due to the importance of rare extreme climatic and biotic events to forest dynamics. Uniform afforestation that initially establishes successfully, but later becomes unstable, is a good example (Fig. 3.4).

3.7 EXAMPLES OF SUCCESSFUL MANAGEMENT OF PROTECTION FORESTS

3.7.1 Minimal tending in Switzerland

Protection is the primary function of Swiss mountain forests. However, silvicultural interventions including timber harvesting are often not profitable, despite the high subsidies that are available. To ensure that only the necessary and effective interventions are subsidized and carried out, a procedure of 'minimal tending' was designed for protection forests with a direct protective function, in a collaborative effort of scientists and practising foresters. This procedure was thoroughly documented (Wasser & Frehner 1996), and many detailed checklists have been provided for each step. This ensures transparent and justifiable decisions.

The procedure follows the steps 2–4 in Box 3.5, as the information collected in step 1 is usually available for Swiss mountain forests. Monitoring is carried out on representative 'indicator plots'. For each area with the same treatment, one indicator plot of 0.25–1.00 ha is delineated. On this indicator plot, the stand condition is recorded and documented (e.g. with photographs), and any relevant changes are monitored.

A key to determine site conditions is provided, and important ecological and silvicultural characteristics of each site type are described. Site-specific target stand conditions are given as minimum and optimum conditions, the parameters including species composition, structural characteristics (vertical and horizontal stand structure, variation in diameter at breast height), stability characteristics of the dominant trees (crown length, coefficient of slenderness), and presence and spatial distribution of the regeneration. In addition, hazard-specific target stand conditions are given, for the natural hazards landslides, erosion, torrent and debris flow, rockfall and snow avalanches. These hazard-specific target stand conditions are further specified, where appropriate, for departure, transit and run-out areas. They include standards for maximum gap size, tree cover, occurrence of windthrow-prone trees, stem density and occurrence of regeneration.

To obtain subsidies for silvicultural interventions, the 'minimal tending' procedure is not mandatory. However, any applied procedure must meet the standards established in 'minimal tending'. An open question is the number of indicator plots; it seems to be difficult to convince foresters to establish and maintain a sufficient number of such plots.

3.7.2 A structural typology in the northern French Alps

The silvicultural decisions in *P. abies* protection forests in the northern French Alps (Chauvin *et al.* 1994; Renaud *et al.* 1994) resemble the procedure in Switzerland, but emphasize stand structure. As an aid in decision making, a structural typology was established, based on detailed terrestrial assessments of the horizontal and vertical stand structures on plots of 50 m × 50 m, on 11 study sites totalling over 300 ha (Chauvin *et al.* 1994). The assessments include qualitative and quantitative variables, the most important being the presence or absence of canopy layers (including the presence of gaps), the basal area and the distribution of the diameter at breast height, and expert judgements on stand resistance to disturbance and future development.

Based on the structural typology, keys have been established that enable a particular stand to be classified as a known structural type. The structural types are then linked to instability types that describe (i) a stand's susceptibility to storm damage, snow break and rime, (ii) its effectiveness to prevent and stop snow avalanches and rockfall, and (iii) any impediments to sufficient

regeneration. In contrast to the Swiss procedure, the term 'instability' is thus used to designate three different stand properties, including (i) its resistance to disturbance, (ii) its effectiveness in protection against natural hazards, and (iii) its resilience after disturbance. For each structural type, silvicultural interventions are proposed. The target stand structure is irregular and clumped, with long internal edges.

To design interventions in a particular stand, the structural and instability types in a protection forest are assessed. A particular stand is divided into units of 50 m × 50 m, and the types are described for each unit. The spatial distribution of 'unstable units' is then used to delineate zones with effective and ineffective protection, and to establish management priorities.

The effect of different silvicultural interventions is closely monitored in several areas that were used to establish the structural typology. This typology is now widely used in forest management.

3.8 SHELTERBELTS

3.8.1 Introduction

Prevention of avalanches and related hazards seeks largely to harness protection conferred by existing forest cover. Closely related is the deliberate planting of trees to ameliorate a hazard, of which shelterbelts are the best known. Man has always made use of the shelter provided by trees if only to get out of the rain or to keep flocks warm during cold nights. Over the last 200 years a much more systematic approach has been taken to the use of trees to shelter agricultural operations and human habitation. The impetus behind this move has been the European agricultural reforms of the eighteenth and nineteenth centuries, particularly in Germany, and the expansion of human agriculture onto land previously not used for crop production (Caborn 1957). These included the prairies of North America, the steppes of Russia and the Ukraine and the heathlands of Denmark. More recently shelterbelts have been planted extensively in parts of China (Song 1991), Japan, New Zealand and Australia (Nuberg 1998).

Trees can provide shelter from the wind, precip-

Area (Fig. 3.8)	Increased	Reduced
Displacement zone	Wind speed Turbulent production	
Cavity	Turbulence Daytime temperature Humidity Lodging and abrasion	Wind speed Night-time temperature
Wake	Daytime temperature Humidity	Wind speed Turbulence Erosion Water loss

Table 3.1 Flow regions of shelterbelt.

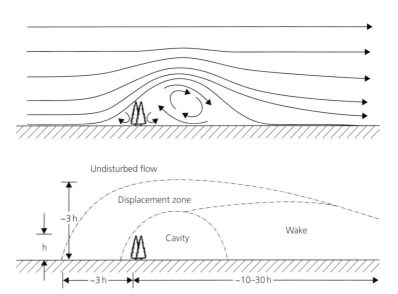

Fig. 3.8 Flow over a shelterbelt.

itation, blowing snow and the sun. Whether shelter is provided by individual trees, groups of trees or rows of trees, the basic aim is to modify the microclimate in the local vicinity. The major influence of trees is in their interception of incoming and outgoing radiation, modification to wind flow patterns and changes to the local humidity. They are also effective at capturing dust and pollution. Recent reviews of current knowledge can be found in Brandle and Hintz (1988) and Palmer *et al.* (1997).

Shelterbelts usually consist of long rows of trees placed perpendicular to the prevailing wind.

The presence of the trees blocking the flow produces a high-pressure region upstream and a low-pressure region downstream. This leads to deflection of the flow upward ahead of the belt and return of the flow behind the belt. The flow can be roughly divided into three regions (Fig. 3.8), the displacement zone, the quiet or cavity zone and the wake or mixing zone (Judd *et al.* 1996). The microclimate in each region is affected differently (Table 3.1). The critical parameters controlling the exact nature of the flow are the height, length and porosity of the shelterbelt. Porosity is difficult to define exactly but is a measure of the resistance of the shelterbelt to the flow and is approximately related to the visual density (optical porosity). Shelterbelts are roughly divided by their optical porosity from dense or impermeable (porosity < 40%), through semi-permeable (40–60%), to open or permeable (>60%). Dense shelterbelts tend to produce a well-defined cavity region with a large amount of shelter very close to the shelterbelt (<10 tree heights) but much less shelter further away. They also produce increased turbulence, increased daytime temperatures and reduced night-time temperatures close to the belt. Semi-permeable belts have no cavity region but a more extensive wake. This produces less dramatic influences but over a larger area downwind (20–30 tree heights). Open shelterbelts may be very ineffective and if they have an open trunk space lead to enhanced wind speeds close to the belt. The length of the belt also affects the area of shelter, with a maximum potential shelter area formed by a triangle from the ends of the belt extending downwind a distance equal to the length of the belt (Brandle & Finch 1991).

Shelter from trees is generally applied to three distinct areas of activity:
1 the shelter of crops;
2 the shelter of livestock;
3 the shelter of buildings and roads.

3.8.2 Shelter of crops

Kort (1988) and Nuberg (1998) provide comprehensive reviews of the influence of shelter on crops. Shelter can provide positive and negative benefits to crops as listed below.

Benefits

• Increase in ambient temperature leading to improvement in germination and growth rates.
• Reduction of moisture loss and control of snow drifting.
• Reduction of mechanical damage leading to improved crop quality.
• Increase in soil organic matter by production of leaf litter.
• Trapping or recycling nutrients.
• Reduction of soil erosion.

• Reduction of crop lodging.
• Promotion of mineralization of soil nitrogen.
• Retention of heat in the soil and air, thereby extending the growing season.
• Reduction of soil acidification in certain soil types.
• Improvement of pollination efficiency for certain crops.
• Control of spray drift.
• Reduction of infiltration of water to groundwater systems preventing the rise of saline water tables.

Disadvantages

• Competition with crops for light, moisture and nutrients leading to reduced crop yields close to belts.
• Increase in lodging due to increased turbulence.
• Reduction in pollination for certain crops.
• Land taken out of production.
• Waterlogging of soil close to dense belts.
• High costs of establishment and management.

3.8.3 Shelter of livestock

Much less work has been carried out on the benefits of shelterbelts for livestock in comparison to crops. Gregory (1995) provides the most comprehensive review, Webster (1997) discusses the energy requirements of animals and Bird (1998) discusses the indirect benefit to livestock through improvements to pasture.

Benefits

• Increased yield of pasture and root crops used for feed.
• Increased animal productivity (milk production and weight gain).
• Reduction in heat loss on clear nights if animals are able to move under trees.
• Reduced mortality in newborn animals and shorn sheep.
• Shade provision reduces overheating and sunburn.
• Increased animal welfare by provision of protection from wind, rain and snow.

- Increased fertility due to better health and provision of more comfortable conditions.
- Increase in range of breeds which can be utilized.
- Shelter for wildlife and game animals (deer, pheasants, etc.)

Disadvantages

- Overcropping close to shelterbelt.
- Spread of disease by concentration of animals close to shelter.
- Increase in insect pests in low wind speed region behind dense belts.
- Reduction in air temperature on clear nights behind belts.

3.8.4 Shelter of buildings and roads

Discussions of the benefits of shelter for buildings and roads can be found in Clarke (1997), DeWalle and Heisler (1988) and Shaw (1988), and are summarized below.

Benefits

- Reduced heat loss during cold weather leading to energy efficiencies.
- Shading in summer providing a reduction in requirements for air conditioning.
- Improved conditions in the vicinity of buildings such as lower wind speeds and increased temperatures.
- Reduction in building damage from driving rain and diurnal temperature fluctuations.
- Increased animal comfort in barns sheltered by trees.
- Roads kept free of snow drifts.
- Visual screening of buildings and roads.
- Reduction in noise levels.
- Capture of pollution such as dust, soot, spray and gases.

Disadvantages

- Take up land.
- Damage to buildings by the roots of trees planted too close.

- Shading by trees reducing the heating benefits of the sun.
- Increased snow build-up from poorly designed or positioned snow belts.

3.8.5 Shelterbelt design

Clearly the benefits of shelterbelts have to be balanced against the disadvantages (Clarke 1997; Blyth 1997). To ensure benefits are maximized careful shelterbelt design is required. Poor design can lead to a potential benefit becoming a disadvantage. In Table 3.2 a summary of the type of shelterbelt to use for different purposes is presented (Hislop *et al.* 1999). In general, semi-permeable belts (porosity 40–60%) are used where a large area of moderate shelter is required. Typical enterprises are the protection of crops and improved pasture. Dense shelterbelts (porosity <40%) are used where intense shelter is required over a short distance, such as for the protection of livestock for lambing and calving and buildings. In some cases it is beneficial to have some of the features of both a semi-permeable and a dense belt. This can be achieved with a hybrid belt in which the upper part of the belt is relatively open and the bottom of the belt is made dense by the planting of shrubs or slow-growing trees.

Generally shelterbelts should be kept as narrow as possible. This is partly to minimize the land used but also to maintain the effective area downstream. Wide forests are found to be very ineffective at providing a large area of shelter (Nageli 1964). However, where snow retention is the objective, a wider belt can provide a larger area of snow accumulation. Multiple shelterbelts have been found to provide no cumulative benefit (Nageli 1964) but rather the increased turbulence behind upwind belts can slightly reduce the effectiveness of subsequent shelterbelts.

3.9 FUTURE MANAGEMENT OF PROTECTION FORESTS

In the past decade, our knowledge about basic ecosystem processes such as nutrient cycling has considerably increased. However, the present state of protection forest management is still far

Table 3.2 Descriptions of shelterbelt types, their impact on windspeeds and their application.

Shelterbelt type	Broad description of features (porosity/ height/length)	Porosity profile	Area of windspeed reduction	Reduction of open windspeed	General application
Windbreak	Semi-permeable As tall as possible As long as necessary	60–40%	20–30 times height of the trees	20–70%	Crops Improved pasture
Windshield	Close to impermeable As tall as necessary As long as necessary	<40%	Up to 10 times height of the trees (max. shelter at 3–5 times the height)	up to 90%	Lambing/calving areas Feeding areas Farm buildings
Hybrid	Lower storey impermeable; upper storey semi-permeable As tall as possible As long as necessary	<40% lower storey 60–40% upper storey	5 times (approx.) height of the lower storey 20–30 times height of the upper storey	up to 90% 20–70%	Where a combination of applications suit both windbreak and windshield shelterbelt types

from a true ecosystem management, and often even further from ecosystem-based management (Schlaepfer 1997) where forests would be managed based on sound ecological knowledge. Instead, we are still mainly treating mountain forest ecosystems as tree assemblies. This is partly justified, because trees are the ecosystem component that is most important in providing effective protection against natural hazards. However, new impacts on forests such as excessive browsing, air pollution and global climate change may impair effective protection not directly by lowering the resistance and resilience of the trees, but by slowly affecting related ecosystem processes. This would be the case, for example, if soil acidification leads to shallower rooting and enhances storm risk. The rapid recovery of many European mountain forests after the devastations of past centuries suggests that these forests are very resilient, although subalpine forests seem to be an exception. Nevertheless, new environmental impacts challenge present forest management, because inconspicuous changes in ecosystem processes will need to be monitored and integrated into management decisions.

In protection forests, managing in a sustainable manner means meeting societal demands for pro-

tection against natural hazards, by maintaining or restoring ecosystem characteristics (components, structures and processes) that are required for long-term effective protection. The management of protection forests is sustainable if it ensures permanent effective protection. Whether present management approaches will achieve this is uncertain. Our understanding of mountain forest ecosystems must therefore be extended. We should find site-specific answers to questions such as 'How can the stability properties of mountain forest ecosystems that are important for effective protection be enhanced, and to what extent?', 'How much deviation from natural conditions in mountain forest ecosystems is possible in the long term, without impairing effective protection?', 'Is higher species diversity an effective means to increase ecological stability properties of mountain forest ecosystems?' and 'How can we integrate other forest uses into the management of protection forests, especially in developing countries?'

3.10 CONCLUSIONS

This chapter has presented an overview of the management of forests that serve as protection against natural hazards. It has focused on protection forests in mountain areas. Such forests are often sensitive ecosystems, and some of them recover slowly from disturbances, including human interventions. Managing protection forests means enhancing or maintaining their resistance to disturbances, including natural hazards, and their ability to recover from disturbances (resilience). Besides effective protection, other uses and benefits of the forest are secondary in protection forests.

Due to many uncertainties in the future development of protection forests, we advocate a management approach that mainly utilizes, or contributes to restore, natural ecosystem components, structures and processes, introducing artificial elements only if required. However, in mountain protection forests high naturalness is only a general approach, and not a goal in itself, because the main goal is effective protection.

No general rules on the management of protection forests can be given. In many cases, a small-scale mosaic of developmental phases is most appropriate to provide effective protection in the long term. However, this depends on the nature of the natural hazards, on site conditions and on the nature of the protective effect, i.e. whether this effect is bound on a particular location of the protection forest or not. In the first case of a direct protective effect, the forest protects underlying assets or people, mainly against avalanches and rockfall, whereas in the second case of an indirect protective effect, the forest protects against hazards starting from whole watersheds, for example floods.

The procedure that we have suggested to manage protection forests (Box 3.5) is flexible enough to be used in various situations. Important features of this procedure are community involvement, reliance on sound and complete information, confinement to interventions that are necessary, effective, justifiable and feasible, a transparent decision-finding process and the implementation of a system to monitor the success.

As reliable ecological knowledge on basic ecosystem processes in protection forests is often lacking, forest managers must rely on their observational skills, which should be developed during professional training, apply adaptive manage-

ment, designing interventions as simple experiments, and document the management decisions, interventions and success. Over the long term, this will increase their understanding of protection forests and help to continuously improve forest management.

ACKNOWLEDGEMENTS

We highly appreciated many helpful comments by Anton Bürgi, John Innes and Josef Senn (all at the Swiss Federal Institute for Forest, Snow and Landscape Research, Birmensdorf).

REFERENCES

Alexander, R.R. (1984) *Natural Regeneration of Engelmann Spruce after Clearcutting in the Central Rocky Mountains in Relation to Environmental Factors.* Research Paper no. 254, Rocky Mountains Forest and Range Experimental Station Fort Collins, Co, USA.

Altwegg, D. (1991) Der gemeinwirtschaftliche Wert von Schutzwäldern. [The socioeconomic value of protection forests.] *Österreichische Forstzeitung* **102**, 22–3.

Bassman, J.H. (1989) Influence of two site preparation treatments on ecophysiology of planted *Picea engelmannii* × *glauca* seedlings. *Canadian Journal of Forest Research* **19**, 1359–70.

Bayerische Staatsforstverwaltung, eds (1994) *Handbuch zur Sanierung von Schutzwäldern im bayerischen Alpenraum [Handbook of Restoration of Protection Forests in the Bavarian Alps].* Bayerische Staatsforstverwaltung, Munich.

Bazzigher, G. (1976) Der schwarze Schneeschimmel der Koniferen (*Herpotrichia juniperi* (Duby) Petak und *Herpotrichia coulteri* (Peck) Bose). [Black snow mould of coniferous trees (*Herpotrichia juniperi* (Duby) Petak and *Herpotrichia coulteri* (Peck) Bose).] *European Journal of Forest Pathology* **6**, 109–22.

Berger, F. (1996) Appreciation of snow avalanches potentialities under forest canopy. In: Wybo, J.L., Therrien, M.C., Guarnieri, F., eds. *The International Emergency Management and Engineering Conference 1996* (TIMEC), Montréal, pp. 221–230. TIEMES/Ecole des Mines de Paris, Paris.

Bird, P.R. (1998) Tree windbreaks and shelter benefits to pasture in temperate grazing systems. *Agroforestry Systems* **41**, 35–54.

Blyth, J. (1997) The economics of shelter provision on farms. In: Palmer, H., Gardiner, B., Hislop, H., Sibbald, A. & Duncan, A., eds. *Trees for Shelter.* Forestry Commission Technical Paper No. 21 pp. 40–5. Forestry Commission, Edinburgh.

Brandle, J.R. & Finch, S. (1991) *How Windbreaks Work.* University of Nebraska Cooperative Extension Bulletin EC-91–1763-B. Lincoln, Nebraska, USA.

Brandle, J.R. & Hintz, D.L. (1988) Windbreak technology: Proceedings of an International Symposium on Windbreak Technology. *Agriculture, Ecosystems and Environment* **22/23**, 598.

Brang, P. (1996) Experimentelle Untersuchungen zur Ansamungsökologie der Fichte im zwischenalpinen Gebirgswald [Experimental studies on seedling ecology of Norway spruce in the intermediate alpine mountain forest]. Dissertation, Federal Institute of Technology Zurich. *Beiheft, Schweizerische Zeitschrift für Forstwesen* 77.

Brang, P. (1997) Ecological niches of seedling establishment in high-elevation forests. *Working Paper, Research Branch, British Columbia Ministry of Forests* 24, 144–53.

Brang, P. (1998) Early seedling establishment of *Picea abies* in small forest gaps in the Swiss Alps. *Canadian Journal of Forest Research* 28, 626–39.

Brassel, P. & Brändli, U.-B. (1999) *Schweizerisches Landesforstinventar: Ergebnisse der Zweitaufnahme 1993–95. [Swiss Forest Inventory: Results of the Second Inventory 1993–95.]* Paul Haupt, Bern.

Bretz Guby, N.A. & Dobbertin, M. (1996) Quantitative estimates of coarse woody debris and standing dead trees in selected Swiss forests. *Global Ecology and Biogeography Letters* 5, 327–41.

Brückner, G. (1986) Man's impact on the evolution of the physical environment in the mediterranean regions in historical times. *Geojournal* **13**, 1–17.

Burschel, P., El Kateb, H., Huss, J. & Mosandl, R. (1985) Die Verjüngung im Bergmischwald. [Regeneration in mixed mountain forests.] *Forstwissenschaftliches Centralblatt* **104**, 65–100.

Caborn, J.M. (1957) *Shelterbelts and Microclimate.* Forestry Commission Bulletin 29. HMSO, Edinburgh.

Cattiau, V., Mari, E. & Renaud, J-P. (1995) Forêt et protection contre les chutes de rochers. [Forest and protection against rockfall.] *Ingénieries: Eau–Agriculture–Territoires* 1, 45–54.

Cerny, A. (1980) Comparing the health condition of natural spruce forests and spruce plantations in the territory of the Czech Socialist Republic. In: Klimo, E., ed. *Stability of Spruce Forest Ecosystems. University of Agriculture, Brno, International Symposium, October 29–November 2, 1979*, pp. 339–342. Institute of Forest Ecology, Faculty of Forestry, University of Agriculture, Brno, Czechoslovakia.

Chauvin, C., Renaud, J.-P., Rupé, C. & Leclerc, D. (1994) Stabilité et gestion des forêts de protection. [Stability and management of protection forests.] *Office National des Forêts. Bulletin Technique* 27, 37–52.

Chen, J.M., Franklin, J.F. & Spies, T.A. (1995) Growing-season microclimate gradients from clearcut edges into old-growth Douglas-fir forests. *Ecological Applications* 5, 74–86.

Clarke, D. (1997) Shelter trees for energy conservation. In: Palmer, H., Gardiner, B., Hislop, H., Sibbald, A. & Duncan, A., eds. *Trees for Shelter*. Forestry Commission Technical Paper No 21. Forestry Commission, Edinburgh.

Coates, K.D., Emmingham, W.H. & Radosevich, S.R. (1991) Conifer-seedling success and microclimate at different levels of herb and shrub cover in a *Rhodo-dendron–Vaccinium–Menziesia* community of south central British Columbia. *Canadian Journal of Forest Research* 21, 858–66.

DeWalle, D.R. & Heisler, G.M. (1988) Use of windbreaks for home energy conservation. *Agriculture, Ecosystems and Environment* **22–23**, 243–60.

Food and Agriculture Organization (FAO) (1997) *State of the World's Forests.* FAO, Rome.

French National Forest Inventory (1998) *Expanding Forests.* French National Forest Inventory, Nogent sur Vemisson, France.

Gigon, A. (1983) Typology and principles of ecological stability and instability. *Mountain Research and Development* **3**, 98–102.

Gigon, A. & Grimm, V. (1997) Stabilitätskonzepte in der Ökologie: Typologie und Checkliste für die Anwendung. [Stability concepts in ecology: typology and a check list for practical use.] In: Fränzle, O., Müller, F. & Schröder, W., eds. *Handbuch der Umweltwissenschaften. [Handbook of environmental sciences.]*, 19 pp. Ecomed, Landsberg.

Gregory, N.G. (1995) The role of shelterbelts in protecting livestock: a review. *New Zealand Journal of Agricultural Research* 38, 423–50.

Grumbine, R.E. (1994) What is ecosystem management? *Conservation Biology* **8**, 27–38.

Gsteiger, P. (1993) Steinschlagschutzwald: Ein Beitrag zur Abgrenzung, Beurteilung und Bewirtschaftung. [Forest protecting against rockfall: a contribution to delineation, decision finding and management.] *Schweizerische Zeitschrift für Forstwesen* **144**, 115–32.

Hamilton, L.S. (1992) The protective role of mountain forests. *Geojournal* **27**, 13–22.

Hamilton, L.S., Gilmour, D.A. & Cassells, D.S. (1997) Montane forests and forestry. In: Messerli, B. & Ives, J. D., eds. *Mountains of the World: A Global Priority*, pp. 281–311. The Parthenon Publishing Group, New York.

Harmon, M.E., Franklin, J.F., Swanson, F.W. *et al.* (1986) Ecology of coarse woody debris in temperate ecosystems. *Advances in Ecological Research* **15**, 133–302.

Hewitt, K. (1997) Risk and disasters in mountain lands. In: Messerli, B. & Ives, J.D., eds. *Mountains of the World: A Global Priority*, pp. 371–408. The Parthenon Publishing Group, New York.

Hislop, A.M., Palmer, H.E. & Gardiner, B.A. (1999) Assessing woodland shelter on farms. In: *Proceedings of 'Farm Woodlands for the Future' Conference.* September 1999. Cranfield University, Silsoe, England.

Holling, C.S. (1996) Surprise for science, resilience for ecosystems, and incentives for people. *Ecological Applications* **6**, 733–5.

Imbeck, H. & Ott, E. (1987) Verjüngungsökologische Untersuchungen in einem hochstaudenreichen subalpinen Fichtenwald, mit spezieller Berücksichtigung der Schneeablagerung und der Lawinenbildung. [Studies on regeneration ecology in a subalpine Norway spruce forest rich of tall herbs, with particular emphasis on snow ablation and avalanche formation.] *Mitteilungen, Eidgenössisches Institut für Schnee- und Lawinenforschung* no. 42. Davos.

In der Gand, H. (1978) Verteilung und Struktur der Schneedecke unter Waldbäumen und im Hochwald [Distribution and structure of the snowpack under trees and in high forest]. In: *Mountain Forests and Avalanches. IUFRO Working Party Snow and Avalanches. Proceedings of the Davos Seminar September 1978*, pp. 97–122. Swiss Institute for Snow and Avalanche Research, Davos.

Ives, J.D. (1985) The mountain malaise: quest for an integrated development. In: Singh, T.V., Kaur, J., Ives, J.D. & Messerli, B., eds. *Integrated Mountain Development*, pp. 33–45. Himalayan Books, Delhi.

Johnson, E.A. (1987) The relative importance of snow avalanche disturbance and thinning on canopy plant populations. *Ecology* **68**, 43–53.

Judd, M.J., Ranpach, M.R. & Finnigan, J.J. (1996) A wind tunnel study of turbulent flow around single and multiple windbreaks, Part I: Velocity fields. *Boundary-Layer Meteorology* **80**, 127–65.

Kimmins, J.P. (1997) *Forest Ecology: A Foundation for Sustainable Management*, 2nd edn. Prentice Hall, Upper Saddle River, New Jersey.

Kirchner, J.W., Finkel, R.C., Riebe, C.S., Granger, D.E., Clayton, J.L. & Megahan, W.F. (1998) Episodic erosion of the Idaho Batholith inferred from measurements over 10-year and 10000-year timescales. In: *EOS, Transactions, American Geophysical Union, 79, F338*. American Geophysical Union Fall Meeting, San Francisco, December 1998.

Kohyama, T. (1984) Regeneration and coexistence of two *Abies* species dominating subalpine forests in central Japan. *Oecologia* **62**, 156–61.

Koppenaal, R.S., Hawkins, B.J., Shortt, R. & Mitchell, A.K. (1995) *Microclimatic Influences on Advance Amabilis Fir Regeneration in Clearcut and Understory Habitats of a Coastal Montane Reforestation Site*. Forest Resource Development Agreement Report no. 223, Forestry Canada and British Columbia Ministry of Forests, Victoria.

Korpel', S. (1995) *Die Urwälder der Westkarpaten [Old-growth Forests in the Western Carpathian Mountains]*. Gustav Fischer, Stuttgart.

Kort, J. (1988) Benefits of windbreaks to field and forage crops. *Agriculture, Ecosystems and Environment* **22–23**, 165–90.

Kusché, W., Schneider, W. & Mansberger, R. (1994) Schutzwaldphasenkartierung aus Luftbildern. [Classification of states in protection forests using remote sensing.] *Centralblatt für das Gesamte Forstwesen* **111**, 23–55.

Langenegger, H. (1979) Eine Checkliste für Waldstabilität im Gebirgswald. [A check list for forest stability in mountain forests.] *Schweizerische Zeitschrift für Forstwesen* **130**, 640–6.

Magill, A.W. & Twiss, R.H. (1965) A guide for recording esthetic and biological changes with photographs. *Research Note* no. 77, Pacific Southwest Forest and Range Experiment Station, Berkeley, California.

Mahrer, F., Bachofen, H., Brändli, U.-B. *et al.* (1988) Schweizerisches Landesforstinventar: Ergebnisse der Erstaufnahme 1982–86. [Swiss Forest Inventory: results of the first inventory 1982–86.] *Berichte, Eidgenössische Forschungsanstalt für Wald, Schnee und Landschaft* no. 305. Birmensdorf.

Mayer, H. & Ott, E. (1991) *Gebirgswaldbau—Schutzwaldpflege: ein waldbaulicher Beitrag zur Landschaftsökologie und zum Umweltschutz [Silviculture in Mountain Forests—Management of Protection Forests: A Silvicultural Contribution to Landscape Ecology and Environmental Protection]*, 2nd revised edn. Gustav Fischer, Stuttgart.

Mayer, H., Zukrigl, K., Schrempf, W. & Schlager, G., eds. (1989) *Urwaldreste, Naturwaldreservate und schützenswerte Naturwälder in Österreich [Old-growth Remnants, Natural Forest Reserves and Natural Forests Worth being Preserved in Austria]*, 2nd edn. Institut für Waldbau, Bodenkundliche Universität Wien.

Mencuccini, M., Piussi, P. & Zanzi Sulli, A. (1995) Thirty years of seed production in a subalpine Norway spruce forest: patterns of temporal and spatial variation. *Forest Ecology and Management* **76**, 109–25.

Mettin, C. (1977) Zustand und Dynamik der Verjüngung der Hochlagenwälder im Werdenfelser Land. [State and dynamics of the regeneration of high-elevation forests in the Werdenfelser Land.] Dissertation, University of Munich.

Meyer-Grass, M. & Schneebeli, M. (1992) Die Abhängigkeit der Waldlawinen von Standorts-, Bestandes- und Schneeverhältnissen. [The relationship between forest avalanches and site, stand and snow parameters.]. In: Fiebiger, G. & Zollinger, F., eds. *Schutz des Lebensraumes vor Hochwasser, Muren und Lawinen [Protection Against Floods, Debris Flow and Avalanches]. Internationales Symposium*

4: Interventions to Enhance the Conservation of Biodiversity

TIMOTHY J. B. BOYLE

4.1 BACKGROUND

The term 'biodiversity' has gained widespread acceptance only since the late 1980s and early 1990s, and has only been used in scientific literature for a little longer (Wilson 1988). The concept was first popularized with the publication of the 'Brundtland Report' of the World Commission on the Environment and Development (WCED 1989), and the subsequent negotiation of the Convention on Biodiversity, as part of the United Nations Conference on the Environment and Development (UNCED), in Rio de Janeiro, in 1992. In recent years, two important principles concerning the conservation of biodiversity have become widely accepted. The first of these is that protected area systems alone cannot adequately conserve the world's biodiversity. This view is clearly stated in the Global Biodiversity Strategy (WRI, IUCN & UNEP 1992), the Global Biodiversity Assessment (UNEP 1995, p. 929) and the work of various other authors (e.g. Reid 1996; Szaro 1996). The second principle is that sustainability can only be achieved through a landscape approach to natural resource management. This is reflected in the enormous amount of attention paid in recent years to how landscape management can contribute to conservation of biodiversity (e.g. Szaro & Johnson 1996), and in the debate over 'integrating or segregating' conservation and management goals at the landscape level (e.g. Boyle & Sayer 1995; Gajaseni 1996).

Taken together, these principles imply that in order to conserve biodiversity and sustainably utilize components of biodiversity, it is necessary to understand the interactions among the ecological and social systems that result in a particular pattern of land cover. Biodiversity represents the major component of ecological systems, and understanding how biodiversity is maintained at the landscape level is critical to managing natural resources sustainably.

Changing patterns of land use have a direct impact on biodiversity. Such changes are driven in part by human population growth and migration, and partly by the demands of the forestry, agricultural and industrial sectors. Thus, both social and ecological systems influence the types of land-use systems, which in turn result in a particular pattern of land cover.

Components of biodiversity provide resources and services for human communities at all scales, and constitute one of the major drivers of the interaction between social and ecological systems. Sustainable management of land-use systems is therefore dependent on the conservation of biodiversity at levels and in spatial patterns which provide optimal values and services to human populations, whilst still maintaining viable ecological systems. The conservation of biodiversity requires knowledge of the three elements shown in Fig. 4.1. We clearly need to understand the impacts of human activities on biodiversity, and to do this we also need to have tools and methodologies to allow us to assess the status of biodiversity. The third component, namely the ability to predict levels of biodiversity at non-sampled points, is also critical. This is because of the complexity of the concept of biodiversity (see below) and the impracticality of measuring levels of biodiversity at every point in the landscape.

This chapter is concerned with the importance of biodiversity in forest ecosystems and what

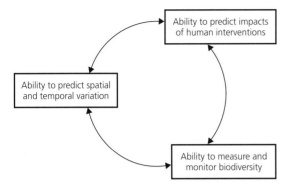

Fig. 4.1 Three critical elements of research for conservation of biodiversity.

actions may be possible and beneficial in conserving forest biodiversity. The role of biodiversity in ecological functioning will be discussed, together with problems and approaches to assessment and monitoring of biodiversity. The impacts of human activities will be reviewed, and possible paths towards improved conservation in managed forests will be proposed.

4.1.1 What is forest biodiversity?

This seemingly simple question is, in fact, very complex. It is complex both in terms of defining biodiversity in an operationally meaningful way, and also in setting limits to what constitutes forest, as opposed to non-forest, biodiversity.

Biodiversity is usually defined in terms related to the diversity of life on earth. Thus, for example, the Convention on Biological Diversity defines biodiversity as:

> . . . the variability among living organisms from all sources including, *inter alia,* terrestrial, marine and other aquatic ecosystems and the ecological complexes of which they are part; this includes diversity within species, between species and of ecosystems (UNEP 1992).

The Global Biodiversity Assessment provides essentially the same definition, and goes on to note that '[it] can be partitioned so that we can talk of the biodiversity of a country, of an area, or of an ecosystem, of a group of organisms, or within a single species' (Bisby *et al.* 1995). Various

other definitions all provide a consistent message, though the wording may differ (e.g. McNeely *et al.* 1990; WRI, IUCN & UNEP 1992).

A common feature of these definitions is the partition of biodiversity into three hierarchical levels, namely those associated with genetic, species and ecosystem diversity. There are also a multitude of other ways to partition biodiversity. It is possible, for example, to characterize biodiversity in terms of processes or functions, and thus to assess trophic diversity, structural diversity or functional diversity. However, although such terms may have great value for certain applications, or for examining particular aspects of biological systems, the genetic–species–ecosystem approach is conceptually simpler and has broader applicability. Nevertheless, a shortcoming of all such definitions is that although they may be conceptually sound, they are most certainly not operationally feasible. It is clearly an impossible task, even in biologically rather simple systems, to assess the numbers and relative frequencies of all component ecosystems, species and genes (alleles) within species that conceptual definitions of biodiversity demand. The implications for assessment and monitoring of biodiversity is discussed further in section 4.2.1.

Turning to the question of what constitutes forest biodiversity, it may be very simplistically considered as that portion of the earth's biodiversity which occurs in forest ecosystems. However, forest ecosystems are, increasingly, found in a matrix of other land uses, especially agriculture. Most of this agricultural land has been derived from forest clearing, and the result is a continuum of landscape patterns, ranging from predominantly forest, with only small agricultural clearings, to a predominantly agricultural landscape in which only small patches of forest survive. As there are ecological interactions among landscape units, it is certainly logical to conclude that small clearings within a forest, which may often resemble natural disturbance patches, contribute to forest biodiversity, but where the threshold occurs as the proportion of forest in the landscape declines is difficult to define. The agricultural land itself may contain forest fragments or individual trees which contribute to conservation of forest biodiversity in two ways. First, they provide habitat for those elements of forest biodiversity which are most tolerant to disturbance and environmental stress, and second they provide the opportunity for movement of genes or individuals, between intact forest areas, of species which may not be able to survive in fragments or an agricultural setting.

4.1.2 Biodiversity and sustainable management

Biodiversity contributes to sustainable management of forests in a variety of ways. Healthy ecosystems require the maintenance of numerous ecosystem processes, such as decomposition and nutrient cycling, carbon and nitrogen fixation, water cycling, etc. All of these processes depend on biotic agents, and therefore on biodiversity. Different organisms may be optimally adapted to mediate these processes under different environmental conditions. Spatial and temporal fluctuations in environmental conditions therefore lead to a greater diversity of organisms contributing to maintaining ecosystem processes. In addition to fluctuations around a mean, there may be directional changes in environmental conditions. The most obvious example is global climate change resulting from increased atmospheric concentrations of 'greenhouse' (i.e. radiatively active) gases. Responses to such directional changes also require high levels of diversity to be maintained within ecosystems in order to provide the evolutionary resources leading to improved adaptation. Finally, components of biodiversity may provide direct benefits to human populations. The most obvious example of this is in the breeding and domestication of many species, especially cereals, legumes and fruits, but also including many other plant and animal species. Many of these species were originally forest species, and many more forest species are utilized without being domesticated, and with only minimal formal management.

The concept of ecosystem health has been adapted from our understanding of human health, and a healthy ecosystem is usually considered to be characterized by two features, namely stability and resilience. Stability refers to the ability of ecosystem processes to withstand the impacts of disturbances, while resilience reflects the capacity to return to previous conditions following disturbances. Although, as mentioned previously, healthy ecosystems depend on biodiversity to maintain ecosystem functions, the relationship between levels of biodiversity and stability are complex. There are numerous examples of empirical studies demonstrating a lack of any simple relationship between levels of biodiversity and stability (e.g. Spitzer *et al.* 1997), and this has led Kimmins (1997) to conclude that there are no 'clear scientific relationships between diversity and ecosystem health'.

The various roles of biodiversity in functioning ecosystems may not always be complementary. For example, increasing yields of useful products from forests may entail simplifying the structure and diversity of such forests—the ultimate case being well-maintained monocultures, including (potentially) monoclonal plantations. The benefits derived from such forests obviously come at a cost in terms of the other roles of biodiversity, especially in relation to adaptiveness to change. Inadequate attention has been given to the trade-offs between the different uses and benefits of biodiversity, and a more detailed system of valuation needs to be developed. Many of the benefits derived from forest biodiversity are 'public goods' which no single owner can claim. These benefits include:

1 Ecological services: the services provided by ecological systems in the form of their role in regulating regional or global climate and freshwater flows, or in carbon sequestration.

2 Existence values: the value people place on maintaining species in their own right, even if they will not use (or view) them.

3 Option values: the value of the option to make use of a component of biodiversity in the future, such as the potential value of species for new pharmaceuticals or as a source of new genetic material for the biotechnology or agriculture industry.

Consequently, although the conservation of forest biodiversity provides benefits to many individuals within a region, nation or globally, only a portion of these benefits can be captured by the landowner faced with a choice between protecting a forest or converting it to another use. This discrepancy between the local costs and benefits of conservation and the broader social costs and benefits is a major force driving the loss of biodiversity. To the extent that mechanisms can be put into place that will return an equitable share of the diffuse global, national or regional benefits of biodiversity to the local custodians of biodiversity (e.g. farmers, local landowners, etc.), the incentive for conservation will be increased.

Some mechanisms that can play important roles in the equitable sharing of benefits at the local, national and global level include: strengthened property rights for smallholders, regulation of access to genetic resources, voluntary contributions to wildlife conservation organizations or through development assistance programmes, and the establishment of user fees or taxes on commodities derived from biodiversity. Most of the economic benefits from biodiversity are local and regional, in the form of various products and services including food, fuel, fibre and clean water, which are a consequence of the role of biodiversity in maintaining healthy and productive ecosystems. But while global benefits are typically smaller, the value forests provide in sequestering carbon is likely to increase significantly in coming decades.

4.1.3 Processes that maintain biodiversity

Ecosystem processes are mediated by components of biodiversity. Due to ecological interactions among organisms, these same processes are largely responsible for maintaining levels of biodiversity. Some processes operate at only one of the different levels of the biodiversity hierarchy, while others may operate at more than one level.

At the genetic level there are four important processes that serve to maintain genetic diversity and structure:

1 *Random genetic drift.* The non-directional changes in genotypic frequencies among generations due to random chance in small populations.
2 *Selection.* Relative differences among genotypes in viability or reproductive success.
3 *Gene flow.* The exchange of genes between populations that differ in genotypic frequencies.
4 *Mating.* The process mediating the recombination and assortment of genes between generations.

Mutation is arguably the single most important genetic process, since it is the origin of all genetic variation—the variation that ultimately results in speciation, and hence species and ecosystem diversity. However, mutation is a process which is usually only expressed over periods of tens of generations, which is beyond the scope of management for many forest species.

At the level of species diversity, processes which serve to maintain diversity include:

1 *Dispersal/migration.* The movement of individuals in space, either at the time of reproduction or thereafter. This process is directly related to gene flow, which represents the movement of gametes in space.
2 *Reproduction.* Impacts on the process of reproduction can have rapid, direct and dramatic consequences. In the case of species with short generation periods, non-overlapping generations or highly specific mutualisms, changes can be particularly devastating.
3 *Regeneration/succession.* An obvious, and highly publicized, consequence of logging is the reduction in area of mature, or 'old-growth' forest, and replacement with forest dominated by pioneer or early successional species.
4 *Trophic dynamics.* Trophic dynamic processes refer to the ways that species from different trophic levels interact. These include pollination, predation and herbivory. As each trophic level is dependent on other levels, impacts on trophic dynamics can be very serious.
5 *Local extinction.* In some cases the dominant process determining change in species composition may be local extinction.

4.2 ASSESSING AND PREDICTING THE STATUS OF BIODIVERSITY

4.2.1 Problems with assessment

As discussed in section 4.1, the conceptual definitions provided in the Convention on Biological Diversity and other sources are of little use in operational assessment of biodiversity as it is impossible to contemplate a comprehensive inventory of biodiversity even in simple systems. A related problem in the measurement of biodiversity lies in the characteristics of the three levels of diversity which create problems in measuring and monitoring. At the genetic and species levels, individual alleles or species are usually quite discrete and recognizable. However, alleles cannot be seen without expensive and relatively sophisticated detection systems. It is also now widely recognized that it is not simply numbers of alleles, but how they combine in forming multilocus genotypes, that determines effective genetic diversity. The huge number of gene loci also makes it impossible to quantify more than a small sample of total genetic diversity. Large numbers are also a problem in quantifying species diversity, and this problem is compounded by the fact that many of the groups of organisms that contribute the most to species diversity are not easy to see because of their small size and/or habitat (e.g. soil microbes). Other species are mobile and have an aversion to being sampled. In contrast to genes and species, ecosystems are often non-discrete, and may require some ordination process for the purposes of identification. However, partly as a consequence of this, the numbers of recognizable ecosystems are much smaller than allele or species numbers. These various characteristics of the three hierarchical levels of biodiversity are summarized in Table 4.1.

Measurements of biodiversity must be understandable and acceptable to natural resource managers and, perhaps as importantly, to the public. One approach to deriving more meaningful measurements of biodiversity is to simplify the concept. Without question, species diversity is conceptually the simplest of the three hierarchical levels of diversity, both because many species are visually distinct and relatively easy to count,

and because species extinction is a dramatic event. By contrast, ecosystems suffer from a lack of distinctness, and changes resulting from human activities may significantly change the characteristics of an ecosystem, but the relative value of two ecosystems may be ambiguous. For example, an 'old-growth' forest and regenerating forest may have highly distinct assemblages of species, but neither may have an intrinsically higher conservation value. It is only when large-scale imbalances in the proportions of different successional stages result from overexploitation, or when certain forest types, such as the coastal forests of Brazil and Ecuador, are in danger of complete elimination that ecosystem loss becomes a major issue.

Consequently, species diversity, or even more frequently, species richness, is often used as a surrogate measure of biodiversity. However, even species counts are extremely resource intensive and expensive, especially if groups of invertebrates or microbes are included in the assessment, as these groups contain immense numbers of species, most of which can only be identified by experts. For this reason much attention has been paid in recent years to the possibility of using limited groups of taxa as surrogate measures of species diversity in a broader range of taxa—the so-called 'indicator species' concept. Research on indicator taxa has thus far yielded conflicting results, even when analyses have been conducted on the same data sets. For example, Lawton *et al.* (1998) report that correlation analyses of diversity within and among eight faunal groups across sites of varying degrees of disturbance indicated that no one group served as an indicator for overall

Table 4.1 Characteristics of different levels of diversity that are of significance to measurement and monitoring.

	Size of units	Numbers of units	Distinctness of units	Ease of conceptualization	Ease of measurement
Genes	Too small	Too large	Mostly OK	Low	Expensive and time-consuming
Species	Many are too small	Too large	Mostly OK	High	Time-consuming, or inaccurate if only a sample
Ecosystems	OK	OK	Low	Medium/low	Difficult if ordination required

faunal diversity. By contrast, Vanclay (personal communication), analysing the same data set but using regression analysis to reveal trends, concluded that birds and butterflies did, in fact, serve as useful indicators of overall diversity. Other studies have examined the use of higher order taxonomic groups such as genera or families, for which numbers are smaller and identification problems less severe. Balmford *et al.* (1996a,b) found that such an approach was effective in assessing overall diversity, and at great cost savings in comparison with species-level assessments.

One reason why conflicting views on the value of indicator taxa have emerged is that the 'target' is a moving one. We know that biodiversity varies widely, in both space and time, and over a range of scales. Global patterns of spatial variation in biodiversity are well understood. There is a trend to increasing species diversity from higher to lower latitudes, and a similar trend from drier to moister ecosystems (e.g. Boyle 1996). However, at a local scale, both the pattern and dynamics of spatial and temporal variations in biodiversity are poorly understood.

For example, the theory of plant succession, supported by studies on population dynamics in natural systems, has indicated that local extinction and re-colonization are normal events for many species. This has led to the concept of 'metapopulations', or populations of populations (Levins 1970). The existence of several populations of a species within a spatial unit implies that the among-population dynamics may be as important as the within-population dynamics in determining the persistence or otherwise of the species. In other words, the number and distribution of populations of a species may be a critical factor in its survival. However, metapopulation dynamics remain poorly understood, with few examples of detailed studies on the implications of disruptions to metapopulations. The contribution of a single population within a metapopulation to the continued persistence of a species will be related not only to its size and genetic diversity, but also to its spatial location within the metapopulation, and the total number of such populations. The conservation value of a particular population is therefore very difficult to quantify. In addition, the relevance of metapopulation

dynamics to long-lived organisms such as trees in the context of rapid global change needs to be elaborated.

4.2.2 Practical approaches to assessment and monitoring

As the concept of biodiversity is hierarchical, it is logical to conclude that any practical approach to assessment and monitoring of biodiversity should adopt a hierarchical approach also. The basic problem, derived from the previous discussion, is that enormous amounts of data need to be collected in the most cost-efficient manner and in a way that allows meaningful analyses relevant to biodiversity. This problem is particularly acute for assessment and monitoring of forest biodiversity, because forests cover very large areas and are also some of the most diverse ecosystems on earth.

Remote sensing is obviously best suited for assessments of the 'higher' levels in the biodiversity hierarchy (i.e. landscapes and ecosystems/forest associations). Satellite imagery, which typically operates at resolutions between 10 and 100 m, is being intensively investigated for its ability to identify and quantify forest resources (e.g. the TREES (Tropical Ecosystem Environment Observations by Satellites) project: Malingreau *et al.* 1994). Preliminary results indicate that, in some cases, data from visible-range and infra-red wavelength detectors (e.g. SPOT (Système Probatoire de l'Observation de la Terre), Landsat) can identify forested and non-forested land with reasonable accuracy, although various types of forest/non-forest interfaces cause classification errors (Sader *et al.* 1990). However, in other regions, especially in the equatorial zone, where cloud cover is frequent and extensive, visible wavelength detectors are of limited value. The use of microwave (radar) detectors (e.g. on the ERS (Earth Remote Sensing Satellite) series of satellites) offers a promising alternative, but classification errors are still a problem with satellite data (Rauste 1998).

Although there is a clear potential for rapid and effective assessment of diversity at the landscape level using satellite imagery, the limits of resolution of satellite data do not permit more detailed information to be obtained. For such purposes, airborne remote sensing will be required, and the same problems encountered by satellite sensors (i.e. limitations of visible wavelength detectors in cloudy areas) apply.

A possible additional advantage of radar remote sensing is that radar can penetrate not only clouds, but also forest canopies, and even soil. Different radar bands, varying in frequency, wavelength and/or polarization, have different properties in terms of penetrability and detection capability (Hoekman 1987). Much work has been done already on the capacity of airborne radar remote sensing for the estimation of forest biomass. Results from this research indicate that:
• The intensity of L-band (24 cm wavelength; 1.25 GHz frequency) backscatter is correlated with above-ground biomass (Dobson *et al.* 1992; Scales *et al.* 1997).
• P-band (68.1 cm wavelength; 440 MHz frequency) backscatter is correlated with tree age, height and diameter (Dobson *et al.* 1992).
• HH polarization return is physically related to crown and trunk biomass, but is also strongly affected by changes in the bottom layer (ground vegetation, surface topography and soil wetness) (Toan *et al.* 1989; Scales *et al.* 1997).
• Especially VH cross-polarization is linked to crown biomass and affected only by tree structure (Toan *et al.* 1989)
• The best results are obtained from using a mixture of incidence angles (Rauste 1998).

These results suggest that airborne radar remote sensing has the potential, at a satisfactory level of resolution, to detect variation in tree height (i.e. the difference between crown surface elevation and ground surface elevation—not detectable by visible wavelength detectors), crown biomass and ground vegetation structure.

As both radar and visible wavelength detectors can be operated simultaneously, the possibility exists for synergism in the use of both (Leckie 1990a,b). Visible wavelength detectors, as well as near infra-red, can be used to detect variability in crown surface colour and texture, which is also expected to be related to forest structural diversity.

If airborne remote sensing techniques are to be useful for rapid assessment of biodiversity, some significant technical problems need to be over-come, including geometric correction, calibration and registration to ground control points. Also, for interpretation of remote sensing data in terms of biodiversity, the relationship between structural diversity and biodiversity would need to be determined for each biome. However, for detection of *change*, the relationship does not need to be determined—data collected at different times will allow an assessment of change in structural diversity. This capacity to monitor changes in diversity over time constitutes a valuable tool for determining the sustainability of forest management practices, and is possibly the simplest and most useful application of airborne remote sensing.

4.2.3 The use of criteria and indicators

Criteria and indicators (C&I) are tools for collecting and organizing information in a way that is useful to assist in evaluating and implementing sustainable forest management. Recent interest in the development and implementation of C&I for sustainable forest management has led to the formulation of a wide range of international, regional and national systems of C&I, both by governments and non-governmental organizations.

Criteria and indicators have proved to be a popular assessment tool, but this popularity has, in itself, led to some confusion resulting from the profusion of 'jargon' terms associated with the large number of proposed systems of C&I. However, C&I remain intuitively attractive precisely because they are frequently, if often subconsciously, used in everyday life. Phenomena such as beauty or safety cannot be measured quantitatively, so we use criteria and indicators to rank, or assess, the status of these phenomena.

C&I have been designed to assess sustainability, including conservation of biodiversity, at a range of spatial scales. National-level C&I, which summarize the status of forest management over the entirety of a country's forest area, have been developed under a number of international initiatives, such as the Helsinki Process, the Montreal Process and the Tarapoto Process. Several systems of C&I which reflect sustainability at a local level, corresponding with individual forest management units (FMUs), have been developed by organizations such as the Rainforest Alliance, the Soil Association and the African Timber Organization.

The assessment of sustainability, including conservation of biodiversity, is essential at both national and local scales. Local-scale assessment is needed because it is the management decisions made in individual FMUs that ultimately determine the fate of forest biodiversity. However, comprehensive assessment is not possible solely at a local scale. For example, the establishment of a system of 'protected areas' is an essential requirement for conservation purposes as one aspect of sustainable forest management (this volume, Chapter 2). It is neither necessary nor possible for each FMU to include protected areas, so national-scale C&I are required to provide an assessment of such factors. Prabhu *et al.* (1996) concluded that the entire question of biodiversity assessment has been avoided in the sets of criteria and indicators proposed to date; instead the emphasis has been to ensure that management conforms with good forest stewardship. This is an understandable and pragmatic first step towards the development of a sustainability assessment tool. Nevertheless the question of how forest management actually impacts on biodiversity still begs an answer.

Human interventions in forests will inevitably affect biodiversity (increasing or decreasing it), so 'sustainability' in the context of conserving forest biodiversity must be considered in relation to a forest management plan. The first requirement for sustainability is therefore the existence of an articulated management plan. Key questions are:
• is the new level of biodiversity stable or continuing to decrease (or increase); and
• is the new level adequate to support all ecological processes (i.e. above a critical threshold level)?

These questions must consider multiple scales both because the definition of biodiversity requires this, and because the processes affecting biodiversity operate at different scales.

C&I are most powerful if they can be used as a tool in 'adaptive' management, by which forest managers (be they corporations, communities or individuals) can regularly assess the quality of their forest management, and adapt management practices accordingly. This implies that members

of the assessment team will not be 'scientific experts', and that conservation of biodiversity will need to be assessed in only a few (±10) person-days. These considerations dictate that the most important characteristic of an effective indicator will be the practicality of assessment in a very short period. This is a serious constraint for assessment of 'conservation', which implies a need to consider temporal dynamics, and for the assessment of biodiversity, which is an extremely broad concept. It is largely a consequence of this constraint that Prabhu *et al.* (1996) concluded, on the basis of field trials in different parts of the world, that most of the currently proposed biodiversity C&I were ineffective.

Stork *et al.* (1997) consequently suggested that *changes in biodiversity may be assessed indirectly through assessment of the processes that maintain and generate biodiversity.* For example, successful pollination is essential for plant species that reproduce sexually. Changes in pollination success will indicate changes in the diversity, abundance or behaviour of pollinating species as well as the structure of future plant (and animal) communities. They rejected the use of indicator species, on the basis that there is considerable doubt about the extent to which extrapolation from one species (or group of species) to another is possible (Landres *et al.* 1988; Lawton *et al.* 1998)—as discussed above. There have been some excellent examples of indicator species/groups for old-growth temperate forests (e.g. Spence *et al.* 1997), but the utility of this concept for tropical forests has yet to be proved.

4.2.4 Prediction at non-sampled points

There is increasing recognition of the need to deal with forest management issues on a landscape-wide basis. This is in contrast with most research and monitoring, which has tended to be at a stand level. Biodiversity must also be dealt with on a landscape-wide basis. There are a number of reasons for this. First, there are processes that contribute to the long-term viability of animal species in a landscape that is relatively large scale. For example, many animal species occur in landscapes as disjunct populations. Their persistence and genetic viability are affected by the metapopulation dynamics that determine the extinction of individuals in a patch and the dispersal of individuals between patches (e.g. Lindenmayer & Possingham 1994).

Secondly, a further and critical premise of landscape ecology is that landscapes are not environmentally homogeneous. Rather, they are composed of often surprisingly broad and complex environmental gradients, which exert a fundamental control on the nature and distribution of biodiversity in a landscape.

Conventional forestry tends to smooth out the inherent environmental heterogeneity found in landscapes by using uniform, 'lowest common denominator' guidelines for planning and management. There is widespread recognition that such an approach is neither necessary nor optimal, but methods need to be developed to facilitate more flexible and environmentally sensitive management.

Nix (1982) proposed that a 'niche' could be generally defined in terms of a species' response to the primary environmental regimes—heat, light, water and nutrients being the main determinants of plant response. Hence there are two fundamental requirements to develop environmentally sensitive management approaches. First, methods are needed to model the primary environmental regimes in a spatially distributed format. Secondly, methods are needed to predict levels and distribution of biodiversity.

The primary environmental regimes can be defined in terms of those processes that determine the distribution and availability of heat, light, water and nutrients. A key point is that these processes occur over a range of space/time scales. At the meso-scale the climatic inputs of energy and precipitation are critical. At the topo-scale, local topography modifies (i) incoming radiation (through the effects of slope, aspect and topographic shading), and (ii) redistributes water through the landscape. The microscale (or stand-level) processes are essentially related to canopy cover which further modifies below-canopy radiation levels. Radiation is of course the main driver of evaporation, and hence a major determinant of soil water status.

A great deal of ecological analysis is directed at the micro- (canopy or stand level) scales, but we need to focus at scales that provide the larger environmental context in order to understand forest responses to management. This means examining the primary environmental regimes at the meso- and topo-scales. Methods are now available to generate reliable spatial estimates of various long-term mean monthly climatic parameters including minimum and maximum temperatures, rainfall, radiation and potential evaporation. These procedures were developed by Hutchinson, Nix and McMahon (see Hutchinson 1991), and have now been successfully applied throughout the world. These models are derived from interpolation of data from the available network of climatic records as a function of longitude, latitude and elevation for temperature and precipitation, and longitude, latitude and cloudiness for radiation. They enable estimates of climate to be made for any location, and when coupled to a suitably scaled digital elevation model (DEM) enable gridded estimates to be generated (as used here, a DEM is a digital model of the topography represented by a regular grid of elevations). These data can then be entered into a GIS (Geographic Information System) as raster data layers.

While meso-scaled methods are now well established, topo-scaled environmental modelling is a more recent development (Mackey 1996). Of particular importance is the use of appropriately scaled DEMs to model (i) surface radiation, (ii) catchment hydrology, and (iii) land form. DEMs enable the effects of slope, aspect and horizon shading to be factored into radiation calculations. This enables a topo-scaled radiation budget to be calculated, including short- and long-wave radiation, thereby providing more precise estimates of air and soil temperatures, photosynthetically active radiation and evaporation. DEMs can also be used to model the flow of water through a catchment and hence to predict the spatial distribution of soil moisture. These models assume, among other things, that subsurface flows track surface slope—an assumption that is not valid in every landscape. They generally work best in humid, erosional landscapes with relatively uniform substrates. Nonetheless, these methods provide the ability to examine quantitatively some of the primary controls on forest productivity at relatively fine spatial resolutions, but over large areas.

The need for accurate predictive information for forest management planning, including the location and shape of conservation areas, is vital. The scale of operations of concessions in many parts of the world makes detailed biodiversity surveying over more than a fraction of the area impractical. The ability to predict species occurrence and habitat type from environmental information that can be obtained from readily available data, GIS and remote sensing would be a valuable management tool. In addition such prediction would also be of use in developing sampling designs for further biodiversity studies and in permanent sample plot location for growth and yield research.

4.3 HUMAN IMPACTS ON BIODIVERSITY AND A FRAMEWORK FOR DECISION-MAKING

4.3.1 Impacts of human activities on biodiversity

Any discussion on the impacts of human activities on biodiversity needs to address a number of issues. First, human activities can result in changes to the extent of forest lands (historically, most often reductions!), or can result in changes to the composition and quality of the forest, while retaining a forest-type land cover. The former type of impact obviously has a major and catastrophic impact on forest biodiversity, especially as the scale of forest loss exceeds some level at which clearings can be considered part of a forested landscape, as discussed above. This process results in fragmentation of forest lands. Impacts that do not result in loss of forest area are many and varied, and for this reason are more difficult to characterize in terms of their consequences for biodiversity. The impacts of fragmentation and other disturbances are discussed in more detail below.

The second issue that needs to be considered is the distinction between underlying threats leading to deforestation or forest degradation, and the immediate agents of such changes. With such a diverse array of global forests, from the tundra–boreal interface to tropical rain forests, it is obviously difficult to discuss threats to forest biodiversity without falling into the trap of over-generalization. However, Soulé (1991) presented an excellent discussion of six classes of threat and the north–south distinction (Fig. 4.2). Soulé distinguished between threats at the genetic, species, community and ecosystem levels, though in many cases the intensity of threat is similar at all levels. In poorer countries, threats due to habitat loss and fragmentation, introduction of exotic species and overexploitation were seen as most severe. By contrast, in richer countries the greatest threats were considered to derive from exotic species, pollution, climate change and habitat fragmentation, while overexploitation was seen as being of much less concern.

As is the case with all such large-scale generalizations, the detailed picture may deviate quite substantially. Schemske *et al.* (1994) evaluated alternative approaches to conservation of rare plants, with particular reference to endangered plant species in the USA. They reviewed the primary threats to 98 species, identifying 14 categories of threat. The most significant threat came from 'development', accounting for more than 20% of the species. 'Collecting' and 'grazing' each accounted for 10% and, in order of importance, these were followed by 'oil, gas, mining', 'water control' and 'trampling'; 'logging'; and 'exotics' and 'off-road vehicles', each at 6%. Only one species was considered to be endangered primarily due to natural causes, i.e. due to no direct human impact.

At first glance, these results would seem to be very different to Soulé's proposed agents of threat in a rich country such as the USA. If 'grazing', 'collecting' and 'logging' are collectively considered to represent 'overexploitation', they account for nearly 30% of all species. By contrast, 'pollution' is not identified as a threat, though elements of 'oil, gas, mining', 'development', 'agriculture' could constitute pollution. Several points need to be made in interpreting this apparent divergence, however. First, Schemske *et al.* reported agents of threat for only one group of organisms, i.e. plants. For other groups, the primary agents may differ quite substantially. For example, groups such as fishes, amphibians and insects may be expected to be much more susceptible to pollution than plants. Second, threats deriving from climate change or habitat fragmentation are much more

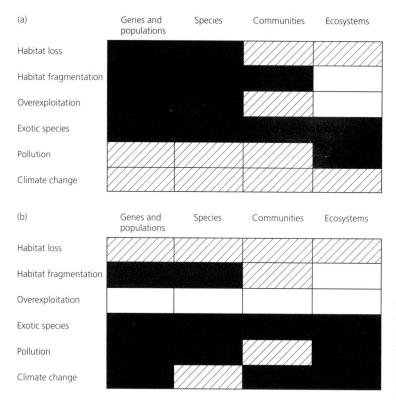

Fig. 4.2 The relationship between threats to biodiversity and different levels in the biodiversity hierarchy for developing (a) and developed (b) nations. ■, Severe threat; ▨, moderate threat; □, low threat. (Adapted from Soulé 1991.)

difficult to identify, as the impacts are often delayed for years, or even decades.

Returning to Soulé's assessment of threats to biodiversity in poorer countries, the combination of 'overexploitation', 'habitat loss' and 'habitat fragmentation' can be interpreted in line with the oft-expressed truism that the underlying cause of deforestation (and consequent loss of forest biodiversity) in tropical countries is poverty. While this may generally be true (but see further discussion, below), the nature of the agents of deforestation may be difficult to ascertain. For example, in reviewing the underlying causes of deforestation in Indonesia, Sunderlin and Resosudarmo (1996) surveyed 13 papers on the topic published since 1990, and noted an evolution in the assumed agents of deforestation, from smallholders to government agencies and the timber industry. They went on to emphasize, however, that there are several agents involved, and interactions among the various agents are also significant. The ten-

dency among researchers, NGOs or donor organizations to identify a single agent as the cause of deforestation, to the exclusion of others, needs to be resisted.

Human activities in forests have the effect of modifying the size and age structure of populations of both plants and animals, and potentially altering levels of biodiversity. When the intensity of disturbance becomes so great that tree cover is removed entirely from parts of the area formerly covered by forest, habitat fragmentation is the result. Ultimately, the major land use may be completely changed from a forest-based to a non-forest-based activity, such as agriculture, leaving only residual trees in the landscape. Such changes to forests are not, of course, simply sequential. At a landscape level, different areas of forest may remain undisturbed, while others may have already become fragmented or have been reduced to residual trees in an agricultural landscape. In particular, disturbance and fragmentation often go hand-in-hand—it is rare to find forest fragments that have not also been affected by human activities within the remnant.

Fragmentation

Most research into the biological effects of ecosystem fragmentation for plants has been of an ecological nature, concentrating on changes in species richness and community dynamics (Meave *et al.* 1991; Young & Mitchell 1994). Much of this work has involved testing various aspects of Island Biogeography Theory (Helliwell 1976). Direct effects of ecosystem fragmentation on plants include loss of populations of species, reductions in remnant population sizes, changes in densities of reproductive individuals, reduced reproductive success and increased isolation of remnant populations. Based on this, the general prediction has been that fragmentation will be accompanied by a loss of species diversity. Recently attention has been drawn to the possible genetic consequences of fragmentation and the implications for plant conservation (Billington 1991; Ledig 1992; Young *et al.* 1996). In the short term, loss of individual genetic diversity (heterozygosity) may reduce individual fitness and, through this, lower population viability. In the longer term, lowered population-level genetic diversity (polymorphism and allelic richness) will limit a species' ability to respond to changing selection pressures, thus reducing its evolutionary potential and an erosion of intraspecific genetic diversity (Young *et al.* 1996).

Research has demonstrated that species loss from forest fragments appears to be nested; that is, there is a predictable sequence of species loss as forest area is reduced (Patterson 1992). For example, understorey insectivorous birds and large top predators seem especially sensitive to fragmentation. Thus the pattern of biodiversity loss in habitat fragments in dynamic landscapes is predictable even though the final cause of extinction itself is generally a result of demographic or environmental stochasticity. Such predictable patterns are useful for setting thresholds for sustainable management. Research targeting the causes, consequences and solutions to population declines is therefore more valuable than research

focusing on small populations (Caughley & Gunn 1996). The decline of species and the fate of biodiversity in dynamic managed systems is firmly grounded in population dynamics.

Our knowledge of the population dynamics of most tropical species is virtually nil. However, there is now a considerable body of ecological theory dealing with the dynamics of spatially structured populations such as those found in fragmented habitats. The 'metapopulation' concept of a set of local populations potentially linked together through dispersal within a landscape has become a leading paradigm in conservation biology (Hanski 1989; Caughley & Gunn 1996). The key feature of metapopulations is that the extinction of local populations is balanced by the process of dispersal and recolonization from other local populations (Levins 1970). The result is that regionally the population persists, and may be thought of as a series of lights with individual lights being switched on and off at random. Various adaptations have been proposed to examine variants in this simple model.

The existence of a metapopulation is dependent upon both dynamics of the landscape and the species' populations (Fahrig & Merriam 1994). Dispersal of individuals may occur between patchy populations, and if balanced by local extinctions the species may exist as an equilibrium metapopulation. Further fragmentation will eventually isolate very small habitat patches which may support small populations separated by distances too great for interfragment movement.

One conclusion from these models is that a metapopulation will be vulnerable to a threshold of habitat availability. When the proportion of forest cleared is greater than this threshold, the rate of local extinction exceeds colonization and the regional population will go extinct. There is a scarcity of empirical data from managed tropical forest ecosystems relating to extinction thresholds.

A second conclusion is that metapopulations may show transient dynamics so that even when this threshold in habitat loss has been exceeded, there will be a time lag until the decline of species is detectable, particularly for species with long generation times. This is extremely important for empirically determining the thresholds of habitat modification as indicators because habitat loss may well exceed the threshold level for regional persistence well before the effects on populations are seen. This implies that extinction is not an instantaneous process and species are 'committed to extinction' at a certain level of habitat loss (Heywood *et al.* 1994).

One problem in studying the impacts of fragmentation on species and genetic diversity are the multitude of factors influencing the outcome. In addition to fragment size, and distance of isolation, discussed above, numerous other factors will affect rates of gene flow (see Volume 1, Chapter 3), and consequently genetic drift. These include:

• *Ecology.* Species adapted to narrow ecological niches, which may therefore tend to be restricted to relatively small isolated patches, even in continuous forest, may be relatively unaffected by fragmentation, especially if their particular microsite types are not directly removed.

• *Reproductive ecology.* For animal-pollinated and -dispersed plants, dispersal, gene flow and recruitment will be reduced if interpopulation distances become greater than the home ranges, or gap-crossing ability, of pollinator or disperser guilds (Powell & Powell 1987).

• *Matrix type.* A species' gap-crossing ability will depend on the nature of the gap. A matrix of grassland is likely to present a very different habitat quality from secondary (scrub) forest, or plantations, to pollinator and disperser guilds.

• *Landscape pattern.* Most research has concentrated on the classical 'island' form of fragmentation. However, in some landscapes, such as in many parts of Central America, surviving fragments are restricted to river valleys, and thus form a dendritic pattern, in which gene flow along fragments may be relatively unaffected, but gene flow across the intervening watershed may be much reduced.

• *Initial density of individuals.* Species which tend to be found at low densities are more likely to be reduced to non-viable effective population sizes, as the intensity of fragmentation increases.

• *Time/variance.* Most studies of fragmentation treat time as a relatively simple factor, such that with increasing time since fragmentation, the impacts of the event will be greater. However, in addition to possibly reducing mean rates of dispersal and outcrossing, the variance of these processes is also likely to increase. While no effective dispersal may occur for a number of years, optimum conditions in one year may permit adequate recruitment to maintain population levels.

Disturbance

Disturbance is a word which can have a variety of connotations (see Volume 1, Chapters 4 and 5). It is therefore important to specify that in this context disturbance refers to any human activity which affects the forest without causing loss of forest cover. Human activities which have the greatest effect in forests usually involve the extraction or harvesting of products from the forest. The variety of products harvested from forests, especially tropical forests, is immense: wood, fruit and other foods, medicinal plants, construction materials, and many more. In general, increased intensity of economic uses of the forest is likely to reduce overall biodiversity, though the relationship is not a simple one.

There have been many studies, in both tropical and temperate/boreal forests, on the impacts of logging on biodiversity. For example, in tropical forests studies have focused on the vegetation (Crome *et al.* 1992), birds (Thiollay 1992; Crome *et al.* 1996), amphibians (Taufik 1998) and insects (Holloway *et al.* 1992; Spitzer *et al.* 1997). Comparable studies in temperate/boreal forests include those by Niese and Strong (1992) on vegetation, Welsh (1988) on birds and Bruce and Boyce (1984) on amphibians and insects. A common, though not universal, conclusion of these studies is that diversity, often measured in terms of the Shannon–Weiner index, tends to increase after logging. One reason for this is that following logging a wider range of environmental conditions is created which offers a greater variety of ecological niches for colonizing species. This is especially true for selective logging regimes in the tropics, where the intensity of logging within a felling area ranges from very heavy (e.g. along and near roads, skidding trails and landings) to zero, where patches of forest were left untouched, particularly on steeper slopes. While some species which require large areas of mature forest conditions will decline dramatically, or disappear from the site, others survive by adapting their behaviour to the new conditions (e.g. Thiollay 1992). Negative impacts are also closely related to the intensity of disturbance. Along ridges, where the intensity of logging is usually highest (Thiollay 1992) or, in the case of illegal selective logging, near villages, or in areas of highest density of the most desirable species (Spitzer *et al.* 1997), far more species are adversely affected.

Lee *et al.* (1996) studied the impact of selective logging on genetic diversity of a range of plant species, including both those which are harvested and those which are not. They recorded losses in genetic diversity for all species, and those losses were greatest for harvested, rare species, and least for non-harvested, common species. Similarly, Tsai and Yuan (1995) recommended that in order to reduce negative impacts on genetic diversity, harvest levels should be adjusted in proportion to the census numbers of mature and submature individuals for each commercial species.

The impacts of non-timber forest product (NTFP) collection are often considered to be more benign than logging impacts (e.g. Panayatou & Ashton 1992), though Hall and Bawa (1993) concluded that there are few, if any, examples of sustainable NTFP collection systems. In a series of comprehensive studies on NTFP collection in the Biligiri Rangan Hills of south India, Hegde *et al.* (1996) demonstrated a negative correlation between the amount of time spent collecting NTFPs and the availability of salaried labour, while Murali *et al.* (1996) found a significant impact on recruitment of the harvested species, especially where harvest levels were highest. The negative relationship between time spent collecting NTFPs and availability of salaried labour is a consequence of the seasonality and unreliability of NTFP collection, where crops in some years may be very small, and also the low value of the crop at the point of collection. Uma Shanker *et al.* (1996) concluded that non-sustainable harvest levels were inevitable given these low value-added benefits accruing to the collectors. There are also numerous other examples of overexploitation of NTFPs (e.g. Browder 1992; Nepstad *et al.* 1992). While most of these studies focus on the direct impacts on the harvested species, Boot

and Gullison (1995) use the example of collection of brazil nut (*Bertholletia excelsa* Hunb. and Bonpl.) to point out that there may be both direct and indirect impacts on non-harvested species. High removal rates of brazil nuts may directly affect squirrels and agouti—the main consumers and seed distributors of the species—and this may have indirect consequences on other plant species if these frugivores substitute other species into their diet. [An economic analysis of NTFPs will be found in Volume 1, Chapter 15.]

4.3.2 Impacts of global climate change

Although the loss of natural forest due to climate change may be balanced by natural regeneration or, more likely, the establishment of plantations on unproductive agricultural land, the biodiversity conservation value of such new forests will, at least initially, be modest. Hobbie *et al.* (1993) considered the likely impact of climate change on genetic and species diversity. They noted that, to accurately predict the consequences of climate change, it is not sufficient to examine only the changes in physical environment, but it is also necessary to consider the changes in resource availability, and changes in disturbance regimes. They concluded that both genetic diversity and species diversity will decline, but that the rate and magnitude of the decline will depend on the existing population structure and population dynamics of different groups of organisms.

Harte *et al.* (1992) point out that changes in the abiotic environment, disturbance regimes and species interactions will result in strong directional selection for most taxa. This will result in an individualistic migratory response to the new environments for each species, leading to reduced effectiveness of pollination and seed dispersal, as plant, pollinator and seed dispersal interactions are disrupted. The new climatic and biotic environments may also promote increased hybridization among related, but previously allopatric, species.

A proposed policy response to climate change is the creation of migratory corridors to overcome existing barriers. Quite apart from the problems associated with creating and managing corridors in the presence of other agents of global change, such as changing land use (mainly involving a reduction in forest land) and pollution, the validity of such an approach is highly questionable, given the individualistic response of different species, as discussed by Harte *et al.* As climate change will result in not only a 'movement' of existing abiotic environments but the creation of novel environments, it is uncertain whether the resultant selective pressures will lead to migration of taxa or evolution *in situ*. In all likelihood, some taxa will respond *in situ*, while others will migrate, though the spatial direction and distance of migration will be different for each species.

Although global circulation models (GCMs) are rapidly increasing in sophistication (Baskin 1993), they are still limited in the ability to integrate the interactions and feedback mechanisms resulting from climate change that are ecologically significant, such as changes in resource availability and disturbance regimes. GCMs also operate at a scale that is unsuitable for prediction at the local level. Consequently, the impact of climate change on forest biodiversity, and appropriate policy responses, is still an area where much additional information is necessary.

4.3.3 A system to assist in decisions to enhance conservation of biodiversity

The area of natural forest land will continue to decline, especially in the tropics, at least in the short to medium term. This may be due partly to population growth and economic development, as discussed above, as well as other factors. Global climate change will also lead to displacement of forest lands, though the impact is expected to be greater at higher latitudes (Hobbie *et al.* 1993). Meanwhile, opportunities for establishing new protected areas in the tropics is rapidly diminishing (Sayer & Wegge 1992).

These trends have led to an assumption that, increasingly, production systems and conservation goals will be integrated. Integration of production and conservation is a form of multiple-use management. Panayatou and Ashton (1992) examined multiple-use management from an economic perspective, and although they noted that greater management skills are required for multiple-use management, and that produc-

tion costs may increase due to potentially higher harvesting costs and the loss of economies of scale, they concluded that 'the case for multiple-use management can often be based on financial analysis alone, without resorting to shadow pricing . . . '.

Although multiple-use management may be justifiable on financial grounds, biodiversity is difficult to value, as noted above. Van Noordwijk *et al.* (1996) constructed a model to evaluate the relative value, in terms of biodiversity conservation, of strategies to segregate or integrate production and conservation, and used data from agroforestry systems in Sumatra, Indonesia, to test the model. The model basically calculates the relative loss of biodiversity per unit increase in productivity, for a range of possibilities from pure conservation to pure production. For systems in which the relative loss of biodiversity is higher at low or intermediate production levels, a segregation policy is optimal, in which production is concentrated on relatively small areas, allowing larger areas to be set aside for conservation. This situation may apply to most food crop-based production systems, due to the requirement of most crops for a high light, low competition environment, even at low production levels. By contrast, integration may be a valid strategy for 'jungle rubber' systems, involving the production of rubber and other crops in modified secondary forest.

The model developed by Van Noordwijk *et al.* also has some limitations. For example, they point out that the impact of increasing productivity has a differential impact on different components of biodiversity. Data from Sumatra indicate that whereas for soil collembola the impact may only be slight, for birds it is much more severe. If all groups of organisms are to be considered, then the choice to integrate or segregate will depend on the most sensitive group, and segregation will generally become much more attractive. Similarly, the choice depends on total productivity possible under pure production systems. Whereas in the 1930s productivity of jungle rubber was comparable to monocultures, now it is only 25–30% of monoculture productivity, which again increases the tendency to segregate.

The conclusions of Van Noordwijk *et al.* would seem to differ from those of Panayatou and Ashton (*op. cit.*). Certainly, Panayatou and Ashton were more concerned with total economic value of the forest than with biodiversity alone. Their studies have indicated that the contribution of biodiversity to total forest value is rather small. Thus, if making management decisions based on total value, the final choice may conflict with what is best for biodiversity. Clearly, the actual decision of when to integrate and when to segregate production and conservation will depend on numerous factors, including not only relative biodiversity values and economic considerations, but also on land tenure systems, regulatory efficiency, administrative flexibility, and so on.

Where segregation of conservation and production objectives is necessary, the identification of optimum areas for conservation is increasingly relying on computer technologies, such as GIS (e.g. Pressey *et al.* 1990; Margules *et al.* 1991; Vane-Wright *et al.* 1991; Ramesh *et al.* 1997). Such approaches may take the form of gap analyses (e.g. Scott *et al.* 1991) or the identification of minimum sets of conservation areas to meet a threshold value of some measurable variable (e.g. Margules *et al.* 1988; Kitching & Margules 1995). In most cases, the measurable variable is species richness, and the threshold value is some arbitrarily fixed proportion of total species richness within a region. Raw measures of species richness may be supplemented with assessments of threat to prioritize area selection or conservation effort (Dickman *et al.* 1993; Smith *et al.* 1993).

Some studies integrating socioeconomic and biophysical data by means of GIS have been undertaken in relation to biodiversity conservation. For example, Fox *et al.* (1994c) investigated the conflict between grazing and conservation of red panda habitat in Langtang National Park in Nepal. A model of grazing impact was developed, which showed that the strong preference for gentle slopes and open canopies in selection of grazing lands coincided with prime red panda habitat. Fox *et al.* (1994a,b) also used GIS to study the farmer decision-making process in relation to biophysical and socioeconomic variables, and temporal change in land use in northern Thailand.

From the earlier discussion, it is clear that whatever the form of forest utilization (logging, NTFP collection, etc.), negative impacts on biodiversity are possible, but equally, management options are available to minimize such consequences. If our understanding of the impacts of any given management activity were sufficiently sophisticated, and if our ability to predict spatial variation in levels of biodiversity were adequate (Fig. 4.1), it would be possible to apply management prescriptions to maintain biodiversity within certain designated limits. Even in less diverse temperate forests we are as yet unable to meet these requirements and are unlikely to be in such a position for the foreseeable future. Consequently, we must find alternatives to ensure the conservation of biodiversity in managed forests.

One option would be to apply uniformly such a benign form of management that conservation of biodiversity is virtually guaranteed. This is unlikely to be acceptable from an economic viewpoint, and is also inefficient as we can assume that there are some combinations of forest type and management action where higher exploitation intensities are possible without unacceptably high losses of biodiversity.

A better option is to utilize a management regime which combines in-built flexibility with frequent monitoring of performance—an 'adaptive' management approach. The concept of adaptive management was first developed in the late 1970s (Holling 1978), and incorporates a 'learning by doing' philosophy. Advocates of adaptive management argue that such an approach increases the rate of knowledge acquisition and enhances information flow among stakeholders, thus creating shared understanding. For these reasons, it has been proposed as an ideal approach for the management of natural resources (e.g. Ludwig *et al.* 1993), where political and economic pressures are likely to promote non-sustainable harvesting levels under normal management regimes. It has especially been recommended for the management of biodiversity (Wieringa & Morton 1996), where the high degree of uncertainty and changing management goals make alternative management approaches difficult. However, in a review of three case studies of adaptive management, McLain and Lee (1996) conclude that the approach often fails due to inadequate attention paid to policy processes, and a frequent tendency among managers to rely too heavily on a modelling approach, ignoring alternative sources of knowledge. However, if these short-comings are addressed, adaptive management offers the best approach to conserving biodiversity in managed systems. It is well suited to planted forests where interventions such as glade and gap formation, extended rotations, thinning to increase stand structure and dead-wooding are all empirically observed to increase biodiversity. For adaptive management to be effective, the monitoring component must not place too heavy an economic or technical burden on the forest manager. Thus, the requirements for the monitoring system match those discussed earlier as being relevant to the assessment of biodiversity using criteria and indicators.

ACKNOWLEDGEMENTS

I would like to acknowledge the contributions made by Drs William Sunderlin, Nick Mawdsley and Brendan Mackey to the ideas presented in this chapter.

REFERENCES

Balmford, A., Green, M.J.B. & Murray, M.G. (1996a) Using higher-taxon richness as a surrogate for species richness: I. Regional tests. *Proceedings of the Royal Society of London B* **263**, 1267–74.

Balmford, A., Jayasuriya, A.H.M. & Green, M.J.B. (1996b) Using higher-taxon richness as a surrogate for species richness: II. Local applications. *Proceedings of the Royal Society of London B* **263**, 1571–5.

Baskin, Y. (1993) Ecologists put some life into models of a changing world. *Science* **259**, 1694–6.

Billington, H.L. (1991) Effect of population on genetic variation in a dioecious conifer. *Conservation Biology* **5**, 115–19.

Bisby, F.A., Coddington, J., Thorpe, J.P. *et al.* (1995) Characterization of biodiversity. In: Heywood, V.H., ed. *Global Biodiversity Assessment*, pp. 21–106. UNEP/Cambridge University Press, Cambridge.

Boot, R.G.A. & Gullison, R.E. (1995) Approaches to developing sustainable extraction systems for tropical forest products. *Ecological Applications* **5**, 896–903.

Boyle, T.J.B. (1996) Biodiversity of Canadian forests, with particular reference to the west coast forests. In: Lawford, R.G., Alaback, P.B. & Fuentes, E., eds. *High-Latitude Rainforests and Associated Ecosystems of the West Coast of the Americas. Ecological Studies 116*, pp. 353–78. Springer-Verlag, New York.

Boyle, T.J.B. & Sayer, J.A. (1995) Measuring, monitoring and conserving biodiversity in managed tropical forests. *Commonwealth Forestry Review* **74**, 20–5.

Browder, J.O. (1992) Social and economic constraints on the development of market-oriented extractive reserves in Amazon rain forests. In: Nepstad, D.C. & Schwartzman, S., eds. *Non-Timber Forest Products from Tropical Forests: Evaluation of a Conservation and Development Strategy. Advances in Economic Botany 9*, pp. 33–42. New York Botanical Garden, New York.

Bruce, R.C. & Boyce, S.G. (1984) Measurements of diversity on the Nantahala National Forest. In: Cooley, J.L. & Cooley, J.H., eds. *Natural Diversity in Forest Ecosystems: Proceedings of the Workshop. Institute of Ecology*, pp. 71–85. University of Georgia, Athens, GA.

Caughley, G. & Gunn, A. (1996) Conservation biology in theory and practice. Blackwell, Cambridge, MA.

Crome, F.H.J., Moore, L.A. & Richards, G.C. (1992) A study of logging damage in upland rainforest in North Queensland. *Forest Ecology and Management* **49**, 1–29.

Crome, F.H.J., Thomas, M.R. & Moore, L.A. (1996) A novel Bayesian approach to assessing impacts of rain forest logging. *Ecological Applications* **6**, 1104–23.

Dickman, C.R., Pressey, R.L., Lim, L. & Parnaby, H.A. (1993) Mammals of particular conservation concern in the Western Division of NSW. *Biological Conservation* **65**, 219–48.

Dobson, M.C., Ulaby, F.T., le Toan, T., Beaudoin, A., Kasischke, E.S. & Christensen, N. (1992) Dependence of radar backscatter on coniferous forest biomass. *IEEE Transactions on Geoscience and Remote Sensing* **30**, 412–15.

Fahrig, L. & Merriam, G. (1994) Conservation of fragmented populations. *Conservation Biology* **8**, 50–9.

Fox, J., Kanter, R., Yarnasarn, S., Elkaisngh, M. & Jones, R. (1994a) Farmer decision making and spatial variables in northern Thailand. *Environmental Management* **18**, 391–9.

Fox, J.J., Krummel, S., Yarnasarn, M., Ekasingh & Podger, N. (1994b) Land use and landscape dynamics in northern Thailand: assessing change in three upland watersheds since 1954. In: Fox, J., ed. *Spatial Information and Ethnoecology: Case Studies from Indonesia, Nepal, and Thailand. East–West Centre Working Paper, Environment Series No. 38*, pp. 27–42. East–West Centre, Honolulu, USA.

Fox, J.P., Yonzon & Podger, N. (1994c) Maps, yaks, and red pandas: using GIS to model conflicts between biodiversity and human needs. In: Fox, J., ed. *Spatial Information and Ethnoecology: Case Studies from Indonesia, Nepal, and Thailand. East–West Centre Working Paper, Environment Series No. 38*, pp. 15–26.

Gajaseni, J. (1996) Biodiversity and agroforestry systems. In: Korpilahti, E., Mikkela, H. & Salonen, T., eds. *Caring for the Forest: Research in a Changing World. Congress Report*, Vol. II. IUFRO XX World Congress, Tampere, Finland. Finnish Forest Research Institute, Tampere, Finland.

Hall, P. & Bawa, K. (1993) Methods to assess the impact of extraction of non-timber forest products on plant populations. *Economic Botany* **47**, 234–47.

Hanski, I. (1989) Metapopulation dynamics: does it help to have more of the same? *TREE* **4**, 113–14.

Harte, J., Torn, M. & Jensen, D.B. (1992) The nature and consequences of indirect linkages between climate change and biological diversity. In: Peters, R., ed. *The Effect of Climate Change on Biodiversity*. Yale University Press, New Haven, CT.

Hegde, R., Suryaprakash, S., Achoth, L. & Bawa, K.S. (1996) Extraction of non-timber forest products in the forests of Biligiri Rangan Hills, India. 1. Contribution to rural income. *Economic Botany* **50**, 243–51.

Helliwell, D.R. (1976) The effects of size and isolation on the conservation value of wooded sites in Britain. *Journal of Biogeography* **3**, 407–16.

Heywood, V.H., Mace, G.M., May, R.M. & Stuart, S.N. (1994) Uncertainties in extinction rates. *Nature* **368**, 105.

Hobbie, S.E., Jensen, D.B. & Chapin, F.S. (1993) Resource supply and disturbance as controls over present and future plant diversity. *Ecological Studies Analysis and Synthesis* **99**, 385–408.

Hoekman, D.H. (1987) Measurements of the backscatter and attenuation properties of forest stands at X-, C- and L-band. *Remote Sensing of Environment* **23**, 397–416.

Holling, C.S. (1978) *Adaptive Environmental Assessment and Management*. John Wiley, New York.

Holloway, J.D., Kirk-Spriggs, A.H. & Chey, V.K. (1992) The response of some rain forest insect groups to logging and conversion to plantation. *Philosophical Transactions of the Royal Society of London B* **335**, 425–36.

Hutchinson, M.F. (1991) Climate analysis in data sparse regions. In: Muchow, R.C. & Bellemay, J.A., *et al. Climatic Risk in Crop Production*, pp. 55–71. CAB International, Wallingford, Oxon.

Kimmins, J.P. (1997) Biodiversity and its relationship to ecosystem health and integrity. *The Forestry Chronicle* **73**, 229–32.

Landres, P.B., Verner, J. & Thomas, J.W. (1988) Ecological use of vertebrate indicator species: a critique. *Conservation Biology* **2**, 316–28.

Lawton, J.H., Bignell, D.E., Bolton, B. *et al.* (1998) Biodiversity inventories, indicator taxa and effects of habitat modification in tropical forest. *Nature* **391**, 72–6.

Leckie, D.G. (1990a) Synergism of synthetic aperture radar and visible/infrared data for forest type discrimination. *Photogrammetric Engineering and Remote Sensing* **56**, 1237–46.

Leckie, D.G. (1990b) Advances in remote sensing technologies for forest surveys and management. *Canadian Journal of Forest Research* **20**, 464–83.

Ledig, F.T. (1992) Human impacts on genetic diversity in forest ecosystems. *Oikos* **63**, 87–108.

Lee, C.T., Wickneswari, R., Norwati, M. & Boyle, T.J.B. (1996) Effect of logging on the genetic diversity of three rainforest species with different life history strategies in a lowland dipterocarp forest. In: *Proceedings of the 2nd Malaysian National Genetics Congress, Kuala Lumpur, November 13–15*, pp. 319–25. Universiti Kebangsaan Malaysia, Kuala Lumpur, Malaysia.

Levins, R. (1970) Extinction. *Lectures on Mathematics in the Life Sciences* **2**, 75–107.

Lindenmayer, D.B. & Possingham, H.P. (1994) *The Risk of Extinction: Ranking Management Options for Leadbeater's Possum Using Population Viability Analysis*. Centre for Resource and Environmental Studies, Australian National University, Canberra.

Ludwig, D., Hilborn, R. & Walters, C. (1993) Uncertainty, resource exploitation, and conservation: lessons from history. *Science* **260**, 17, 36.

Mackey, B.G. (1996) The role of GIS and environmental modelling in the conservation of biodiversity. In: *Proceedings of the Third International Conference/ Workshop on Integrating GIS and Environmental Modeling CD-ROM, January 21–25*. US National Center for Geographic Information and Analysis, Santa Fe, New Mexico, USA.

McLain, R.J. & Lee, R.G. (1996) Adaptive management: promises and pitfalls. *Environmental Management* **20**, 437–48.

McNeely, J.A., Miller, K.R., Reid, W.V., Mittermeier, R.A. & Werner, T.B. (1990) *Conserving the World's Biological Diversity*. IUCN, Gland, Switzerland/WRI, CI, WWF-US, World Bank, Washington, DC.

Malingreau, J.-P., Aschbacher, J., Achard, F., Conway, J., De Grandi, F. & Leysen, M. (1994) TREES ERS-1 Study '94: Assessment of the Usefulness and Relevance of ERS-1 for TREES. In: *Proceedings of the First ERS-1 Pilot Project Workshop*. Toledo/Spain, 22–24 June 1994. ESA SP-365, October 1994. European Space Agency, Frascati, Italy.

Margules, C.R. & Kitching, I.J. (1995) Assessing priority areas for biodiversity and protected area networks. In: Boyle, T.J.B. & Boontawee, B., eds. *Measuring and Monitoring Biodiversity in Tropical and Temperate Forests* pp. 355–64. CIFOR, Kuala Lumpur.

Margules, C.R., Nicholls, A.O. & Pressey, R.L. (1988) Selecting networks of reserves to maximize biological diversity. *Biological Conservation* **43**, 663–76.

Margules, C.R., Pressey, R.L. & Nicholls, A.O. (1991) Selecting nature reserves. In: Margules, C.R. &

Austin, M.P., eds. *Nature Conservation: Cost Effective Biological Surveys and Data Analysis*, pp. 90–7. CSIRO, Melbourne.

Meave, J., Kellman, M., MacDougall, A. & Rosales, J. (1991) Riparian habitats as tropical forest refugia. *Global Ecology and Biogeography Letters* 1, 69–76.

Murali, K.S., Uma Shanker, Uma Shaanker, R., Ganeshaiah, K.N. & Bawa, K.S. (1996) Extraction of non-timber forest products in the forests of Biligiri Rangan Hills, India. 2. Impact of NTFP extraction on regeneration, population structure, and species composition. *Economic Botany* 50, 252–69.

Nepstad, D.C., Brown, I.F., Luz, L., Alechandra, A. & Viana, V. (1992) Biotic impoverishment of Amazonian forests by tappers, loggers and cattle ranchers In: Nepstad, D.C. & Schwartzman, S., eds. *Non-timber Forest Products from Tropical Forests: Evaluation of a Conservation and Development Strategy. Advances in Economic Botany 9*, pp. 1–14. New York Botanical Garden, NY.

Niese, J.N. & Strong, T.F. (1992) Economic and biodiversity trade-offs in managed northern hardwoods. *Canadian Journal of Forestry Research* 22, 1807–13.

Nix, H.A. (1982) Environmental determinants and evolution in Terra Australis. In: *Evolution of the Flora and Fauna of Arid Australia*, pp. 47–65. Peacock Publications, Sydney.

Panayatou, T. & Ashton, P.S. (1992) *Not by timber alone: Economics and Ecology for Sustaining Tropical Forests.* Island Press, Washington, DC.

Patterson, B.D. (1987) The principle of nested subsets and its implications for biological conservation. *Conservation Biology* 1, 323–34.

Powell, A.H. & Powell, G.V.N. (1987) Population dynamics of male euglossine bees in Amazonian forest fragments. *Biotropica* 19, 176–9.

Prabhu, R., Colfer, C.J.P., Venkateswarlu, P., Tan, L.C., Soekmadi, R. & Wollenberg, E. (1996). *Testing Criteria and Indicators for Sustainable Management of Forests: Final Report of Phase I.* CIFOR, Indonesia.

Pressey, R.L., Bedward, M. & Nicholls, A.O. (1990) Reserve selection in mallee lands. In: Noble, J.C., Joss, P.J. & Jones, G.K., eds. *The Malle Lands: A Conservation Perspective*, pp. 167–78. CSIRO, Melbourne.

Ramesh, B.R., Menon, S. & Bawa, K.S. (1997) A vegetation based approach to biodiversity gap analysis in the Agastyamalai region, Western Ghats, India. *Ambio* 26, 529–36.

Rauste, Y. (1998) *Multi-mode Radarsat data in detailed forest mapping in West Africa.* Abstract posted at internet site: http://198.103.176.13/ENG/ADRO/Symposium/Abstracts/abstracts/616i.html.

Reid, W.V. (1996) Beyond protected areas: changing perceptions of ecological management objectives. In: Szaro, R.C. & Johnson, D.W., eds. *Biodiversity in*

Managed Landscapes. Theory and Practice, pp. 442–53. Oxford University Press, Oxford.

Sader, S.A., Stone, T.A. & Joyce, A.T. (1990) Remote sensing of tropical forests: an overview of research and applications using non-photographic sensors. *Photogrammetric Engineering and Remote Sensing* 56, 1343–51.

Sayer, J.A. & Wegge, P. (1992) Biological conservation issues in forest management. In: Blockhus, J.M., Dillenbeck, M., Sayer, J.A. & Wegge, P., eds. *Conserving Biological Diversity in Managed Tropical Forests*, pp. 1–4. IUCN, Gland, Switzerland.

Scales, D., Keil, M., Schmidt, M., Kux, H. & dos Santos, J.R. (1997) Use of multitemporal ERS-1 SAR data for rainforest monitoring in Acre, Brazil, within a German/Brazilian cooperation project. In *Proceedings of the International Seminar on the Use and Application of ERS SAR in Latin America*. Viña del Mar, Chile, 25–29 November 1996. European Space Agency, Frascati, Italy.

Schemske, D.W., Husband, B.C., Ruckelshaus, M.H., Goodwillie, C., Parker, I.M. & Bishop, J.G. (1994) Evaluating approaches to the conservation of rare and endangered plants. *Ecology* 73, 584–606.

Scott, J.M., Csuti, B. & Ciacco, S. (1991) Gap analysis: assessing protection needs. In: Hudson, W.E., ed. *Landscape Linkages and Biodiversity*. Island Press, Washington DC.

Smith, P.J., Pressey, R.L. & Smith, J.E. (1993) Birds of particular conservation concern in the Western Division of NSW. *Biological Conservation* 69, 315–38.

Soulé, M.E. (1991) Conservation tactics for a constant crisis. *Science* 253, 744–50.

Spence, J.R., Langor, D.W., Hammond, H.E.J. & Pohl, G.R. (1997) Beetle abundance and diversity in a boreal mixed wood forest. In: Watt, A.D., Stork, N.E. & Hunter, M., eds. *Forests and Insects*, pp. 287–301. Chapman & Hall, London.

Spitzer, K., Jaros, J., Havelka, J. & Leps, J. (1997) Effect of small-scale disturbance on butterfly communities of an Indochinese montane rainforest. *Biological Conservation* 80, 9–15.

Stork, N.E., Boyle, T.J.B., Dale, V. *et al.* (1997) *Criteria and Indicators for Assessing the Sustainability of Forest Management: Conservation of Biodiversity.* CIFOR Working Paper 17. Center for International Forestry Research, Bogor, Indonesia.

Sunderlin, W.D. & Resosudarmo, D.P. (1996) *Rates and Causes of Deforestation in Indonesia: Towards a Resolution of the Ambiguities.* CIFOR Occasional Paper. Center for International Forestry Research, Bogor, Indonesia.

Szaro, R.C. (1996) Biodiversity in managed landscapes: principles, practice and policies. In: Szaro, R.C. & Johnson, D.W., eds. *Biodiversity in Managed Land-*

scapes. Theory and Practice, pp. 727–70. Oxford University Press, Oxford.

Szaro, R.C. & Johnson, D.W., eds. (1996) *Biodiversity in Managed Landscapes: Theory and Practice.* Oxford University Press, Oxford.

Taufik, A. (1998) *Study on the impact of selective logging on the abundance of anurans in the rain forests of Central Kalimantan, Indonesia.* PhD Thesis, University of Edinburgh.

Thiollay, J.-M. (1992) Influence of selective logging on bird species diversity in a Guianan rain forest. *Conservation Biology* **6**, 47–63.

le Toan, T.A., Beaudoin, J., Riom & Guyon, D. (1989) Relating forest biomass to SAR data. *IEEE Transactions on Geoscience and Remote Sensing* **30**, 403–11.

Tsai, L.M. & Yuan, C.T. (1995) A practical approach to conservation of genetic diversity in Malaysia: genetic resource areas. In: Boyle, T.J.B. & Boontawee, B., eds. *Measuring and Monitoring Biodiversity in Tropical and Temperate Forests*, pp. 207–18. Center for International Forestry Research, Bogor, Indonesia.

Uma Shanker, Murali, K.S., Uma Shaanker, R., Ganeshaiah, K.N. & Bawa, K.S. (1996) Extraction of non-timber forest products in the forests of Biligiri Rangan Hills, India. 3. Productivity, extraction and prospects of sustainable harvest of amla *Phyllanthus emblica* (Euphorbiaceae). *Economic Botany* **50**, 270–9.

UNEP (United Nations Environment Programme) (1992) Convention On Biological Diversity. UNEP Na.92–7807, Nairobi, Kenya.

UNEP (United Nations Environment Programme) (1995) In: Heywood, V. & Watson, R., eds. *Global Biodiversity Assessment.* UNEP, Nairobi, Kenya.

Vane-Wright, R.I., Humphries, C.J. & Williams, P.H. (1991) What to protect?—systematics and the agony of choice. *Biological Conservation* **55**, 235–54.

Van Noordwijk, M., van Schaik, C.P., de Foresta, H. & Tomich, T.P. (1996) Segregate or integrate nature and agriculture for biodiversity conservation. In: *Proceedings of the Global Biodiversity Forum, November 1–2, Jakarta, Indonesia.* IUCN, Gland, Switzerland.

WCED (World Commission on the Environment and Development). (1989) *Our Common Future.* Oxford University Press, Oxford.

Welsh, D.A. (1988) Meeting habitat needs of non-game forest wildlife. *Forestry Chronicle* **64**, 262–6.

Wieringa, M.J. & Morton, A.G. (1996) Hydropower, adaptive management and biodiversity. *Environmental Management* **20**, 831–40.

Wilson, E.O. (1988) *Biodiversity.* National Academy Press, Washington, DC.

WRI (World, Resources Institute), IUCN (The World Conservation Union) and UNEP (United Nations Environment Programme). (1992) *Global Biodiversity Strategy.* World Resources Institute, Cambridge, UK.

Young, A.G. & Mitchell, N. (1994) Microclimate and vegetation edge effects in a fragmented podocarp–broadleaf forest in New Zealand. *Biological Conservation* **67**, 63–72.

Young, A.G., Boyle, T.J.B. & Brown, A.H.D. (1996) The population genetic consequences of habitat fragmentation for plants. *Trends in Ecology and Evolution* **11**, 413–18.

Part 2
Sustainable Wood Production

The world's forests must continue to supply wood in steadily increasing quantities. The fundamentals of ensuring that supply is sustainable is at the heart of sound forest management and harks back to the forester's dictum 'of always aiming to improve the soil'. It is with no apology that the first three chapters of Part 2 take a strong soil-centred approach. It is the one dimension of the forest ecosystem that bad management can irreparably harm: you can do little about the climate but quite a lot about the soil in which trees grow. Bob Powers (Chapter 5) provides an overview of the whole theme, which nicely complements Attiwill and Weston's account of forest–soil processes (Volume 1 Chapter 7), and provides a foundation on which Larry Morris builds (Chapter 7) as he works through forest soils issue by issue for achieving sustainability.

Sandwiched between these two chapters (Chapter 6) is a short account of the contribution silviculture plays in enhancing yield through good husbandry. Although Lee Allen touches on the main issues only in summary form, the tools available to the forester are presented.

The two final chapters of Part 2 examine how foresters should approach the threat posed by insect pests and fungal diseases and what they can do about them. Both Hugh Evans' and John Gibbs' fine overviews are slanted towards forest plantations partly because British forestry is almost wholly plantation based and partly because the expense of establishing plantations requires that such an investment should be protected. The accounts found in Chapters 8 and 9 are, however, more generally applicable and provide the extra dimension of sustainable management: paying due regard to protecting the valuable resource.

Omitted from Part 2 is the role of genetics and tree improvement, though in both Chapters 5 and 6, this important dimension of sustained yield is fully acknowledged. As Lee Allen notes 'the key to operational productivity is to use the best genetic material available'. Also omitted is formal consideration of silvicultural systems, and the forest protection threat that fire, climatic hazards and mammal damage represent. This Handbook is not primarily focused on forestry practices and operations themselves, and these themes are readily accessible in other texts.

5: Assessing Potential Sustainable Wood Yield

ROBERT F. POWERS

5.1 INTRODUCTION

5.1.1 Aims of this chapter

Society is making unprecedented demands on world forests to produce and sustain many values. Chief among them is wood supply, and concerns are rising globally about the ability of forests to meet increasing needs. Assessing this is not easy. It requires a basic understanding of the principles governing forest productivity: how wood yield varies with tree and stand development; the implications of rotation length, utilization standards and silvicultural treatment; and effective techniques for judging site potential and detecting changes in it. This chapter reviews these concepts and suggests workable approaches for assessing a site's ability to sustain wood production.

5.1.2 The sustainability problem

Exploitation has dominated our use of forests for 10 000 years. Of the 6.2 billion ha of forests and woodlands thought to exist at the start of the Holocene, between one-fifth and one-third have been lost to soil erosion, conversion to agriculture, excessive fuelwood gathering and livestock grazing, and desertification from poor land-use practices (Postel & Heise 1988; Waring & Running 1998). Concerns have persisted from the Chou Dynasty in 1127 BC (Hermann 1976) to the present that societal demands on forests are excessive and are compromising the land's capacity for sustaining multiple resources.

5.1.3 Sustainability of what, and for whom?

Society values forests both for the products they produce, such as lumber, fuelwood, lichens, herbs and mushrooms; and for the functions they serve, including watershed protection, runoff moderation, fish and wildlife habitat, recreation and aesthetics (see Volume 1, Chapters 12, 14 and 15; Volume 2, Chapters 2–4 and 10–12). More recently, high value has been placed on the role that forests play in the global carbon cycle (Bouwman & Leemans 1995; Landsberg & Gower 1997). Concerns about sustained production often focus on wood products. But sustainability also applies to other forest properties and functions. Environmentalists, concerned that ecological values have suffered from overemphasis on wood extraction, call for more conservative management practices that reduce wood harvest and preserve or restore other ecological values (Drengson & Taylor 1997).

Such arguments carry strong emotional force and sway public views on how forests should be managed for sustainability. A 'green advocacy' has gained such momentum that it has spawned a cottage industry to certify what is, and is not, 'sustainable forestry' (Anonymous 1995). Many in the private forestry sector are sceptical of third-party 'green certification' where criteria may be based more on speculation than on science (Berg & Olszewski 1995). In general, leading forest scientists agree that timber harvesting, *if carried out carefully enough to preserve potential productivity*, need not compromise other ecosystem values (Attiwill 1994; Kimmins 1996; Nambiar 1996). Yet, ignoring green certification could limit markets for industrial wood.

As currently practised, green certification standards aimed at protecting multiple forest values often are so conservative that wood harvests are less than could be sustained had the sites been severely degraded (Kimmins 1996). Reduced harvests must be accompanied either by reduced demand for wood, or by the substitution of other products. Both are at odds with global realities. Well-intended notions that mineral resources can 'save forests' by substituting for wood when such conversions are powered by fossil fuels make ecological nonsense as long as people are part of the ecosystem (Smith *et al.* 1997). Substitutes manufactured from non-renewable resources demand far more energy than wood products and have a much greater capacity for pollution. Furthermore, reduced wood production by industrialized nations creates a strong incentive for other countries to accelerate forest harvesting beyond sustainable levels to reap the rewards of a global market, irrespective of green certification (Kimmins 1996).

5.1.4 Whose problem is it?

Although forest product consumption is related closely to gross domestic product, high wood use is not restricted to technologically advanced nations. Per capita demands for paper and other wood-based products are mounting in countries where increases in both population and standards of living spell higher demands for wood (Anonymous 1997; Sutton 1999). Recent projections show that the global demand for wood increases by 70–80 million m^3 annually—a volume equivalent to British Columbia's entire allowable harvest in 1993 (Kimmins 1996). Paper consumption is expected to rise in developing countries by 4% to 6% annually, reaching approximately 120 million m^3 by 2010 (Anonymous 1997; Brown *et al.* 1997). Fuelwood deficits will occur in arid and semiarid regions.

Coupled with this is a projected decline in global forest area of over 16 million ha (about 0.4%) annually (Anonymous 1997). While forested area has stabilized in many industrialized nations, it has declined elsewhere (see Volume 1, Chapter 1). As wood becomes scarce, prices rise and consumers in industrialized nations choose alternative products such as brick, concrete, and

steel or aluminium (Binkley 1997). But such substitutes carry energy and pollution costs to society that are more than 10 times greater than wood (Sutton 1999). All signs point to crisis, and appeals have been made to international leaders not only to take stock of their existing forest resources, but also to develop effective indices of forest sustainability (Canadian Forest Service 1995). Sustainable yield is everyone's problem. It cannot be ignored.

5.1.5 Searching for solutions

Even where forested area has stabilized, increasing demands from a fixed base lead to shorter rotations and greater utilization. Can yields be sustained? Or will high rates of removal outstrip the site's ability to sustain growth? Questions raised as early as the nineteenth century (Ebermeyer 1876) await convincing answers.

Stocking and competition control

Each forest site has an inherent set of resources as determined by soil, climate and topography. These resources can be partitioned in many ways, but the general focus of forest management for wood yield has been in finding cost-effective ways of directing these resources to achieve the greatest gain in merchantable wood at the least expense to the manager. Obviously, minimal management would produce low-cost yield. However, yields per unit land area also would be low because some units would be under-stocked, others overstocked. Understocking means that some of the site's potential productivity would be spent on weeds or tree species of marginal value. Overstocking means that intertree competition would keep trees relatively small with a greater proportion of fixed carbohydrate used for maintenance respiration and a lesser proportion used for growth. Trees would be weakened from competition for moisture, light and nutrients. As stresses build, such trees and stands are susceptible to catastrophic losses from insects, diseases, wildfire and the vagaries of climate.

To date, forest management's greatest contributions to improved yield are in the fields of reforestation, weed control and thinning (Smith *et al.*

1997). Reliable regeneration methods, especially planting, have reduced the lag between forest harvest and the establishment of a new stand. Enlightened methods of site preparation and vegetation control have improved survival and allowed young trees to capture fresh sites quickly. Timely applications of thinning have reduced mortality and focused site resources on the most valuable trees with little loss in overall stand productivity. Thinning builds the vigour and mechanical strength of individual trees and their resistance to insects, disease and windthrow. Removing weaker trees in advance of mortality also reduces fuel build-up and the wildfire risk common to unmanaged stands. Because thinning can alter stand composition by removing undesirable trees and by creating canopy gaps, it also creates opportunities for natural regeneration in advance of final harvest. Collectively, stocking and competition control practices have had a monumental impact on the practice of forestry.

But prompt regeneration, vegetation control and thinning cannot improve inherent site quality. Other than manipulating canopy architecture to the degree that site resources are focused more completely on desirable trees so that merchantable sizes are reached faster, the site's productive potential is not improved. While stocking and competition control are absolutely critical to capturing a site's potential for merchantable wood yield, they will not improve that potential.

Genetic salvation?

Genetic improvement has revolutionized agriculture, but parallel gains are not as likely in forestry. Time works against us. Errors in choosing genetic strains of agricultural crops generally can be corrected the following year—an impossible practice with long-lived forest trees. Under warm temperate and subtropical conditions, yield gains of from 14% to 23% per generation—perhaps more—are believed possible (Nambiar 1996). But, in contrast with modern agriculture where drought can be eliminated with irrigation, forests generally are subject to the natural stresses of climate that limit genetic expression. Substantive gains reported under favourable glasshouse conditions

are rarely found in the field (Matheson & Cotterill 1990), and large absolute responses to genetic improvement may be restricted to sites with the most favourable growing conditions (Zobel & Talbert 1984).

Genetic progress to date has centred on selecting for better adaptation by weeding out poor performers and by producing hybrids for special-purpose plantings. Selections favouring disease resistance and higher wood quality have been impressive. Yet, little has been done to alter the fundamental mechanisms by which CO_2, nutrients and water are processed to synthesize biomass. Opportunities exist for improving nutrient retranslocation within the tree to make them less dependent on uptake from the soil (Libby 1987). Possibilities for improved drought tolerance and photosynthetic efficiency have been described but have not been achieved (Matheson & Cotterill 1990; Boyle *et al.* 1997).

Current selection strategies simply may mean that stands reach growth ceilings sooner and that a higher proportion of net primary productivity is allocated to merchantable parts of the tree. Reaching merchantable sizes sooner suggests shorter rotation periods and greater rates of nutrient removal from the site. Accelerated harvesting may not be sustained without help from management to maintain potential productivity.

Is soil management a key?

Increasingly, soil management is seen as the underpinning of sustainable forest productivity (Powers *et al.* 1990; Dyck *et al.* 1994; Kimmins 1996; Nambiar 1996). Given climatic constraints, forest growth is primarily limited by the soil's ability to supply moisture, air and nutrients in proportion to a tree's potential demand. Yields generally are raised by treatments that enhance soil quality relative to the principal limiting factors (Chapter 6). However soil improvement treatments will not be applied everywhere. High economic returns on such investments are limited to regions of the world with mild climates and favourable moisture regimes. While soil improvement always is an option, our first responsibility is to protect or enhance the land's existing capacity to grow wood. Have we altered this? Can changes be detected? What levels are sustainable relative to management goals?

5.2 SUSTAINABLE WOOD YIELD

Before we can assess the land's potential for sustainable wood yield we should clarify the terms and principles governing forest growth and measurement. Volume 1 of this handbook covers ecological and physiological principles in detail, but a brief review is useful here.

5.2.1 Definitions

Productivity

The primary measure of forest productivity is the rate at which organic compounds are produced from CO_2 through photosynthesis. Depending on moisture, nutrient and temperature constraints, this *gross primary productivity* (GPP) is a linear function of light interception by the forest canopy (Cannell 1989). It usually is expressed as Mg ha^{-1} year^{-1} (or g m^{-2} year^{-1}). GPP is not measured easily and not all of it is converted to growth. As much as two-thirds of GPP is respired back to the atmosphere as the 'cost' of constructing and maintaining living cells. The absolute amount of respired carbon is greater in older stands because the mass of living tissue also is greater. Respiration rates are sensitive to temperature, and essentially double with each 10°C rise (Landsberg & Gower 1997). Thus, respiration rates are higher in summer than in winter and are greatest in warm, humid climates.

The difference between the amount of CO_2 fixed into organic compounds and the amount respired is termed *net primary productivity* (NPP). For forest trees, NPP varies between 0.37 and 0.5 of GPP, but averages a relatively constant 45% (Schulze *et al.* 1977; Ryan *et al.* 1996; Landsberg & Waring 1997). The NPP rate usually is expressed as the annual amount of biomass produced per unit area. NPP is apportioned mainly into the structural materials of roots and mycorrhizae, foliage and reproductive parts, branches and boles, roughly in that order of priority (Waring & Schlesinger 1985; Landsberg & Gower 1997). Difficulty in measuring below-ground conditions

limits our understanding of true distributions under field conditions, and most of our knowledge rests on observed changes above ground.

Like respiration, NPP increases with temperature and moisture, but at a faster rate. Therefore, NPP rates generally are highest under moist, tropical conditions where above-ground NPP hovers near 30 Mg ha^{-1} year^{-1} (Leith 1975), and may briefly reach 60 Mg ha^{-1} year^{-1} in fully stocked stands (Binkley *et al.* 1992). Rates generally are half to two-thirds lower at the middle latitudes (Grier *et al.* 1989), and fall to only 5 Mg ha^{-1} year^{-1} where mean annual air temperatures approach freezing (Van Cleve & Powers 1995). As leaf area or mass increases, a greater proportion of NPP is shifted to wood growth, reaching about 50% once crown mass has stabilized (Waring & Schlesinger 1985). In general, maximum wood production occurs at a one-sided leaf area–ground area ratio (LAI) of about 3 (Landsberg & Gower 1997), although the optimal ratio varies with canopy architecture and the photosynthetic efficiency of leaves (Beadle 1997).

Presumably, LAIs greater than 3 impose the risk of periodic water or nutrient stresses on most sites or lower net carbon assimilation from lower- and inner-crown leaves shaded below their photosynthetic compensation point. Thus, very high LAIs may lead not to increased growth, but to increased respiration and mortality (Waring & Schlesinger 1985). Some of the highest leaf areas measured are not in the tropics, but on middle latitude sites of moderate productivity (Beadle 1997). Wood growth rates will increase with leaf area if water, nutrients and temperatures are not limiting, and if sunlight penetrates the canopy sufficiently that all leaves have a positive carbon balance.

Site quality

'Site' defines an area in terms of its environment, and 'quality' is a relative measure of the site's productive capacity to grow forest vegetation. Together, they refer to the potential of a land unit for tree growth when the land is stocked fully. Full stocking occurs when a site is at its 'foliar carrying capacity'. That is, when the forest has attained the amount of leaf biomass that it is capable of supporting for an extended period. Foliar carrying capacity occurs shortly after crown closure. It is reached early if trees are growing closely together or later if trees are spaced widely. This carrying capacity is a 'potential' determined by climate, soil and topography factors interacting upon a particular forest genotype. It fluctuates about a long-term mean because of annual vagaries in climate and occasional outbreaks of defoliating insects or diseases when trees are under stress.

Unfortunately, foliar biomass is difficult to measure directly and non-destructively, but LAI is a popular and convenient surrogate in many ecological studies, and has correlated linearly with tree growth response to silvicultural treatment (Della-Tea & Jokela 1991). In practice, foliar biomass, LAI and NPP are not the usual measures of site quality. The historical emphasis on wood has made bole wood volume production per annum the conventional standard. Although bole wood ranks relatively low as a sink for photosynthate (Waring & Schlesinger 1985; Landsberg & Gower 1997), bole volume is measured with relative ease. And given that bole wood has a high societal value, its acceptance as a practical measure of site quality is understandable.

Site quality is neither static nor immutable. In the long term, it will aggrade or degrade with changes in climate and stage of soil development (Jenny 1980; Van Cleve & Powers 1995). Changes triggered by the anthropogenic production of air pollutants may affect site quality quickly, particularly in Europe where high atmospheric depositions of SO_x, NO_x and NH_x reportedly have reduced soil pH and base cation status, and raised aluminium solubility (Van Breemen 1990). Site quality also can be altered rapidly for better or worse by management (Powers *et al.* 1990; Nambiar 1996). Examples include fertilization, irrigation, drainage, or tilling to modify limiting site factors; and losses in soil fertility or aeration through careless management.

Yield

Defining 'yield', the amount of wood available from a forest for human use, depends on purpose. In many regions, forests are the principal source of fuel, and branches and twigs are considered a component of yield. Regardless, interest centres mainly on bole wood. Therefore, yield is defined here as the total amount of bole wood available for harvest at a given time (Avery & Burkhart 1983). Yield is usually expressed as volume per unit area ($m^3 ha^{-1}$), but also as biomass ($Mg ha^{-1}$). It is what remains of GPP not lost in respiration, sequestered in foliage, branches or roots, or consumed by animals or disease. Because yield is a cumulative function of stand growth—and because stand growth varies by age, stocking and species—yields will vary over time for a given stand. Also, forest tree species differ genetically in their responses to shade, temperature, moisture, soil physical and chemical properties, and biotic pests. Therefore, adjacent stands of the same age, but stocked with genetically dissimilar trees, will produce different yields. Also because a site's growth potential is conditioned by climatic and edaphic properties, sites differing in climate or soil will differ in potential yields.

Sustained yield

A non-declining, continual supply of wood from a forest over decades or centuries defines a 'sustained yield'. Individual stands within a forest are not appropriate units for assessing sustained yield because such stands would need a perfect balance of multiple age classes occupying equivalent areas and site qualites to ensure regular harvesting of a constant yield without interruption, an ideal not occurring in nature (Smith *et al.* 1997). Sustained yield rests on two requirements. One is the certainty of timely forest regeneration. This simple fact is what separates forest management from forest exploitation. Without a *commitment to successful regeneration*, sustainable yield is impossible. Appropriate regeneration methods ensure rapid, complete reforestation following harvest. Methods vary with tree species, climate and management objectives, but all are an integral part of any sustained yield silvicultural system (Smith *et al.* 1997). The other requirement of sustained yield is that harvesting rates for a forest are balanced by growth.

Sustained yield can be confusing. Yield can be differentiated into total or merchantable wood production. Historically, some managers have justified high harvesting rates in older stands because low growth rates there were equalled or surpassed by high growth rates in much younger (but submerchantable) stands. In mixed stands, one species may have greater commercial value than others. Selective harvesting of the more favoured species will inevitably alter forest composition so that merchantable yields are not sustained. Sustained yield requires deliberate and persistent management. It is not achieved easily.

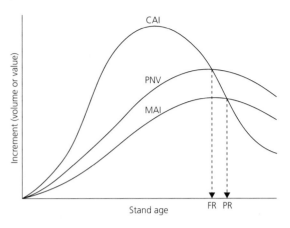

Fig. 5.1 Generalized relationship between current annual volume increment (CAI), mean annual volume increment (MAI) and present net value (PNV) of stands or individual trees. Financial rotations (FR) generally occur sooner than physical rotations (PR). (After Evans 1992.)

Rotation

Central to sustained yield is the 'rotation', a concept that applies both to individual trees and to stands. A rotation is the period between tree establishment and the age at which the oldest tree is harvested. Intermediate harvests, or 'thinnings', can occur at any point, but the period of rotation is the age attained by the oldest tree at harvest. As will be seen, rotation length is a management decision that depends on biology, economics and ecological principles.

Rotation lengths traditionally are of two types: 'physical' and 'financial' (Smith *et al.* 1997). Physical rotations are aimed at maximizing yield per unit time, which occurs arithmetically at the culmination of mean annual volume increment of bole wood. This is the point of inflection in the sigmoid growth curve that typifies tree or stand development over time, and is mathematically equivalent to the intersection between current annual volume increment and mean annual volume increment. The age of a physical rotation is indicated by 'PR' in Fig. 5.1. In the tropics this usually occurs at 10–15 years, but in temperate and boreal forests it is much later (40 and 70 years) (Hamilton & Christie 1971; Plonski 1974; Clutter *et al.* 1983; Evans 1992). The range in mean annual increment (MAI) at culmination is remarkably similar on average sites of tropical and temperate latitudes (Table 5.1). Lower MAIs for boreal species reflect low yields of natural, unmanaged stands. With periodic thinnings, MAIs would be greater and culmination age would be extended (Smith *et al.* 1997).

Harvesting at the culmination of MAI assures

the maximum yield of wood per unit time. But for slow-growing trees, this rarely is the most profitable rotation strategy. Rather, it is most appropriate for fast-growing, high-value species where costs of production and alternative interest rates of capital are relatively low (Evans 1992). Naturally, biological growth continues much longer, but at a decreasing rate. Financial rotations are determined by the maximum monetary return on a capital investment in silviculture. The higher the compound interest rate on the investment, the shorter the rotation. Generally, financial rotations are less than physical rotations—although they could be longer if there is exceptional value in older trees (such as extremely valuable veneer). They occur at the age of intersection between the stand's present net value and current annual increment ('FR', Fig. 5.1). Should the interest rate for the financial rotation be zero, rotation lengths will be the same.

A third view of rotation length that has gained momentum is that of 'ecological rotation' (Kimmins 1974), in which the harvest interval varies by the time needed for the ecosystem to recover fully from the last harvest. For example, multiple short rotations—or possibly those involving very high rates of organic matter

Table 5.1 Examples of stand conditions at culmination of mean annual volume increment (MAI) for boreal, temperate, and tropical forest tree species on sites of average yield class based mainly on plantation data. (From Hamilton & Christie 1971; Plonski 1974; Evans 1992.)

Species	Age at culmination (yr)	Cumulative volume ($m^3 ha^{-1}$)	Volume MAI ($m^3 ha^{-1} yr^{-1}$)	Top height (m)
Boreal forests				
Betula papyrifera	40	134	3	13
Picea mariana	65	169	2	10
Pinus banksiana	35	156	4	12
Pinus strobus	65	277	4	16
Populus tremuloides	50	342	7	21
Temperate forests				
Abies grandis	55	1100	20	31
Fagus sylvatica	85	680	8	28
Picea abies	75	900	12	25
Picea sitchensis	55	880	16	26
Pinus nigra	60	720	12	22
Pinus sylvestris	75	600	8	22
Populus×euramericana	35	350	10	33
Pseudotsuga menziesii	55	880	16	30
Tropical forests				
Eucalyptus camaldulensis	15	120	8	–
Eucalyptus grandis	9	162	18	–
Gmelina arboria	10	200	20	–
Pinus caribaea	16	320	20	–
Pinus patula	16	288	18	–
Swietenia macrophylla	30	420	14	–

removal or soil disturbance—may remove more nutrients than the site is capable of restoring by the next rotation. Or, more time may be needed for stands to develop ecological conditions of particular social value (structures, flora and fauna characterizing late seral stages of forest development). Ecological rotation strategies recognize that sustained wood yield hinges on more complex factors than suggested by the simple marriage of historical growth trends and recent inventories. Put simply, the ecological rotation is the basis for sustainable yield for wood and myriad other forest products and values. Our understanding of how forest management affects wood production over multiple rotations is strong. Unfortunately, our understating of long-term effects of forest management on other values is weak. Therefore, the discussion that follows will centre on what we know best: wood yield.

5.2.2 Key concepts

Inherent potential productivity

Wood yield depends partly on site quality and partly on management. Many important factors such as genotype, stocking, most biotic pests, and certain soil physical and chemical properties are under silvicultural control. Others, such as climate and topographic features affecting solar radiation and precipitation, are not. These latter factors (climate, soil, and topography) define 'inherent potential productivity' for a site. In short-rotation tropical plantations, the upper limit for inherent, potential productivity varies a few Mg, by about 10 Mg bole wood ha^{-1} $year^{-1}$ in the first decade (Lugo *et al.* 1988). For the rest of the world, it is less. This inherent potential for productivity exists irrespective of tree stocking. It

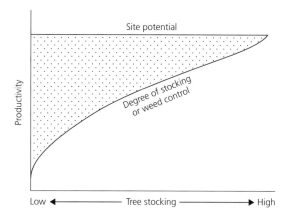

Fig. 5.2 Relationship between degree of stocking (or weed control) and the potential productivity inherent to a site. Improved stocking or weed control captures a greater proportion of the site's inherent potential.

is realized when the unit of land is stocked at its foliar carrying capacity, and is represented by the flat portion of the curve in Fig. 5.2. Whether or not this potential is realized depends on management.

Actual and inherent potential productivity

'Actual productivity' is the current rate of stand growth. Poor stocking, weed competition and damage from pests commonly hold forest production below its inherent site potential. If stands are understocked with trees (the stand is below foliar carrying capacity for that species), only a fraction of the potential productivity is realized. This occurs when stands are very young, trees are spaced widely or thinned too heavily, or much of the site is occupied by weeds. Competition for water and nutrients by adapted weeds can have a substantial impact on early tree growth (Nambiar & Sands 1993), particularly on droughty, infertile sites (Powers & Reynolds 1999b). Even on very productive sites, as much as 50% of stand growth may be lost to weed competition through the first 20 years if trees are widely spaced (Oliver 1990). Long-term projections of early and repeated vegetation control in pine plantations suggest 100% improvements in volume production on the droughtiest sites after 50 years, but only 12% improvements on more fertile and better watered

sites (Powers & Reynolds 2000). The nearer that tree stocking approaches full site occupancy (foliar carrying capacity), the greater the proportion of potential productivity that will be captured by the trees (Fig. 5.2).

Assuming that site quality is not altered, greater management investments in stocking, pest control and genetic improvement will increase yield and capture the site's potential earlier. Figure 5.3 illustrates this principle. In Fig. 5.3a, potential productivity is constrained by inherent soil properties. However, actual productivity is less than that because of low stocking. In Fig. 5.3b, stocking is improved either by replanting to a higher density with a better-adapted species, through timely weed control, or because the trees have grown older and larger. With stocking no longer a constraint, the stand has reached the potential productivity permitted by its inherent soil properties. Replanting also is a chance for genetic improvement, meaning that the stand achieves full stocking more rapidly and reaches potential productivity sooner. However, the productive potential remains constrained by soil.

Altering site potential

A site's productive potential is *malleable*—especially when productivity is limited by certain soil properties. Soil management practices that erase physical or chemical limitations will raise potential productivity to a new level (Chapter 6; Nambiar 1996). This is illustrated in Fig. 5.3c by treatments that improve soil quality to a point where productivity is now constrained by climate. Depending on treatment, such gains may be short-lived. Gains in volume growth of 30% are common from a single application of N fertilizer, but the effect usually dissipates by 10 years (Ballard 1984; Allen 1990). Part of this gain is through improved nutrition that increases foliar biomass, but part also is due to improved water-use efficiency (the amount of carbon fixed per unit of water transpired) (Mitchell & Hinckley 1993; Powers & Reynolds 1999). Assuming a deficiency exists, effects of phosphorus fertilization last longer (Ballard 1984).

Fertilizing repeatedly with multiple nutrients

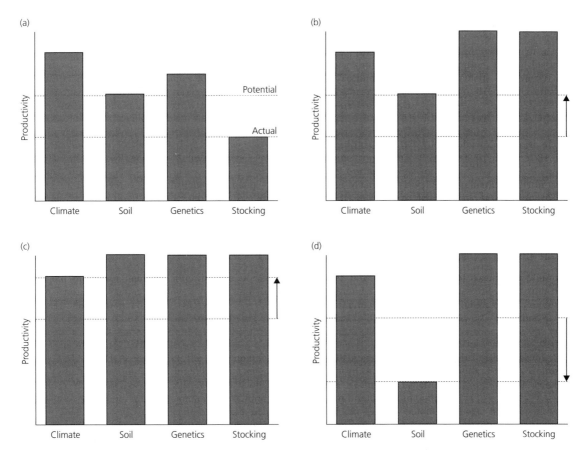

Fig. 5.3 Relationship of climate, soil, tree genetics and tree stocking to potential and actual site productivity. (a) Poor stocking holds actual productivity below the potential as limited by soil factors. (b) Potential productivity (as limited by soil) is achieved through improved stocking. (c) Soil improvements raise potential productivity to the limit set by climate. (d) Soil degradation leads to a loss in potential productivity. (From Powers *et al.* 1996.)

at rates proportional to stand demand may alter site quality fundamentally, conceivably doubling volume growth through the first 50 years if soil moisture is not severely limiting (Powers & Reynolds 2000). Alternatively, poor management practices leading to soil degradation from erosion, compaction or nutrient drain can reduce potential productivity from its inherent level (Powers *et al.* 1990) (Fig. 5.3d). Overcoming site, stocking or genetic constraints requires capital and intervention. Consequently, many managers operate at lower, less costly levels of intensity. Regardless, an important element of sustained productivity lies in protecting or enhancing the soil resource.

Growth and the partitioning of biomass

Yield assessment is helped by a working knowledge of how forests and individual trees develop and how dry matter is partitioned above and below ground. Given freedom from disturbance, biomass production of stands (or individual trees) follows a general pattern of increase from establishment, to maximal rates near crown closure (or attainment of a dominant canopy position) (Fig. 5.4). Essentially, this is a function of leaf area per unit ground area (or per tree) because primary productivity is related linearly to canopy light interception (Cannell 1989). This linear trend seems unaffected by water or nutrient stress

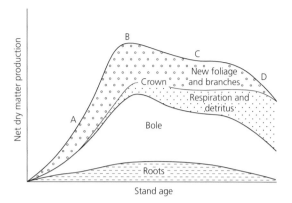

Fig. 5.4 Generalized pattern of net primary productivity assimilation into principal tree components of a forest stand. Major phases are: A, rapid increases in assimilation rates as crowns expand when stand is below leaf area carrying capacity; B, peak productivity at crown closure; C, period of maturity — leaf area is fixed and an increasing proportion of gross assimilation is used as maintenance respiration; D, rapid decline as stand senesces from natural causes. (Modified from Waring & Schlesinger 1985.)

(Della-Tea & Jokela 1991; Powers & Reynolds 1999b) because water or nutrient limitations affect a site's foliar carrying capacity and, within temperature constraints, its rate of photosynthesis. The first priority for photosynthate is cell respiration. Rates increase with gains in living tissues, essentially doubling with each 10°C rise in temperature, and can account easily for half or more of gross primary productivity (Waring & Schlesinger 1985). Carbohydrate remaining beyond respiratory needs is partitioned into biomass (NPP).

Production rates (and yields) are low per unit ground area when forest stands are very young and foliage mass has yet to reach site carrying capacity (Fig. 5.4). Much of the carbon assimilated in this phase is directed to foliage and branch production. Total production rates rise exponentially as crown mass expands during Phase A, and then become sigmoid as the stand reaches foliar carrying capacity in Phase B (Switzer & Nelson 1972; Waring & Schlesinger 1985). During Phase B, nutrient uptake peaks and leaf area and mass stabilize. Sites characterized by favourable climate

and soil conditions can carry more foliar mass than poorer sites.

Once foliar carrying capacity is reached, crown mass is stable through Phase C, other than for variation due to storm damage, seasons of unusual climate, outbreaks of defoliating insects, or damage to stems or root systems. Because leaf mass is stable, new foliage produced in the upper crown leads to senescence and abscission of older foliage and a surge in litterfall. An increasing proportion of the photosynthate produced by the foliage in Phase C is spent on maintaining respiring tissues accumulating in branches, boles and roots. Root growth is not as well documented as above-ground growth, but fine root production is believed secondary only to maintenance respiration and perhaps new leaf production in its demand for carbohydrate (Oliver & Larson 1990). The poorer the site, the greater the proportion of NPP allocated to fine root production, and roots may account for 40% or more of the total NPP (Vogt *et al.* 1997). Conversely, treatments that improve water or nutrient availability apparently shift a higher proportion of NPP away from roots to other parts of the tree (Vogt *et al.* 1997). Roots grow through the soil volume by following paths of least resistance, such as ped faces, existing root channels, animal borings and other regions of moist but aerated, low-strength soil. Eventually, roots will exploit the full volume of available soil. Shallow soils will be occupied sooner, deeper soils later. To a large extent, root growth probably parallels the pattern of foliage production.

As trees grow larger, more photosynthate is required to maintain living tissues. This leads ultimately to Phase D, marked by a gradual decline in net wood growth. Eventually, respiration/production imbalances, nutrient stress, increased hydraulic resistance in tall trees, and crown loss from wind abrasion lead to sharp declines in vigour and growth (Waring & Schlesinger 1985; Ryan *et al.* 1997). Phase D marks a period of senescence that ends in mortality. For further discussion, see Volume 1, Chapters 8 and 9.

Biomass and nutrient accumulation

Because foliar biomass is fixed once the stand

reaches foliar carrying capacity, wood increments depicted in Fig. 5.4 accumulate on the bole so that total bole biomass surpasses the biomass in foliage and limbs. Beyond crown closure, tree boles will accumulate as much as 10 times more biomass than in the crown (Kimmins *et al.* 1985). This dry matter difference belies the fact that nutrient concentrations are disproportionately higher in foliage than in wood, meaning that the nutrient *content* of crowns and boles can be similar. Figure 5.5 shows a typical trend for the distribution of N in the standing biomass of a developing forest.

Potential sustainable yield

To date, no international consensus exists on the definition of forest sustainability (Sullivan 1994; Jaggi & Saandberg 1997; Nambiar & Brown 1997). Largely, this is due to conflicts between ecocentric and anthropocentric views and the forces at play in wood-based economies. The concept that wood harvests should centre only on trees facing imminent mortality (Camp 1997) is naïve, given economic and political realities and the paucity of evidence that more aggressive wood harvests cannot be sustained. Binkley (1997) has shown that global forest area would have to be increased by 1.1 billion ha (one-third) to meet current wood demand if all forests were managed under current concepts of green certification (MAI = $0.7\,m^3\,ha^{-1}$ year^{-1}). Switzerland's policy of allowing more silvicultural flexibility (Jaggi & Sandberg 1997) seems more practical. The Swiss model strives to find a balance between ecology, economy and culture. It aims at ensuring a steady flow of diversified forest products and a more stable local forest economy, but it clearly is labour intensive. Furthermore, the Swiss approach cannot meet the internationally rising need for raw wood.

Assuming an average MAI of $10\,m^3\,ha^{-1}$—increments quite common to temperate and tropical forests (Table 5.1)—today's global demand for wood could be met simply by increasing the area of planted forests from its current 3% of the world's forest area to 5% (Binkley 1997). But demand continues to rise. By mid-century, global wood production may need to rise by at least 2 billion m^3 (Sutton 1999). At an MAI of $10\,m^3$ ha^{-1}, this projects to a new managed forest area twice the size of Nigeria or British Columbia. Given international pressures to de-emphasize timber harvests in native forests, new highly productive plantations are the only hope for meeting this shortfall (Nambiar 1999). Most likely, new plantations will be extended into grassland, shrubland and agricultural regions of marginal value. We must recognize that sustained, high wood production has a legitimate place on the forest management continuum between farm forestry and wilderness management (Nambiar 1996).

Texts exist on procedures for determining maximum sustainable yield (Assmann 1970; Clutter *et al.* 1983), but conventional approaches rest on the assumption that site potential is static. Once a site's inherent potential productivity is recognized, some or all of it may be captured in a rotation. However, if harvesting affects the site to the degree that soil properties are degraded (Fig. 5.3d), that potential will not be realized in subsequent rotations unless the soil has been restored to its original condition. The quantity of wood that can be removed per unit time without lowering the site's inherent potential productivity is defined as 'potential sustainable yield', and the time needed to do so is the ecological rotation. But for a variety of reasons (lack of historical

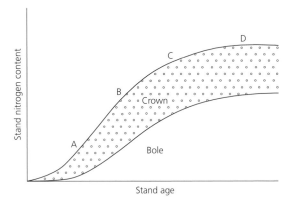

Fig. 5.5 Generalized pattern of nitrogen accumulation in above-ground biomass of a developing stand—derived from Fig. 5.4.

records, changes in rotation length, genotype and silvicultural practices) we must recognize that there is no immutable reference point for assaying potential sustainable yield (Nambiar 1999). The challenge is to develop analytical methods for guiding appropriate management.

5.3 PROTECTING THE RESOURCE

5.3.1 A nutritional balancing act

The question of nutrient drain

Rotation length and intensity of utilization determine the amount of biomass and nutrients removed from a site over time. How this relates to the concept of ecological rotation can be demonstrated from Fig. 5.5. Two consecutive harvests made at stage B of stand development would remove considerably more biomass than a single harvest at stage D during the same time-span. Assuming that soil compaction or erosion does not occur with more frequent entry, and that soil organic matter is not depleted severely by more frequent site preparation and warmer soil temperatures, the mere act of removing more wood per unit time should not affect the site's productive potential.

Nutrient removal is another matter, particularly if utilization standards are high. Nutrients, especially N, are removed at very high rates when crowns are harvested along with boles (Fig. 5.5). Switzer and Nelson (1972) estimated that three 20-year rotations of whole-tree harvests in southern pine plantations would remove three-quarters more biomass, more than double the N, but roughly the same amount of calcium as one 60-year rotation. Whether or not this degrades potential sustainable yield depends on quantity of nutrient removal, the existing fertility of the soil and the rate of nutrient replenishment (Chatper 7). To sustain the site's inherent potential productivity, rotation length would be determined by the time it takes for the site to re-establish its inherent productivity.

Productive sites generally have fertile soils. Frequent nutrient removals on such sites should have no effect on potential productivity as long as nutrient supply meets forest demand. Natural rates of nutrient input through precipitation,

biological N fixation and mineral weathering presumably are higher on productive sites with favourable climate than on poorer sites where rooting volume, soil moisture or temperature are limiting. Therefore, short rotations and high utilization standards will surely degrade the potential productivity of poorer sites.

There is no sound evidence that moderate rotations coupled with low-intensity site preparation practices have a detrimental effect on soil fertility and sustainable yield on most sites. Declines that have occurred are associated with coarse-textured soils ranking low in organic matter and nutrient retention properties (Powers *et al.* 1990; Morris & Miller 1994; Nambiar 1996). Even there, growth declines are not associated with harvest removals *per se*, but with extreme site preparation practices (intense slash burning, heavy equipment, topsoil removal) that reduce soil rooting volume and deplete soil organic matter and nutrients. Alternatives to these practices are well known and practicable (Chapters 6 and 7).

Avoiding nutritional problems

A key to maintaining soil fertility in sustained yield planning is to balance nutrient losses with nutrient replacement (Kimmins 1974; Van Miegroet *et al.* 1994). Two pathways are recognized. In the more intensive approach, nutrient losses associated with harvest and site preparation are minimized where possible, and—where serious enough—replaced with fertilization. This can be costly. For example, if soil N availability declines below the level needed to achieve the site's foliar carrying capacity, potential yield will be reduced as well. However, N is leached readily if it is applied at rates exceeding plant uptake or immobilization. This means that fertilization would have to be carried out periodically throughout the rotation to avoid nutrient stress. Fertilization rates would not be constant, but would vary with forest demand, being low prior to crown closure, maximal at crown closure, and low again during the closed stand phase (Fig. 5.4). Timing and rates would be determined by regular analyses of foliar nutrient status. The second pathway is to set harvest schedules by the time needed for the ecosystem to restore itself nutritionally (Kimmins 1974). This approach (to be discussed later) involves process simulation using computer models of varying complexity (Proe *et al.* 1994).

5.3.2 Soil physical changes

While much attention has been paid to nutritional aspects of management practices on sustainable yield, soil physical aspects are equally important. Mass wasting and surface erosion obviously have catastrophic consequences because natural restoration takes centuries or millennia. But beyond this, the most serious problem is compaction. Compaction degrades potential site productivity by the loss of soil macroporosity. Substantive loss in macroporosity reduces soil aeration, water infiltration and available water-holding capacity, and increases physical resistance to root penetration (Sands *et al.* 1979; Powers *et al.* 1990, 1998). Loss of aeration also shifts soil microbial activity to bacterially dominated processes that slow organic matter decomposition and nutrient mobilization. Reductions in such obligate aerobes as fungi reduce mycorrhizal symbiosis and can lead to phosphorus deficiency.

Effects of compaction on forest growth are not always obvious. For example, trees established on compacted landings and skid trails often grow at lower stocking and have less competition from weed species. Consequently, early growth can be equivalent to that of trees on less-compacted areas which generally grow at greater stocking densities and with higher levels of weed competition (Powers 1999). In time, differences between tree growth on compacted and less-compacted areas become even less apparent because of the localized nature of compacted units and the large edge effect.

Many forest managers assume that soil compaction is a surface effect that dissipates with time through such natural processes as frost heaving. However, frost heaving is not an important process at the lower latitudes, and compaction may be irreversible without human intervention. Even where frost heaving does occur, compaction may persist for a quarter century (Froehlich *et al.* 1985).

In a unique experiment, Tiarks and Haywood (1996) compared first and second rotation responses of *Pinus elliotii* to three site preparation treatments on a silt loam soil in Louisiana. Treatments were established in 1960 following the harvesting of a natural stand of pines. Logging slash was burned, and sites received either no further treatment ('Check'), disking, or disking plus bedding. Trees were planted and measured regularly, and the plantations were harvested in 1983. Growth analysis showed that trees were substantially taller in the disking plus bedding treatment. Slash was burned again on all plots, but mechanical site preparation was not repeated. The same plots were replanted and trees were measured regularly through year 10.

Comparing 10-year growth for first and second rotation stands showed that trees in the Check treatment grew consistently faster in the second rotation than either of the mechanically site-prepared treatments. Measurements of soil strength at 10 years in the second rotation revealed that a compaction pan existed beneath the two mechanically site prepared treatments and had persisted for 33 years (Fig. 5.6). The pan was continuous below 20 cm in disked plots, and was similar but discontinuous where soil had been shaped by plough into furrows and mounds ('bedding'). Soil strengths below 20 cm often were at or above 2 MPa, meaning that root growth was likely reduced (Sands *et al.* 1979). The net effect was that roots in the second rotation were confined to a shallow, impoverished soil zone which largely was depleted of available phosphorus in the first rotation, causing incipient deficiency in the second. Similar effects of mechanical treatment were shown for first and second rotation plantations on sandy soils in Australia where logs had been skidded by tractor (Sands *et al.* 1979). The cumulative effect of mechanical traffic can seriously alter soil physical properties at depths not normally probed by a spade, and effects will persist without management intervention.

5.4 DETERMINING SUSTAINABLE YIELD

5.4.1 Assessing site potential

Inherent productivity

A key to developing a sustainable yield strategy is to know the inherent productivity of the forest

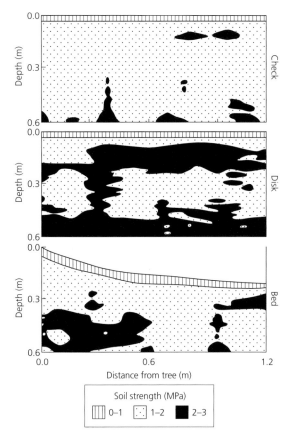

Fig. 5.6 Penetrometer measurements of soil strength on burned only (Check), burned and disked (Disk) and burned, disked and bedded (Bed) treatments 13 years after site preparation in the second rotation. Measurements taken at 10-cm intervals from planted trees. (From Tiarks & Haywood 1996.)

and how it is affected by various management options (Fig. 5.3). Both tree and stand volumes can be estimated non-destructively using established mensurational methods (Avery & Burkhart 1983), and data may exist from previous rotations for tracking stand development to final harvest. Provided that management continues in a similar way from one rotation to the next, that yield trends show no decline, and that site quality is not altered by factors beyond management control, historical records offer a reasonable basis for planning sustainable yield. Lacking historical records, stand inventories must be made, records kept of growth and yield, and trials established

that encompass appropriate silvicultural options (Beets *et al.* 1994). Results would provide a basis for evaluating effects of management on productivity at an operational scale. Powers *et al.* (1994) suggest methods for establishing such trials with statistical reliability.

Direct measures

Site index, the height that the largest trees in a stand attain at some specific age, is the traditional estimator of site quality and potential productivity (Smith *et al.* 1997). But as Avery and Burkhart (1983) have noted, site index has many problems:

1 Stand age often is difficult to determine, and small errors can compound to larger errors in the site index estimate.

2 The concept is suited mainly for even-aged, pure stands.

3 Stand density measures are not considered.

4 Site index is assumed to be constant, yet height growth can change dramatically with climatic cycles.

5 Rarely can the site index of one species be translated into the site index of another.

Site index alone is too imprecise a measure of productivity potential or productivity change. In summary, site index simply amalgamates all the historical events that have shaped the height growth pattern for a particular stand. It is not reliable for projecting the effect of new site or stand conditions (Waring & Running 1998). Nor is it practicable for sites lacking older trees with no history of suppression.

Obviously, the best measure of sustained wood yield is yield itself—bole wood production at harvest. Differences in bole wood yields in successive rotations of the same age indicate changes in potential sustainable yield. Results would be precise and seemingly convincing, but there are problems. Converting unmanaged, natural stands to plantations almost always leads to yield improvement simply because stocking is controlled throughout the life of the stand. Even in consecutive rotations of plantations, similar yields could result from improved competition control or genetic composition in the current rotation that masks a true decline in site quality

(Powers *et al.* 1990; Burger 1996). Such confounding is especially likely in short rotation strategies because factors that temporarily override soil impacts are most effective when stands are young. Climatic differences between consecutive rotations also will affect yield outcomes. In general, yield records from consecutive rotations are not safe indicators of sustainable yield unless yields clearly are declining. Evans & Masson (2000) (this volume, Chapter 18) report a careful study over three rotations of pine and found no significant decline in yield despite no fertilizer application and more adverse climatic conditions.

Squire *et al.* (1985) largely overcame this problem by collecting seeds from one rotation to produce tree seedlings for the next (thereby keeping genotype constant), then treating the second-generation stand similarly to the first both on the same and on matched sites (keeping silvicultural treatments constant and short-term climate similar). Careful comparisons of stand development in consecutive or concurrent rotations indicated that site productivity was sustained if logging slash was retained. Nambiar (1996) concluded that the effect was due mainly to nutrient availability and that productivity could be sustained from one rotation to the next by combining weed control with fertilization. Such trials, of course, are expensive in capital and time.

Another problem is that precise measures of stand productivity are not obtained easily. Bole wood volume for a stand is determined from site-specific equations that estimate volume from an easily measured variable such as bole diameter, doing this for all trees in a stand inventory, and summing the bole volumes for the whole stand. Developing such equations requires either felling trees so that measurements may be taken on the ground, or by measuring upper bole diameters and heights so that volume can be estimated non-destructively. Several procedures have been developed to accomplish this (Avery & Burkhart 1983). But because bole form can change with different degrees of competition and different positions of the live crown, mathematical equations must be robust enough to account for form changes that normally occur throughout the life of a stand. Done diligently, periodic volume estimation is an excellent way to track the expression of productivity in a stand. Individual tree and stand-growth simulators have been developed for more general use, and can be parameterized for specific site and tree conditions (Landsberg & Gower 1997), and offer sizeable advantages over former ways of estimating growth and yield. The problem is, if changes have or are occurring, empirical methods such as direct measurement and growth simulation offer no insight as to the causes.

Less direct measures

A fundamental measure of yield potential is a site's carrying capacity for foliar biomass, and anything altering this alters sustainable yield. Therefore, periodic measures of foliar mass in the stable period following canopy closure (Fig. 5.4, phase C) should provide powerful, physiologically based indices of productive potential. Because foliar mass is linked directly with NPP, foliar-based productivity inferences require no knowledge of average tree size, age or stocking (Waring & Running 1998). Unfortunately, foliar mass is difficult to measure directly and non-destructively, but LAI is a popular and convenient surrogate. Instruments for measuring light transmitted through the canopy are the most common way of estimating LAI, although leaf area–sapwood area allometric relationships also are used (Beadle 1997; Landsberg & Gower 1997; Waring & Running 1998).

While leaf area meters are versatile, they underestimate mass if foliage is clumped. Cherry *et al.* (1998) have shown that very accurate measures of true LAI can be made if meters are calibrated for specific stands. Foliage mass also changes over a season and, to be useful, sampling should be standardized seasonally along permanent transects. If allowances are made for changes due to thinning, long-term trends in LAI should provide an effective index of stand productivity. Sapwood area measurements also offer a physiological basis for estimating foliar mass, but destructive sampling such as increment coring is needed to estimate sapwood area if the leaf area–sapwood area ratio is known. Otherwise, the relationship must be established by intensive and destructive sampling of whole trees. Leaf area–sapwood area

Table 5.2 Typical soil quality threshold standards used by the USDA Forest Service to indicate detrimental changes in soil productivity. In general, at least 85% of the activity area must be within threshold standards. (Modified from Powers *et al.* 1998.)

Nature of disturbance	Threshold value for detrimental disturbance in surface horizons
Area affected	Total area of detrimental disturbance should not exceed 15% of the activity area, exclusive of roads and landings
Altered wetness	Area becomes perennially flooded or drained, and the natural function and value of the land is lost
Erosion	Sheet or rill erosion exceeds estimated natural rates of soil formation over a rotation
Soil cover	Effective cover on less than 50% of the area, or as modified by slope
Organic matter	Forest floor missing from 50% or more of the area
Infiltration	Reduced by 50% or more
Compaction	Bulk density increase of 15% or more, or reduction of 10% or more of total porosity
Rutting and puddling	Ruts to at least 15 cm deep for more than 3 m
Detrimental burning	Forest floor consumed, mineral soil reddened
Displacement	Removal of half or more of the A horizon over an area at least 6 m² and 1 m wide

relationships vary with species, bole position, stand age, stocking density and possibly site quality. Therefore, ratios established elsewhere may not be appropriate for the stand of interest. Regardless, both LAI and leaf area–sapwood area methods should be seen as indices which are probably two or more levels removed from direct estimates of productive potential, such as bole wood production. Calibrated properly, they should provide good assessments of the trend in stand productivity. But as with direct tree measures, causal factors are not explicit.

Indirect measures

Burger (1996) called for an unbiased measure of soil quality and potential site productivity that is independent of stocking and genetic influences on current growth. Because soil is a major factor determining potential productivity, and because it is readily affected by management and largely independent of current stand conditions, soil-based variables have been recommended as alternative indices of sustainable productivity (Canadian Forest Service 1995; Burger 1996; Powers *et al.* 1998). Accordingly, the United States Forest Service has adopted a programme for monitoring soil condition as a surrogate for site quality. Each of the geographical regions of the USA is developing operationally practicable sampling criteria for detecting soil changes that

would lower site quality and potential sustainable yield over a rotation (Powers *et al.* 1998).

Soil quality is assumed to reflect inherent potential productivity, but standards are not meant to be precise estimators. Rather, they are used as 'early warning signals' that something may be impairing potential sustainable yield. Standards vary from region to region, but the general concept for 10 threshold soil conditions used by the United States Forest Service is described in Table 5.2. Shortcomings of such standards are that subsoil conditions are not addressed, correlations have not been made with site potential, and important processes are not well integrated.

Are more precise indices feasible? Van Miegroet *et al.* (1994) presented a more detailed list of fundamental, soil-based variables that could be measured to develop precise indices of soil quality status. Repeated sampling on given sites would offer valid measures of soil quality change, and might suggest possible causes. For example, substantially increased cation leaching and exchangeable acidity, and decreased pH would suggest acidification from atmospheric deposition. Increases in soil nitrate would suggest that the cause was N deposition as noted near European feedlots (Van Breemen 1990). Along with static measures (soil chemistry, moisture-holding capacity, biotic diversity) are process-based measures (decomposition, mineralization, immobi-

Table 5.3 Partial list of soil properties, processes, and methods of assessment that relate to site quality and potential sustainable yield. (Modified from van Miegroet *et al.* 1994.)

Soil property	Key processes	Assessment
Soil fauna	Detrital fragmentation/mixing Soil aeration	Pitfall traps and counting Heat extractions and counting
Soil microbiota	Decomposition Nutrient transformation Soil aggregation	Dilution plate Incubation/fumigation Respiration
Forest floor	Energy substrate Decomposition Nutrient cycling	Gravimetric/chemical analysis Decomposition in litterbags Isotopic techniques
Mineralization potential	Nutrient release Nutrient transformations	Incubations/exchange resins Solution lysimetry
Reaction	Proton exchange (H and Al) Nutrient availability	pH Exchangable acidity
Organic matter	Energy substrate Nutrient supply/retention Water retention	Combustion Fractionation Spectroscopy/^{13}C NMR
Fertility	Nutrient availability	Kjeldahl analysis Chemical extractions
Oxygen content	Soil biotic activity Redox reactions	Soil gas analysis Redox potential measures
Water input	Recharge Availability to roots	Infiltrometry Hydraulic conductivity
Water content	Uptake by plants/soil biota Nutrient availability	Gravimetric analysis Neutron scattering Resistance blocks/TDR
Porosity/strength	Gas diffusion Water availability Root impedence	Bulk density Pore size analysis Mechanical resistance

lization, leaching). Typical measures are indicated in Table 5.3. If tailored to a particular climatic regime and soil type, periodic readings would provide a reliable index of site quality trends and inference into possible causes. However, sampling complexities and the cost of detailed analyses relegate such detailed soil-based indices to only a few intensively studied research sites. Findings must be interpreted carefully. For example, soil organic N, carbon and microbial biomass were appreciably lower following 30 years of continuous weed control in a pine forest than in adjacent, control plot soils (Busse *et al.* 1996). Rather than indicating a decline in soil fer-

tility from weeding, this simply suggests that long-term weeding slows the rate at which soil fertility aggrades.

Recognizing this, Powers *et al.* (1998) called for a simple set of integrative measures for extensive soil quality monitoring in forests of the US. To be effective, such measures must: (i) reflect physical, chemical, and biological soil processes important to sustained productivity; (ii) integrate multiple properties and processes; (iii) be operationally practicable on a variety of sites; and (iv) be sensitive to the overriding conditions of climate. They proposed that penetrometer measurements of soil strength, anaerobic incubations of soil N availability and (indirectly) microbial biomass, and physical signs of soil invertebrate activity (biopores, faecal aggregates) serve as first-approximation surrogates for more detailed (and costly) measures of soil physical, chemical and biological properties.

These biologically relevant, integrative measures are particularly useful for operational monitoring. Soil strengths above 2 MPa indicate increasing stress to root growth (Taylor *et al.* 1966), and root growth essentially ceases beyond 3 MPa (Sands *et al.* 1979). Figure 5.6 (p. 118) indicates how soil strength can detect degraded soil quality and forest productivity over two rotations on a pine site in Louisiana (Tiarks & Haywood 1996). Nitrogen mineralized anaerobically correlates well with site quality in such diverse regions as Australia, California and Maine (Powers *et al.* 1998), and with field rates of N mineralization (Adams & Attiwill 1986).

Although direct correlations between soil invertebrates and forest productivity have yet to be established, their significance in ecosystem processes is unquestioned (Van Cleve & Powers 1995) and is recognized particularly well in Europe (Shaw *et al.* 1991). Soil invertebrates are difficult to measure directly, but the functional products of their activity—biopores, faecal aggregates, and stable soil aggregates—are not (Powers *et al.* 1998). Findings from such soil sampling would not be interpreted in an absolute sense. Rather, they would serve as baselines for measuring soil quality trends at each particular site. As the concept of soil quality evolves, so do standards for effective monitoring.

Process simulators

Mathematical models based on physiological processes and the site factors that affect them offer a promising means for estimating how forests grow under differing management regimes. For a general discussion see Volume 1, Chapters 8–10. Early process-based simulation models were based on the simple premise that a single factor (a nutrient, light or soil moisture) was the principal determinant of tree or stand behaviour. Thus, they were inherently unrealistic. More sophisticated models involving foliar mass as a determinant of growth

have emerged in the last two decades (Landsberg & Waring 1997). Growth projections may follow assumptions on how photosynthate is partitioned into biomass components according to light interception and canopy temperatures. Nutritional or water factors are handled through less dynamic submodels. For example, BIOMASS (McMurtrie *et al.* 1992) is a stand-level simulator that models photosynthetic and respiratory processes as a function of air temperature. Soil properties interact through the assumption that assimilation is a linear function of foliar N concentration and that water deficits control stomatal behaviour (Landsberg & Gower 1997).

Other 'biogeochemical' simulators are driven by soil nutrient or water availability submodels that affect foliar production and assumed partitioning of carbon, and are a large step forward in modelling the concept of an ecological rotation. Century—originally a grassland model modified recently for forests (Metherell *et al.* 1993)—is a soil-based model of organic matter decomposition that is tied loosely with NPP (Landsberg & Gower 1997). Recognizing the need to couple below- and above-ground processes and conditions, researchers have combined model components to try and simulate a more realistic picture of ecosystem processes. For example, G'DAY (McMurtrie *et al.* 1992) combines elements of Century with BIOMASS, a physiological model of forest vegetation, to account for short-term changes due to rises in atmospheric CO_2. In turn, such combined models have been further modified to simulate longer-term processes (Murty *et al.* 1996). The most advanced models marry both elements. For example, Forest BGC (Running & Coughlan 1988) is a landscape-scale ecosystem model based on leaf area as influenced by climate, leaf water status and soil N availability. It simulates the carbon cycle by modelling photosynthesis, respiration and the partitioning of NPP into above- and below-ground components. Characteristics of these and other simulation models are outlined in Table 5.4 and are discussed in more detail elsewhere (Proe *et al.* 1994; Landsberg & Gower 1997; Homann *et al.*, 2000).

One should recognize that all models are abstractions of reality because our understanding of most site processes is imperfect. Therefore,

Table 5.4 Examples of mathematical process models that simulate forest growth and their attributes. From Homann *et al.* (in press), Landsberg and Gower (1997), Landsberg and Waring (1997) and Proe *et al.* (1994).

Model	Calibrating variables	Timestep	Reference
BIOMASS	Climatic	Monthly	McMurtrie *et al.* 1992
Century–Forest	Climatic, Soil	Monthly	Metherell *et al.* 1993
FOREST-BGC	Climatic	Yearly	Running & Coughlan 1988
G'DAY	Climatic, Soil	Monthly	Comins & McMurtrie 1993
LINKAGES	Climatic	Yearly	Post & Pastor 1996
NuCM	Climatic, Soil	Daily–monthly	Liu *et al.* 1991
PnET	Climatic	Monthly	Aber & Federer 1992
3-PG	Climatic, Soil	Monthly	Landsberg & Waring 1997

many critical processes affecting water and nutrient availability are modelled implicitly using assumed relationships with standard growth functions. Biogeochemical models attempt a more realistic approach, but their outputs still rest on many assumptions and uncertainties awaiting further research. Recently, Homann *et al.* (2000) compared soil chemistry predictions from several calibrated biogeochemical models with independent data not used in the calibration. Errors between observed and predicted values averaged between 24% and 56%, with individual discrepancies as great as 1000%.

Practical problems with the more advanced simulators are that many of the calibration parameters must be estimated because they are not directly measurable, and that calibration may be operationally impracticable. In their validation efforts with FOREST-BGC, Milner *et al.* (1996) voiced the need for better soil data and the need for other modifiers of stockability for marginal sites and ecotones. They also expressed frustration at configuring even-age stand simulators to make reasonable projections for multiaged stands. Uncertainties surrounding nutrient dynamics and the overriding significance of climate led Landsberg and Waring (1997) to consider only climatic variables, soil water-holding capacity and stand leaf area in their physiological model of forest productivity, 3-PG. Recent tests of 3-PG predictions against 30-year growth data from Australian and New Zealand plantations showed excellent correlations (Landsberg & Waring 1997).

The history of simulation models mirrors the evolution in our understanding of how ecosystems function. They serve an extremely useful scientific purpose for integrating knowledge and testing hypotheses, but to date they cannot be extrapolated with confidence to specific sites or new situations. However, models can be used as a 'gaming' tool to develop first approximations of sustainable yield strategies if reasonable calibration data are available. In the absence of site-specific data, tabulations such as those by Kimmins *et al.* (1985) may provide first approximations. Also the availability of climatic and leaf area data from remote sensing furthers the ease of parameterization. The key for the general acceptance of simulation models, though, is that they be logical, easily calibrated, validated, and upgraded as needed. Easily parameterized, physiologically based growth models such as 3-PG seem exceptionally promising. Coordinated programmes like the North American LTSP study (Powers *et al.* 1990) will help to validate them.

5.4.2 Scheduling the harvest

Area regulation

Assuming management strategies that protect site quality, one classical way of managing for sustained yield is through the *area method of regulation* (Smith *et al.* 1997). It occurs when a forest contains multiple even-aged stands that collectively comprise all age classes in a rotation. Individual stands may be of varying extent—the only criterion being that each age class occupies equivalent ground area. *If* site quality is similar for all stands; *if* they are tended as needed to assure adequate stocking and vigour; *if* harvesting from a particular age class is balanced by ingrowth from the next youngest age class; *if* site quality and gene pools have not been degraded; and *if* no catastrophic events occur from fire, flood, wind or pestilence—then sustained yield has been achieved. Despite all the conditional 'ifs', area regulation is a popular means for achieving sustained yield in even-aged management systems. It is particularly useful if the area in submerchantable age classes is recognized for other values, such as wildlife habitat or fuelwood production. But it seems unlikely that all the 'if' conditions can be met.

Volume regulation

Another management strategy for sustained yield is the *volume method of regulation* (Smith *et al.* 1997). Unlike area regulation, it does not require a balance of age classes occupying equal ground areas. Its main characteristic is that the forest contains an appropriate distribution of size classes that describe an 'inverse J-shaped' (negative exponential) curve. In such a regulated forest, harvesting would occur in all age classes in order to maintain stocking that ensures healthy, orderly and predictable growth. Also ideally, there would be a ready market for all sizes of harvested materials. Rarely does this occur. Markets for small, young trees often are marginal or non-existent, and only the largest and oldest trees have high value. Although it is difficult under these conditions, sustained yield may still be achieved, provided that the volume removed in larger trees is met by accelerated growth in smaller size classes through various silvicultural treatments such as weeding, thinning and fertilization. Volume regulation demands more knowledge of stand conditions and more silvicultural input than area regulation. Therefore, it is the more technically demanding of the two. However, it is a more flexible system because it frees the forester from rigid adherence to area regulation, and is equally appropriate in even- or uneven-aged management strategies.

A special case often applies to plantations where the objective is to maximize wood yield over time. In Britain, this requires thinning at an intensity of 70% of the maximum mean annual volume increment for that yield class and, of course, to fell the crop at the age of maximum mean annual increment (Fig. 5.1). If the productivity of the successor plantation is expected to be greater than the current, maximum mean annual increment will occur earlier in the successor rotation (e.g. FR in Fig. 5.1) and a shorter rotation may be justified for the current plantation (Evans 1992).

Infrequent stand entries, low levels of extraction, and just enough ground disturbance to secure regeneration as proposed for 'ecological forestry' (Drengson & Taylor 1997) equate to yields that presumably can be sustained indefinitely, although they may be less than the site's yield potential. Conditions suitable for the biotic diversity characterizing late seral forests would be maintained, and a continuous forest cover would enhance many watershed values.

5.5 SYNTHESIS AND SUMMARY

Society places high value in wood. Therefore, sustaining or enhancing wood yields from one rotation to the next should be both an ethical and economic goal of forest managers. Achieving this depends on management practices that sustain or enhance the site's foliar carrying capacity for the forest species of interest. This capacity, varying by genotype, is set by climate, topography and soil. Of these, only soil is amenable to management. Therefore, sustained productivity and yield hinge on good soil management.

Recent innovations in harvesting technology, the conversion of natural stands to plantations, rises in air pollution, and the spectre of global climate change compound the difficulty in assessing potential sustainable yield. Practicable methods are needed for monitoring trends in site potential. Direct observations of growth and yield give true measures of stand performance, but are labour intensive. Such measures are conditioned by stand history and, if changes occur, offer little information on possible causes. Trends in foliar mass or area can be monitored non-destructively by periodic surveys of LAI once canopies have closed. If changes are noted, causes may be inferred from changes in soil properties and processes revealed through soil quality monitoring. Long-term field studies of multiple treatments representing forest management strategies are needed both to determine the true impact of management on potential sustainable yields, and to calibrate or validate mathematical models meant to simulate fundamental site processes affecting yield. Despite their recognized imperfections, computer simulators may be our best means for projecting the sustainability of any forest management system and come closest to the concept of *ecological rotation*.

REFERENCES

Aber, J.D. & Federer, C.A. (1992) A generalized, lumped-parameter model of photosynthesis, evapotranspiration and net primary production in temperate and boreal forest ecosystems. *Oecologia* **92**, 463–74.

Adams, M.A. & Atttiwill, P.M. (1986) Nutrient cycling and nitrogen mineralization in eucalypt forests of south-eastern Australia. II. Indices of nitrogen mineralization. *Plant and Soil* **92**, 341–62.

Allen, H.L. (1990) Manipulating loblolly pine productivity with early cultural treatment. In: Gessel, S.P., Lacate, D.S., Weetman, G.F. & Powers, R.F., eds. *Sustained Productivity of Forest Soils*, pp. 301–17. University of British Columbia Faculty of Forestry, Vancouver.

Anonymous (1995) The certified forest: what makes it green? *Journal of Forestry* **93**, 1–41.

Anonymous (1997) The global fibre supply study (GFSS). In: *Eighteenth Session, North American Forestry Commission, Ashville, North Carolina*. Food and Agricultural Organization of the United Nations. W 3158/E.

Attiwill, P.M. (1994) Ecological disturbance and conservative management of eucalypt forests in Australia. *Forest Ecology and Management* **63**, 301–46.

Avery, T.E. & Burkhart, H.E. (1983) *Forest Measurements*, 3rd edn. McGraw-Hill, New York.

Ballard, R. (1984) Fertilization of plantations. In: Bowen, G.D. & Nambiar, E.K.S., eds. *Nutrition of Plantation Forests*, pp. 327–60. Academic Press, New York.

Beadle, C.L. (1997) Dynamics of leaf and canopy development. In: Nambiar, E.K.S. & Brown, A.G., eds. *Management of Soil, Nutrients and Water in Tropical Plantation Forests*. Monograph no. 43, pp. 169–212. Australian Centre for International Agricultural Research, Canberra.

Beets, P.N., Terry, T.A. & Manz, J. (1994) Management systems for sustainable productivity. In: Dyck, W.J., Cole, D.W. & Comerford, N.B., eds. *Impacts of Forest Harvesting on Long-Term Site Productivity*, pp. 219–46. Chapman & Hall, London.

Berg, S. & Olszewski, R. (1995) Certification and labeling. A forest industry perspective. *Journal of Forestry* **93**, 30–1.

Binkley, C.S. (1997) Preserving nature through intensive plantation forestry: the case for forestland allocation with illustrations from British Columbia. *Forestry Chronicle* **73**, 553–9.

Binkley, D., Dunkin, K.A., DeBell, D. & Ryan, M.G. (1992) Production and nutrient cycling in mixed plantations of *Eucalyptus* and *Albizia* in Hawaii. *Forest Science* **38**, 393–408.

Bouwman, A.F. & Leemans, R. (1995) The role of forest soils in the global carbon cycle. In: McFee, W.W. & Kelly, J.M., eds. *Carbon Forms and Functions in Forest Soils*, pp. 503–25. Soil Science Society of America, Madison, WI.

Boyle, T.J.B., Cossalter, C. & Griffin, A.R. (1997) Genetic resources for plantation forestry. In: Nambiar, E.K.S. & Brown, A.G., eds. *Management of Soil, Nutrients and Water in Tropical Plantation Forests*. Monograph no. 43, pp. 25–64. Australian Centre for International Agricultural Research, Canberra.

Brown, A.G., Nambiar, E.K.S. & Cossalter, C. (1997) Plantations for the tropics—their role, extent and nature. In: Nambiar, E.K.S. & Brown, A.G., eds. *Management of Soil, Nutrients and Water in Tropical Plantation Forests*. Monograph no. 43, pp. 1–23. Australian Centre for International Agricultural Research, Canberra.

Burger, J.A. (1996) Limitations of bioassays for monitoring forest soil productivity: rationale and example. *Soil Science Society of America Journal* **60**, 1674–8.

Busse, M.D., Cochran, P.H. & Barrett, J.W. (1996) Changes in ponderosa pine site productivity following removal of understory vegetation. *Soil Science Society of America Journal* **60**, 1614–21.

Canadian Forest Service (1995) *Criteria and Indicators for the Conservation and Sustainable Management of Temperate and Boreal Forests*. The Montreal Process. Canadian Forest Service, Natural Resources Canada, Quebec.

Cannell, M.G.R. (1989) Physiological basis of wood production: a review. *Scandanavian Journal of Forest Research* **4**, 459–90.

Cherry, M., Hingsgton, A., Battaglia, M. & Beadle, C. (1998) Calibrating the LI-COR 2000 for estimating leaf area index in eucalypt plantations. *Tasforests* **10**, 75–81.

Clutter, J.L., Fortson, J.C., Pienaar, L.V., Brister, G.H. & Bailey, R.L. (1983) *Timber Management: A Quantitative Approach*. John Wiley & Sons, New York.

Comins, H.N. & McMurtrie, R.E. (1993) Long-term response of nutrient-limited forests to CO_2 enrichment: equilibrium behavior of plant–soil models. *Ecological Applications* **3**, 666–81.

Della-Tea, F. & Jokela, E.J. (1991) Needlefall, canopy light interception, and productivity of young intensively managed slash and loblolly pine stands. *Forest Science* **37**, 1298–313.

Drengson, A.R. & Taylor, D.M., eds. (1997) *Ecoforestry. The Art and Science of Sustainable Forest Use.* New Society Publishers, Gabriola Island, BC.

Dyck, W.J., Cole, D.W. & Comerford, N.B., eds. (1994) *Impacts of Forest Harvesting on Long-Term Site Productivity.* Chapman & Hall. London.

Ebermeyer, E. (1876) *Die gesamte Lehte der Waldstreu mit Rucksicht auf die chemische Statik den Waldbaues.* Springer, Berlin.

Evans, J. (1992) *Plantation Forestry in the Tropics.* Oxford University Press, London.

Evans, J. & Masson, P. (2000) Sustainable plantation forestry: a case study of wood production and environmental management strategies in the Usutu Forest, Swaziland. In: Evans, J., ed. *The Forests Handbook*, Vol. 2, pp. 357–70. Blackwell Science, Oxford.

Froehlich, H.A., Miles, D.W.R. & Robbins, R.W. (1985) Soil bulk density recovery on compacted skid trails in central Idaho. *Soil Science Society of America Journal* **49**, 1015–17.

Grier, C.C., Lee, K.M., Nadkarni, N.M., Klock, G.O. & Edgerton, P.J. (1989) *Productivity of Forests of the United States and its Relation to Soil and Site Factors and Management Practices: A Review.* General Technical Report PNW-GTR-222. USDA Forest Service, Pacific Northwest Research Station, Portland, OR.

Hamilton, G.J. & Christie, J.M. (1971) Forest management tables (metric). *Forestry Commission Booklet* No. 34. Her Majesty's Stationery Office, London.

Hermann, R.K. (1976) *Man and Forests—A Prodigal Relation.* Northwest Area Foundation Series, pp. 29–51. School of Forestry, Oregon State University, Corvallis.

Homann, P.S., McKane, R.B. & Sollins, P. (2000) Belowground processes in forest-ecosystem biogeochemical simulation models. In: Boyle, J.B., Powers, R.F., eds. *Forest Soils and Ecosystem Sustainability.* Elsevier, Amsterdam.

Jenny, H. (1980) *The Soil Resource. Origin and Behavior.* Springer-Verlag, New York.

Kimmins, J.P. (1974) Sustained yield, timber mining, and the concept of ecological rotation: a British Columbia view. *Forestry Chronicle* **50**, 27–31.

Kimmins, J.P. (1996) Importance of soil and role of ecosystem disturbance for sustained productivity of cool temperate and boreal forests. *Soil Science Society of America Journal* **60**, 1643–54.

Kimmins, J.P., Binkley, D., Chatarpaul, L. & de Catanzaro, J. (1985) *Biogeochemistry of Temperate Forest Ecosystems: Literature on Inventories and Dynamics of Biomass and Nutrients.* Information Report PI-X-47E/F. Petawawa National Forestry Institute, Canadian Forestry Service, Chalk River, Ontario.

Landsberg, J.J. & Gower, S.T. (1997) *Applications of Physiological Ecology to Forest Management.* Academic Press. London.

Landsberg, J.J. & Waring, R.H. (1997) A generalized model of forest productivity using simplified concepts of radiation-use efficiency, carbon balance and partitioning. *Forest Ecology and Management* **95**, 209–28.

Leith, H. (1975) Modeling the primary productivity of the world. In: Leith, H. & Whitaker, R.H., eds. *Primary Production of the Biosphere*, pp. 237–62. Springer-Verlag, New York.

Libby, W.J. (1987) Do we really want taller trees? Adaptation and allocation as tree-improvement strategies. *The HR MacMillan Lectureship in Forestry*, The University of British Columbia, Vancouver. 16p.

Liu, S., Munson, R., Johnson, D.W. *et al.* (1991) Application of a nutrient cycling model (NuCM) to northern mixed hardwood and southern coniferous forests. *Tree Physiology* **9**, 173–82.

Lugo, A.E., Brown, S. & Chapman, J. (1988) An analytical review of production rates and stemwood biomass of tropical forest plantations. *Forest Ecology and Management* **23**, 179–200.

McMurtrie, R.E., Comins, H.N., Kirschbaum, M.U.F. & Wang, Y.P. (1992) Modifying existing forest to take account of effects of elevated CO_2. *Australian Journal of Botany* **40**, 657–77.

Matheson, A.C. & Cotterill, P.P. (1990) Utility of genotype–environment interactions. In: Nambiar, E.K.S., Squire, R., Cromer, R., Turner, J. & Bordman, R., eds. *Management of Water and Nutrient Relations to Increase Forest Growth.* Forest Ecology and Management **30**, 159–74.

Metherell, A.K., Harding, L.A., Cole, C.V. & Parton, W.J. (1993) *Century Soil Organic Matter Model Environment. Technical Documentation Agroecosystem, Version 4.0.* Great Plains System Research Unit Technical Report 4. US Department of Agriculture, Agriculture Research Service, Fort Collins, CO.

Milner, K.S., Running, S.W. & Coble, D.W. (1996) A biophysical soil-site model for estimating potential productivity of forested landscapes. *Canadian Journal of Forest Research* **26**, 1174–86.

Mitchell, A.K. & Hinkley, T.M. (1993) Effects of foliar nitrogen concentrations on photosynthesis and water use efficiency in Douglas-fir. *Tree Physiology* **12**, 403–10.

Morris, L.A. & Miller, R.E. (1994) Evidence of long-term productivity change as provided by field trials. In: Dyck, W.J., Cole, D.W. & Comerford, N.B., eds. *Impacts of Forest Harvesting on Long-Term*

Site Productivity, pp. 41–80. Chapman & Hall, London.

Murty, D., McMurtrie, R.E. & Ryan, M.G. (1996) Declining forest productivity in aging forest stands: a modeling analysis of alternative hypotheses. *Tree Physiology* **16**, 187–200.

Nambiar, E.K.S. (1996) Sustained productivity of forests is a continuing challenge to soil science. *Soil Science Society of America Journal* **60**, 1629–42.

Nambiar, E.K.S. (1999) Pursuit of sustainable plantation forestry. *South African Forestry Journal* **189**, 45–62.

Nambiar, E.K.S. & Sands, R. (1993) Competition for water and nutrients in forests. *Canadian Journal of Forest Research* **23**, 1955–68.

Oliver, C.D. & Larson, B.C. (1990) *Forest Stand Dynamics*. McGraw-Hill, New York.

Oliver, W.W. (1990) Spacing and shrub competition influence 20-year development of planted ponderosa pine. *Western Journal of Applied Forestry* **5**, 79–82.

Plonski, W.L. (1974) *Normal Yield Tables (Metric) for Major Species in Ontario*. Division of Forests, Ontario Ministry of Natural Resources, Toronto.

Post, W.M. & Pastor, J. (1996) LINKAGES—an individual-based forest ecosystem model. *Climate Change* **34**, 253–61.

Postel, S. & Heise, L. (1988) Reforesting the Earth. *Worldwatch Paper 83*. Worldwatch Institute, Washington, DC.

Powers, R.F. (1999) On the sustainable productivity of planted forests. *New Forests* **17**, 263–306.

Powers, R.F. & Reynolds, P.E. (1999) Ten-year responses of ponderosa pine plantations to repeated vegetation and nutrient control along an environmental gradient. *Canadian Journal of Forest Research* **29**, 1027–38.

Powers, R.F. & Reynolds, P.E. (2000) Intensive management of ponderosa pine plantations: sustainable productivity for the 21st century. *Journal of Sustainable Forestry* **10**, 249–55.

Powers, R.F., Alban, D.H., Miller, R.E. *et al.* (1990) Sustaining site productivity in North American forests: problems and prospects. In: Gessel, S.P., Lacate, D.S., Weetman, G.F. & Powers, R.F., eds. *Sustained Productivity of Forest Soils*, pp. 49–79. University of British Columbia Faculty of Forestry, Vancouver.

Powers, R.F., Mead, D.J., Burger, J.A. & Ritchie, M.W. (1994) Designing long-term site productivity experiments. In: Dyck, W.J., Cole, D.W. & Comerford, N.B., eds. *Impacts of Forest Harvesting on Long-Term Site Productivity*, pp. 247–86. Chapman & Hall, London.

Powers, R.F., Tiarks, A.E., Burger, J.A. & Carter, M.C. (1996) Sustaining the productivity of planted forests. In: Carter, M.C., ed. *Growing Trees in a Greener World: Industrial Forestry in the 21st Century*, pp. 97–134. School of Forestry, Wildlife and Fisheries, Louisiana State University, Baton Rouge.

Powers, R.F., Tiarks, A.E. & Boyle, J.R. (1998) Assessing soil quality: practicable standards for sustainable forest productivity in the United States. In: Davidson, E., Adams, M.B. & Ramakrishna, K., eds. *The Contribution of Soil Science to the Development of and Implementation of Criteria and Indicators of Sustainable Forest Management*. SSSA Special Publication 53, pp. 53–80. Soil Science Society of America, Madison, WI.

Proe, M.F., Rauscher, H.M. & Yarie, J. (1994) Computer simulation models and expert systems for predicting productivity decline. In: Dyck, W.J., Cole, D.W. & Comerford, N.B., eds. *Impacts of Forest Harvesting on Long-Term Site Productivity*, pp. 151–86. Chapman & Hall, London.

Running, S.W. & Coughlan, J.C. (1988) A general model of forest ecosystem processes for regional applications. I. Hydrologic balance, canopy gas exchange and primary production processes. *Ecological Modelling* **42**, 125–54.

Ryan, M.G., Hubbard, R.M., Pongracic, S., Raison, R.J. & McMurtrie, R.E. (1996) Foliage, fine-root, woody-tissue and stand respiration in *Pinus radiata* in relation to nitrogen status. *Tree Physiology* **16**, 333–43.

Ryan, M.G., Binkley, D. & Fownes, J.H. (1997) Age-related decline in forest productivity: patterns and process. *Advances in Ecological Research* **27**, 213–62.

Sands, R., Greacen, E.L. & Gerard, C.J. (1979) Compaction of sandy soils in radiata pine forests. I. A penetrometer study. *Australian Journal of Soil Research* **17**, 101–13.

Schulze, E.-D., Fuchs, M.I. & Fuchs, M. (1977) Spatial distribution of photosynthetic capacity and performance in a mountain spruce forest of northern Germany. *Oecologia* **29**, 43–61.

Shaw, C.H., Lundqvist, H., Moldenke, A. & Boyle, J.R. (1991) The relationships of soil fauna to long-term forest productivity in temperate and boreal ecosystems: processes and research strategies. In: Dyck, W.J. & Mees, C.A., eds. *Long-Term Field Trials to Assess Environmental Impacts of Harvesting*. FRI Bulletin no. 161, pp. 39–77. Ministry of Forests, Forest Research Institute, Rotorua, New Zealand.

Smith, D.M., Larson, B.C., Kelty, M.J. & Ashton, P.M.S. (1997) *The Practice of Silviculture: Applied Forest Ecology*. John Wiley, New York.

Squire, R.O., Farrell, P.W., Flinn, D.W. & Aeberli, B.C. (1985) Productivity of first and second rotation stands of radiata pine on sandy soils. *Australian Forestry* **48**, 127–37.

Sutton, W.R.J. (1999) Does the world need planted forests? *New Zealand Journal of Forestry* 24–9.

Switzer, G.L. & Nelson, L.E. (1972) Nutrient accumulation and cycling in loblolly pine (*Pinus taeda* L.) plantation ecosystems: the first twenty years. *Soil Science Society of America Proceedings* **36**, 143–7.

Taylor, H.M., Robertson, G.M. & Parker, J.V. Jr (1966) Soil strength–root penetration relations for medium to coarse-textured soil materials. *Soil Science* **102**, 18–22.

Tiarks, A.E. & Haywood, J.D. (1996) Site preparation and fertilization effects on growth of slash pine for two rotations. *Soil Science Society of America Journal* **60**, 1654–63.

Van Breemen, N. (1990) Deterioration of forest land as a result of atmospheric deposition in Europe: a review. In: Gessel, S.P., Lacate, D.S., Weetman, G.F. & Powers, R.F., eds. *Sustained Productivity of Forest Soils*, pp. 40–8. University of British Columbia Faculty of Forestry, Vancouver.

Van Cleve, K. & Powers, R.F. (1995) Soil carbon, soil formation, and ecosystem development. In: McFee, W.W. & Kelly, J.M., eds. *Carbon Forms and Functions in Forest Soils*, pp. 155–200. Soil Science Society of America, Madison, WI.

Van Miegroet, H., Zabowski, D., Smith, C.T. & Lundkvist, H. (1994) Review of measurement techniques in site productivity studies. In: Dyck, W.J., Cole, D.W. & Comerford, N.B., eds. *Impacts of Forest Harvesting on Long-Term Site Productivity*, pp. 287–362. Chapman & Hall, London.

Vogt, K., Asbjornsen, H., Ercelawn, A., Montagnini, F. & Valdes, M. (1997) Roots and mycorrhizas in plantation ecosystems. In: Nambiar, E.K.S. & Brown, A.G., eds. *Management of Soil, Nutrients and Water in Tropical Plantation Forests*. Monograph no. 43, pp. 247–96. Australian Centre for International Agricultural Research, Canberra.

Waring, R.H. & Running, S.W. (1998) Forest ecosystems. In: *Analysis at Multiple Scales*, 2nd edn. Academic Press, New York.

Waring, R.H. & Schlesinger, W.H. (1985) *Forest Ecosystems. Concepts and Management*. Academic Press, New York.

Zobel, B. & Talbert, J. (1984) *Applied Forest Tree Improvement*. John Wiley, New York.

6: Silvicultural Treatments to 1

H. LEE ALLEN

6.1 INTRODUCTION

Worldwide consumption of forest products is expected to continue to rapidly increase during the twenty-first century. By contrast, the land base used for wood production is expected to decline because of population pressures, environmental concerns and the lack of adequate management by many landowners. Even today, removals equal or exceed growth rates in some areas. These changes are coming at a time when there is a developing recognition that the potential productivity of forests in many regions can be much higher than is currently realized (Bergh et al. 1998a; Sampson & Allen 1999). With investments in appropriate management systems, growth rates exceeding $25\,m^3\,ha^{-1}\,year^{-1}$ for pines are biologically possible and can be financially attractive for a broad range of site types in temperate, subtropical and tropical regions. Production rates can be substantially higher ($>50\,m^3\,ha^{-1}\,year^{-1}$) for eucalyptus in humid subtropical and tropical regions (Evans 1992).

Historically, the practice of silviculture has focused on controlling the composition, quantity and structure of forest vegetation and the maintenance of site quality. As forest plantations have become important sources of fibre, fuel and structural material, this custodial role has given way to active intervention to improve both plant and soil resources. Forest managers are finally recognizing that intensive plantation silviculture is like agronomy; both the plant and the soil need to be actively managed to optimize production. Silvicultural treatments including soil tillage, vegetation control, fertilization, fire and thinning can dramatically affect soil resource availability.

The key to optimizing pr⌷   ⌴e the best genetic material avail⌴   ⌴o provide resources in quantities suffic   ⌴ to allow the trees' genetic potential to be realized.

Most forest investments are of a long-term nature; consequently, forest landowners have tended to focus on minimizing the costs associated with stand establishment and tending, particularly in boreal and temperate forests characterized by slow growth rates. This approach has resulted in millions of planted hectares with growth rates that are substantially less than potential. Recognition that production can be increased dramatically has prompted many landowners to invest in intensive silvicultural regimes to increase production. While per hectare costs increase with intensive plantation culture, the resultant increases in production rates coupled with the reduction in area needed to grow a given amount of wood may substantially reduce production costs per ton of wood produced (Allen et al. 1998, Fig. 6.1).

Clearly, effective manipulation of both genetic and soil resources are essential for cost-effective and environmentally sustainable forest production. However, effective manipulation of the genetic and site resources requires an understanding of what resources limit forest production, how these resources are affected by silvicultural treatment, and how crop trees may differ in their ability to acquire and utilize these resources. Fortunately, our understanding of the physiological and ecological processes regulating forest production has greatly increased during the last decade. Applying the fundamental concept that 'leaves grow trees and resources grow leaves' is a powerful approach to understanding and managing

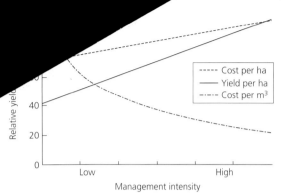

Fig. 6.1 Effects of increasing management intensity on relative yield, cost per hectare and production cost per ton of wood produced.

forest production. This chapter briefly reviews the physiological and ecological bases for forest production (see also Chapter 8 in Volume 1 of this handbook) and then reviews contemporary silvicultural practices with a focus on how their application may impact resource availability and productivity.

6.2 ECOPHYSIOLOGICAL BASIS FOR FOREST PRODUCTION

6.2.1 Production–leaf area relationships

It is now generally accepted that much of the variation in biomass and wood production can be accounted for by variation in light interception. (see Chapter 8 in Volume 1 of this handbook; Linder 1987; Cannell 1989; Gower *et al.* 1994; Landsberg & Gower 1997). Light interception is principally a function of the amount of leaf area and the duration of leaf area display. Differences in individual tree crown architecture and stand canopy structure can also affect light interception. Empirical data from field studies with Scots pine, Norway spruce (Axelsson & Axelsson 1986; Linder 1987; Bergh *et al.* 1998b), radiata pine (Linder *et al.* 1987; Benson *et al.* 1992; Snowdon & Benson 1992), southern pines (Vose & Allen 1988; Colbert *et al.* 1990; Albaugh *et al.* 1998) and *Eucalyptus globulus* (Pereira *et al.* 1994) have shown that leaf area and consequently wood production

are below optimum levels in many areas of the world. Low nutrient availability is a principal factor causing suboptimal levels of leaf area in many areas (Linder 1987; Vose & Allen 1988; Colbert *et al.* 1990; Albaugh *et al.* 1998). Low soil water availability, high vapour pressure deficits, high temperatures and elevated ozone levels have also been shown to adversely affect leaf area production and/or retention (Benson *et al.* 1992; Teskey *et al.* 1987; Hennessey *et al.* 1992; Stow *et al.* 1992; Pereira *et al.* 1994).

The variation in production per unit of leaf area (generally defined as the slope of the wood production–leaf area relationship and referred to as growth or leaf area efficiency) can also contribute to the variation in production. Growth efficiency can vary due to differences in photosynthetic efficiency, respiration and partitioning to various biomass components. Improved nutrient and water availability has been shown to increase photosynthetic efficiency (Linder 1987; Murthy *et al.* 1996), and above-ground productivity proportionally more than below-ground productivity in stand-level studies (Vogt *et al.* 1986; Gower *et al.* 1992; Haynes & Gower 1995; Albaugh *et al.* 1998). On an individual site basis, changes in growth efficiency generally do not contribute as much to changes in productivity as to changes in leaf area. However, when species (e.g. loblolly pine in the south-east USA or radiata pine in Chile) are planted across large areas with substantial regional differences in precipitation and temperature, regional variation in growth efficiency may be substantial (Sampson & Allen 1999).

6.2.2 Limiting resources

To effectively ameliorate resource limitations with silvicultural treatments, forest managers must be able to identify what resources (e.g. light, water, nutrients and oxygen) are limiting the production of crop trees. A resource limitation develops when a stand's production potential, given the level of all other resources, cannot be achieved because of a lack of available supply of that resource (Liebig's Law of the Minimum or the lowest hole in the bucket analogy). A stand's potential use ('need' or 'requirement') for a resource will vary with stand development. In addition, resource availability varies spatially, within a tree, a stand, and across sites and also temporally, within a day, a growing season, and over a rotation. These spatial and temporal variations in potential use and supply make the identification and ranking of importance of the resources that limit production a challenge. Because forest production is generally considered on an annual basis and at the stand level, it is very likely that more than one resource (e.g. water or different nutrients) will limit production.

Low soil water availability has been considered by many to be the principal resource limiting forest productivity (Gholz *et al.* 1990). This is probably true for seedlings that exploit a limited soil volume (Dougherty & Gresham 1988) or for mature trees growing on shallow soil with limited rooting volume. However, once roots have fully exploited most soils, low soil water availability is typically no longer the principal limitation to growth. Evidence for the secondary importance of water includes the widespread growth responses found in fertilizer trials (Binkley *et al.* 1995) and the modest responses to water additions as compared with nutrient additions in trials that have included both (Linder 1987; Albaugh *et al.* 1998; Bergh *et al.* 1998b). Clearly, in most humid temperate and boreal forest regions, nutrient limitations are more limiting to leaf area and consequently production than water. Fortunately, nutrient limitations are easier and less costly to ameliorate than water limitations.

Why are nutrient limitations so widespread? Consideration of a stand's potential nutrient use and nutrient supply indicates a large disparity between nutrient needs and supply in many forest plantation systems (Allen *et al.* 1990). Typically, nutrient availability is high following harvesting and site preparation, as these disturbances provide suitable conditions for rapid decomposition and release of nutrients from the accumulated forest floor and slash material. This is the assart effect described by Kimmins (1997). In young stands, use of nutrients by crop trees is quite small owing to their small size, low leaf area and lack of site occupancy. However, as leaf area development and stand growth accelerates, nutrient use increases rapidly. At the same time, the supply of readily available nutrients is being

rapidly sequestered within the accumulating tree biomass and forest floor. Furthermore, as the canopy closes, the environmental conditions conducive to rapid organic matter decomposition and soil N mineralization are no longer present. Consequently, a stand's nutrient requirement for maximum growth generally exceeds soil supply (particularly for N) at or even before canopy closure (Fig. 6.2). As nutrient supply diminishes, leaf area production and in turn growth becomes regulated (and limited) by the available nutrient pool (Chapter 7). It is not surprising then, that the majority of field trials in intermediate-aged stands have shown strong responses to nutrient additions. In young stands, the development of nutrient limitations is still possible when levels of available nutrients (particularly phosphorus (P) and boron (B)) in the soil are low and the soil volume exploited by roots is small. In addition, as other silvicultural treatments (e.g. vegetation control and/or tillage) are used to improve water availability, crop tree growth and use of nutrients will be increased at young ages. Fertilization may then be needed to sustain rapid growth on all but the most fertile sites.

During extended periods of drought when environmental conditions are otherwise suitable for tree growth (e.g. areas with Mediterranean climates or subtropical or tropical savannahs), low soil water availability and/or high vapour pressure deficits will limit leaf production and

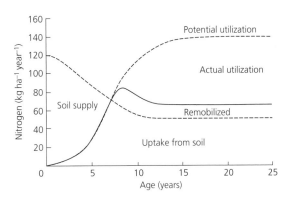

Fig. 6.2 Conceptual relationships between soil supply, stand potential use, and stand actual use and age for nitrogen.

productivity (Linder *et al.* 1987; Benson *et al.* 1992; Pereira *et al.* 1994). During the rainy season, stands on these same sites may be strongly limited by nutrients.

Lack of soil aeration can also be a major factor limiting production on sites where soils are saturated for even a short period during the growing season. Such poorly drained soils are very common on coastal and river flood plains and comprise a substantial amount of land. With appropriate drainage (water control), these sites are potentially very productive because soil water limitations are almost non-existent.

6.3 SILVICULTURAL TREATMENTS

Silvicultural treatments can have either positive, or, if inappropriately applied, negative effects on:
- site resource availability;
- the allocation of resources to crop trees; or
- the ability of crop trees to acquire and utilize site resources.

The most obvious method for ameliorating a resource limitation is to add more of that resource. For example, nutrients can be added by fertilization and water can be added with irrigation. However, significant impacts can be made on resource availability without additions. On many sites, increasing the availability of a site's existing resources by changing the quantity and/or quality of soil volume can have as much, or more, impact on production as adding resources.

6.3.1 Water control

Water control (drainage) can have major effects on the physical, chemical and biological properties and processes of the soil volume, through its impacts on the water table and depth of aeration (Chapter 7; Morris & Campbell 1991). The magnitude and duration of growth responses to water control depends on soil type, species, intensity and depth of ditching, and the level at which water is maintained. Gains of up 3–10 m in site index are possible with drainage to 50–60 cm on many poorly drained soils of the south-eastern USA (Morris & Campbell 1991). Similar gains have been reported for bogs in the boreal forest.

On soils with poor internal water movement (poorly structured clays, organic soils), ditch spacing of 100 m or closer may be needed. Survival and early growth of planted seedlings is such a problem on poorly drained soils that in addition to drainage, bedding or mounding is almost universally applied on these site types. If drainage is not used, bedding or mounding is essential to establish most species.

6.3.2 Tillage

Tillage (cultivation) treatments such as bedding, disking, subsoiling, ripping, or combination ploughing can have major effects on the physical, chemical and biological properties and processes of the soil volume through their impacts on soil depth, structure, strength, and aeration (Chapter 7; Morris & Lowery 1988). Tillage treatments can be very effective at ameliorating soil physical limitations (e.g. reducing soil strength and increasing aeration porosity) whether they occur naturally or result from past soil damage from equipment trafficking. Each of the tillage treatments provides a different volume and configuration of tilled soil. Understanding whether aeration and/or soil strength restrict root growth and where these restrictions occur is essential to effectively use these treatments. For example, bedding is typically used on poorly drained soils where aeration is a problem. By contrast, subsoiling is used where high soil strength in a subsurface horizon restricts root growth.

In addition to the direct effects of better soil aeration on root growth, aeration can also improve nutrient availability by increasing microbial activity and therefore mineralization rates. Increased aeration will also favour more oxidized and available forms of various nutrient elements. The positive effects of subsoiling on soil depth and consequently rooting volume can dramatically increase the amount of water and/or nutrients available to crop trees. In addition to individual tree growth gains, tillage treatments typically improve seedling survival and stand uniformity resulting in substantial gains in yield and value.

Height growth responses to bedding range from 1 to 3 m and are usually achieved early in the rotation (Gent *et al.* 1986). Responses are typically the greatest on poorly drained clays and diminish as soil texture grades from clay to loam to sand and as drainage improves. Growth responses to bedding do occur on well-drained sites, although gains of 1 m or more are limited to poorly structured clayey soils with thin surface horizons. Application of ripping, subsoiling and combination ploughing has recently increased. Ripping on shallow rocky soils has resulted in dramatic increases in survival and growth of planted pines (Fallis & Duzan 1998). Growth responses following subsoiling or combination ploughing are mixed, with responses very dependent on soil texture, structure, mineralogy and the timing of treatment. Gains of 1–1.5 m have been reported on well-drained and poorly structured clay soils or soils with a hardpan within 40 cm of the surface.

For effective manipulation of soil resources, it is critical to remember that the need for and effectiveness of tillage treatments is very dependent on soil and site conditions. Tillage treatments need to be applied at the appropriate soil moisture conditions. If a soil is too wet, more damage than good may be done. If a soil is too dry, the power requirements and therefore costs can exceed benefits. Potential losses of soil to erosion must also be considered during any tillage operation. Areas with slopes and loamy surface soils are particularly susceptible to erosion losses. If tillage is undertaken on these sites, care must be taken to till on the contour and leave strips of undisturbed soil.

6.3.3 Fire

Prescribed fire is widely used to reduce logging debris and slash after harvest to facilitate planter access and to reduce wildfire hazard. Burning can also have significant positive or negative impacts on resources, particularly nutrient availability. Hot site preparation fires have the potential to result in substantial volatile losses of nitrogen (200–400 kg ha^{-1}) and sulphur and ash losses of phosphorus and micronutrients (Flinn *et al.* 1979a; Vose & Swank 1993). On the positive side, fire may result in a short-term increase in the availability of many nutrients, especially potassium and phosphorus. The rapid early growth of many plantations following site preparation burning is the result of this increased availability of nutrients as well as control of competing vegetation.

6.3.4 Non-crop vegetation control

The allocation of site resources to crop trees can be increased by reducing the use of site resources by non-crop vegetation by chemical or mechanical means. Prior to planting, both mechanical (e.g. chopping, shearing, piling and tillage) and/or chemical site preparation treatments are widely used where woody vegetation such as sprouts and seedlings of tree species, or waxy leaved brush species (e.g. ericaceous shrubs), rapidly become competitors with the planted species. Over the last two decades, there has been a shift to greater use of chemical treatments as new chemicals and tank mixes have been registered that are more effective at reducing the sprouting of undesirable species, have less risk of erosion and soil damage, allow for the treatment of large areas in a short period of time, and can be less expensive. Mechanical treatments are still used where the use of herbicides is restricted and/or where debris removal or the need for soil tillage dictates the use of mechanical methods. Where competing woody vegetation is a problem and it is effectively controlled, substantial gains in survival, individual tree growth and stand yield have been realized (Clason 1993; Glover & Zutter 1993; Zutter *et al.* 1995). Analysis indicates that a negative exponential relationship exists between final crop tree yield and number of hardwood stems at a young age (Glover & Zutter 1993; Fig. 6.3).

Effective reduction of non-crop vegetative regrowth does not end with site preparation. In most subtropical and tropical areas, effective grass, herbaceous and brush control during the first year(s) of plantation establishment is absolutely essential for good survival and rapid early growth of the planted species. This is true for replanting of previously forest sites or afforestation of pastureland. In temperate and boreal regions, survival and early growth can also be dramatically improved. Several very effective pre- and post-emergent herbicides are presently available and their use has resulted in early

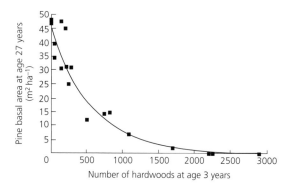

Fig. 6.3 Relationship between 27-year basal area yield and number of hardwood stems at age 3 for a loblolly pine plantation in the south-eastern USA. (From Glover & Zutter 1993.)

height growth gains from 1 to 2 m across a broad range of site and climatic conditions (Fortson *et al.* 1996; Lauer *et al.* 1993, Lowery *et al.* 1993; Richardson 1993). Even where height growth gains have not been observed, diameter growth gains can be substantial (Wagner *et al.* 1999). On most sites, the early growth gains found with weed control do not continue to increase over time; long-term results indicate that early gains have been maintained with treated and non-treated areas showing parallel growth projectories (Lauer *et al.* 1993; Mason & Milne 1999; Miller *et al.* 1995). This type of response has been described as Type B (Morris & Lowery 1988) or Type I (Snowdon & Waring 1984). On other sites, the early growth responses have been partially lost (Allen & Lein 1998) with time (Type C or Type III (Richardson 1993) responses). Clearly, an understanding of the resources that limit production and the effectiveness of the vegetation control treatments in ameliorating these limitations over the short and long term is needed to make projections.

Recent research has focused on the timing of vegetation control (Wagner *et al.* 1999), the quantity and composition of vegetation that needs control (Glover & Quicke 1999; Lauer & Glover 1999), and the size of the area needing vegetation control to optimize crop tree growth (Mason & Kirongo 1999). Generally spot or band treatments that control vegetation within a 4-m^2 area immediately surrounding the planted seedling have

been found to be quite effective for the first year. Depending on the composition and quantity of the competition vegetation, plantation spacing and the growth rate of the crop trees, additional release treatments may be needed in subsequent years to optimize growth.

6.3.5 Thinning

Thinning is a very common practice in plantations, particularly where products such as sawtimber, veneer or poles are desired rather than just pulpwood or fuelwood. One of the major objectives for thinning can be to increase the growth and value of the remaining crop trees. In this regard, thinning is very similar to vegetation control where the objective is to allocate site resources to crop trees. In overstocked stands, thinning is also essential to put a stand in a condition to respond to other treatments such as fertilization or pruning.

It is interesting to note that foresters typically consider light a major limitation to individual tree growth, especially in fully stocked stands. It is generally thought that crop trees respond to thinning because they receive more light. In fact, in fully stocked stands with low leaf areas ($<2.5\,m^2\,m^{-2}$), a situation typical of the south-eastern USA, crop tree responses to thinning are small. In stands with low leaf areas, the limitation for individual tree (and stand) production is not light but the ability to capture light (Gillespie *et al.* 1994). Low leaf areas are due to chronic nutrient limitations. What little responses crop trees exhibit to thinning may be due to the reallocation of the site's limited nutrient supply to the remaining crop trees. However, in these situations, nutrients will need to be added to increase leaf area and consequently growth.

In practice, thinning is frequently undertaken with objectives other than increasing crop tree growth. These objectives include providing intermediate wood flow and/or cash flow, reducing losses due to density-dependent mortality, reducing the susceptibility to insects, especially bark beetles (Nebeker *et al.* 1984), increasing forage production, and improving habitat for certain wildlife species. As a result, thinning prescriptions and practices are highly variable and range from a one-time mechanical removal of every other row in a plantation to careful selection of crop trees through a series of two or three selective thinning cycles.

6.3.6 Nutrient additions

Research has shown that nutrient limitations are very widespread and that nutrient additions are needed on most sites to even come close to achieving optimum rates of production. Much attention has been focused on ameliorating the gross and highly visual deficiencies of P and B. In many areas of the world, plantations of eucalyptus and pines routinely receive fertilizer applications that provide the equivalent 20–40 g of P and 1–2 g of B per seedling immediately before or soon after planting (Boomsma & Hunter 1990; Jokela *et al.* 1991; Herbert 1996). Where labour costs are low, fertilizers are typically applied by hand either on the surface, in holes, or in shallow trenches surrounding the seedlings. Where labour costs are high, fertilizers are applied by tractor (band or broadcast) or broadcast from the air. The sources of P that are used include diammonium phosphate (DAP), triple superphosphate (TSP), rock phosphate or NPK blends (e.g. 10:10:10). The benefits of early P fertilization on P-deficient soils typical of many coastal plain regions and soils derived from sandstones and rhyolites have long been recognized (Turner & Lambert 1988). Long-term volume gains range from 3 to over $10 \, m^3 ha^{-1} \, year^{-1}$ on severely deficient sites (Ballard 1972; Flinn *et al.* 1979b; Pritchett & Comerford 1982; Gent *et al.* 1986; Turner & Lambert 1988). Because the response to applications of $40–50 \, kg \, ha^{-1}$ P may last for 20 or more years (Fig. 6.4), P fertilization on deficient sites is viewed by many as an improvement in site quality, a Type A (Morris & Lowery 1988) or Type II (Snowdon & Waring 1984) response.

Identification of stands in need of early fertilization has been based on landscape/soil type, soil and foliar tests, and experience. With the advent of effective vegetation control, it is now apparent that early fertilization will improve growth even on sites that have not previously been considered nutrient deficient. The critical value for soil P below which a fertilizer response is typically seen

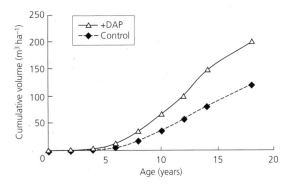

Fig. 6.4 Cumulative volume growth during the first 18 years following planting for a loblolly pine plantation with and without $280 \, kg \, ha^{-1}$ diammonium phosphate (DAP) applied at the time of planting.

is 5–8 p.p.m. depending on soil type and extraction procedure. Critical values for foliar P vary by species and range from 0.08 to 0.12% (Jokela *et al.* 1991; Judd *et al.* 1996).

By time of canopy closure, a plantation's potential to use nitrogen typically outstrips the available soil supply resulting in restricted leaf area development and growth (Allen *et al.* 1990; Fig. 6.2). Not surprisingly, strong and consistent responses have been found following N or N + P fertilization in mid to late rotations stands in the south-eastern USA (Allen *et al.* 1990), the Pacific North-west (Chappell *et al.* 1991), Australasia (Boomsma & Hunter 1990) and northern Europe (Möller 1992). Growth gains averaging 25–35% ($2–5 \, m^3 ha^{-1} \, year^{-1}$) over a 6- to 10-year period following single applications of 200 kg/ha N or 200 kg/ha N plus 25 kg/ha P are typical. In the south-east USA, both N and P are deficient with responses to additions of N + P much greater than the additive effects (synergistic) of the two elements applied alone (Allen *et al.* 1990). Mid to late rotation fertilization is typically a very attractive financial investment, with internal rates of return exceeding 15% or more. Until recently, fertilization has been delayed until after the first or second thinning to maximize returns on investment. This is now changing as the need for nutrients earlier in a plantation's life is recognized and increasing wood values have made even early fertilization a very attractive financial investment.

In the south-eastern USA alone, over 400 000 ha are treated annually in midrotation stands. This area is greater than the sum of midrotation fertilization in all other regions of the world.

In intermediate-aged stands, little response is typically observed to additions of P alone except on very P deficient sites as indicated by low foliar P concentrations and very low leaf areas. Applications of DAP at ≥200 kg ha^{-1} are typically used as the remedial treatment in these stands. Additions of N alone can actually be detrimental to growth on these severely P deficient sites.

The diagnostic techniques for identifying intermediate-aged stands that will be biologically responsive to fertilization have undergone substantial revision. In the past, stand attributes (e.g. basal area and site index), soil/landscape type, foliar concentrations and experience were used together to identify responsive stands and prescribe the appropriate elements and rates to apply. Unfortunately, the ability to predict the responsiveness of an individual stand was limited. This situation has now changed with the application of recent research that quantified the linkages among stand productivity, leaf area and nutrient availability (e.g. Albaugh *et al.* 1998). Differences between a stand's current leaf area and its potential leaf area are now being used to estimate responsiveness to nutrient additions. Leaf area of a fully stocked stand (basal area >20 m^2 ha^{-1}) should be 3.5 or greater; otherwise, the stand is probably in need of nutrients. The further leaf area is below this value, the greater the probability and magnitude of response.

6.4 INTERACTIONS AMONG TREATMENTS

Where two silvicultural treatments affect resource availability and/or the allocation of resources to the crop trees in a similar manner, their combined effects may be less than the sum of each individual effect (less than additive). For example, disking not only improves the rooting environment, but may also reduce competing hardwood vegetation. The expectation would be that the combination of disking and chemical control of hardwoods would yield less than the additive effects of the two treatments applied

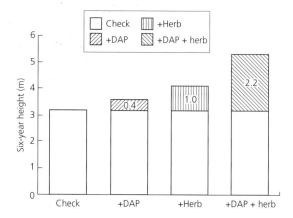

Fig. 6.5 Cumulative height of 6-year loblolly pine in the south-eastern USA following vegetation control (2 years' banded application of Velpar (herb)) and/or fertilization (280 kg ha^{-1} diammonium phosphate (DAP)) applied at the time of planting.

alone. By contrast, the effects of effective hardwood control and fertilization may be more than additive as, without control, hardwoods will also respond to fertilizer resulting in increased competition for light and water resources with pine crop trees (Fig. 6.5).

It is clear that obtaining optimum plantation production will require the use of integrated systems that couple intensive management of site resources and genetics. Much research is now underway to understand the interactions among silvicultural treatments and genetics. The beneficial effects of improved genetics and intensive culture appear to be at least additive (McKeand *et al.* 1997). Clonal forestry also brings new challenges. Procedures for selecting and deploying clones must be optimized with an understanding of how to effectively take advantage of genetic–environment interactions that are so often found with clones.

6.5 FUTURE OPPORTUNITIES

The potential growth rates for forest plantations are very high, much higher than commonly thought just a few years ago. In the last 5 years, our expectations have increased dramatically. Our challenge now is to develop and implement the appropriate silvicultural systems to realize this potential in a cost-effective and environmentally sustainable way. To be successful will require a basic understanding of how resource availability limits forest production and how crop trees may differ in their ability to acquire and utilize these resources. Key challenges from a resource management perspective include: understanding the relative contributions of water and nutrient limitations to stand productivity across a range of site and stand developmental conditions, assessing resource limitations that are constantly changing, identifying sites where nutrients other than N and P are limiting growth, developing the appropriate dosage–frequency prescriptions to ameliorate these limitations, pest and disease issues with intensive culture, understanding the impacts of intensively managed plantations within a landscape context, developing people capable of making the site-specific prescriptions, and securing the capital sufficient to implement those prescriptions.

REFERENCES

Albaugh, T.J., Allen, H.L., Dougherty, P.M., Kress, L.W. & King, J.S. (1998) Leaf-area and above- and below-ground growth responses of loblolly pine to nutrient and water additions. *Forest Science* **44**, 317–28.

Allen, H.L. & Lein, S. (1998) Effects of site preparation, early fertilization, and weed control on 14-year old loblolly pine. *Proceedings of the Southern Weed Science Society* **51**, 104–10.

Allen, H.L., Dougherty, P.M. & Campbell, R.G. (1990) Manipulation of water and nutrients—practice and opportunity in southern U.S. pine forests. *Forest Ecology and Management* **30**, 437–53.

Allen, H.L., Weir, R.J. & Goldfarb, B. (1998) Investing in wood production in southern pine plantations. *Paper Age* (April 1998), 20–1.

Axelsson, E. & Axelsson, B. (1986) Changes in carbon allocation patterns in spruce and pine trees following irrigation and fertilization. *Tree Physiology* **2**, 189–204.

Ballard, R. (1972) Influence of a heavy phosphate dressing and subsequent radiata pine response on the properties of a Riverhead clay soil. *New Zealand Journal of Forest Science* **2**, 202–16.

Benson, R.E., Myers, B.J. & Raison, R.J. (1992) Dynamics of stem growth of Pinus radiata as affected by water and nitrogen supply. *Forest Ecology and Management* **52**, 117–38.

Bergh, J., McMurtrie, R.E. & Linder, S. (1998a) Climatic factors controlling the productivity of Norway spruce: a model-based analysis. *Forest Ecology and Management* **110**, 127–39.

Bergh, J., Linder, S., Lundmark, T. & Elfving, B. (1998b) The effect of water and nutrient availability on the productivity of Norway spruce in northern and southern Sweden. *Forest Ecology and Management.* **110**, 51–62.

Binkley, D., Carter, R. & Allen, H.L. (1995) Nitrogen fertilization practices in forestry. In: Bacon, P., ed. *Nitrogen Fertilization in the Environment*, chap. 11. Marcel Dekker, New York.

Boomsma, D.B. & Hunter, I.R. (1990) Effects of water, nutrients, and their interactions on tree growth, and plantation forest management practices in Australasia: a review. *Forest Ecology and Management* **30**, 455–76.

Cannell, M.G.R. (1989) Physiological basis of wood production: a review. *Scandinavian Journal of Forestry Research* **4**, 459–90.

Chappell, H.H., Cole, D.W., Gessel, S.P. & Walker, R.B. (1991) Forest fertilization research and practice in the Pacific Northwest. *Fertilizer Research* **27**, 129–40.

Clason, T.R. (1993) Hardwood competition reduces loblolly pine plantation productivity. *Canadian Journal of Forest Research* **23**, 2133–40.

Colbert, S.R., Jokela, E.J. & Neary, D.G. (1990) Effects of annual fertilization and sustained weed control on dry matter partitioning, leaf area, and growth efficiency of juvenile loblolly and slash pine. *Forest Science* **36**, 995–1014.

Dougherty, P.M. & Gresham, C.A. (1988) Conceptual analysis of southern pine plantation establishment and early growth. *Southern Journal of Applied Forestry* **12**, 160–6.

Evans, J. (1992) *Plantation Forestry in the Tropics.* Clarendon Press, Oxford.

Fallis, F. & Duzan, H.W., Jr (1998) Effects of ripping (deep subsoiling) on loblolly pine plantation establishment and growth—nineteen years later. In: *Proceeding of the Eighth Biennial Southern Silvicultural Research Conference*. General Technical Report SRS-1, pp. 525–9. USDA Forest Service, Washington, DC.

Flinn, D.W., Hopmans, P., Farrell, P.W. & James, J.M. (1979a) Nutrient loss from burning Pinus radiata logging residue. *Australian Forest Research* **9**, 17–23.

Flinn, D.W., Moller, I. & Hopmans, P. (1979b) Sustained growth responses to superphosphate applied to established stands of *Pinus radiata. New Zealand Journal of Forest Science* **9**, 201–11.

Fortson, J.C., Shiver, B.D. & Shackelford, L. (1996) Removal of competing vegetation from established loblolly pine plantations increases growth on Piedmont and Upper Coastal Plain sites. *Southern Journal of Applied Forestry* **20**, 188–93.

Gent, J.A., Allen, H.L., Campbell, R.G. & Wells, C.G. (1986) Magnitude, duration, and economic analysis of loblolly pine growth response following bedding and phosphorus fertilization. *Southern Journal of Applied Forestry* **10**, 124–8.

Gholz, H.L., Ewel, K.C. & Teskey, R.O. (1990) Water and forest productivity. *Forest Ecology and Management* **30**, 1–18.

Gillespie, A.R., Allen, H.L. & Vose, J.M. (1994) Effects of canopy position and silvicultural treatment on the amount and vertical distribution of foliage in individual loblolly pine trees. *Canadian Journal of Forest Research* **24**, 1337–44.

Glover, G.R. & Quicke, H. (1999) Growth response of loblolly pine, sweetgum, and water oak in a pine-hardwood density study. *Canadian Journal of Forest Research* **29**, 968–78.

Glover, G.R. & Zutter, B.R. (1993) Loblolly pine and mixed hardwood stand dynamics for 27 years following chemical, mechanical, and manual site preparation. *Canadian Journal of Forest Research* **23**, 2126–32.

Gower, S.T., Vogt, K.A. & Grier, C.C. (1992) Carbon dynamics of Rocky Mountain Douglas-fir: influence of water and nutrient availability. *Ecological Monographs* **62**, 43–65.

Gower, S.T., Gholz, H.L., Nakane, K. & Baldwin, V.C. (1994) Production and allocation patterns of pine forests. In: Gholz, H.L., Linder, S. & McMurtrie, R.E., eds. *Environmental Constraints on the Structure and Productivity of Pine Forest Ecosystems: A Comparative Analysis. Ecological Bulletin* **43,** 115–35.

Haynes, B.E. & Gower, S.T. (1995) Belowground carbon allocation in nonfertilized and fertilized red pine plantations in northern Wisconsin. *Tree Physiology* **15**, 317–25.

Hennessey, T.C., Dougherty, P.M., Cregg, B.M. & Wittwer, R.F. (1992) Annual variation in needlefall of a loblolly pine stand in relation to climate and stand density. *Forest Ecology and Management* **51**, 329–38.

Herbert, M. (1996) Fertilizers and eucalypt plantations in South Africa. In: Attiwill, P.M. & Adams, M.A., eds. *Nutrition of Eucalypts*, pp. 303–25. CSIRO Publishing, Australia.

Jokela, E.J., Allen, H.L. & McFee, W.W. (1991) Fertilization of southern pines at establishment. In: Duryea, M. & Dougherty, P., eds. *Forest Regeneration Manual*, chap. 14. Kluwer Academic Publishers, Netherlands.

Judd, T.S., Attiwill, P.M. & Adams, M.A. (1996) Nutrient concentration in *Eucalyptus*: a synthesis in relation to differences between taxa, sites, and components. In: Attiwill, P.M. & Adams, M.A., eds. *Nutrition of Eucalypts*, pp. 123–53. CSIRO Publishing, Australia.

Kimmins, J.P. (1997) *Forest Ecology*, 2nd edn. Prentice Hall, New York.

Landsberg, J.J. & Gower, S.T. (1997) *Application of Physiological Ecology to Forest Management.* Academic Press, London.

Lauer, D.K. & Glover, G.R. (1999) Stand level response to occupancy of woody shrub and herbaceous vegetation. *Canadian Journal of Forest Research* **29**, 979–84.

Lauer, D.K., Glover, G.R. & Gjerstad, D.H. (1993) Comparison of duration and method of herbaceous weed control on loblolly pine response through midrotation. *Canadian Journal of Forest Research* **23**, 2116–25.

Linder, S. (1987) Responses to water and nutrients in coniferous ecosystems. In: Schulze, E.D. & Wolfer, H.Z., eds. *Potentials and Limitations of Ecosystems Analysis. Ecology Studies*, 61, pp. 180–202. Springer-Verlag, Berlin.

Linder, S., Benson, M.L., Myers, B.J. & Raison, R.J. (1987) Canopy dynamics and growth of *Pinus radiata*. I. Effects of irrigation and fertilization during a drought. *Canadian Journal of Forest Research* **17**, 1157–65.

Lowery, R.F., Lambeth, C.C., Endo, M. & Kane, M. (1993) Vegetation management in tropical forest plantations. *Canadian Journal of Forest Research* **23**, 2006–14.

McKeand, S.E., Crook, R. & Allen, H.L. (1997) Genetic stability on predicted family responses to silvicultural treatments in loblolly pine. *Southern Journal of Applied Forestry* **21**, 84–9.

Mason, E.G. & Kirongo, B. (1999) Responses of radiata pine clones to varying levels of pasture competition in a semiarid environment. *Canadian Journal of Forest Research* **29**, 934–9.

Mason, E.G. & Milne, P.G. (1999) Effects of weed control, fertilization, and soil cultivation on growth of *Pinus radiata* at midrotation in Canterbury, New Zealand. *Canadian Journal of Forest Research* **29**, 985–92.

Miller. J.M., Busby, R.L., Zutter, B.R., Zedaker, S.M., Edwards, M.B. & Newbold, R.A. (1995) Response of loblolly pine to complete woody and herbaceous control: projected yields and economic outcomes—the COMPROJECT. In: *Proceedings of the Eighth Biennial Southern Silvicultural Research Conference.* General Technical Report SRS-1, pp. 81–9. USDA Forest Service Washington D.C.

Möller, G. (1992) The Scandanavian experience in forest fertilization research and opportunities. In: Chappell, H.N., Weetman, G.F. & Miller, R.E., eds. *Forest Fertilization: Sustaining and Improving Nutrition and Growth of Western Forests*, chap. 23. Institute of Forest Resources Contribution 72. University of Washington, Morris, LA.

Morris, L.A. & Campbell, R.G. (1991) Soil and site potential. In: Duryea, M. & Dougherty, P., eds. *Forest*

Regeneration Manual, chap. 10. Kluwer Academic Publishers, The Netherlands.

Morris, L.A. & Lowery, R.F. (1988) Influences of site preparation on soil conditions affecting stand establishment and tree growth. *Southern Journal of Applied Forestry* **12**, 170–8.

Murthy, R., Dougherty, P.M., Zarnoch, S.J. & Allen, H.L. (1996) Effects of elevated CO_2, nitrogen and water on net photosynthesis and foliar nitrogen concentration of loblolly pine trees. *Tree Physiology* **16**, 537–46.

Nebeker, T.E., Hodges, J.D., Karr, B.K. & Moehring, D.M. (1984) *Thinning Practices in Southern Pines – With Pest Management Recommendations*. USDA Forest Service Technical Bulletin 1703. United States Department of Agriculture, Washington D.C.

Pereira, J.S., Madeira, M.V., Linder, S., Ericsson, T., Tomé, M. & Araújo, M.C. (1994) Biomass production with optimised nutrition in *Eucalyptus globulus* plantations. In: Pereira, J.S. & Pereira, H., eds. *Eucalyptus for Biomass Production*, pp. 13–30. Commission of the European Communities Brussels.

Pritchett, W.L. & Comerford, N.B. (1982) Long-term response to phosphorus fertilization on selected Southeastern Coastal Plain soils. *Soil Science Society of America Journal* **46**, 640–4.

Richardson, B. (1993) Vegetation management practices in plantation forests of Australia and New Zealand. *Canadian Journal of Forest Research* **23**, 1989–2005.

Sampson, D.A. & Allen, H.L. (1999) Regional influences of soil available water, climate, and leaf area index on simulated loblolly pine productivity. *Forest Ecology and Management* **124**, 1–12.

Snowdon, P. & Benson, M.L. (1992) Effects of combinations of irrigation and fertilisation on the growth and above-ground biomass production of *Pinus radiata*. *Forest Ecology and Management* **52**, 87–116.

Snowdon, P. & Waring, H.D. (1984) Long-term nature of growth responses obtained to fertilizer and weed control applied at planting and their consequences for forest management. In: *Proceedings IUFRO Symposium on Site and Productivity of Fast Growing Plantations*, 2, pp. 701–11. South African Forest Research Institute, Pretoria, South Africa.

Stow, T.K., Allen, H.L. & Kress, L.W. (1992) Ozone impacts on seasonal foliage dynamics of young loblolly pine. *Forest Science* **38**, 102–19.

Teskey, R.O., Bongarten, B.C., Cregg, B.M., Dougherty, P.M. & Hennessey, T.C. (1987) Physiology and genetics of tree growth response to moisture and temperature stress: an examination of the characteristics of loblolly pine (*Pinus taeda* L.). *Tree Physiology* **3**, 41–61.

Turner, J. & Lambert, M.J. (1988) Long-term effects of phosphorus fertilization on forests. In: Cole, D.W. & Gessel, S.P., eds. *Forest Site Evaluation and Long-Term Productivity*, chap. 13. University of Washington Press, Seattle.

Vogt, K.A., Grier, C.C. & Vogt, D.J. (1986) Production, turnover, and nutrient dynamics of above- and below-ground detritus of world forests. *Advances in Ecology Research* **15**, 303–77.

Vose, J.M. & Allen, H.L. (1988) Leaf area, stemwood growth and nutrient relationships in loblolly pine. *Forest Science* **34**, 547–63.

Vose, J.M. & Swank, W.T. (1993) Site preparation burning to improve southern Appalachian pine-hardwood stands: aboveground biomass, forest floor mass, and nitrogen and carbon pools. *Canadian Journal of Forest Research* **23**, 2255–62.

Wagner, R.G., Mohammed, G.H. & Noland, T.L. (1999) Critical period of interspecific competition for northern conifers associated with herbaceous vegetation. *Canadian Journal of Forest Research* **29**, 890–7.

Zutter, B.R., Miller, J.H., Zedaker, S.M., Edwards, M.B. & Newbold, R.A. (1995) Response of loblolly pine plantations to woody and herbaceous control – eighth-year results of the region-wide study – the COMPROJECT. In: *Proceedings of the Eighth Biennial Southern Silvicultural Research Conference*. General Technical Report SRS-1, pp. 75–80. USDA Forest Service, Washington, DC.

7: Sustainable Management of Soil and Site

LAWRENCE A. MORRIS

7.1 INTRODUCTION

There exists no universally agreed upon definition of sustainable soil management. Most
authorities are concerned with the capacity of
forests to produce goods and services (*sensu*
Powers *et al.* 1990; Morris & Miller 1994;
Kimmins 1996; Malcolm & Moffat 1996;
Nambiar 1996) and define sustainable soil
management in a manner similar to the
following:

> Sustainable soil management provides for
> the continued or improved capacity of the
> soil to provide an environment suitable for
> plant root growth and to supply water and
> essential plant elements at rates and in ratios
> necessary to meet tree growth demands and
> to produce a nondiminishing supply of wood
> and other forest products.

These production-centred definitions of sustainability are subject to criticism. Drengson and
Taylor (1997) have unflatteringly referred to such
definitions as 'sustainable exploitation'. These
authors have a point. While sustained production
is one important aspect of sustainable soil management, it is important to recognize that soil has
other important functions. For instance, recalcitrant organic compounds in the soil are responsible for long-term storage of carbon (C) that helps
reduce atmospheric C and the potential for global
warming. Soils have the capacity to bind contaminants and serve as filters for the water we drink.
Soils provide habitat for a myriad of organisms
which is significant in itself, but also important
from an anthropocentric viewpoint because these
organisms are important in C and nutrient
cycling and some have the capacity to break

bonds of toxic compounds and convert them into
harmless compounds.

Despite criticism of production-focused definitions of sustainability, measures of forest production provide a tractable means for determining
when soil change is unacceptable (for a discussion
see Morris & Miller 1994; Worrell & Hampson
1997). Thus, in this chapter, discussion is focused
on developing an understanding of soil requirements for tree growth and how characteristics
related to these requirements can be degraded or
improved by management activities. Other issues
related to sustainable soil and site management
are also discussed, but this discussion is within
the context of the capacity of a soil to function as
an environment for plant growth.

7.2 REQUIREMENTS FOR SUSTAINABLE SOIL MANAGEMENT

From a plant growth standpoint, the soil supplies
13 essential plant nutrients (N, P, K, Ca, Mg, S, Fe,
B, Mn, Cu, Zn, Mo and Cl) and water in an environment that provides anchorage and in which
roots can function. With careful control, such an
environment can be created in an aerated solution. Maintaining such a system long enough to
grow a tree would require constant monitoring
and enormous inputs—but it could be done. Such
a system would have little room for error. If essential elements were not added in the correct
amounts or if aeration was suspended, the system
would fail.

By contrast, productive soil is an environment
for root growth that provides adequate water
supply and enough air-filled pore space to allow
oxygen to diffuse to, and carbon dioxide to diffuse

away from, growing roots. It contains large amounts of stored nutrients and releases these nutrients to the soil water where they are available for plant uptake at rates and in ratios needed to support plant growth. This environment is maintained by activities of soil fauna that burrow, dig, transport and bind individual soil particles into a complex matrix of individual particles, aggregates of individual particles, and a range of microscopic to macroscopic pores. In the absence of human influences that degrade the soil through excessive nutrient removal, accelerated soil erosion, physical damage through use of heavy equipment or introduction of contaminants, a productive soil can sustain forest growth indefinitely.

Clearly, a completely artificial system that requires constant input cannot be considered sustainable. Nor can we define sustainability in such a manner so as to preclude any effect on soils and/or removal of products for human use. Rather, sustainable soil management must be evaluated within the context of management objectives and the capacity of the soil to recover following some impact. From the standpoint of the potential of soil to support forest growth, specific criteria for sustainable soil management are as follows:

1 *The volume of soil with physical characteristics suitable for root growth must be maintained.* Limitations to this volume are a function of mechanical resistance; the movement of oxygen to, and carbon dioxide away from, root surfaces; water availability and soil depth. Adverse effects on soil physical conditions that directly reduce the total volume of soil which roots can exploit or the ability of roots to proliferate within rooted soil volume will result in productivity loss.

2 *Nutrients must be supplied at non-diminishing rates and in ratios appropriate for plant growth.* Even the most intensively managed forests rely on the soil as a source of most nutrients. Thus, sustainable management requires that the capacity of the soil to store and release nutrients must be maintained. To ensure this long-term supply of essential plant elements, a balance between removals and inputs from atmospheric deposition and mineral weathering must exist.

3 *Contamination by acids, metals, salts or pesticides through management activities or the abuse of forests for waste treatment must be avoided.* Clearly, management should not result in the accumulation of contaminants beyond levels that impact plant growth. Also, since contamination may also occur as a result of long-range transport of contaminants to the forest (e.g. burning of fossil fuels), sustainable soil management may require correction for impacts not directly associated with forest management.

4 *Processes involving a broad range of soil biota are responsible for maintaining nutrient supply and favourable soil physical conditions and these processes must be maintained.* Sustainable management requires that important soil processes are not degraded during forest management to the extent that biotic communities cannot function and recover to conditions similar to those in natural forests should active management cease.

This last point requires some elaboration. Worldwide there has been a trend to remove areas of forest from production and reserve them for other uses (Chapter 2). The inevitable consequence of this is to increase production pressures on industrial and privately owned forestlands that are managed primarily for wood production. The high levels of wood production that will be required from these lands cannot be met without significant inputs and some change in soil characteristics. The requirement that soil processes are not degraded during forest management does not mean that no change in process occurs, only that if active management were suspended, the site would return to the original level of productive capacity supported by a functioning soil system. Examples of soils that have been degraded beyond this point include millions of hectares of agricultural land in Africa, where poor agricultural practices have resulted in excessive nutrient removal and continued erosion that is slowly reducing these soils to almost inert systems (Eswaran *et al.* 1997). It is worth noting that, for these lands, establishing or maintaining a forest cover is considered an important component of sustainable soil management (Eswaran *et al.* 1993).

It should also be clearly stated that sustainable soil management is only one component of sustainable forest management. Although most definitions of sustainable forest management include

Table 7.1 Criteria for sustainable management of forests proposed by three different initiatives. (Note that criteria related to sustained production as well as to maintaining soils are included as a separate criterion in all three initiatives.) (Adapted from Brand 1998.)

No.	Montreal Process (non-European boreal and temperate forests)	Helsinki Process (Europe)	Amazonian Process (tropical forests)*
1	Conservation of biological diversity	Maintenance and appropriate enhancement of forest resources and their contribution to global carbon cycles	Socioeconomic benefits
2	Maintenance of the productive capacity of forest ecosystems	Maintenance of forest ecosystem health and vitality	Policies and legal–institutional framework for sustainable development of forests
3	Maintenance of forest ecosystem health and vitality	Maintenance and encouragement of productive functions of forests (wood and non-wood)	Sustainable forest production
4	Conservation and maintenance of soil and water resources	Maintenance, conservation and appropriate enhancement of biological diversity in forest ecosystems	Conservation of forest cover and biological diversity
5	Maintenance of forest contribution to global carbon cycles	Maintenance and appropriate enhancement of protective functions in forest management (notably soil and water)	Conservation and integrated management of water and soil resources
6	Maintenance and enhancement of long-term multiple socioeconomic benefits to meet the needs of society	Maintenance of other socioeconomic functions and conditions	Science and technology for the sustainable development of forests
7	Legal, institutional and economic framework for forest conservation and sustainable management		Institutional capacity to promote sustainable development in Amazonia

* National level criteria.

conservation of soil resources, other components such as protection of water resources, maintenance of biological diversity, contributions to global ecology and societal benefits are equally important (Table 7.1).

7.3 SOIL CHARACTERISTICS AFFECTING ROOT GROWTH, WATER AND NUTRIENT SUPPLY

7.3.1 Soil as an environment for root growth

Three physical factors are directly related to root growth:

1 mechanical resistance to penetration;
2 soil water potential;
3 gas diffusion rates that control oxygen availability and concentrations of toxic metabolites near roots.

Changes in these physical factors also influence root growth indirectly through their effect on microbial activity, decomposition rate, nutrient transformations, activity of organisms that promote the formation of soil structure, and temperature.

Large continuous macropores provide a path for root growth even in soils with high mechanical resistance. In the absence of such pores, root growth depends upon the ability of roots to enter and expand small pores. Mechanical resistance, measured as the force required to push a metal probe through the soil at a specified rate, provides the most commonly used approximation of mechanical resistance to root penetration.

If other limitations do not occur, root growth and extension is greatest at minimal resistance and generally declines linearly with increased mechanical resistance up to a maximum resistance above which growth ceases or is confined to large pores. Tree root growth restriction begins at resistances as low as 0.5 MPa and stops at resistances between 2.0 MPa and 3.0 MPa The relationship between penetrometer resistance and root growth is relatively consistent across a range of plant species and soil textures (Fig. 7.1).

Water availability is a second factor directly affecting root growth. Costantini *et al.* (1996) studied root elongation of Caribbean pine (*Pinus caribaea*) under conditions of low mechanical resistance. Across the range of soil moisture potentials ranging from near field capacity to dry soils, root elongation declined linearly to near zero. Ludovici and Morris (1997) and Torreano and Morris (1998) evaluated root growth of loblolly pine (*Pinus taeda* L.) seedlings grown in rhizotrons filled with an artificial medium of fine sand and fritted clay that provided low resistance at all moisture contents (maximum resistance <0.5 MPa). Maximum root growth occurred near field capacity and declined linearly as water

potential declined to –2.2 MPa. (Fig. 7.2). Note that although the relationship was linear in both of these studies, absolute growth rates varied. Ludovici and Morris (1997) reported lower growth rates at equivalent potentials than those reported by Torreano and Morris (1998), apparently because of nutrient limitations caused by competing plant species. In the former study, reductions in water availability were partly due to competing plant water use while in the latter study no competing plants were present.

Excess soil water does not directly injure tree roots; roots grow vigorously in well-oxygenated solution. Adverse effects of poor aeration are the result of insufficient air-filled pores to transfer oxygen to respiring roots and carbon dioxide and potentially toxic wastes away from roots. Diffusion of oxygen through soils is 3×10^6 slower in water-filled than air-filled pores (Drew 1979), thus, the reduction in macropore volume (pores that drain under gravity and that are air filled at field capacity) results in decreased diffusion and is a major consequence of forest harvest or other activities that compact the soil. Generally, oxygen diffusion rates are sufficient to support uninhibited root growth in soils with air-filled porosity greater than 10–12% for both crop and tree species (Vomocil & Flocker 1961; Mukhtar *et al.* 1988; Simmons & Pope 1988), leading to a generally accepted critical value for aeration porosity (air-filled porosity at field capacity) of 10%. Activities that reduce aeration porosity below 10% can

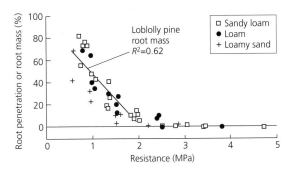

Fig. 7.1 Root penetration of cotton (*Gossypium hirsutum* L.) on soils of three textures (symbols) as a function of resistance to penetration. Root mass of loblolly pine in a coarse-textured soil is superimposed on these data. (Redrawn from Taylor *et al.* 1966; Torreano 1992.)

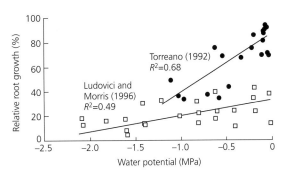

Fig. 7.2 Relative elongation of loblolly pine roots in coarse-textured soils of low mechanical resistance as a function of soil water potential. (Source: Ludovici & Morris 1997.)

be expected to reduce root growth and site productivity.

7.3.2 Nutrient supply

Essential elements are absorbed by plants largely in their inorganic forms from the soil solution (although some elements can be absorbed in soluble low molecular weight organic compounds). Factors affecting the release of each essential element are unique. Generally, however, essential elements can be grouped into those that:

1 occur largely as a component of organic matter and whose availability is determined largely by processes affecting organic matter decomposition;

2 are elements held on the soil exchange complex in large amounts and whose availability tends to be controlled by exchange reactions;

3 are elements that occur as solid compounds and whose availability is largely determined by conditions that affect the solubility of these compounds (Table 7.2).

Decomposition and mineralization

A complex process of decomposition and element mineralization occurs before elements bound in organic matter are released into soil water in inorganic forms (Fig. 7.3). Fungi that have the capability of attacking intact cell walls first invade litter falling to the forest floor. Following this fungal invasion, compounds that protect the plant tissue from pathogen and insect attack are leached, and larger soil fauna begin feeding. This process breaks large tissues into small pieces with high surface area that are more easily decomposed by soil microbes. It is during this stage of decomposition, as the amount of carbon available as an energy source declines relative to limiting nutrients in the tissue (usually N), that net mineralization occurs and nutrients are released into soil water where they are available for use by higher plants.

Table 7.2 Grouping of essential plant nutrients by the major soil factors and processes affecting their availability for root uptake.

Nutrient (available forms in soil solution)	Retention and availability in soil	Comment(s)
N (NH_4^+, NO_3^-)	Mineralization from organic matter and immobilization in microbial biomass	Most N is stored in organic forms, availability depends on microbial activity, substrate (C:N ratio), temperature and moisture
S (SO_4^{2-}), Cl^-	Anion exchange and specific adsorption	Sulphur is specifically adsorbed by Fe and Al hydroxides, organic matter can hinder sorption by coating or competing for sites, oxidation and reduction reactions are important
K^+, Ca^{2+}, Mg^{2+}; Al^{3+}, H^+	Cation exchange	Substantial quantities held on cation exchange slowly replaced by mineral weathering; as pH decreases and Al and H on cation exchange increases, cation reserves will decrease
P ($H_2PO_4^-$, HPO_4^{2-}), B (H_3BO_3, $B(OH)_4^-$, $H_2BO_3^-$), Mo (MoO_4^{2-}, $HMoO_4^-$)	Solubility	Relatively insoluble oxyanions; solubility is strongly pH dependent because of potential to form precipitation products with other species in solution
Fe^{2+}, Fe^{3+}, Mn^{2+}, Zn^{2+}, Cu^{2+}	Solubility and chelation	Micronutrients that occur in low concentrations, pH-dependent formation of low solubility hydroxides, may form more soluble organic chelates; oxidation and reduction reactions important

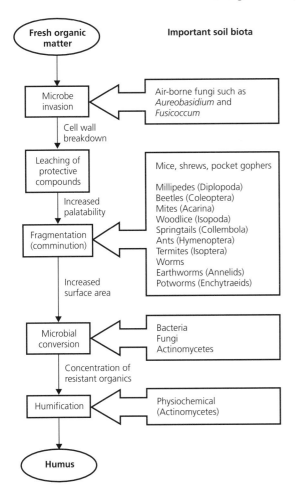

Fig. 7.3 Steps in the decay of leaf and branch litter falling to the forest floor and major groups of organisms associated with each step.

As illustrated in Fig. 7.3, this process involves the activities of numerous soil organisms and, potentially, could be affected by changes that affect even one species of organism. Changes in soil temperature and moisture affect overall faunal and microbial activity and can shift habitat suitability such that different groups of organisms are favoured. The introduction of contaminants can eliminate sensitive species. Such shifts in populations or reductions in diversity of soil biota could have ramifications for element supply and forest sustainability. Studies that have eliminated groups of organisms (fungi, bacteria or soil fauna) have shown that decomposition and mineralization are reduced (Coleman *et al.* 1996). However, a relatively high level of redundancy appears to exist in the system and reduction in the activity or loss of a single species seems unlikely to affect long-term nutrient supply. We do not have enough information to know the limits of system flexibility.

Exchange capacity

Ca, K and Mg belong to a group of elements that are held on charged soil surfaces in major quantities. Plant availability of these elements is controlled by the total number of charged sites available in the soil and the ratios at which different elements occur on these sites. Two types of exchange sites are significant. The first is associated with crystalline inorganic particles (primarily clays) and results from the substitution of elements of different valence into the structure of the particle. The charges associated with these exchange sites are permanent features of the soil. The second type of exchange site is associated with organic matter or coatings of inorganic oxides. These materials can develop charged surfaces capable of retaining ions through the disassociation of H^+ and OH^-. The degree that this occurs depends upon the concentrations of H^+ in soil solution (pH) and, thus, this type of exchange is termed variable charge or pH-dependent charge.

Long-term changes in the supply of nutrients from the exchange complex can occur in one of three ways.

1 through a change in the total number of potential exchange sites, such as might occur in soils that have a major portion of exchange associated with organic matter;

2 through a change in soil acidity that affects the number of variable charge sites;

3 through a change in the relative amount of elements on the exchange sites.

For example, cultivation or fertilization of coarse-textured (sandy) soils with N can stimulate organic matter decomposition that could lead to reductions in exchange sites associated with organic matter, ultimately leading to reduced availability of Ca, K or other exchangeable nutrients. Acidic deposition can alter both the number of pH-dependent exchange sites and the proportion of sites occupied by non-essential H^+ and Al compounds.

Fixation and solubility

The capacity of many soils to supply the macronutrient P and essential micronutrients is largely determined by the relative solubility of compounds containing these elements. For P, the concentration in soil water is a function of the solubility of P-containing precipitation products with Fe, Al and Ca, giving rise to the well-known relationship between soil solution P and pH illustrated in Fig. 7.4. At low pH, P in solution is sorbed to Fe and Al oxide surfaces; at high pH, P in solution will form low-solubility products with Ca. Studies of P availability in forests have shown these reactions to be greatly affected by interactions with soluble organic acids formed during the decomposition of forest floor litter. For instance, Fox *et al.* (1990) showed that P released from spodic horizons of slash pine forests (*Pinus elliottii* Engelm.) was greatly increased by a number of low molecular weight organic acids. Comerford and Skinner (1989) similarly demonstrated that addition of the low molecular weight organic acids oxalate and citrate increased P availability in radiata pine (*Pinus radiata* D. Don) forests in New Zealand. Thus, we can also conclude that changes in species composition and associated changes in the production of these organic acids can influence P supply.

Micronutrient cations (Fe^{3+}, Mn^{2+}, Cu^{2+} and Zn^{2+}) are generally unavailable at high pH because they form low-solubility products with major base cations in solution. The micronutrients B and Mo occur as oxyanions (anions formed in combination with oxygen) that have low solubility at low pH. As is the case with P, the availability of these nutrients is greatly affected by the concentrations of soluble organics in solution. They can bind with two or more organic groups to form stable ring structures that allow them to remain in solution at pH values above those at which the formation of insoluble precipitation products would normally have occurred. This increases availability to plants.

All of the aforementioned nutrients occur in organic matter and are subject to the same type of mineralization process illustrated for N in Fig. 7.3; however, for these nutrients we do not normally think of mineralization as the process that controls availability to plants. This is because material mineralized from organic matter and released into soil solution is subject to the aforementioned solubility reactions and, regardless of how much mineralization occurs, soil solution concentrations are controlled by these reactions. In sandy soils with low exchange and few minerals, mineralization may be the controlling process and the supply of P and micronutrients will be subject to the same sorts of perturbations as N and may be affected by human impacts in similar ways.

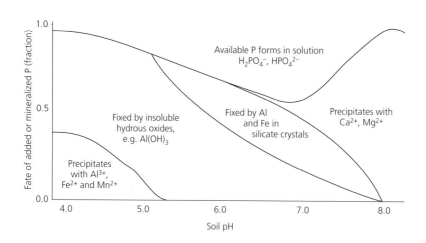

Fig. 7.4 The relationship between soil pH and P availability in mineral soils. (Redrawn from Buckman & Brady 1969.)

Organic matter

Because of its role as a reservoir and source of nutrients, energy source for soil biota, stabilizer of soil structure and contributor to water-holding capacity, maintenance of soil organic matter is often considered a requisite feature of sustainable soil management. In all systems, soil organic matter content tends to converge to a constant that represents a dynamic equilibrium among the addition of fresh organic material to the soil pool and decomposition, leaching or erosion from the soil. The quality and composition of organic matter additions, climate, soil water regime, clay content and availability of essential plant nutrients all contribute to determining this equilibrium content.

The role that organic matter plays in soil productivity is crucial (see Volume 1, Chapter 7); however, despite its potential contribution to many factors affecting soil productivity, the direct relationship between total soil organic matter and tree-growth-based measures of site productivity is poor. Experiments on agricultural crop productivity have been continuously conducted at Rothamsted for over 150 years and have been recently summarized by Poulton (1996). These long-term results show that with adequate fertilizer inputs and low erosion rates, yields on plots with low organic matter can equal those with high levels of organic matter maintained through the application of farmyard manure. However, Poulton points out that under similar fertilization

regimes, growth is greater on soils with organic matter. Equivalent long-term resear not available in production forests. Instea researchers have attempted to evaluate the influence of organic matter and other soil properties by sampling growth and numerous soil properties on many sites and then estimating the importance of specific soil properties to growth through correlation techniques. Carmean (1975) summarized more than 150 such attempts to establish relationships between soil properties, including soil organic matter measurements and forest site quality in the USA, between 1950 and the early 1970s. Although organic matter content was an important variable in a number of studies, few of the relationships were strong. This does not imply that organic matter is unimportant, only that such approaches ignore the complex interactions among soil properties. Recently, there have been significant efforts to develop meaningful ways to separate total soil organic matter into biologically active pools that are better related to productivity and non-reactive pools that are poorly associated with productivity (Wander *et al.* 1994; Sanchez 1998). Although some progress has been made, it seems unlikely that even direct measurements of an active pool are likely to produce consistent results over a range of forest conditions.

The conceptual response of conversion of forest to agriculture and subsequent abandonment on carbon content is presented in Fig. 7.5. For managed forests, an increase in soil carbon will occur immediately following harvest due to slash

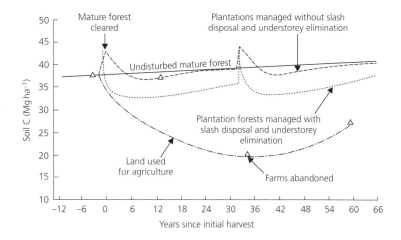

Fig. 7.5 Theoretical changes in soil carbon content of a clay-textured Ultisol of the south-eastern USA resulting from harvest of mature forests, agricultural use and abandonment to return to forest as compared with the effects of forest harvest. Data from Van Lear *et al.* (1995) (symbols) are plotted on the theoretical relationship.

147

ᴴ higher
ᴄʰ is
d

…all reduction early in
…st begins to grow and
…er, soil carbon should
…evels. By contrast, clear-
…ersion to agriculture is
…reduced organic matter
…ᴄe to a lower level that rep-
…um between carbon inputs
…nd decomposition of these
materials. Conᴗ…ᴗn of old-growth forests, particularly in cooler climates where organic matter accumulations in forest floor and soil are large, will likely lead to a reduction in soil organic matter content. Generally, these decreases will be small. Where forests are established on old agricultural lands, soil organic matter will increase towards a new equilibrium that may approach that of old-growth forests.

A variety of data support the generally small impact of harvesting on soil organic matter illustrated in Fig. 7.5. In a review, Johnson (1992) concluded that there is no general trend toward reduced organic matter following forest harvest in temperate and warm climates, and that when changes were measured they were small (<10%). Recently, this same author (Johnson and Todd 1997) evaluated long-term soil organic matter on a site in Tennessee subject to two levels of biomass removal in harvest (bole-only removal with branches and tops left on site or complete above-ground harvest and removal from the site). Fifteen years following harvest there were no significant differences in soil organic matter between harvest treatments or with the non-harvested reference stand.

Of course, exceptions to this generally small impact exist. Organic matter in soils with low clay contents is not stabilized into recalcitrant pools. Nambiar (1996) showed that on such soils organic matter can be decreased by as much as 40% during the 6-year period between felling one rotation of radiata pine and crown closure of the next rotation; it was not clear to what level the soil organic matter would recover by the end of the rotation. Carlyle (1993) similarly showed that organic matter declined quickly following timber harvest on sandy soils. Weed control and slash and litter removal accelerated these declines. Thus, at least on some sites, there exists a poten-

tial for forest harvest and regeneration activities to lead to decreased soil organic matter. Experience in agriculture indicates that either fertilizer or organic matter additions are required to maintain productivity when this occurs. Both of these are viable options for forest management. Additionally, slash management, rotation length and competition control can be adjusted to promote accumulation of soil organic matter on sensitive sites (Costantini *et al.* 1997).

7.4 TREE GROWTH AS EVIDENCE FOR SUSTAINABILITY

Because tree growth is determined by rooting conditions, nutrient supply and water availability, measures of a single variable (where a single variable limits growth) or models that combine variables affecting tree growth should be useful means for evaluating sustainable soil management. Process models capable of predicting changes in soil conditions and forest growth response to these changes have been developed and calibrated for a number of ecosystems. Among the best known are the FORECAST model and its FORCYTE predecessors developed in British Columbia. This model predicted forest growth response to natural or management-induced changes in nutrient cycling (Kimmins *et al.* 1996). LINKAGES (Pastor & Post 1985), which linked forest floor and soil C and N cycles to N availability and tree growth, is another example of a process-based model. Productivity indices that relate tree growth and productivity to the measures of selected soil variables have also been developed (Gale & Grigal 1990). Such indices can be used to assess the potential for productivity change where information on soil conditions is available, but tree growth is not. Nonetheless, while these models are useful for predicting long-term trends, tree growth remains the best integrator of soil conditions and productivity.

In theory, measurement of net primary productivity would provide the soundest basis for evaluating changes in site productivity. In practice, this measurement has seldom been employed in studies of forest growth because of the time, expense and error in its determination.

Instead, most long-term studies in forestry rely on the growth of a target tree species (usually a commercially valuable species) as a means of assessing changes in site productivity. Site index or height growth is the most commonly used measure of growth because it is easily measured and is less sensitive to changes in other stand conditions, particularly stocking, than other measures (Morris & Miller 1994).

Well-documented studies of tree height growth over several rotations are limited. Evans (1996) provides data on height growth for three rotations of weeping pine (*Pinus patula*) in Swaziland. On granite-derived soils, third rotation height exceeded height of either first or second rotations. On gabbro-derived soil, height growth decreased between the first and second rotations. Because these plantations were not fertilized, the author attributed this decline to reduced P availability in the slowly weathered gabbro soils. This study is particularly instructive because it underscores how differences in a specific soil nutrient can affect sustainability.

In the absence of data on growth of successive rotations, comparisons of tree height among different soil treatments within the same rotation have been used to assess sustainability of specific practices. This must be done with caution. Many management activities have short-term impacts on soil conditions that are ephemeral and poorly related to sustainable soil and site management. For instance, herbicide applications that reduce herbaceous plant competition will often result in more rapid early growth in plantations. When such competition control does not result in accelerated erosion or nutrient loss the increased growth will be short term. As illustrated in Fig. 7.6a, when measured at almost any plantation age, height of trees in herbicide-treated plots will be taller than in non-herbicide-treated plots. This does not mean soil and site productivity was increased by herbicide application. Instead, elimination of herbaceous competition increased the allocation of available site resources to trees and increased early growth. This increase was maintained throughout the rotation but, because the trees did not continue to increase in growth relative to the non-herbicide trees, we can not interpret this as a change in long-term productivity.

This growth pattern exemplifies but one of a number of patterns of tree growth response to forest management activities illustrated in Fig.

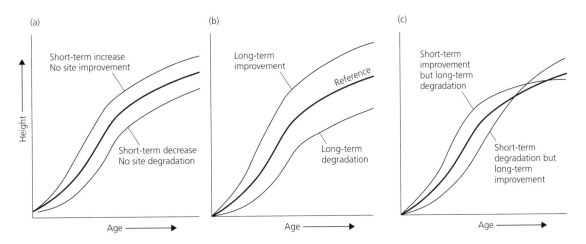

Fig. 7.6 Patterns of tree height growth representative of different types of disturbance and management inputs: (a) short-term increases or decreases in productivity resulting from ephemeral changes in soil conditions that do not affect sustainability; (b) changes in productivity that result from long-term improvements or degradation of soil conditions; (c) transient changes in productivity resulting from ephemeral differences that mask long-term changes in soil properties. Reduced growth observed at the end of the rotations in (b) and (c) indicate lack of sustainable soil management. (Adapted from Hughes *et al.* 1979.)

7.6. Patterns of reduced growth illustrated in Fig. 7.6b and c are representative of long-term changes in site productivity because they demonstrate growth rates that diverge from the undisturbed reference at the end of the rotation period. Unfortunately, few studies of soil conditions and forest growth provide repeated measures of forest growth over long enough periods to clearly identify the growth response pattern. Without long-term repeated measurement of growth pattern it is easy to mistake short-term changes with long-term sustainability. Thus, many of the inferences we make regarding long-term soil impacts are speculative.

7.5 MANAGEMENT IMPACTS ON THE ROOTING ENVIRONMENT

7.5.1 Compaction and rutting

Harvesting and other forest management activities can have direct and indirect effects on soil conditions that affect root growth and soil processes. In particular, the use of heavy equipment to harvest trees can lead to soil compaction. Increases in resistance to root penetration, reduced macroporosity and gas transfer potential, and altered moisture regimes all occur on compacted sites. Where puddling (the loss of soil structure) also occurs as a result of machine trafficking, self-compaction on drying and loss of pore continuity can contribute to even greater damage.

Numerous studies have measured tree growth on soils that were artificially compacted in the laboratory or operationally compacted during forest harvest. For most conditions (but not all) reduced tree growth occurs on compacted soils (Hatchell *et al.* 1970; Froehlich *et al.* 1986; Helms & Hipkin 1986). Reductions in growth reported for trees planted or naturally regenerated in operationally compacted skid trails and logging decks typically range from 5 to 20% when measured during the early part of the rotation. In the US Pacific Northwest, 6–12% reductions in basal area growth of 16-year-old ponderosa pine (*Pinus ponderosa* Engelm.) planted on skid trails have been reported for a sandy clay loam soil (Froehlich 1979) and a 13% reduction in height of 15-year-old

ponderosa pine (Helms & Hipkin 1986). In the US southeast, Perry (1964) showed that growth of 26-year-old loblolly pine (*Pinus taeda* L.) planted on skid trails was reduced by 52%. Similarly large reductions have been reported for 5-year-old pine planted on compacted skid trails in this region by Lockaby and Vidrine (1984). The height of trees planted on severely compacted skid trails was 60% less than for trees planted on undisturbed areas.

Exceptions to the general pattern of reduced growth on compacted soils exist, but are rare. Where compaction is associated with no change or an increase in tree growth it is usually the result of interactions with other factors affecting tree growth, such as increased contact of roots with mineral soil and improved planting or reduced competition from competing plant species (e.g. Clayton *et al.* 1987). Examples also exist where measurable compaction occurs but these compacted soils remained within a favourable range of bulk density and no reductions in tree growth are observed (Miller *et al.* 1996).

The degree to which compaction damage poses a threat to sustainable soil management is a function of the severity and depth of damage, the extent of damage (area) and the speed at which soils can recover from damage. On average, ground-based harvesting equipment affects from a low of about 10% to as much as 60% of the harvested area, with most investigators reporting from 15% to 30% of the harvested area in primary or secondary skid trails where the potential for soil damage is greatest (Miller *et al.* 2000). Thus, the potential for productivity loss is high. Within damaged areas, rates of recovery vary widely. In climates where surface soils are subject to freezing and thawing, recovery can occur quickly and may take only 1–2 years (Mace 1971; Holman *et al.* 1978). Here compaction may reduce initial growth rates but appears unlikely to pose a threat to long-term soil sustainability, because the soils recover to predisturbance levels prior to the next harvest. By contrast, soils in warmer climates not subject to freezing and thawing cycles or lacking in shrink-swell clays may take decades to recover from severe compaction damage. Perry (1964) estimated that Ultisols in North Carolina could take as long as 40 years to recover from severe logging damage. Grecean and Sands (1980) similarly stated that on sandy soils of South Australia, 50-years-old skid trails were still compacted relative to surrounding soils. Although repeated long-term measures of tree growth response on these compacted areas are not available, it is reasonable to expect that growth on these sites will remain lower than in non-impacted areas indicative of long-term degradation (Fig. 7.6b).

Generally, risks to sustainable management posed by soil physical damage are low where compaction is limited to the surface soil and/or the ameliorating effects of freezing and thawing, wetting and drying, insect activity and root growth are large. Deep compaction resulting from the use of heavy equipment when soils are wet, and therefore weak, poses a significant risk to sustainable production, particularly when associated with degradation of soil structure (puddling) in warm climates with low shrink-swell soils (Fig. 7.7).

7.5.2 The role of tillage

Tillage (cultivation) has a role in sustainable management of soils both as an ameliorative practice to correct damage associated with harvest and as a means for overcoming natural soil and site limitations. Examples of tillage treatments that are commonly employed in plantation management are bedding, disk harrowing, subsoiling and spot cultivation. Each of these treatments can improve soil physical conditions; however, the nature of improvement and soil conditions under which treatments are most appropriate varies (Morris & Lowery 1988). The raised mound created during bedding increases large pore space and lifts the seedlings; it has particular value where root growth is limited by high water tables and poor aeration (Gent *et al.* 1983). However, because surface soils are concentrated in the bedded zone, nutrient availability in the vicinity of roots of newly planted seedlings is also increased by bedding (Morris & Pritchett 1982; Attiwill *et al.* 1985; Page-Dumroese *et al.* 1989). Disk harrowing, which turns the soil and incorporates surface debris to depths of 15–25 cm, can reduce bulk density, decrease resistance to penetration and increase macropore and total soil porosity in

a

b

Fig. 7.7 Severe rutting caused by harvesting with ground-based equipment when soils were saturated (a) and nutrient displacement of organic matter and nutrients during preparation of some sites for planting (b) are two management activities that have been demonstrated to reduce forest productivity. Note that ameliorative treatments are possible but they add to the cost of plantation establishment.

surface soils (Gent *et al.* 1984) and result in increased root growth in the tilled zone (Morris & Lowery 1988). Improvements in nutrient availability also occur following disking as a result of the stimulatory effects of organic matter incorporation and exposure of fresh surfaces to microbial activity. Unlike these two treatments, which only influence surface soil conditions, subsoiling has the potential to ameliorate root restrictive conditions at depths up to 1 m. When undertaken with appropriate equipment under dry conditions that promote fracturing, subsoiling can increase macroporosity, create continuous pores and reduce resistance to root penetration. Finally, spot cultivation, which is fairly new to forestry, combines the benefits of all of these treatments. Surface soils are concentrated in a mound as they are on bedded sites, and subsoils are fractured and deep root restrictions ameliorated as they are on subsoiled sites.

The benefits of soil tillage for correcting soil damage have been shown for several combinations of site type and tillage treatments. For instance, soil compaction and puddling resulting from rutting is a major problem wherever poorly drained soils are harvested by ground-based equipment such as in the Lower Coastal Plain of the USA during wet periods. Natural recovery of soils from harvesting damage has been estimated to take from 15 to 20 years on these sites (Hatchell & Ralston 1971), or almost a full rotation. On such sites, bedding in combination with NPK fertilization has been shown to correct most soil physical damage and return productivity to levels observed for non-trafficked areas within 4 years (Aust *et al.* 1998). On well-drained soils in the Piedmont of North Carolina, Gent *et al.* (1984) showed that disking could return the surface 15 cm of skid trails compacted during harvest to bulk density, macroporosity and saturated hydraulic conductivity comparable to preharvest conditions. Twelve years after planting of these sites, loblolly pine height was 31% greater in skid trails that were disked harrowed than on non-disked skid trails (Stewart 1995).

While the short-term benefits of such treatments are clear, long-term benefits are less clear. Many responses to treatment seem to fall into responses illustrated in Fig. 7.6a. Trees planted after these treatments achieve an early growth advantage that is maintained throughout the rotation, but the advantage or response does not increase; thus, this is not a long-term change in site productivity. Such growth responses have been reported for *Pinus elliottii* and *P. taeda* in the USA following bedding (Hughes *et al.* 1979), bedding and disk harrowing (Haywood 1980; Tiarks 1982; McKee & Wilhite 1986), and following bedding and subsoiling for *P. radiata* in New Zealand (Mason *et al.* 1988).

The pattern of the response to such tillage treatments may be due to the reasons for the observed initial response. Responses would tend to be temporary where benefits of tillage are largely caused by:

• decreased plant competition;
• accelerated occupation of rooting volume without an increase in absolute rooting volume;
• short-term increases in N mineralization;
• beneficial changes in thermal regimes or water infiltration and distribution.

Where amelioration of soil physical conditions which restricted total rooting volume were largely responsible for initially poor growth, responses appear to be long term. Thus, bedding of sandy sites does not appear to have long-term benefits because much of the initial benefit is associated with ephemeral increases in nutrient mineralization and competition control. Responses to bedding may be long term on fine-textured soils where beds provide increased aerated volume during wet periods of the year.

7.6 EROSION

Unplanned or poorly conducted forest harvest, especially when such harvests are conducted without plans for continuous management, can accelerate rates of soil erosion. Erosion affects sustainability in several ways: it reduces the depth of soil available for root exploitation and effective rooting depth, it degrades physical characteristics and decreases the quality of the soil as a root growth environment, it reduces the water-holding capacity of a site and the water available to support plant growth, and it results in reductions in nutrient storage and availability. While reductions in nutrient storage and availability can, potentially, be compensated for by fertilization, other erosion-associated changes are difficult to correct through intervention.

Well-known examples of non-sustainable loss of soil due to erosion caused by poor-quality row-crop farming include the Piedmont of southeastern USA, where cotton farming led to loss of between 10 and 20 cm of surface soil (Trimble 1974) and contributed to the eventual collapse of this industry. On these soils, it has been estimated that each centimetre of topsoil lost through erosion results in reduced corn yields by 2.5% (Langdale *et al.* 1979). Similar relationships exist for other regions. For instance, Lal (1998a) reviewed studies of soil erosion and its effect on crop productivity in the topics. Loss of 30 cm of topsoil through simulated erosion generally reduced crop yields by 10% to 50%. Larney *et al.* (1995) predicted yield reductions of 120% for a loss of 15 cm of topsoil for crops in Alberta. Such relationships between yield decline and soil erosion are crop and soil specific. For the tropics, Lal (1985) showed small reductions in maize yield at cumulative soil loss less than 2 mm, with proportionately more yield reduction as cumulative soil loss increased beyond 2 mm depth (Fig. 7.8). The aforementioned estimates by Larney were for sensitive soil and demanding crops; for better soils and less demanding crops yield reductions for a given amount of erosion would be lower.

Forest management-induced increases in soil erosion are generally much less than those observed under food crop production. Except in mountainous regions, soil erosion in natural and managed forests is generally very low (Table 7.3). Generally, rates of erosion in managed forests average less than 1 Mg ha^{-1} yr^{-1} throughout temperate and tropical areas. For instance, in the southeastern USA, maximum rates of erosion following forest harvest and mechanical preparation are 14.2 Mg ha^{-1} yr^{-1} (Yoho 1980) on erosive sites and these rates only occur for the first postdisturbance year.

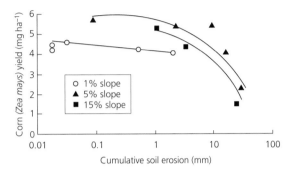

Fig. 7.8 The relationship between corn (*Zea mays*) yield and cumulative soil loss for a tropical soil. Similar relationships undoubtedly exist for forests, but because of the low erosion typical of most forests, erosion is at the low end of these curves where differences are difficult to detect. (*Redrawn* from Lal (1985) used with permission.)

Table 7.3 Soil loss from managed forests during the first postdisturbance year under a range of harvest and regeneration practices that include forest floor disturbance.

Location and management activity	Sediment yield (Mg ha^{-1} yr^{-1})	Reference
Southeastern USA		Yoho 1980
Undisturbed forests	trace–0.7	
Forest thinning	trace–0.1	
Annual burning	0.7–17.7	
Careful clearcut	0.1–0.4	
Careless clearcut	3.0	
Mechanical site preparation	12.5–14.2	
Wet tropics (Nigeria)		Lal 1986b
Undisturbed forest	trace	
Manual harvest	trace	
Manual harvest and tillage	3.78	
Shear harvest	0.97	
Bulldozer clearing with root rake	0.28	
Bulldozer clearing with root rake and tillage	4.19	
Wet tropics (Indonesia)		Wiersum 1985
Undisturbed forest	0.3	
Trees removed	0.4	
Trees and undergrowth removed	0.8	
Trees undergrowth and forest floor debris removed	43.2	

Very few data exist that can be directly used to estimate the effects of erosion on forest productivity at the low rates typically observed (Elliot *et al.* 1998). They are within the rates considered acceptable for sustained agricultural production, but these agricultural rates are based on cropping that includes annual fertilizer inputs and tillage. For agronomic crops, relationships between soil erosion and productivity are less significant for more intensively managed systems that receive fertilization (Thomas *et al.* 1989), and we can expect the same to be true for forests. Thus, erosion is unlikely to be a major factor in intensively managed forests in areas of low relief or moderately rolling topography, where erosion rates are low, organic matter conserved and fertilizer additions are common.

In contrast to areas of low relief, forest harvest and regeneration activities can result in severe erosion in mountainous regions. Reductions in root strength coupled with increased soil water and disturbance of the forest floor can lead to rill erosion and surface gullying as well as increased landslide frequency. Where surface gullying occurs, erosion can be severe. Klock (1982) reviewed the effects of harvesting methods on erosion in central Washington State. Salvage logging in fire-damaged Douglas fir (*Pseudotsuga menziesii* (Mirb.)Franco) forests resulted in the disturbance of 75% of the area. Most soil loss was in rills and gullies that, when averaged across the site, was equivalent to a loss of 3.8 cm of surface soil. In a bioassay, growth of Douglas fir and ponderosa pine (*Pinus ponderosa* Laws.) seedlings associated with this level of surface soil loss ranged from 15 to 75% and 14 to 105%, respectively, of undisturbed controls. Rill and gully erosion is considered to be a major contributor to growth declines and site deterioration under teak (*Tectona grandis*) in India and Indonesia, because of the lack of understorey development and the burning of the litter layer which exposes the mineral soil to intense rain (Bell 1973; Perhutani 1992), and in steep lands of Australia (Costantini *et al.* 1997).

Both road construction and harvesting can contribute to increased landslide frequency in mountainous areas. Road construction changes the geometry of the slope while root decay following harvest leads to reduced soil strength during the regeneration phase of the forest (5–10 years) (Sidle 1991). In a survey of landslide frequency in the western USA following unusually heavy rainfall, Dyrness (1967) found that 34 of 47 mass wasting events were directly associated with roads and 8 of 47 events occurred in harvested areas. Only 5 of 47 events occurred in undisturbed forests. Obviously, areas within landslides will have significantly reduced productivity.

7.7 BALANCING INPUTS AND OUTPUTS OF NUTRIENTS

7.7.1 Nutrient balance in undisturbed and managed forests

As illustrated in section 7.3.2, sustainable soil management requires that the supply of nutrients in available forms must be maintained. In the short term, changes in rates of release can occur independent of the total content of soil-stored nutrients. In the long term, sustainability requires a balance between the direct export of nutrients during harvest or accelerated loss of nutrients by management activities and the addition of nutrients through natural processes or human intervention. There is no such thing as a perfectly balanced system. Even undisturbed forests tend to accumulate some nutrients and export others (Table 7.4). However, they tend to accumulate nutrients that are limiting growth and export nutrients that are available in excess. As most forest systems tend to be N limited, there is a tendency to accumulate N in undisturbed forests. Whether or not other nutrients are accumulated or leaked from the system depends upon the mineral composition of the soil and the underlying geology. For instance, both K and Ca tend to be lost from forested watersheds in the Douglas fir forest in Oregon studied by Sollins *et al.* (1980) because of the high concentrations of these cations in primary minerals. In slash pine forests of Florida studied by Riekerk (1983), K tends to be accumulated in the system. The quartz sands which form the parent material for these soils are low in K and K is one of the nutrients that tends to be deficient on these sites (Allen, Chapter 6, this volume).

Table 7.4 Balance between annual nutrient input in precipitation and soluble export for undisturbed forests across a range of forest and climatic conditions.

Forest type (location)	Nutrient input ($kg\,ha^{-1}\,yr^{-1}$)					Reference
	N	P	K	Ca	Mg	
Douglas fir (British Columbia)						Feller & Kimmins (1984)
Input	4.0	<0.1	1.0	7.0	1.0	
Export	0.8	<0.1	1.5	27.2	4.1	
Gain(+) or loss(−)	+3.2	0	−0.5	−20.2	−3.1	
Douglas fir (Oregon, USA)						Sollins *et al.* 1980
Input	2.0	0.3	0.9	3.6	1.2	
Export	1.5	0.7	9.5	123.1	8.6	
Gain(+) or loss(−)	+0.5	−0.4	−8.6	−119.5	−7.4	
Spruce plantation (Germany)						Cole & Rapp 1981
Input	21.8	0.50	3.7	12.6	2.6	
Export	14.9	0.02	2.1	13.5	3.7	
Gain(+) or loss(−)	+6.9	+0.48	+1.6	−0.9	−1.1	
Mixed oak–birch (UK)						Cole & Rapp 1981
Input	5.8	0.2	3.3	6.9	5.4	
Export	12.6	0.2	8.3	59.8	6.0	
Gain(+) or loss(−)	−6.8	0	−5.0	−52.9	−0.6	
Northern hardwoods (New Hampshire, USA)						Whittaker *et al.* 1979
Input	5.8	0.1	1.1	2.6	0.7	
Export	2.3	trace	1.7	11.7	2.8	
Gain(+) or loss(−)	+3.5	+0.1	−0.6	−8.5	−2.1	
Slash pine (Florida, USA)						Riekerk 1983
Input	7.5	1.1	2.3	4.8	3.1	
Export	2.6	0.1	0.4	2.4	1.9	
Gain(+) or loss(−)	+4.9	+1.0	+1.9	+2.4	+1.2	
Mixed (Panama)						Golley *et al.* 1975
Input	–	1.0	9.5	29.3	4.9	
Export	–	0.7	9.3	163.2	43.6	
Gain(+) or loss(−)	–	+0.3	+0.2	−133.9	−38.7	

Harvest removals

Removal of nutrients accumulated in wood, branch and leaf material during forest harvest has the potential to deplete soil nutrients at a rate that exceeds natural inputs. This is particularly true for forests managed for short rotations and high levels of utilization. Comprehensive reviews of removal rates for various types of forest harvest and utilization are available. Results for several representative forests are summarized in Table 7.5. From Table 7.5 it is clear that periodic removal of stems from forests is sustainable from a nutrient balance standpoint—for most systems, atmospheric inputs of N, P and S over the period between harvests can replace nutrients removed during harvest. Soil reserves of K, Ca and Mg are relatively high and, although information on the release of these nutrients from weathering of primary materials is not very accurate, sufficient quantities of these nutrients are available for many rotations. Some exceptions exist. For instance, Ca and K content of many tropical soils is low and removals in harvests in intensive short-rotation forestry can exceed estimated atmospheric and weathering inputs and reduce availability. Nambiar (1996) provides an example for plantations in Brazil. In this example rose gum grown in 8- to 12-year rotations was projected to have insufficient K after only one rotation. For this site, only minor levels of forest harvest could be considered sustainable without replacement of Ca. Another example is provided by Nykvist (1998), who discusses how harvesting can lead to Ca limitations in tropical forests in Malaysia.

Because forests tend to accumulate nutrients more rapidly when they are young, shortening the rotation lengths between harvests results in greater nutrient removal over selected periods. Also, removal of branch and foliage through more complete utilization will result in increased removals at each harvest. The amount of and frequency of biomass removal that can occur before long-term balance becomes negative varies greatly (Table 7.5); however, short-rotation forestry that includes removal of stems, branches and foliage (whole-tree harvest) is likely to result in negative balances for at least one nutrient and is not sustainable unless nutrient additions through organic amendment or fertilization is a component of forest management.

Slash disposal

Disposal of slash left on the site following harvest is common on forested sites being prepared for planting. Such removal allows for easier soil tillage prior to planting and can improve the speed and quality of planting. It can also contribute to unsustainable nutrient balances. Two types of slash disposal are common: mechanical displacement of slash from the planting surface and burning. Mechanical displacement of slash is accomplished by using a tractor-mounted rake or

Table 7.5 Comparison of nutrient removals in harvest and as a consequence of slash disposal during preparation of the harvested site for regeneration across a range of forest and climatic conditions.

Forest (location)	Age (yr)	Activity	Nutrient removals (kg ha^{-1} (kg ha^{-1} yr^{-1}))					Reference
			N	P	K	Ca	Mg	
Spruce (Sweden)	18	Conventional stem harvest	20 (1.1)	3 (0.2)	18 (1.0)	16 (0.9)	–(–)	Tamm (1969)
		Whole-tree harvest	163 (9.1)	25 (1.4)	106 (5.9)	157 (8.7)		
	50	Conventional stem harvest	217 (4.3)	20 (0.4)	248 (5.0)	278 (5.6)		
		Whole-tree harvest	842 (168)	80 (1.6)	442 (8.8)	463 (9.2)		
Scots pine (UK)	50	Conventional stem harvest	72 (1.4)	9.6 (0.2)	55 (1.1)	–(–)	–(–)	Miller & Miller (1991)
		Whole-tree harvest	248 (5.0)	25.0 (0.5)	118 (2.4)	–(–)	–(–)	
Douglas fir (Washington)	53	Conventional stem harvest	478 (9.0)	56 (1.1)	225 (4.2)	–(–)	–(–)	Bigger & Cole (1983)
		Whole-tree harvest	728 (13.7)	96 (1.8)	326 (6.2)	–(–)	–(–)	
Loblolly pine plantation (North Carolina)	22	Conventional stem harvest	57.0 (2.5)	4.6 (0.2)	35.4 (1.6)	51.5 (2.3)	14 (0.6)	Tew *et al.* (1986)
		Whole-tree harvest	180.4 (8.2)	18.6 (0.8)	89.4 (4.1)	178.0 (8.1)	35 (1.6)	
		Chop and burning of slash*	46.0 (2.9)					
		Windrowing of slash*†	654 (29.7)	41 (1.9)	118 (5.4)	429 (19.5)	74 (3.4)	
Slash-longleaf pine (Florida)	40	Conventional stem harvest	59 (1.5)	5 (0.1)	20 (0.5)	80 (2.0)	17 (0.4)	Morris & Pritchett (1982)
		Whole-tree harvest	110 (2.8)	10 (0.3)	36 (0.9)	118 (3.0)	27 (0.7)	
		Windrowing of slash‡	373 (9.3)	17.8 (0.5)	27.3 (0.7)	163.1 (4.1)	41 (1.0)	Morris *et al.* (1983)
Gmelina (Nigeria)	10	Thinning 25% of stems	66 (6.6)	9.0 (0.9)	–(–)	124 (12.4)	15.1 (1.5)	Nwoboshi (1983)
		Stem coppice w/standards	384 (38.4)	184.9 (18.5)	–(–)	1135 (113.5)	281 (28.1)	
		Whole-tree harvest	959 (95.9)	371.5 (37.2)	–(–)	2424 (242.4)	615 (61.4)	

* Following stem-only harvest.

† Note that windrowing does not remove nutrients from the site but displaces them from much of the soil surface.

‡ Following stem-only harvest with limbs left at a delimbing gate.

blade to push slash into long rows or smaller circular piles (Fig. 7.7). This creates a reasonably debris-free surface for tillage or for planting. Although the material is not moved from the site, nutrients localized in windrows are not as available as if they were left distributed across the site. Morris *et al.* (1983) showed that the amounts of N, P and K displaced into windrows during slash disposal on a Coastal Plain site in the south-eastern USA exceeded removals during stem-only harvest by sixfold and represented more than 10% of the site's nutrient reserves. Similar results have been reported for Monterey pine (*Pinus radiata* D. Don) plantations in New Zealand (Webber 1978) and loblolly pine (*Pinus taeda* L.) plantations in the Piedmont of the south-eastern USA (Tew *et al.* 1986) and can probably be generalized to other areas where such techniques are used. Few long-term results of tree growth response to slash and soil displacement during site preparation are available. Reductions in plant competition and accelerated mineralization following such disturbance (Vitousek *et al.* 1992) often increase seedling growth during the early portion of a rotation (Smethurst & Nambiar 1990; Allen *et al.* 1991). Only later in the rotation may reductions in nutrient availability and growth occur (growth pattern indicated in Fig. 7.6c) that indicate such practices are not sustainable (Fox *et al.* 1988; Morris & Miller 1994; Powers *et al.* 1994).

Burning, another means for slash disposal, is employed throughout the world. Losses of N and S, both of which volatilize at temperatures reached in fire, can be significant and can contribute to negative balances of these nutrients. In boreal forests where debris and forest floor accumulations are large, more than $500\,\mathrm{kg\,ha^{-1}}$ of N can be lost during slash burning (Feller & Kimmins 1984). In the tropics, similarly high losses of N can occur when significant amounts of vegetation are left on the site following commercial harvest or when a site is cleared for agriculture (Lal 1986a; Holsher *et al.* 1996). Typically, burning is less severe and N and S losses smaller following harvest of managed forests because of lesser accumulation of slash and litter. Prescribed burning during the rotation can also contribute to N loss. In the south-eastern USA, where such burns are common, they do not usually have a measurable effect on soil nutrient content (Wells *et al.* 1979); however, they have been implicated in productivity declines in other areas (e.g. Bell 1973).

Runoff and leaching

Investigations of water quality changes following forest harvest and regeneration of forests leave little doubt that water quality can be adversely affected by harvesting and regeneration activities through increased nutrient concentrations. Concentrations of dissolved nutrients have been found to increase in watershed runoff in temperate conifers (Fredriksen *et al.* 1975; Feller & Kimmins 1984), temperate hardwoods (Likens *et al.* 1970; Van Lear *et al.* 1985) and in tropical forests (Malmer 1996). Generally, losses of N are considered among the most important in temperate systems. Additionally, high amounts of P can be exported in runoff in tropical systems because of the tendency for rapid peak flows and limited contact between runoff and soil surfaces (Malmer 1996). However, except in instances where regrowing vegetation has been kept from the site, such as by repeated herbicide control (e.g. Likens *et al.* 1970), increased loss tends to be of a relatively short 1–3 years' duration, after which conditions return to near predisturbance levels. When considered over the course of the 20 or more years established forests remain on the site, the accelerated loss during these establishment years tends to be small in comparison with other nutrient inputs (e.g. atmospheric, weathering) and exports (e.g. harvest removals).

7.7.2 Nutrient additions

Reviews of site sustainability tend to focus on harvest and associated management activities considered to have potentially negative impacts. Several harvest- and regeneration-associated practices have the potential to improve productivity and contribute to sustainable soil management.

P fertilization at planting

Phosphorus fertilization is an example of a nutrient addition that can contribute to sustainable soil management. On P-deficient sites, fertilization at planting can improve growth throughout a full rotation (Pritchett & Comerford 1982) and into a second rotation. For instance, Gentle *et al.* (1986) report *P. radiata* growth response to fertilization with two forms of phosphorus at age 16 in both the initial rotation and subsequent rotation. During the first rotation, height and volume accumulation curves increased throughout the rotation indicative of a change in site productivity that is more than ephemeral, such as that illustrated in Fig. 7.6b. While this rapid growth may not continue without continued fertilization, there is little reason to believe that if fertilization were stopped the site productivity would drop below initial levels. The general persistence of response is due to the large amounts of applied P (50–100 kg ha^{-1}) in comparison to removals in forest harvest (5–70 kg ha^{-1}). Thus, exceptions to the rule that P fertilization results in long-term growth responses are rare (e.g. Mead & Gadgil 1978).

N fertilization

Nitrogen is typically applied to forest stands at or near the time of crown closure when N demand is near its maximum and when the flush of N mineralization associated with harvest and stand establishment (Vitousek *et al.* 1992) is over. Unlike P fertilization, which has been shown to improve soil P availability and site productivity for an entire rotation or more, a single application of N is usually considered to result in only short-term improvements in soil N supply. Following fertilization, available soil N can be increased 10-fold to 100-fold; however, this increase is ephemeral and little detectable difference in available or mineralizable soil N occurs a year or more after application. (Johnson *et al.* 1980; Whynot & Weetman 1991). Most of the longer 3-year to 10-year growth response reported following N fertilization is due to recycling of N assimilated during this short period of increased available soil N. Consequently, growth response to N fertilization typically follows the ephemeral improvement pattern illustrated in Fig. 7.6a.

Miller (1981) speculated that the potential for N additions to improve site productivity beyond the current rotation was likely only when the amount of N added was large in relation to the original capital of the site. Managed forests typically contain between 2000 and 5000 kg ha^{-1} N in the soil (Morris & Miller 1994). If 200 kg ha^{-1} of N is applied as fertilizer, as is typical for most operational fertilization, and an average of 50% is retained in the forest floor and vegetation components not removed during harvest, then this application rate corresponds to only a 2–5% change in total N storage. Because only a small portion of this additional N is mineralized in any year, it is unlikely to result in large changes in productivity. Cases where long-term improvements in growth followed a single fertilizer application appear to be restricted to cases where very large amounts of N were added in an organic form such as sewage sludge (Miller *et al.* 1992).

In contrast to single applications of inorganic N, where the potential to alter the long-term capacity of soil to supply N is limited, there is potential for repeated applications of inorganic N at low rates to increase productivity, leading to an increase in soil organic matter N mineralization and productivity that can last beyond the current rotation. However this will require fertilization well above current operational practices. Thus, we can conclude that increases in forest growth associated with N fertilization are sustainable to the extent that social and economic conditions allow for continued N fertilization.

Cations and micronutrients

Generally, Ca, Mg and K applied to forests as inorganic fertilizer are retained within the soil exchange complex where they can contribute to increased soil supply and recycled within the ecosystem. Long-term responses to single applications are the norm and to the extent that additions of these nutrients replace losses, they will contribute to sustainable soil management. Most micronutrients are bound quite tightly to the soil, and fertilizer additions exceed utilization by trees and removal during harvest. Consequently, micronutrient fertilization is also likely to lead to long-term increases in forest growth and sustainable soil management in areas where micronutrient deficiencies occur.

Waste recycling

Land application of sludge created during treatment of municipal or industrial wastewater has been applied to forestland for many years. In many, if not most, cases, short-term increases in forest productivity have been demonstrated. Henry (1986) studied the effects of pulp and paper mill sludge applied alone or in combination with municipal sludge on growth of Douglas fir, noble fir (*Abies nobilis* (Dougl.) Lind.), white pine (*Pinus strobus* L.) and cottonwood (*Populus* spp.) seedlings. Response was variable, but seedling growth was generally decreased by application of primary mill sludge with high carbon-to-nitrogen ratios and increased following application of low carbon-to-nitrogen ratio secondary mill sludge or primary mill sludge–municipal sludge mixtures. Bockheim *et al.* (1988) found that growth of thinned 28-year-old red pine (*Pinus resinosa* Ait.) was increased by a single application of a mixed primary–secondary pulp mill sludge applied at intermediate (70 Mg ha^{-1}) rates with growth response decreasing at higher rates of sludge application. In the US South, McKee *et al.* (1986) showed that growth response of 1-, 3-, 8- and 28-year-old loblolly pine to a single application of textile mill sludge and combinations of sludge and municipal sludge depended on the age of the stand, carbon-to-nitrogen ratio of the sludge and the method of application. Incorporation of textile mill sludge into soils of young stands improved growth even when carbon-to-nitrogen ratios were high. By contrast, the growth of older trees was improved only by the addition of textile sludge with low carbon-to-nitrogen ratios and higher nitrogen (N) availability.

The long-term benefits of such applications are not well documented; however, Harrison *et al.* (1995) showed that single applications of municipal sludge can result in increases in organic matter, N and P measured up to 15 years after application. These long-term increases in organic matter and nutrient content were associated with long-term increases in annual basal area growth increment typical of the increase pattern illustrated in Fig. 7.6b (Henry & Cole 1998). Modelling efforts for hardwood forest in the north-eastern USA indicate that 4 Mg ha^{-1} of anaerobically digested municipal sludge could be applied to forests at 3-year intervals without compromising drinking water quality. Projections indicate that wood yield would increase by 25% and soil organic matter would double from about 40 Mg ha^{-1} to 80 Mg ha^{-1} at a newly established steady-state content (Crohn 1995).

Wood and coal ash has been used for many years as a liming agent and source of potassium (K) for agricultural crops. In Europe, wood ash has been routinely applied to peat soils. In the short term, application of wood ash at rates of 30 t ha^{-1} can decrease tree growth and negatively impact soil fauna through large increases in soil pH and high osmotic potential due to high concentration of soluble salts (Lundkvist 1983). In the long term, application of ash is beneficial and can play a role in balancing inputs and outputs for K, Ca and Mg in managed forests (Vance 1996). Andersson and Lundkvist (1989) reviewed studies of wood ash application in Europe. In most cases, application of wood ash to forests increased forest growth. It is noteworthy that his response followed the pattern of an ephemeral degradation but long-term improvement illustrated in Fig. 7.6c. On many sites, ash application reduced short-term growth but over longer, >30-year periods, ash application increased growth.

7.8 CONTAMINATION

Contamination of soils through activities directly associated with forest management is rare, but not unknown. For example, boron toxicity has been reported for several forests as a result of overapplication of this micronutrient. Smith and Whitton (1975) reported B toxicity in radiata pine produced by airborne and soil-applied ash, and Neary *et al.* (1975) reported B toxicity in red pine irrigated with high B wastewater. Application of waste and wastewater with high salt or Na contents has also been reported (Benyon *et al.* 1996). Generally, however, contaminant-associated risks to soil and site sustainability are the result of activities outside the direct sphere of forest management. Risks may be associated with regional- or global-scale transport phenomena such as the transport and deposition of acids, or may be more local in nature such as resulting from metal contamination from mining. Unfortunately, these phenomena play an increasing role in determining productivity and long-term sustainability of soil management.

7.8.1 Soil acidification

When plants adsorb cations such as Ca^{2+} and K^+ from the soil solution, or anions such as NO_3^-, electroneutrality must be retained. Positive charges of the cations must be balanced by formation of organic anions and H^+ is exuded from the roots to the soil. The accumulation of positively charged Ca^{2+}, Mg^{2+} and K^+ and negatively charged anions such as NO_3^- or SO_4^{2-} in growing forests is unequal. As a result, there is an acid production under growing forests. If the forest is burned, base cations are returned to the soil and there is no net long-term acidification. However, if biomass is removed, as during harvest, acidity generated by biomass accumulation endures.

One measure of soil sustainability is the degree to which balance between acid production and acid neutralization is achieved (Mayer 1998). Generally, acid produced through biomass accumulation is relatively small and much of it can be neutralized by base production during weathering of parent material. Consequently, acid generation solely by biomass accumulation and harvest is not generally considered to pose a threat to sustainable soil management. However, as illustrated in Table 7.6, forests in industrialized areas are affected by anthropogenic activities such as the burning of fossil fuels that have increased the deposition of the strong acid anions NO_3^- and SO_4^{2-} and associated H^+. Only a portion of this acid loading can be neutralized through the release of base cations by mineral weathering, and excess H^+ inputs displace base cations from the cation exchange complex, affecting the capacity of the soil to supply Ca, Mg and K to growing plants. The degree to which this occurs depends on soil texture-related exchange characteristics of the soil as well as on factors such as the extractable soil Al, SO_4^{-2} adsorption potential, and the selectivity coefficients of individual cations (Binkley *et al.* 1989). Soils with high concentrations of exchangeable bases will tend to be less affected by acid additions (better buffered)

than those with low contents of exchangeable bases (see Volume 1, Chapter 11, for further discussion).

Over the very long term, sustainable soil management requires neutralization of the acid loads imposed by harvest removals and atmospheric input. This may only be possible by liming and replacement of base cations by fertilization in soils that are subject to acid deposition and large removals of biomass (Mayer 1998).

7.8.2 Pesticides

The use of pesticides over extensive forest areas is generally restricted to cases where commercially significant forests are threatened by epidemic outbreaks of pests (see also Evans, Chapter 8, this volume). Often these pests are introduced (e.g. gypsy moth, tussock moth, spruce budworm) and the potential for major economic loss is large. The application of pesticides over large areas required to control such outbreaks can reduce overall biological activity and change the structure of faunal and microbial communities. For instance, Hastings *et al.* (1989) described changes in soil and litter fauna following application of lindane for the control of southern pine beetle in North Carolina. Mites, collembola and other arthropods did not return to pretreatment levels for at least 2 years. In a study of soil biology following the application of 2,4,5-T, in the Black Forest of Germany, Mittmann (1991) reported that mite populations were only slightly affected but enchytraeid and nematode populations were significantly reduced by application. Soil microflora was also reduced for a short period following application.

A number of herbicides such as glyphosate, hexazinone, 2,4-D, picloram, triclopyr and imazapyr are commonly used in plantation forestry to prepare sites for planting or to control plant competition during establishment. Two to three herbicide applications are typically applied over a 2–3-year period in intensively managed plantations. Although differences in mode of action and toxicity exist among these herbicides, in general they do not appear to reduce the overall activity of soil organisms when applied at rates and intervals used in forest management; however, they clearly shift populations (Eijsackers & Van de Bund 1980;

Table 7.6 Acid budgets for forests in Germany. (Source: Mayer 1988).

	Significance	Coniferous forests $(kmol\,H^+ha^{-1}yr^{-1})$	Deciduous forests $(kmol\,H^+ha^{-1}yr^{-1})$
Generation of acidity			
Atmospheric deposition	Strong acids of HNO_3 and H_2SO_4 have the potential to reduce soil pH to very low levels	2.0–7.0	1.0–4.0
Acid production by cation accumulation in biomass	If biomass is removed during harvest, this results in long-term contribution to soil acidity	0.4	0.2
Mineralization of organic matter coupled with NO_3 or SO_4-leaching	Except under unusual conditions, leaching of mineralized NO_3 and SO_4 is small	0–1.3	0–1.3
Dissociation of carbonic acid	Dissociation of HCO_3 in equilibrium with CO_2 concentrations in soil atmosphere is pH-dependent and is not significant in very acid (pH<4.5) soils		
Total loading to soil		2.4–8.4	1.2–5.2
Consumption (fate) of acidity			
Buffering by exchangeable Ca and Mg	Acidity buffered by soil exchange results in long-term reductions in base saturation and increases in H^+ and Al on exchange complex.	−1.0 sandy soils −7.0 clay soils	−1.0 sandy soils −7.0 clay soils
Mineral weathering	Acidity balanced by weathering does not affect long-term nutrient supply or soil conditions	−0.2 to −1.0	−0.2 to −1.0
Specific adsorption of SO_4	Acidity associated with SO_4 adsorption is neutralized when the SO_4 is sorbed on soil surfaces. If sulphate deposition decreases, this adsorbed SO_4 can be desorbed and contribute to acidity delaying recovery	Negligible	Negligible
Leakage in water		0.4–7.2 (sandy) 0–1.2 (clayey)	0.2–4.0 sandy 0.2–4.0 (clayey)

Greaves & Malkoney 1980). The significance of these shifts is not known, but they are generally transient and do not appear to constitute a threat to long-term soil productivity.

7.8.3 Metals and salts

Contamination of forests soils through excess of a single element and direct toxicity or through creation of imbalanced nutrition is almost always a result of activities associated with industrial pro-

cessing, mining and smelting operations or oil production. Sudbury Canada is a well-known example of the devastating effects that smelting operations can have on forests. A 40 square mile (100 km²) area around the Sudbury complex is almost completely devoid of vegetation and trees are virtually absent in a 140 square mile area (360 km²). The direct burning of foliage by SO_2 emissions, as well as acidification of soil and the toxic accumulations of Cu, contribute to tree death and decline (Winterhalder 1984; Gunn *et al.*

1995). Upper Silesia in Poland is littered with waste piles from Pb and Zn smelting which are devoid of vegetation due to high concentrations of soluble metals and salinity (Daniels *et al.* 1999). Larger scale dieback of spruce forests in Poland has been tied to the accumulation of metals and acidification (Drozd *et al.* 1998). In Finland, Helmisaari *et al.* (1995) documented reductions in growth of Scots pine (*Pinus sylvestris* L.) forests in the vicinity of a Cu–Ni smelter. In the USA, Beyer (1988) described forest mortality on Blue Mountain, Pennsylvania, resulting from Zn toxicity and erosion-associated stress as a result of Zn smelting. Damage occurred up to 19 km from the smelter. Similarly contaminated areas exist in forests almost anywhere in the world where mining occurs, including other areas in the USA, Canada (Wotton *et al.* 1986), Portugal (Mench *et al.* 1999), Korea (Jung & Thorton 1996), Spain (Vaamonde *et al.* 1999) and Russia (Stepanov & Cheren'kova 1990).

Subtle changes in biological processes occur even where obvious biological collapse and direct tree injury do not occur. Tolerance of specific soil biota to metals varies and those organisms normally associated with development of mor humus seem to be most tolerant of high metal concentrations. Bacterial populations, earthworms and soil arthropods associated with mixing all decrease in metal-contaminated sites. There can be an overall reduction in decomposition and a change in litter from a mull to a mor type (Beyer 1988). The long-term effects of such changes are not clear but the implication is that productivity will slowly be reduced due to the slower nutrient turnover associated with mor litter types.

Waste brine is typically generated in association with petroleum production, and extensive areas of the US south and elsewhere that were formerly forested have been contaminated by high salt and Na concentrations (Newbold *et al.* 1988). When the source of salt is removed and soils are easily flushed by precipitation, recovery can be rapid, taking a few years (Auchmoody & Walters 1988). However, reforestation of sites where soil structure has been severely degraded can require leaching of salts, introduction of organic matter to increase soil structure and selection of salt-tolerant species (Dyer & Farrish 1997).

Even 'clean' energy production, can have negative impacts on forest growth. For instance, trees within the spray drift zone of a cooling tower of a geothermal electric generating plant in California (Glaubig & Bingham 1985) and in Italy (Bussotti *et al.* 1997) have been damaged by boron toxicity. Thus, it is clear that soil and site productivity can be degraded by nutrient and metal contaminants and that sustainable management of soil and site will include remediation of contaminated soils.

7.9 SOIL BIOLOGY

Examples of collapse of soil biotic systems exist, but they are rare except in association with salt or metal contamination as discussed in the previous section. Lal (1986b) describes an example of system collapse following application of the pesticide Furadan. This pesticide is a cholinesterase inhibitor and has broad toxicity. Two applications of it to control stem borer in corn killed nearly all earthworms, termites and other soil fauna, resulting in a collapse of soil structure and the beginning of a downward spiral of degradation due to decreased infiltration and increased erosion. Another, less dramatic example of direct biological effects is provided by Salminen and Haimi (1996), who demonstrated that European white birch (*Betula pendula*) growth and N concentrations of leaves were reduced in sodium pentachlorophenol (PCP) contaminated forest soils due to reductions in enchytraeid and collembola populations and reduced microbial biomass.

Fauna are key regulators of nutrient cycling through the breakdown of large organic debris into smaller pieces (comminution) and mixing of mineral and organic matter, grazing, predation on heterotrophic microbes and production of fecal matter that provides a microsite for other fauna and microbes. Disturbance associated with forest management affects soil biota, but results vary from site to site and it is difficult to draw general conclusions. The most basic of management activities, harvesting, has been evaluated in a number of studies and has been shown to affect faunal populations differently depending on the ecosystem. Studies by Huhta *et al.* (1969) in Finland, Blair and Crossley (1988) in North Carolina and Bird and Chatarpaul (1986) in

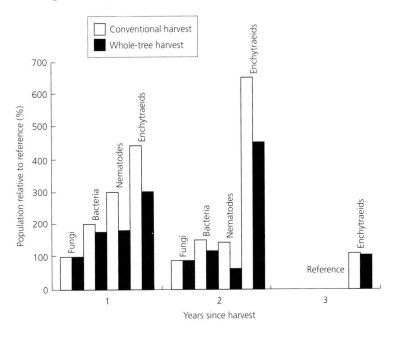

Fig. 7.9 Effects of conventional harvest that leaves foliage and branch residues on site or whole-tree harvest that removes most residues on soil biology. (Adapted from Shaw *et al.* 1991.)

Ontario have all shown that microarthropod densities in harvested forests are lower than in uncut stands for up to 8 years following harvest. Enchytraeid populations, which are particularly important in mor humus of boreal forests, appear to respond favourably to the increased slash immediately following harvest of these forests (Fig. 7.9), but these soft-bodied species appear sensitive to changes in salt concentration and populations may be reduced following fertilization (Huhta *et al.* 1986). Both of these changes appeared short-lived, with populations returning to levels near those of non-harvested controls within several years. Entry *et al.* (1986) studied microbial response to timber harvesting in the Rocky Mountains of Montana. They found few differences in microbial biomass or the ratio of fungal to bacterial mass as a result of harvest, but burning residues resulted in reductions in biomass and a shift toward bacteria. Few long-term comprehensive studies of soil biology following harvest have been done. In one such study, Houston *et al.* (1998) evaluated the effects of clearcut harvesting on general measures of biological activity such as microbial biomass and respiration rates as well as on faunal community structure. They were unable to detect significant differences between non-harvested and harvested stands for any of these measures. Bengtsson *et al.* (1997) evaluated the effect of logging residue addition or removal on populations of macroarthropods and enchytraeids 15–18 years following treatment. They found that residue removal changed the quantity of fauna and the number of some groups but no species or group disappeared.

Microbial community response to forest management activities is also variable. Populations of bacteria and fungi can be expected to respond to changes in temperature and moisture conditions of the microenvironments they inhabit and the availability of specific substrates they require; the duration of response can be expected to correlate well with the duration of these changes. Where moisture and temperature is more favourable following harvest, such as immediately following clearcut harvest in Sweden, overall increases in microbial populations may occur (Fig. 7.9).

Compaction can markedly alter soil fauna and microbial communities. Fungi are obligate aerobes and reduced oxygen availability in compacted soils will reduce the overall activity of fungi (Smeltzer *et al.* 1986) and the formation of mycorrhizae (Amaranthus *et al.* 1996), as well as reduce overall respiration rates as microbial respiration is shifted to less energy efficient metabolic pathways (Dulohery *et al.* 1996). Further, at least in some cases, compaction can increase populations of fungi such as *Fusarium* that are tree pathogens (Smeltzer *et al.* 1986). Undoubtedly, these microbial effects contribute to the reduced tree growth under compacted soil conditions.

Considerable information on the effect of fire on soil biology is available. Immediately following fire, populations of soil fauna and microorganisms are reduced (Jurgensen *et al.* 1979) or, in extreme cases, eliminated from the upper portion of the mineral soil (Cerri *et al.* 1985). Recovery of microorganisms is rapid. Many faunal species also recover quickly but populations of those fauna that utilize specific forest floor habitat destroyed by fire will require extended periods to recover.

In contrast to the generally short-term response of soil biology to harvest, liming appears to have long-term effects on soil fauna. In particular, prolonged increases in earthworm populations occur following liming of acidic soils (Persson 1988), and improvements in soil physical characteristics and nutrient availability associated with increased activity of this group of fauna occur.

There is little agreement as to the significance of these changes. In some cases, reductions in faunal populations have been associated with reduced litter decomposition rates such as for reduced mite populations described by Blair and Crossley (1988), but, as these investigators noted, even here a causal effect was not established. Perhaps we can conclude that management activities that do not create disturbances beyond the scope and duration of possible natural disturbances are unlikely to negatively affect long-term sustainability through detrimental impacts on soil biology. Harvest, soil tillage and fertilization are no more severe than insect-induced mortality, windthrow piles or burning in their effects on soil biology and would not be expected to reduce productivity even though they may alter community composition. By contrast, contamination or soil physical damage may results in shifts in biology that have lasting effects.

7.10 CONCLUSIONS

Disturbance is a feature of all forests. Hurricanes,

fire, cyclical outbreaks of insects are a part of natural forest systems and all lead to soil impacts. We do not believe these impacts lead to long-term loss of sustainability even though they lead to change. It is a common mistake to equate a measurement of soil change with a loss in sustainability.

It has been argued (see for example Maser 1994) that intensive forest management, particularly plantation management, is not sustainable. Certainly, characteristics of tree plantations differ from those of natural forests. Indeed, many aspects associated with natural forests are lost upon conversion to plantation management. Tree species diversity is reduced. Understorey species composition is altered and while species diversity as measured by several diversity indexes may be increased, rare species may be lost. Plantations are also considered candidates for insect and disease outbreaks. In a recent review of sustainability of plantation forestry, Evans (1999) describes more than a dozen examples of insect or disease outbreaks associated with plantation forestry. While these issues are important, they are not directly related to the potential of the soil and site to sustain productivity.

One of the key characteristics of sustainable soil management is the degree to which management is integrated with the social and economic fabric of each country. In industrialized nations, mechanized harvesting is the norm. As shown in this chapter, the impacts of such harvesting on nutrient removals and soil physical conditions can be large. If the same socioeconomic conditions that support such mechanized harvests also support the use of fertilizers to replace nutrients and require tillage or other ameliorative treatments to correct soil damage, site productivity can be maintained. By contrast, where conditions will not support the use of ameliorative treatments, greater levels of protection and reduced dependence on ameliorative treatments are required if sustainability is to be achieved.

In summary, we know that soils and forests are resilient. Sites that have been cleared, farmed and abandoned have returned to become productive forests. Nevertheless, many of our activities either intentional, such as the use of heavy machinery for forest harvest, or unintentional, such as long-term transport and deposition of acids, have been shown to decrease soil and site productivity. Without greater attention to protection of these resources we may find that significant efforts will be required to maintain current levels of productivity with less opportunity for increasing productivity to meet increasing world demands.

REFERENCES

Allen, H.L., Morris, L.A. & Wentworth, T.R. (1991) Productivity comparisons between successive loblolly pine rotations in the North Carolina Piedmont. In: Dyck, W.J. & Mees, C.A., eds. *Long-Term Field Trials to Assess Environmental Impacts of Harvesting. Proceedings of the IEA/BE T6/A6 Workshop, Florida, USA.* Bulletin 161, pp. 125–36. New Zealand Forest Research Institute, Rotorua.

Amaranthus, M.P., Page-Dumroese, D.S., Harvey, A., Cazares, E. & Bednar, L.F. (1996) *Soil Compaction and Organic Matter Affect Conifer Seedling Nonmycorrhizal and Ectomycorrhizal Root Tip Abundance and Diversity.* USDA Forest Service Research Paper. PNW-RP-494. Pacific Northwest Forest and Range Experiment Station, Portland, OR.

Andersson, F.O. & Lundkvist, H. (1989) Long-term Swedish field experiments in forest management practices and site productivity. In: Dyck, W.J. & Mees, C.A., eds. *Research Strategies for Long-Term Site Productivity. Proceedings of the IEA/BE A3 Workshop.* New Zealand Forest Research Institute Bulletin 152, pp. 125–37.

Attiwill, P.M., Turvey, N.D. & Adams, M.A. (1985) Effects of mound-cultivation (bedding) on concentration and conservation of nutrients in a sandy podzol. *Forest Ecology and Management* 11, 97–110.

Auchmoody, L.R. & Walters, R.S. (1988) Revegetation of a brine-killed forest site. *Soil Science Society of America Journal* 52, 277–80.

Aust, W.M., Burger, J.A., McKee, W.H., Scheerer, G.A., Jr & Tippett, M.D. (1998) Bedding and fertilization ameliorate effects of designed wet-weather skid trails after four years for loblolly pine (*Pinus taeda*) plantations. *Southern Journal of Applied Forestry* 22, 222–6.

Bell, T.I.W. (1973) Erosion in Trinidad teak plantations. *Commonwealth Forestry Review* 52, 223–33.

Bengtsson, J., Persson, J. & Lundkvist, H. (1997) Long-term effects of logging residue addition and removal on macroarthropods and enchytraeids. *Journal of Applied Ecology* 34, 1014–1022.

Benyon, R.G., Myers, B.J., Theiveyanathan, S. *et al.* (1996) Effects of salinity on water use, water stress and growth of effluent-irrigated *Eucalyptus grandis.*

In: *Proceedings of the 14th Land Treatment Collective Meeting, September 30–October 1, Canberra,* pp. 169–74. CSIRO Forestry and Forest Products, Canberra.

Beyer, W.N. (1988) Damage to the forest ecosystem on Blue Mountain from zinc smelting. In: *Proceedings of the University of Missouri's Annual Conference on Race Substances in Environmental Health,* pp. 249–62. University of Missouri, Columbia, MO.

Bigger, C.M. & Cole, D.W. (1983) Effects of harvest intensity on nutrient losses and future productivity in high and low productivity red alder and Douglas-fir stands. In: *Proceedings of the Symposium on Forest Site and Continuous Productivity.* USDA Forest Services General Technical Report PNW-163, pp. 167–87. Pacific Northwest Forest and Range Experiment Station, Portland, OR.

Binkley, D., Driscoll, C.T., Allen, H.L., Schoeneberger, P. & McAvoy, D. (1989) Acidic deposition and forest soils. In: *Ecological Studies 72,* pp. 12–24. Springer-Verlag, New York.

Bird, G.A. & Chatarpaul, L. (1986) Effect of whole-tree and conventional forest harvest on soil microarthropods. *Canadian Journal of Zoology* 64, 1986–93.

Blair, J.M. & Crossley, D.A. (1988) Litter decomposition, nitrogen dynamics and litter microarthropods on a southern Appalachian hardwood forest 8 years following clearcutting. *Journal of Applied Ecology* 25, 683–98.

Bockheim, J.G., Benzel, T.C., Rui-Lin, L. & Theil, D.A. (1988) Sludge increases pulpwood production. *Biocycle* 29, 57–9.

Brand, D.G. (1998) Criteria and indicators for the conservation and sustainable management of forests: progress to date and future directions. *Biomass and Bioenergy* 13, 247–53.

Buckman, H.O. & Brady, N.C. (1969) *The Nature and Properties of Soils.* Macmillan, New York.

Bussotti, F.E., Cenni, A., Cozzi & Ferretti, M. (1997) The impact of geothermal power plants on forest vegetation: a case study at Travale (Tuscany, central Italy). *Environmental Monitoring and Assessment* 45, 181–94.

Carlyle, J.C. (1993) Organic carbon in forested sands soils: properties, processes and the impact of forests management. *New Zealand Journal of Forest Science* 23, 390–402.

Carmean, W.H. (1975) Forest site quality in the United States. *Advances in Agronomy* 27, 209–69.

Cerri, C.C.B., Volkoff, B. & Eduardo, B.P. (1985) Efieto do desmatamento sobre a biomassa microbiana em latossolo amarelo da Amazonia. *Revista Brasilera Cien Solo* 9, 1–4.

Clayton, J.L., Kellogg, G. & Forester, N. (1987) *Soil Disturbance–Tree Growth Relations in Central*

Idaho Clearcuts. USDA Forest Service Research Note INT-372. USDA Forest Service Intermountain Forest Experiment Station, UT.

Cole, D.W. & Rapp, M. (1981) Elemental cycling in forest ecosystems. In: Riechle, D.E., ed. *Dynamic Processes in Forest Ecosystems*, pp. 341–409. Cambridge University Press, London.

Coleman, D.C. & Crossley, D.A., Jr (1996) *Fundamentals of Soil Ecology*. Academic Press, San Diego.

Comerford, N.B. & Skinner, M.F. (1989) Residual phosphorus solubility for acid, clayey, forested soil in the presence of oxalate and citrate. *Canadian Journal of Soil Science* 69, 111–17.

Costantini, A.H., So, H.B. & Doley, D. (1996) Early *Pinus caribaea* var. hondurensis root development 1. Influence of matric suction. *Australian Journal of Experimental Agriculture* 36, 839–47.

Costantini, A.H., Dunn, G.M. & Grimmet, J.L. (1997) Towards sustainable management of forest plantations in south-east Queensland. II. Protecting soil and water values during second rotation Pinus plantation management. *Australian Forestry* 60, 226–32.

Crohn, D.M. (1995) Sustainability of sewage sludge land application to northern hardwoods. *Ecological Applications* 5, 53–62.

Daniels, W.L., Stuczynski, T.I. & Chaney, R.L. (1999) Selection of grass species and amendments for revegetation of Pb/Zn smelter wastes. In: *Proceedings of the 5th International Conference on Biogeochemistry of Trace Elements, July 11–15, Vienna*, vol. II, pp. 868–9. International Society for Trace Element Research, Vienna, Austria.

Drengson, A. & Taylor, D. (1997) An overview of ecoforestry. In: Drengson, A. & Taylor, D., eds. *Ecoforestry, the Art and Science of Sustainable Forest Use*, pp. 17–32. Ecoforestry Institute, Glendale, OR.

Drew, M.C. (1979) Plant responses to anaerobic conditions in soil and solution culture. *Current Advances in Plant Sciences* 36, 1–14.

Drozd, J. *et al.* (1998) *The Degradation of Soils in Devastated Ecosystems of the Karkonosze Mountains and Possibilities of the Prevention*. Wroclaw, PTSH, 125.

Dulohery, C.J., Morris, L.A. & Lowrance, R. (1996) Assessing forest soil disturbance through biogenic gas fluxes. *Soil Science Society of America Journal* 60, 291–8.

Dyer, J.M. & Farrish, K.W. (1997) Reforestation of a brine contaminated oil field site. *Land and Water* July/August 38–40.

Dyrness, C.T. (1967) *Mass Soil Movements in the H.J. Andrews Experimental Forest*. USDA Forest Service Research Paper PNW-42. Pacific Northwest Forest and Range Experiment Station, Portland, OR.

Eijsackers, H. & Van de Bund, C.F. (1980) Effects on soil fauna. In: Hance, R.J., ed. *Interactions Between Herbicides and the Soil*, pp. 255–305. Academic Press, New York.

Elliot, W.J., Page-Dumroese, D. & Robichaud, P.R. (1998) The effects of forest management on erosion and soil productivity. In: Lal, R., ed. *Erosional Impact on Soil Quality*. CRC Press, Boca Raton, FL.

Entry, J.A., Stark, N.M. & Lowenstein, H. (1986) Effect of timber harvesting on microbial biomass fluxes in a northern Rocky Mountain forest soil. *Canadian Journal of Forest Research* 16, 1076–81.

Eswaran, H.S., Virmani M. & Spivey, L.D., Jr (1993) Sustainable agriculture in developing countries: constraints, challenges and choices. In: *Technologies for Sustainable Agriculture in the Tropics*. American Society of Agronomy Special Publication 56, pp. 7–24. Maidson, WI.

Eswaran, H., Almaraz, R., Reich, P. & Zdruli, P. (1997) Soil quality and soil productivity in Africa. *Journal of Sustainable Agriculture* 10, 75–94.

Evans, J. (1996) The sustainability of wood production from plantations: evidence over three successive rotations in the Usutu Forest, Swaziland. *Commonwealth Forestry Review* 75, 234–9.

Evans, J. (1999) *Sustainability of Forest Plantations: The Evidence*. The Department for International Development, London.

Feller, M.C. & Kimmins, J.P. (1984) Effects of clearcutting and slash burning on streamwater chemistry and watershed nutrient budgets in southwest British Columbia. *Water Resources Research* 20, 29–40.

Fox, T.R., Morris, L.A. & Maimone, R.A. (1988) The impact of windrowing on the productivity of a rotation age loblolly pine plantation. In: *Fifth Biennial Southern Silviculture Research Conference*. US Forest Services General Technical Report SO-74, pp. 133–139. USDA Forest Service Southeast Experiment Station, New Orleans, LA.

Fox, T.R., Comerford, N.B. & McFee, W.W. (1990) Kinetics of phosphorus release from spodosols: effects of oxalate and formate. *Soil Science Society of America Journal* 54, 1763–7.

Fredriksen, H.A., Moore, D.G. & Norris, L.A. (1975) The impact of timber harvest, fertilization and herbicide treatment on streamwater quality in western Oregon and Washington. In: Bernier, B. & Winget, C.H., eds. *Forest Soils and Forest Land Management*, pp. 283–313. Laval University Press, Quebec.

Froehlich, H.A. (1979) Soil compaction from logging equipment: effects on growth of young ponderosa pine. *Journal of Soil Water Conservation* 34, 276–8.

Froehlich, H.A., Miles, D.W.R. & Robbins, R.W. (1986) Growth of young *Pinus ponderosa* and *Pinus contorta* on compacted soils in central Washington. *Forest Ecology and Management* 15, 285–94.

Gale, M.R. & Grigal, D.F. (1990) Development of a soil- and root-based productivity index model for trembling aspen. In: Dixon, R.K., Meldahl, R.S., Ruark, G.A. & Watten, W.C., eds. *Process Modeling* of Forest Growth Responses to Environmental Stress, pp. 303–9. Timber Press, Portland.

Gent, J.A., Jr, Ballard, R. & Hassan, A.E. (1983) The impact of harvesting and site preparation on the soil physical properties of lower Coastal Plain soils. *Soil Science Society of America Journal* **47**, 173–7.

Gent, J.A., Ballard, R., Jr, Hassan A.E. & Cassel, D.K. (1984) Impact of harvesting and site preparation on physical properties of Piedmont forest soils. *Soil Science Society of America Journal* **48**, 173–177.

Gentle, S.W.F.R., Humphreys & Lambert, M.J. (1986) An examination of a *Pinus radiata* fertilizer trial fifteen years after treatment. *Forest Science* **32**, 822–9.

Glaubig, B.A. & Bingham, F.T. (1985) Boron toxicity characteristics of four northern California endemic tree species. *Journal of Environmental Quality* **14**, 72–7.

Golley, F.B., McGinnis, J.T., Clements, R.G., Child, G.I. & Duever, J.J. (1975) *Mineral Cycling in a Tropical Moist Forest Ecosystem.* University of Georgia Press, Athens.

Greaves, M.P. & Malkoney, H.P. (1980) Effects on soil microflora. In: Hance, R.J., ed. *Interactions Between Herbicides and the Soil*, pp. 223–53. Academic Press, New York.

Grecean, E.L. & Sands, R. (1980) Compaction of forest soils: a review. *Australian Journal of Soil Research* **18**, 163–89.

Gunn, J.M., Conroy, W.E., Lautenbach, W.E. *et al.* (1995) From restoration to sustainable ecosystems. In: *Restoration and Recovery of an Industrial Region.* Springer Verlag, New York. 335–44.

Harrison, R.B., Henry, C.L., Cole, D.W. & Dongsen Xue. (1995) Long-term changes in organic matter in soils receiving applications of municipal biosolids. In: McFee, W.W. & Kelly, L.M., eds. *Carbon Forms and Functions in Forest Soils*, pp. 139–53. Soil Science Society of America, Madison, WI.

Hastings, F.L., Brady, U.E. & Jones, A.S. (1989) Lindane and fenitrothion reduce soil and litter mesofauna on Piedmont and Appalachian sites. *Environmental Entomology* **18**, 245–50.

Hatchell, G.E. & Ralston, C.W. (1971) Natural recovery of surface soils disturbed in logging. *Tree Planter's Notes* **22** (2), 5–9.

Hatchell, C.E., Ralston, C.W. & Foil, R.R. (1970) Soil disturbances in logging. *Journal of Forestry* **68**, 772–5.

Haywood, J.D. (1980) *Planted Pines do not Respond to Bedding on an Acadia-Beauregard-Kolen Silt Loam Site.* USDA Southern Forest Experiment Station Research Note SO-259. USDA Forest Service Southern Forest Experiment Station, New Orleans, LA.

Helmisaari, H.-S., Derome, J., Fritze, H. *et al.* (1995) Copper in Scots pine forests around a heavy metal smelter in southwestern Finland. *Water, Air and Soil Pollution* **85**, 1727–32.

Helms, J.A. & Hipkin, C. (1986) Effects of soil com-

paction on tree, in a California ponderosa pine plantation. *Western Journal of Applied Forestry* 121–4.

Henry, C.L. (1986) Growth response, mortality and foliar nitrogen concentrations of four tree species treated with pulp and paper and municipal sludges. In: Cole, D.W., Henry, C.L. & Nutter, W.L., eds. *The Forest Alternative for Treatment and Utilization of Municipal and Industrial Wastes*, pp. 258–65. University of Washington Press, Seattle.

Henry, C.L. & Cole, D.W. (1998) Use of biosolids in the forest: technology, economics and regulations. *Biomass and Bioenergy* **13**, 269–77.

Holman, G.T., Knight, F.B. & Struchtemeyer, R.A. (1978) *The Effects of Mechanized Harvesting on Soil Conditions in the Spruce–Fir Region of North-Central Maine.* University of Maine. Life Sciences and Agricultural Experimental Station Bulletin 751. Orono, ME.

Holsher, D., Moller, R.F., Denisch, M. & Folster, H. (1996) *Nutrient Cycling in Agroecosystems* **47**, 49–57. The Netherlands.

Houston, A.P.C., Visser, S. & Lautenschlager, R.A. (1998) Microbial processes and fungal community structure in soils from clear-cut and unharvested areas of two mixed wood forests. *Canadian Journal of Botany* **76**, 630–40.

Hughes, J.H., Campbell, R.G., Duzan, H.W. & Dudley, C.S. (1979) *Site Index Adjustments for Intensive Forest Management Treatments at North Carolina.* Weyerhaeuser Forest Research Technical Report 042–1404/79/24. Weyehauser Co., NC.

Huhta, V.M., Nurminen & Valpas, A. (1969) Further notes on the effect of silvicultural practices on the fauna of coniferous forest soil. *Annales Zoologici Fennici* **6**, 327–34.

Huhta, V.M., Hyvonen, R., Koskenniemi, A., Vilkamaa, P., Kaasalainen, P. & Sulander, M. (1986) Response of soil fauna to fertilization and manipulation of pH in coniferous forests. *Acta Forestica Fennici* **195**, 30 pp.

Johnson, D.W. (1992) The effects of forest management on soil carbon storage. *Water, Air and Soil Pollution* **64**, 83–120.

Johnson, D.W., Edwards, N.T. & Todd, D.E. (1980) Nitrogen mineralization, immobilization and nitrification following urea fertilization of a forest soil under field and laboratory conditions. *Soil Science Society of America Journal* **44**, 610–16.

Johnson, D.W. & Todd, D.E., Jr (1997) Effects of harvesting intensity on forest productivity and soil carbon storage in a mixed oak forest. In: Lal, R., Kimble, J. M., Follet, R.F. & Stewart, B.A., eds. *Management of Carbon Sequestration in Soil*, pp. 351–63. CRC Press, Boca Raton.

Jung, M.C. & Thorton, I. (1996) Heavy metal contamination of soils and plants in the vicinity of a lead-zinc mine, Korea. *Applied Geochemistry* **11**, 53–9.

Jurgensen, M.F., Larsen, M.J. & Harvey, A.E. (1979)

Forest Soil Biology–Timber Harvesting Relationships: A Perspective. USDA Forest Service General Technical Report INT-69. Intermountain Forest and Range Experiment Station, Ogden, UT.

Kimmins, J.P. (1996) Importance of soil and role of ecosystem disturbance for sustained productivity of cool temperate and boreal forests. *Soil Science Society of America Journal* **60**, 1643–54.

Kimmins, J.P. & Scoullar, K.A. (1995) Incorporation of nutrient cycling in the design of sustainable, stand-level, forest management systems using the ecosysteem management model FORCAST and its output format FORTOON. In: Nilsson, L.O., ed. *Nutrient Uptake and Cycling in Forest Ecosystems.* Ecosytem Research Report No. 13. UNR 15405.

Klock, G.O. (1982) Some effects of soil erosion on forest and range soil productivity. In: *Determinants of Soil Loss Tolerance*, pp. 53–65. American Society of Agronomy and Soil Science Society of America, Madison WI.

Lal, R. (1985) Soil erosion and its relation to productivity in tropical soils. In: Swaify, S.A., Moldenhauer, W.C. & Lo, A., eds. *Soil Erosion and Conservation*, pp. 237–61. Soil Conservation Society of America, Ankeny, IO.

Lal, R. (1986a) Conversion of a tropical rainforest: agronomic potential and ecological consequences. *Adances in Agronomy* **39**, 173–259.

Lal, R. (1986b) Soil surface management in the tropics for intensive land use and sustained production. *Advances in Soil Science* **5**, 1–109.

Lal, R. (1998a) Soil quality and sustainability. In: Lal, R., Blum, W.H., Valentine, C. & Stewart, B.A., eds. *Methods of Assessment of Soil Degradation. Advances in Soil Science*, pp. 17–30. CRC Press, Boca Raton.

Lal, R. (1998b) Soil erosion impact on agronomic productivity and environmental quality. *Critical Reviews in Plant Sciences* **17**, 3419–64.

Langdale, G.W., Box, J.E., Leonard, R.A., Jr, Barnett, A.P. & Fleming, W.G. (1979) Corn yield reduction in eroded southern Piedmont soils. *Journal of Soil Water Conservation* **34**, 226–8.

Larney, F.J., Izaurralde, R.C., Janzen, H.H., *et al.* (1995) Soil erosion crop productivity relationships for six Aberta soils. *Journal of Soil Water Conservation* **50**, 87–91.

Likens, G.E., Bormann, F.H., Johnson, N.M., Fisher, D.W. & Pierce, R.S. (1970) Effects of forest cutting and herbicide treatment on nutrient budgets in the Hubbard Brook ecosystem in New Hampshire. *Ecological Monographs* **40**, 23–47.

Lockaby, B.G. & Vidrine, C.G. (1984) Effects of logging equipment traffic on soil density and growth and survival of loblolly pine. *Southern Journal of Applied Forestry* **8**, 109–12.

Ludovici, K.H. & Morris, L.A. (1997) Competition-induced reductions in soil water availability reduced pine root extension rates. *Soil Science Society of America Journal* **61**, 1196–202.

Lundkvist, H. (1983) Effects of clear-cutting on the enchytraeids in a Scots pine forest soil in central Sweden. *Journal of Applied Ecology* **20**, 873–85.

Mace, A.C., Jr (1971) *Recovery of Forest Soils from Compaction by Rubber-tired Skidders.* Minnesota Forestry Research Notes 226. University of Minnesota, MN.

McKee, W.H., Jr, McLeod, K.W., Davis, C.E., McKevlin, M.R. & Thomas, H.A. (1986) Growth response of loblolly pine to municipal and industrial sewage sludge applied to four ages on upper Coastal Plain sites. In: Cole, D.W., Henry, C.L. & Nutter, W.L., eds. *The Forest Alternative for Treatment and Utilization of Municipal and Industrial Wastes*, pp. 272–81. University of Washington Press, Seattle.

McKee, W.H. Jr & Wilhite, L.P. (1986) Loblolly pine response to bedding and fertilization varies by drainage class on lower Atlantic Coastal Plain sites. *Southern Journal of Applied Forestry* **10**, 16–21.

Malcolm, D.C. & Moffat, A.J. (1996) Forestry, soils and sustainability. In: Taylor, A.G., Gordon, J.E. & Usher, M.B., eds. *Soils, Sustainability and Natural Heritage*, pp. 194–209. Scottish Natural Heritage, HMSO, Edinburgh.

Malmer, A. (1996) Hydrological effects and nutrient losses of forest plantation establishment on tropical rainforest land in Sabah. *Malaysian Journal of Hydrology* **174**, 129–48.

Maser, C. (1994) *Sustainable Forestry. Philosophy, Science and Economics.* Delray Press, St Lucie Beach, FL.

Mason, E.G., Cullen, A.W.J. & Rijkse, W.C. (1988) Growth of two *Pinus radiata* stock types on ripped and ripped/bedded plots at Karioi forest. *New Zealand Journal of Forest Science* **18**, 287–96.

Mayer, R. (1998) Soil acidification and cycling of metal elements: cause–effect relationships with regard to forestry practices and climatic changes. *Agriculture, Ecosystems and Environment* **67**, 145–52.

Mead, D.J. & Gadgil, R.L. (1978) Fertilizer use in established radiata pine stands in New Zealand. *New Zealand Journal of Forest Science* **8**, 105–34.

Mench, M., DeKoe, T., Vansgronsveld, J., Bussiere, S., Boisson, J. & Masson, P. (1999) Remediation of the Hales mine spoil by inactivation and phytostabilisation: study in lysimeters. In: Wenzel, W.W., Adriano, D.C., Alloway, H.E. *et al.* eds. *5th International Conference on the Biogeochemistry of Trace Elements. July 11–15 1999, Vienna*, pp. 914–15. International Society for Trace Element Research, Vienna, Austria.

Miller, H.G. (1981) Forest fertilization: some guiding concepts. *Forestry* **54**, 157–67.

Miller, H.G. & Miller, J.D. (1991) Energy forestry—the nutritional equation. In: Aldhous, B.A., ed. *Wood for Energy—the Implications for Harvesting Utilisation and Marketing.* Institute of Chartered Forestry, Edinburgh.

Miller, R.E., Boyle, J.R., Harvey, A.E., Ballard, T.M., Palazzi, L.M. & Powers, R.F. (1992) Fertilizers and other means to maintain long-term productivity of western forests. In: Chappell, H.N., Weetman, G.F. & Miller, R.E., eds. *Forest Fertilization: Sustaining and Improving Nutrition and Growth of Western Forests.* Institute of Forest Resources Contribution 73, pp. 203–222. University of Washington, Seattle, WA.

Miller, R.E., Scott, W. & Hasard, J.W. (1996) Soil compaction and conifer growth after tractor yarding at three Coastal Washington locations. *Canadian Journal of Forest Research* **26**, 225–36.

Miller, R.E., Colbert, S.A. & Morris, L.A. (2000) *Effects of Heavy Equipment on Physical Properties of Soils and on Long-term Productivity: A Review of Literature and Current Research.* NCASI Technical Bulletin. New York, National Council for Air and Stream Improvement, Research Triangle Park, NC.

Mittman, H.W. (1991) Reactions of mite populations to the influence of environmental chemicals in a beech-wood floor. In: Schuster, R. & Murphy, P.W., eds. *The Acari: Reproduction, Development and Life History Strategies*, pp. 495–6. Chapman & Hall, New York.

Morris, L.A. & Lowery, R.F. (1988) Influence of site preparation on soil conditions affecting stand establishment and tree growth. *Southern Journal of Applied Forestry* **12**, 170–8.

Morris, L.A. & Miller, R.E. (1994) Evidence for long-term productivity changes as provided by field trials. In: Dyck, W.J., Cole, D.W. & Comerford, N.B., eds. *Impacts of Forest Harvesting on Long-Term Site Productivity*, pp. 41–80. Chapman & Hall, London.

Morris, L.A. & Pritchett, W.L. (1982) Nutrient storage and availability in two managed pine flatwoods forests. In: Coleman, S., Mace, A. & Swindel, B., eds. *Impacts of Intensive Forest Management Practices Symposium Proceedings, March 9–10, 1982*, pp. 17–23. School of Forest Resources and Conservation, University of Florida, Gainesville, FL.

Morris, L.A., Pritchett, W.L. & Swindel, B.F. (1983) Displacement of nutrients into windrows during site preparation of a pine flatwoods forest. *Soil Science Society of America Journal* **47**, 591–4.

Mukhtar, S.J.L., Baler & Kanwar, R.S. (1988) *Soil Aeration and Crop Growth Response to Excess Water.* American Society of Agricultural Engineering Paper 88–2611. American Society of Agricultural Engineers, St. Joseph, MI.

Nambiar, E.K.S. (1996) Sustained productivity of forests is a continuing challenge to soil science. *Soil Science Society of America Journal* **60**, 1629–42.

Neary, D.G., Schneider, G. & White, D.P. (1975) Boron toxicity in red pine (*Pinus resinosa*) following municipal wastewater irrigation (needle tip discoloration, foliar analysis). *Proceedings of the Soil Science Society of America* **39**, 981–2.

Newbold, R.A., Lockaby, B.G. & Thompson, C.E. (1988) Conversion of forestland to oil and gas production in northern Louisiana: an evaluation. *Journal of Soil and Water Conservation* **43**, 331–3.

Nwoboshi, L.C. (1983) Potenital impacts of some harvesting options on nutrient budgets of a *Gmelina* pulpwood plantation in Nigeria. In: Ballard, R. & Gessel, S.P., eds. *IUFRO Symposium Forest Site and Continuous Productivity.* US Forest Service General Techical Report PN W-163, pp. 212–17. Pacific Northwest Forest and Experiment Station, Portland, OR.

Nykvist, N. (1998) Logging can cause a serious lack of calcium in tropical rainforest ecosystems: an example from Sabah, Malaysia. In: Schulte, A. & Ruhiyat, D., eds. *Soils of Tropical Forest Ecosystems: Characteristics, Ecology and Management*, pp. 7–91. Springer-Verlag, Berlin.

Page-Dumroese, D.S., Jurgensen, M.F., Graham, R.T. & Harvey, A.E. (1989) *Soil Chemical Properties of Raised Planting Beds in a Northern Idaho Forest.* USDA Forest Service Research Paper INT 419, 7p. USDA Intermountain Research Station, Ogden, UT.

Pastor, J. & Post, W.M. (1985) *Development of a Linked Productivity–Soil Process Model.* ORNL/TM-9519. Oak Ridge National Environmental Laboratory, Environmental Science Division, Oak Ridge, TN.

Perhutani, P. (1992) Teak in Indonesia. In: H.Wood, H., ed. *Proceedings of a Regional Seminar on Teak in Asia.* FORSPA Publications 4. FAO, Bangkok.

Perry, T.O. (1964) Soil compaction and loblolly pine growth. *Tree Planter's Notes* **67**, 9.

Persson, T. (1988) Effects of liming on the soil fauna in forests—a literature review. *Swedish Environmental Protection Board Report* **3418**, 47–92.

Poulton, P.R. (1996) The Rothamsted long-term experiments: are they still relevant? *Canadian Journal of Plant Sciences* **76**, 559–71.

Powers, R.F., Alban, R.E., Miller, A.E. *et al.* (1990) Sustaining site productivity in North American forests: problems and prospects. In: Gessel, S.P., Lacate, D.S., Weetman, G.F. & Powers, R.F., eds. *Sustained Productivity of Forest Soils, Proceedings of the 7th North American Forest Soils Conference,* University of British Columbia, Vancouver, pp. 49–79.

Powers, R.F., Mead, D.J., Burger, J.A. & Ritchie, M.W. (1994) Designing long term site productivity experiments. In: Dyck, W.J., Cole, D.W. & Comerford, N.B., eds. *Impacts of Forest Harvesting on Long-Term*

Site Productivity, pp. 247–86. Chapman & Hall, London.

Pritchett, W.L. & Comerford, N.B. (1982) Long-term response to phosphorus fertilization on selected southeastern Coastal Plain soils. *Soil Science Society of America Journal* **46**, 640–4.

Riekerk, H. (1983) Impacts of silviculture on flatwoods runoff, water quality and nutrient budgets. *Water Resources Bulletin* **19**, 73–9.

Salminen, J. & Haimi, J. (1996) Effect of pentachlorophenol in forest soil: a microcosm experiment for testing ecosystem responses to anthropogenic stress. *Biology and Fertility of Soils* **23**, 182–8.

Sanchez, F.G. (1998) Soil organic matter and soil productivity: searching for the missing link. In: *The Productivity and Sustainability of Southern Forest Ecosystems in a Changing Environment*, pp. 543–56. Springer-Verlag, New York.

Shaw, C.H., Lundkvist, H., Moldenke, A. & Boyle, J.R. (1991) The relationship of soil fauna to long-term productivity in temperate and boreal ecosystems: processes and research strategies. In: Dyck, W.J. & Mees, C.A., eds. *Long Term Field Trials to Assess Environmental Impacts of Harvesting. Proceedings of the IEA/BE T6/A6 Workshop. Forest Research Institute Bulletin 161, Rotorua, NZ*, pp. 39–77. New Zealand Forest Research Institute, Rotorua.

Sidle, R.C. (1991) A conceptual model of changes in root cohesion in response to vegetation management. *Journal of Environmental Quality* **20**, 43–52.

Simmons, G.L. & Pope, P.E. (1988) Influence of soil water potential and mycorrhizal colonization on root growth of yellow-poplar and sweetgum seedlings grown in compacted soil. *Canadian Journal of Forest Research* **18**, 1392–6.

Smeltzer, D.L.K., Bergdahl, D.R. & Donnelly, J.R. (1986) Forest ecosystem responses to artificially induced soil compaction II. selected soil microorganism populations. *Canadian Journal of Forest Research* **16**, 870–82.

Smethurst, P.J. & Nambiar, E.K.S. (1990) Effects of slash and litter management on fluxes of nitrogen and tree growth in a young *Pinus radiata* plantation. *Canadian Journal of Forest Research* **20**, 1498–507.

Smith, R.E. & Whitton, J.S. (1975) Note on boron toxicity in a stand of radiata pines in Hawkes Bay. *New Zealand Journal of Science* **18**, 109–14.

Sollins, R.C., Grier, F.M., McCorison, K., Cromack, Jr & Fogel, R. (1980) The internal cycles of an old growth Douglas-fir ecosystem in western Oregon. *Ecological Monographs* **50**, 261–85.

Stepanov, A.M. & Cheren'kova, T.V. (1990) Study of forest biocenoses near a copper smelting plant. *Biological Bulletin of the Academy of Sciences of the USSR* **16**, 212–18.

Stewart, C.C. (1995) *Harvesting and site preparation*

effects on soil physical properties 12 years after establishment of a loblolly pine plantation. MSc Thesis, North Carolina State University, Raleigh, NC.

Tamm, C.O. (1969) Site damage by thinning due to removal of organic matter and plant nutrients. In: *Proceedings of the IUFRO Meeting on Thinning and Mechanization, Stockholm*, pp. 175–9. Royal College of Forestry, Stockholm, Sweden.

Taylor, H.M., Roberson, G.M. & Parker, J.J., Jr (1966) Soil strength-root penetration relations for medium to coarse-textured soil materials. *Soil Sciences* **102**, 18–22.

Tew, D.T., Morris, L.A., Allen, H.L. & Wells, C.G. (1986) Estimates of nutrient removal, displacement and loss resulting from harvest and site preparation of a *Pinus taeda* plantation. *Forest Ecology and Management* **15**, 257–67.

Thomas, P.J., Simpson, T.W. & Baker, J.C. (1989) Erosion effects on productivity of Cecil soils in the Virginia Piedmont. *Soil Science Society of America Journal* **53**, 928–33.

Tiarks, A.E. (1982) Effect of site preparation and fertilization on slash pine growing on a good site. In: *Proceedings of the Second Biennial Southern Silvicultural Research Conference, November 1982, Atlanta, GA. USDA Forest Services General Technical Report SE-24*, pp. 34–9. USDA Forest Service Southern Forest Experiment Station, New Orleans, LA.

Torreano, S.J. (1992) *Effects of soil water availability, aeration and soil mechanical impedance on loblolly pine* (Pinus taeda L.) *root development*. PhD Dissertation. The University of Georgia, Athens.

Torreano, S.J. & Morris, L.A. (1998) Loblolly pine root growth and distribution under water stress. *Soil Science Society of America Journal* **62**, 818–27.

Trimble, S.W. (1974) *Man-induced Soil Erosion on the Southern Piedmont 1700–1970*. Soil Conservation Society, Ankeny, Iowa.

Vaamonde, C.M.L., Alvarez, E., Fernandez & Marcos, M.L. (1999) Heavy metals in a mine dump in Galicia NW Spain. In: *II Proceedings of the 5th International Conference on Biogeochemistry of Trace Elements, July 11–15, Vienna*, pp. 976–7. International Society for Trace Element Research, Vienna, Austria.

Vance, E.D. (1996) Land application of wood-fired and combination boiler ashes: an overview. *Journal of Environmental Quality* **25**, 937–44.

Van Lear, D.H., Douglas, J.E., Cos, S.K. & Augspurger, M.K. (1985) Sediment and nutrient export in runoff from burned and harvested pine watershed in the South Carolina Piedmont. *Journal of Environmental Quality*. **14**, 169–74.

Van Lear, D.H., Kapeluck, P.R. & Parker, M.M. (1995) Distribution of carbon in Piedmont soil as affected by loblolly pine management. In: McFee, W.W. & Kelly,

L.M., eds. *Carbon Forms and Functions in Forest Soils*, pp. 489–502. Soil Science Society of America, Madison WI.

Vitousek, P.M., Andariese, S.W., Matson, P.A., Morris, L. & Sanford, R.L. (1992) Effects of harvest intensity, site preparation and herbicide use on soil nitrogen transformations in a young loblolly pine plantation. *Forest Ecology and Management* **49**, 277–92.

Vomocil, J.A. & Flocker, W.J. (1961) Effect of soil compaction on storage and movement of soil air and water. *Transactions of the American Society of Agricultural Engineering* **4**, 242–6.

Wander, M.M., Traina, S.J., Stinner, B.R. & Peters, S.E. (1994) Organic and conventional management effects on biologically active soil organic pools. *Soil Science Society of America Journal* **58**, 1130–9.

Webber, B. (1978) Potential increases in nutrient requirements of *Pinus radiata* under intensified management. *New Zealand Journal of Forest Science* **8**, 146–59.

Wells, R.E., Campbell, L.F., Debano, C.E. *et al.* (1979) *Effects of Fire on Soil*. USDA Forest Service General Technical Report WO-7. USDA Forest Service, Washington DC.

Whittaker. R.H., Likens, G.E., Bormann, F.H., Eaton, J.S. & Siccama, T.G. (1979) The Hubbard Brook ecosystem study: forest nutrient cycling and element behavior. *Ecology* **60**, 203–70.

Whynot, T.W. & Weetman, G.F. (1991) Repeated fertilization effects on nitrogen fluxes measured by sequential coring. *Soil Science Society of America Journal* **55**, 1101–11.

Wiersum, K.F. (1985) Effects of various vegetation layers in an *Acacia auriculiformis* forest plantation on surface erosion in Java, Indonesia. In: El-Swaify, S., Moldenhauer, W.C. & Lo, A., eds. *Soil Erosion and Conservation*, pp. 79–89. Soil Conservation Society of America, Ankeny, IA.

Winterhalder, K. (1984) Environmental degradation and rehabilitation in the Sudbury area. *Laurentian University Review* **16** (2), 15–47.

Worrell, R. & Hampson, A. (1997) The influence of some forestry operations on the sustainable management of forest soils—a review. *Forestry* **70**, 61–85.

Wotton, D.L., Jones, D.C. & Phillips, S.F. (1986) The effect of nickel and copper deposition from a mining and smelting complex on coniferous regeneration in the boreal forest of northern Manitoba. *Water, Air and Soil Pollution* **31**, 349–58.

Yoho, N.S. (1980) Forest management and sediment production in the South—a review. *Southern Journal of Applied Forestry* **4**, 27–36.

8: Management of Pest Threats

HUGH F. EVANS

8.1 THE AIM OF INTERVENING IN MANAGED FORESTS

The ecological principles that determine the diversity and abundance of herbivores and other invertebrates associated with trees are discussed in Chapter 6 in Volume 1 of this handbook. Balancing forces acting at various trophic levels tend to sustain quite complex populations of invertebrates within a given forest or woodland community such that resource availability is not usually a limiting factor. However, changes in one or more of the attributes can arise either naturally or by human influence. Indeed, there are few forest ecosystems worldwide that have not been substantially altered by human activity. This is particularly true in Europe and especially in Britain where there is virtually no 'natural' forest remaining. With such massive intervention as the 'norm', it is pertinent to consider how such management can result in pest infestations and lead to the need to develop regimes for both pest management and, at the same time, maintenance or enhancement of wider biodiversity.

8.1.1 Outbreaks of invertebrate pests: setting thresholds

In relation to invertebrate pressures on trees, there is no universal definition of an outbreak population. This is particularly the case in forestry where the tolerance of trees to attack may enable large populations of invertebrates to build-up without leading to management intervention. Thus, rather than refer to outbreaks of particular pests, it is more pertinent to consider

damage thresholds, irrespective of the population size that is required to exceed that threshold. Naturally, once the threshold is defined, the question of reducing population size becomes important but, even then, the quantitative basis for intervention may be imprecise and measured more by damage reduction than by pest population reduction *per se*. Losses arising from the activities of invertebrates may be simple to quantify, especially when the parameter of concern is tree death, as is often the case for the more destructive bark beetles of Europe (e.g. *Ips typographus* attacking spruce (Christiansen & Bakke 1988)) and North America (e.g. *Dendroctonus frontalis* (Flamm *et al.* 1988) and *Dendroctonus ponderosae* (Raffa 1988)) attacking pines. In other cases, such as green spruce aphid, *Elatobium abietinum*, although dramatic losses of foliage can be observed frequently, the effects on tree growth are difficult to quantify, but can be serious (Straw & Fielding 1997).

Given the difficulties in determining what level of growth loss or tree mortality is economically significant, it is not always straightforward to determine an appropriate course of management action. However, in economic as well as environmental terms, intervention for pest management must follow from a defined need to manage that pest, a step that must be based on quantitative parameters as much as possible. This requires knowledge of pest outbreaks and of the damage thresholds that, when exceeded, lead to the need for intervention.

8.1.2 Great Britain as an example of a region where pest infestation has resulted in active intervention

Pest status and the setting of thresholds for management will differ both locally and regionally, reflecting ecological and socioeconomic variation. Britain, as an island and a country where tree cover is low but expanding, provides one example of regional variability in both the types of pest present and on methods of intervention. Table 8.1 lists some of the invertebrates that have reached pest status and have required active intervention in Britain during the last century. It is immediately apparent that the majority of the pest species are exotic to Britain, a consequence of both the nature of our tree flora and also the potential for new species to be carried to the country with international trade since Britain imports around 90% of its timber needs. This also illustrates that, although island status provides some interesting insights into the effects on the variety of flora and fauna, the significant impacts on forestry relate more to its influence on the rate of appearance and range of forest pests. Recent establishments of damaging forest pests have emphasized three principal aspects of island status, which would apply to both physical and ecological islands, spatially separated from a point of origin.

• Pests may pose a greater hazard in new locations than in their countries of origin. This principle is based on the supposition that most herbivores are coevolved both with the host tree species (including site and climatic conditions typical for that species) and with the spectrum of natural enemies associated with that tree. They will therefore tend towards the endemic state and rarely result in damaging infestations (naturally, with exceptions). It follows therefore that new associations may result in greater impacts than in the original host range. Table 8.2 lists some of the pests that fall into this category.

• Provision of a wider range of host tree species may encourage invertebrate populations, originating from both native and exotic insect sources, to reach outbreak levels. Pine beauty moth in Britain illustrates this very well. Prior to the extensive planting of the exotic lodgepole pine,

Pinus contorta, in Scotland from the 1960s onwards, *Panolis flammea*, although present in most of Britain, was at very low densities on Scots pine, *Pinus sylvestris*, its native host. During the late 1970s, very high densities of the moth were recorded in northern Scotland, leading to quite extensive tree mortality (Stoakley 1987). Infestations were restricted to lodgepole pine stands and were most severe on particular provenances of the tree, demonstrating a significant effect of host suitability and tree selection by adult moths, particularly in relation to ratios of α- and β-pinenes in foliage (Leather 1987). Active intervention, including aerial application of pesticides and changes in planting practice (mixtures of lodgepole pine and Sitka spruce, use of more resistant lodgepole pine provenances), has been necessary to prevent wholesale tree losses (Evans *et al.* 1991; Leather 1992).

• A further example is provided by the recent increase in densities and distribution of Asian longhorn beetle, *Anoplophora glabripennis* (Coleoptera: Cerambycidae) in China. Increases in populations from previously innocuous levels have been attributed to extensive planting of particularly susceptible species of poplar, *Populus dakuanensis*, *Populus* × *canadensis* and *Populus nigra* 'Italica', all of which are damaged by the beetle, resulting in large losses of trees (Gao *et al.* 1993). The beetle species has since been intercepted with increasing frequency in international trade and has now established and is causing damage in the USA (Chicago and New York) (Haack *et al.* 1997; Cavey *et al.* 1998).

The appearance and impacts of new pests, whether exotic in origin or as a result of fortuitous switching to a new host plant, may be exacerbated by the likelihood that natural enemies normally associated with pests may be absent. Evidence for this supposition is, however, often anecdotal and it is difficult to assess the effects quantitatively. Barbosa and Schaefer (1997) assessed this hypothesis in an analysis of invasions by four species of lymantriid moth in North America and concluded that there was little evidence to support the view that impacts were greater in the absence of the full spectrum of natural enemies. By contrast, there are examples of new pest infestations whose severity has been clearly linked to

Table 8.1 Some examples of insects that have reached pest status in Britain. (Based on Evans 1997.)

Insect species	Order/Family	Status	Origin
Douglas fir seed wasp *Megastigmus spermatrophus*	Hymenoptera/ Torymidae	Can be serious in a poor seed year	Exotic North America
Douglas fir woolly aphid *Adelges cooleyi*	Homoptera/ Adelgidae	First noted in 1913. Shoot distortion of Douglas fir	Exotic North America
Larch budmoth *Zeiraphera diniana*	Lepidoptera/ Tortricidae	Periodically causes severe defoliation of Scots and, particularly, lodgepole pines	Exotic Europe
Pine looper moth *Bupalus piniaria*	Lepidoptera/ Geometridae	First major outbreak in 1953, then periodically since then, especially on lodgepole pine sites	Native Exotic host
Pine beauty moth *Panolis flammea*	Lepidoptera/ Noctuidae	Major infestations on lodgepole pine in Scotland (from mid 1970s)	Native Exotic host
Large pine weevil *Hylobius abietis*	Coleoptera/ Curculionidae	Increasing problem, linked to felling and restocking of conifer crops	Native Native and exotic hosts
Great spruce bark beetle, *Dendroctonus micans*	Coleoptera/ Scolytidae	Established in early 1970s—outbreaks in Wales, bordering counties of England, Lancashire and Kent	Exotic
Large larch bark beetle *Ips cembrae*	Coleoptera/ Scolytidae	Minor outbreaks in Scotland and northern England. Occasionally on spruce	Exotic
Six-toothed bark beetle *Ips sexdentatus*	Coleoptera/ Scolytidae	Secondary pest on all pine species	Exotic
Winter moth *Operophtera brumata*	Lepidoptera/ Geometridae	Major defoliator on oak but more recently causing localized outbreaks on Sitka spruce	Native Exotic host
Horse chestnut scale *Pulvinaria regalis*	Homoptera/ Coccidae	New scale insect pest on horse chestnut, lime and sycamore, first recorded in 1964	Exotic
European spruce sawfly *Gilpinia hercyniae*	Hymenoptera/ Diprionidae	Minor records from 1906 but major outbreak period from 1968 to 1975 in mid Wales	Exotic
Web-spinning larch sawfly *Cephalcia lariciphila*	Hymenoptera/ Pamphiliidae	First record in 1953, then major outbreak from 1972 to 1979 in Wales and many forests in England.	Exotic
Knopper gall wasp *Andricus quercuscalicis*	Hymenoptera/ Cynipidae	First record in 1961. Knopper galls on acorns reduce oak seed production in poor mast years	Exotic
Gypsy moth *Lymantria dispar*	Lepidoptera/ Lymantriidae	Extinct in Britain in 1907. Asian strain poses higher risk than European; small infestations current in north London	Native extinct Current population is exotic

Table 8.2 Some examples of pests that have been more damaging in new regions than in their regions of origin.

Pest	Order/Family	Impact in region of origin	Impact in new region
Pine shoot beetle *Tomicus piniperda*	Coleoptera/ Scolytidae	Europe: generally secondary causing blue stain of felled wood	USA: severe shoot pruning of high value Christmas trees
Pinewood nematode *Bursaphelenchus xylophilus*	Nematoda/ Aphelenchoididae	North America: generally innocuous. Tree mortality only on exotic Scots pine 'off-site'	Japan, China, Korea: rapid wilting and wholesale death of susceptible pine species, linked to high temperatures
Woodwasp *Sirex noctilio*	Hymenoptera/ Siricidae	Europe, North America: little damage, secondary	New Zealand, Australia: tree mortality due to absence of natural enemies and presence of exotic conifer hosts
Balsam woolly aphid *Adelges piceae*	Homoptera/ Adelgidae	Europe: little impact on silver fir (*Abies alba*)	North America: crown dieback and tree death of several species of fir
Beech coccus *Cryptococcus fagisuga*	Homoptera/ Coccidae	Europe: associated with fungus *Nectria coccinea* var. *faginata*. Some damage and loss	Eastern North America: severe combined effects of scale and fungus on American beech which is poorly defended compared with European beeches

establishment without their normal natural enemies. European spruce sawfly, *Gilpinia hercyniae* (Hymenoptera: Diprionidae), was introduced to Canada in the 1930s (Bird & Elgee 1957) and also reached damaging proportions in Britain in the 1960s, having been present in the latter country since early this century (Billany 1978). Extensive tree damage was observed in both situations and, in both cases, a baculovirus (nucleopolyhedrovirus) was introduced, either deliberately (Canada: Bird & Elgee 1957), or by natural appearance and spread (Wales: Entwistle *et al.* 1983). Populations of the sawfly were quickly reduced to very low levels and remain innocuous in both countries. Another sawfly, web-spinning larch sawfly, *Cephalcia lariciphila* (Hymenoptera: Pamphiliidae), also reached outbreak population levels in Britain during the 1970s. Populations were subsequently regulated to very low densities by an accidentally intro-

duced parasitoid *Olesicampe monticola* (Hymenoptera: Ichneumonidae) (Billany & Brown 1980; Billany *et al.* 1985), thereby restoring a specific host–parasitoid relationship that is a normal factor in the natural range of the sawfly in Europe.

8.1.3 Pest risk assessment and the problems of exotic pests

Movement of forest insects in international trade is now commonplace, although not all will be pests in either the original or the receiving country. The process of assessing the risks from this potential transfer route is known as Pest Risk Assessment (PRA) and is a procedure that is normally carried out in the receiving country. Although there is no single international standard for carrying out PRAs, basic methodology based on the International Plant Protection Convention (IPPC) of the Food and Agriculture Organization

(FAO) is being used increasingly and refined by plant protection organizations worldwide. In particular, the European and Mediterranean Plant Protection Organization (EPPO) has developed a variant of the FAO scheme that uses a combination of datasheets and a questionnaire to assess risks (EPPO 1993). Despite these advances in methodology, risk assessment is still dominated by qualitative rather than quantitative descriptions, although this is not a major constraint to identification of high-risk pests and for the development of risk mitigation measures.

PRA uses standard risk terminology in that it assesses the three principal components of risk.

• *Hazard.* Identify the hazard posed by the potential quarantine organism in its existing range. This includes assessment of damage, using data on financial, environmental and socioeconomic aspects, as well as relationships to the biology and ecology of the organism.

• *Likelihood of the hazard being realized.* The key element here is assessment of the full range of pathways over which the organism could be moved, both naturally and, particularly, in international trade. Each pathway should be assessed against the likelihood of survival and transfer of the organism and of the volume of trade along the pathway. Finally, the likelihood of establishment and potential for damage should be assessed, taking particular account of potential tree species, climate, presence or absence of natural enemies and other regional factors, such as forest management practice.

• *Risk.* This is the product of hazard and likelihood of the hazard being realized. Thus, a high hazard organism, causing serious damage in its country of origin, may actually pose a low overall risk if there are few effective pathways for transfer and establishment. The opposite might apply for relatively low hazard, but nevertheless damaging, organisms that have many effective pathways.

PRA procedures are essential components in determining whether quarantine measures should be applied to control the movement of wood between countries. Regional Plant Protection Organizations (RPPOs) determine the precise measures to be used to reduce risks and, thus, must be well aware of pest infestations interna-

tionally. Introduction of restrictions to free trade must be justified under World Trade Organization rules and, thus, require sound scientific justification for the measures taken. Regular contact between RPPOs aims to standardize procedures, so that quarantine measures are the minimum necessary to provide plant health safeguards without adversely affecting trade. Within Europe, the EPPO works closely with the European Union (EU) to develop pest listings and procedures to enable trade to continue, both within the region and with the rest of the world. EU rules are based on a combination of named pests, relevant to the entire Community, and also more defined lists of both pests and commodities so that plant health procedures can concentrate on traded items that act as substrates on which named pests could be carried. Similar procedures are used by other RPPOs.

A recent PRA has been carried out by the EU on pinewood nematode, *Bursaphelenchus xylophilus* (Nematoda: Aphelenchoididae), which is causing serious damage to pines and other conifer genera in Japan, China and Taiwan (Evans *et al.* 1996). This organism represents an example of a greater hazard in the receiving country than in the native range (North America) of the pest. Indeed, it was some years after pines showing intense wilt symptoms were noted in Japan that the causal organism was recognized to be *B. xylophilus* (Mamiya 1988). A further complication in assessing risk from *B. xylophilus* is the fact that it has a close association with an insect vector, which research has shown to be predominantly adult beetles in the genus *Monochamus* (Coleoptera: Cerambycidae) (Warren & Linit 1992). Pathway analysis carried out by Evans *et al.* (1996) is reproduced in Fig. 8.1. The greatest potential for carriage is linked to green wood that is capable of supporting the beetle through to emergence in the receiving country; this coincides with the categories of wood that constitute the greatest volumes of trade, particularly from North America to the rest of the world. It was concluded that *B. xylophilus* and vectors in the genus *Monochamus* are high-risk quarantine organisms for the EU and that mitigation measures, involving heat treatment of wood to reach a core temperature of 56°C for 30 minutes, should

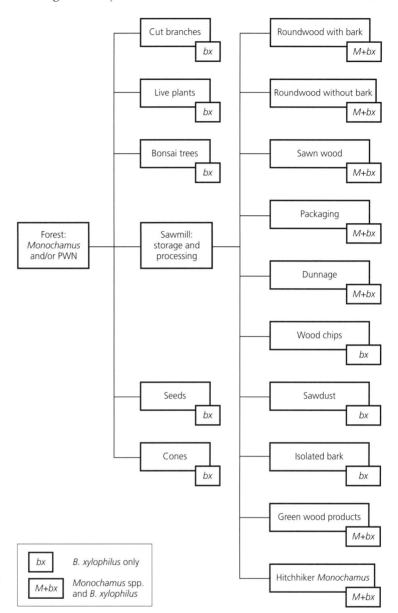

Fig. 8.1 Pathway analysis for potential transportation of pinewood nematode (PWN), *Bursaphelenchus xylophilus* and its vector *Monochamus* spp. in wood. (From Evans *et al.* 1996.)

be applied to wood originating in countries having the nematode (Evertsen *et al.* 1991; Task Force on Pasteurization of Softwood Lumber 1991).

A PRA has also been carried out on Asian longhorn beetle, *A. glabripennis*, in view of its impact in the USA and the fact that it has been intercepted with increasing frequency in several countries, including Canada and the UK (H. F. Evans &

A. McLeod, unpublished). Measures to reduce the risk of carriage of this pest in international trade include a requirement for freedom from grub holes caused by the feeding activity of larvae within the wood and which could provide evidence for the presence of the beetle or, alternatively, that the wood has been kiln dried, using heat, to <20% moisture content (MC).

8.2 ESTABLISHING THE PRINCIPLES OF PEST MANAGEMENT: INTEGRATED PEST MANAGEMENT

Chapter 6 in Volume 1 of this handbook deals with many of the general principles governing pest and disease dynamics. This section now builds on those principles to develop the science of pest management, particularly the principles of integrated pest management (IPM). It has already been noted that determining whether an invertebrate should be classified as a pest is a subjective process and may vary from location to location, depending on the damage threshold established for that pest. The process of pest management commences with an appraisal of the problem and, ideally, should include a detailed assessment of the reasons for the outbreak. In this way, it may be possible to break the cycle that has led to the problem and develop longer-term, stable strategies for reducing numbers to endemic levels. This is the underlying principle behind IPM, which seeks to develop pest management strategies that are ecologically based and make balanced use of a range of control measures, avoiding where possible the use of chemical pesticides (Kogan 1986). At the core of IPM is the establishment of an economic injury level (EIL) or damage threshold, which is tested by monitoring (scouting) for the presence of the pest or symptoms of the pest. Although it is convenient to use the term EIL, it is less strictly relevant to modern forest practice, where multipurpose objectives include parameters that are not easy to quantify in economic terms. EIL, as used in this chapter, is taken to mean any damage that may require pest management, even if the purpose is not primarily economic. On the basis of survey results, decisions are made on whether to intervene and on the methods to be used for intervention. Further complication arises if more than one pest has to be considered, especially if the pests differ substantially in biology or timing. This concept was well summarized by Kogan (1998) in a recent review of IPM (Fig. 8.2). It is a fortunate characteristic of pest management in forestry that, in the majority of situations, pest management is concerned with only one or very few pests at a given location and therefore the implementation within Fig. 8.2 can

be simplified. Indeed, many of the headline techniques in Fig. 8.2 are used increasingly in forestry but can be concentrated on a single pest species. IPM is therefore a central tenet in forest pest management and there are a number of successes in its use for the major pest species.

8.2.1 The biology and ecology of the pest as a basis for integrated pest management

Many of the principles that determine the population densities of invertebrates were considered in Chapter 6 in Volume 1 of this handbook, and these can be used to assess imbalances in dynamics that lead to overexploitation of resources and hence exceed the EIL.

At the heart of IPM is the requirement for detailed knowledge of the biology and ecology of the pest. In particular, the factors that determine how the pest locates and colonizes a suitable host tree are fundamental to the dynamics of the pest and, hence, to enable assessment of its potential to cause damage to the tree. In the majority of cases, it will be the adult stage that is mobile and capable of selecting the precise host tree and oviposition site within that tree. Clearly, the sites for oviposition and later feeding on the tree will depend on the particular guild of herbivore. Table 8.3 summarizes some of the attributes of different feeding guilds and the various strategies within the various life stages of each guild.

8.2.2 Host selection

One of the common factors, irrespective of guild, that determines the initial probability of pest infestation is the ability of adult stages to detect suitable cues from a putative host tree and to oviposit accordingly. Therefore, in developing management strategies it is important to understand what cues are used for oviposition and to avoid creating conditions for enhancement of those cues. This usually requires rather detailed study of each species that poses a pest threat. For example, European pine sawfly, *Neodiprion sertifer*, responds to the size of tree and needle length as well as presence of high terpene concentrations in Scots pine (Zumr & Stary 1993). These authors speculated that the sawfly was trading off the

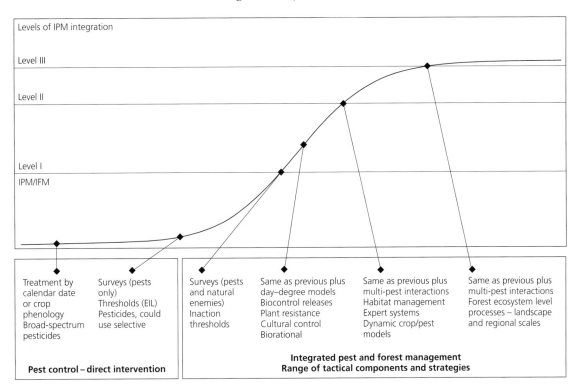

Fig. 8.2 The components of integrated pest management (IPM) and integrated forest management (IFM) in relation to forest pests and their longer-term management. (Based on Kogan 1998.) Reproduced with permission, from *Annual Review of Entomology*, Volume 43 © 1998 by Annual Reviews. www.AnnualReviews.org

suitability of foliage for larval growth against plant defensive qualities conferred by high terpene concentrations in order to gain 'enemy-free' space arising from lower numbers of parasitoids on these trees. Other pine-feeding herbivores (pine beauty moth, *Panolis flammea*, and larch bud moth, *Zeiraphera diniana*) showed strong oviposition preferences to monoterpene profiles of lodgepole and Scots pines, but the response varied with tree age (Trewhella *et al.* 1997). In particular, *P. flammea* preferred northerly provenances of lodgepole pine when presented with young trees but switched preferences to southerly provenances in older trees. This preference was linked to the differing monoterpene profiles of the provenances in relation to tree age. Ultimately, knowledge of adult oviposition strategies can provide clues for the development of both monitoring and pest management regimes, particularly in the development

of longer-term methods of control using factors such as enhanced tree resistance to attack (Thompson & Pellmyr 1991).

Larval feeding can also be linked directly to tree suitability, both nutritionally and defensively, and to regional temperature and humidity regimes that will determine rates of development. Many of the potential trade-offs between nutrition and defence were discussed in Chapter 6 in Volume 1 of this handbook, and these are but two of the complex interactions, including the effects of natural enemies, that influence the growth and mortality rates of the herbivore populations that attack the trees.

8.2.3 Tree suitability and initial rates of invertebrate colonization and growth

Studies of plant effects at the guild level enable a more detailed appraisal of mechanisms to be

Table 8.3 Feeding guilds of forest invertebrates in relation to oviposition and feeding habits.

Herbivore guild	Examples	Adult behaviour	Oviposition sites	Larval feeding strategy
Bark feeders	Scolytidae (bark beetles)	Attraction to suitable trees and use of aggregation pheromones	In maternal galleries under bark, usually in batches	Single or group feeding, often accompanied by fungal associate
Wood borers	Cerambycidae (long-horn beetles)	Attraction to suitable trees	In slit cut in bark, usually singly	Single feeding, initially under bark, then into wood
	Siricidae (wood wasps)	Attraction to stressed trees	Ovipositor 'drills' into outer sapwood. Single or multiple eggs in each drill, depending on tree suitability	Injection of fungal spores assists larval feeding in wood
Chewers	Lepidoptera (range of moths) Hymenoptera (range of sawflies) Coleoptera (adult and larval bettles)	Adults detect suitable host trees	Bark, buds or leaves, depending on ultimate feeding sites of larvae and, in some cases, adults	Normally on foliage, may specialize on young or older foliage or flowers
Suckers	Homoptera (range of aphids and scale insects)	Adult dispersal mainly. Settle and sample rather than remote detection	Both egg laying and birth of nymphal stages On bark, buds, leaves or roots	Feeding by insertion of stylets into phloem, virtually anywhere on tree, but usually foliage
Gall formers	Hymenoptera Diptera	Adult dispersal and specialist selection of host trees	Eggs often in buds or flower heads	Feeding by larvae results in tissue proliferation and gall formation

carried out. Bark beetles, in particular, have been studied extensively in relation to plant parameters that impinge at scales ranging from individual trees, through the forest stand to wider landscape features. The selection of suitable host trees by bark beetles is initiated by adults locating trees through a combination of visual and olfactory stimuli (Coulson 1979; Wood 1982). The limiting factor is the degree of mobilization of defences at the bark interface when the adult beetles make their initial attacks. Certainly, in conifers, there may be local defences that are either preformed and/or induced (Berryman 1988). Preformed resins, produced as soon as beetles invade bark and cambial tissues, are composed of secondary compounds such as monoterpenes, sesquiterpenes and resin acids, the bulk of resin flow occurring within the first 8 hours after invasion (Nebeker *et al.* 1992). This alone may be sufficient to deter low-density beetle attacks, but for many of the so-called mass attack bark beetles, the combination of numbers of beetles and associated fungi will overcome initial defences and result in secondary, or induced, defences as soon as the cambial layer is breached (Berryman 1972). Induced defences are more directed than preformed and arise from morphological and bio-chemical changes such as necrosis, the develop-ment of layers of protective cells (callousing) and the production of new phenolic and monoterpene compounds, thus providing both physical and chemical partitioning of the affected area (Paine *et al.* 1997).

Vigorously growing, healthy trees have a full suite of both preformed and induced defences that are normally capable of overcoming low-density attacks by bark beetles. For example, Norway and Sitka spruces have both preformed defences, in the form of lignified stone cells and resin pro-duction, and induced defences comprising the dynamic wound response that constrains attack zones to the immediate point of entry (Wainhouse *et al.* 1997). This combination alone is sufficient to provide protection against weak fungal pathogens (Wainhouse *et al.* 1997) and, to a great extent, against the bark beetle *Dendroctonus micans* that uses individual rather than mass attack tree-invasion strategies (Wainhouse *et al.* 1990; Wainhouse & Beech-Garwood 1994). Mass attack bark beetles, such as *Dendroctonus punc-tatus*, *D. frontalis* and *D. rufipennis* in North America and *Ips typographus* in Europe, often sustain background populations in weakened or recently dead trees where the suite of defences are poor, thus enabling relatively low numbers of colonizing beetles to establish broods (Kulhavy & Miller 1989). Such a strategy results in local increases in bark beetle populations that may reach densities sufficient to overcome healthy living trees, although there is still a relationship with overall tree vigour. In essence, tree defences have a finite limit above which beetles will suc-cessfully invade and establish broods. Thus, a threshold density of bark beetle adults and fungal associates has been recognized for many of the major pest species (Christiansen 1985; Mulock & Christiansen 1986; Christiansen *et al.* 1987; Raffa & Smalley 1995). The pattern is similar in all cases: adults initially colonizing a tree emit an aggregation pheromone to attract other adults thus increasing the density towards the critical threshold. Once the threshold is exceeded and there is no further host resistance, pheromone production is curtailed and no further aggregation takes place (Birgersson & Bergstrom 1989). This

mechanism is an adaptation by the bark beetle to reduce the potential effects of intraspecific com-petition within the tree that would certainly arise if colonizing beetles continued to arrive after the optimal number of beetles had established broods. Raffa and Berryman (1983) investigated the at-tack dynamics of *D. ponderosae* on lodgepole pine, *Pinus contorta*, and demonstrated a thresh-old density of 40 broods m^{-2} to overcome tree defences. However, if densities exceeded 80 broods m^{-2}, severe intraspecific competition took place leading to reduced brood size and breeding success. The authors determined that beetle reproduction and survival was optimal at densities around 62 broods m^{-2}. Similarly, in a study of the population dynamics of southern pine beetle, *D. frontalis*, it was shown that com-petition for resources could be quite significant both during oviposition (contest competition) or during larval feeding within the gallery systems (scramble competition) (Reeve *et al.* 1998).

The discussion on bark beetles above provides an example of the dynamic nature of herbivore–tree interactions, reflecting the fact that both the herbivore and the tree have evolved mechanisms to increase the probability of survival. Interven-tion by man can dramatically affect the dynamic balance between the various organisms, so that pest outbreaks may become more commonplace if general tree health is compromised by anthro-pogenic influences.

8.3 WHY ARE SOME INSECTS PESTS?

This is the key question that, once answered, can be used to develop appropriate IPM techniques. Discussion will refer back to earlier sections and will develop general principles, with selected examples to illustrate principles that will be elaborated in the remainder of the chapter. Once the key factors are recognized, the process of pest management proceeds through a number of steps to assess the risk of a damaging population build-ing up and to determine the rate and extent of population build-up that may exceed the damage threshold. Decisions can then be made on whether to intervene, either directly over the short term or indirectly by longer term changes in forest management (as part of integrated forest management, IFM) to return populations below the EIL.

8.3.1 Risk factors in relation to pest population size

Application of knowledge of pest biology at a range of spatial scales can provide the data necessary to develop risk rating and management options for a given pest or range of pests (Steele *et al.* 1996). Such an approach is at the core of IPM and, with inclusion of tree provenance, site preparation and other silvicultural practices, can be extended into IFM (equivalent to integrated crop management (ICM) in agriculture). IFM is an holistic approach to enable the forest manager to integrate a range of management options to achieve the single or multiple management aims of a particular forest. Conceptually, risk rating is based on information of the relationships between tree susceptibility (the converse of tree resistance), site factors, climate, anthropomorphic influences and previous history of infestation. These are summarized in Table 8.4, which outlines the principal components affecting tree susceptibility to attack; these include both natural and man-influenced factors. There are clearly cross-links between the various factors, but it is worth noting that, ultimately, invertebrate exploitation, by definition, takes place at the invertebrate–host plant interface. This will be dependent on the availability of various niches on individual trees, even though the other factors from regional to local scales will influence the precise characteristics of the tree itself. IFM therefore offers the opportunity to influence several of the factors associated with tree genus and species and, depending on the underlying site and form of silvicultural practice, can utilize a suite of secondary characteristics to maintain invertebrate populations below the EIL. Thus, in managed forests, the process of pest management requires anticipation of the potential problems and should include preparatory management as early as the forest nursery stage.

Speight and Wainhouse (1989) discussed the role of silvicultural techniques in pest management and distinguished the relative values of the process for each stage from nursery/postplanting, through stand management/forest hygiene to felling and postharvest protection. Such an approach recognizes that each stage of forest growth and succession carries different risks from invertebrates and, once these risks are recognized, can lead to different forms of forest management. Ironically, it is often the converse to active preventative management that illustrates the link between forest vulnerability and stand structure.

8.3.2 Spruce budworms in North America: a detailed example to illustrate the principles of pest dynamics and their links to forest structure and management

The links between long-term changes in forest structure and the incidence of outbreak pest populations are well illustrated by spruce budworms in North America. Although the two principal spruce budworm species (eastern spruce budworm, *Choristoneura fumiferana* (Mattson *et al.* 1988) and western spruce budworm, *Choristoneura occidentalis* (Talerico 1983)) have specific host plant preferences, they are biologically very similar and for many of their ecological attributes can be considered as one entity. These moths are severe pests of mixed conifer forests in Canada and northern USA. In relation to host preferences and infestation dynamics, although eastern spruce budworm attacks a wide range of conifer species, the frequency and severity of outbreaks tends to be associated with forests dominated by mature balsam fir, *Abies balsamea* and, to a lesser extent, white spruce, *Picea glauca.* (Witter *et al.* 1983). The dynamics of the moth are governed by the dispersal capacity of the adult females such that, during epidemics, they migrate *en masse* to alight up to 600 km from their take-off points (Dobesberger *et al.* 1999). One of the key factors in determining where the mass migration alights is the predominance of mature trees above the general canopy level, often associated with managed stands where the moth has been suppressed by application of insecticides or where fire prevention policies have altered the natural uneven-aged stand structure (Blais 1985). Analysis of the trends in outbreak frequency indicates that epidemic populations appear to have occurred more frequently during the twentieth

Table 8.4 Scale and other factors acting on invertebrate/tree association in relation to potential management of forest pests.

Factor	Linked factors	Effects on trees and invertebrate associates
Geographical influences	Latitude Altitude Topography	Linked to factors below, but particularly in relation to natural tree species mix, growth rates and range of invertebrate species associates at the landscape scale
Site	Underlying rock type Soil type	Nutrition Soil depth Nutrition Stability Moisture retention pH
Climate	Rainfall Wind Sun Temperature	Soil moisture content; drought Windblown trees Photobiology effects on growth Growth rates (trees and invertebrates) Stress
Tree genus and species	Provenance Growth Regeneration capacity Defence	Growth Susceptibility Form—silvicultural and niche provision for invertebrates Size—silvicultural and plant architecture Age—rotation length and availability for colonization Ability to regenerate Natural stocking density Chemical characteristics
Silvicultural practice	Monoculture Mixture Site preparation Nutrition Weed control Thinning Age structure Retention of old trees	Genetic uniformity Uniform growth and defence (linked to provenance) Clonal selection Genetic mix Specific and clonal susceptibility Establishment Growth and defence Growth Growth and defence Niche provision for invertebrates Niche extension Biodiversity enhancement Old-growth specialists

(Increasing management influence →)

century than over the previous hundred years: 23 outbreaks during the period up to 1980 compared with only nine in the previous 100 years (Blais 1983). Changes in forest management, including selective harvesting, the introduction of regimes to reduce the incidence of fires and the use of insecticides to control incipient populations of the moth, appear to have contributed to this phenomenon (Mattson *et al.* 1988; Bergeron *et al.* 1998).

The key factor for *C. fumiferana* is the trend towards spruce–balsam fir dominated stands, particularly including mature trees, a combination that is known to pose the highest risk of spruce budworm attack and population increase (Blais 1983). However, while there appears to be little doubt that such changes in forest management have been accompanied by more frequent and sustained outbreaks of spruce budworms, including the western spruce budworm, *C. occidentalis*, it is by no means clear how the dynamics of the moths are affected in practice. Such a finding illustrates the difficulties in understanding and interpreting the key processes, even when good data sets of the dynamics of the moths over time are available. For example, the classic 'Green River' study that was used to develop the first life tables for spruce budworm (Morris & Miller 1954) has been used to support quite different hypotheses on the dynamics of the moth over historical time. Blais (1983) used the data to support the hypothesis that outbreaks of the moth have increased in severity and frequency during the twentieth century. Royama (1984, 1997), on the other hand, re-analysed the data to show that outbreak frequency has not changed and the apparent cyclicity observed by Blais could be explained by natural enemy-induced variation around the mean trend of population change over time.

Before considering how information on the dynamics of spruce budworms can be used to develop hazard rating and management systems, some consideration will be given to their life histories (Fig. 8.3). Host tree selection and exploitation by larvae is not determined entirely by adult choice and oviposition patterns. Although the adult stages will select the more mature trees in a forest block for oviposition (Mattson *et al.* 1988), they exhibit both local and migratory patterns of oviposition behaviour. Thus, a proportion of eggs are laid soon after emergence, at a time when females are poor fliers, ensuring that local exploitation of potentially suitable host material continues, but not necessarily to the point of overexploitation and tree death. Indeed, the primary host of *C. fumiferana*, balsam fir, can continue to set buds for several years despite complete defoliation, thus sustaining a continuous outbreak for many generations in a single area (Royama 1984). Once the female moths have matured they show strong tendencies for flight, often exhibiting mass migrations for considerable distances, particularly if local defoliation levels are very high (Régnière & Lysyk 1995). Landing and host selection appears to be associated with the presence of dominant trees above the canopy, thus favouring the more mature balsam fir dominated forests that are now typical of ecosystems

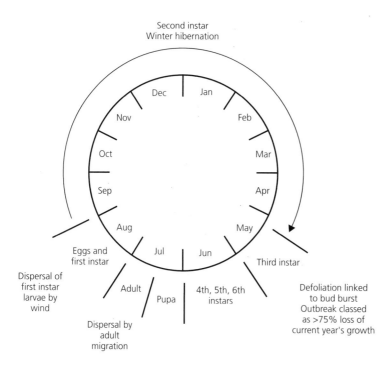

Fig. 8.3 The life history of spruce budworms (*Choristoneura fumiferana* and *Choristoneura occidentalis*).

exploited by spruce budworm (Régnière & Lysyk 1995). Oviposition takes place on these favoured trees, but the newly hatched larvae do not feed and tend to migrate, by wind movement on silken threads, between trees. This may take place over considerable distances and is influenced by local and regional patterns of wind direction and strength. Indeed, this non-feeding stage is the most vulnerable in the life cycle and losses of around 85% are common between the egg and third instar stage (Mattson *et al.* 1988). Overwintering takes place in silken hibernacula when larvae are in the second instar stage. These emerge in the following spring just prior to bud burst and there is therefore a requirement for close synchrony with the biological development of suitable host trees (Shepherd 1985; Lawrence *et al.* 1997).

Although mortality in the early stages is high, it is not the main regulatory factor in determining the classic regular and long-lived cycles associated with spruce budworm outbreak dynamics. Royama (1997) analysed various datasets, including the Green River data, on spruce budworm dynamics and concluded that the key mortality factors were natural enemies acting in a density-dependent way on the third instar to pupal stages. Interestingly, the closely related jack pine sawfly, *C. pinus*, has a much shorter outbreak cycle than the two spruce budworms (10 years compared to 30 years), and this has been attributed to the dependence of second instar larvae of the former on availability of pollen for spring feeding (Nealis & Lomic 1994). Lack of synchrony with pollen production is thought to make the larvae more vulnerable to the actions of natural enemies, especially parasitoids, some of which are shared with spruce budworms, thus reducing the length of time between outbreak cycles.

Hazard rating for spruce budworms

Further complexity in interpretation of spruce budworm dynamics arises when considering interactions with other biotic agents and abiotic factors acting on tree health and availability for colonization. Risk rating for a given pest must therefore take such factors into account. A good example that illustrates how interactions can influence risks of infestations and provide guidance on management practices is a risk rating system for central Idaho forests that incorporates factors and ecological influences related to 11 different biotic and abiotic agents—Douglas fir beetle (*Dendroctonus pseudotsugae*), mountain pine beetle (*Dendroctonus ponderosae*), western pine beetle (*Dendroctonus brevicomis*), spruce beetle (*Dendroctonus rufipennis*), Douglas fir tussock moth (*Orgyia pseudotsugata*), western spruce budworm (*Choristoneura occidentalis*), dwarf mistletoes (*Arceuthobium* spp.), root and butt rots (*Heterobasidion annosum* and *Phaeolus schweinitzii*), honey fungus (*Armillaria* spp.) and wildfire (Steele *et al.* 1996). Within the natural conifer forests of Idaho, the individual and combined interactions of these agents has been assessed in relation to presence and condition of principal tree species; Douglas fir (*Pseudotsuga menziesii*), Engelmann spruce (*Picea engelmannii*), grand fir (*Abies grandis*), lodgepole pine (*Pinus contorta*), ponderosa pine (*Pinus ponderosa*), subalpine fir (*Abies lasiocarpa*), western larch (*Larix occidentalis*) and whitebark pine (*Pinus albicaulis*).

Steele *et al.* (1996) assessed a wide range of published hazard rating models to produce a composite system that appears to work well within the specific confines of the Central Idaho ecosystem. The Stand Hazard Rating Form is reproduced in Box 8.1. Each score is derived from specific instruction sheets for the named pests, diseases and for fire hazard. When interactions between factors are likely to have a significant effect on total hazard rating, these are shown in the table and the adjustments to the scores indicated. The specific Rating Guide, which illustrates clearly how the combined influences of the primary host tree species can be rated to develop an overall hazard score hazard, for *C. occidentalis* is reproduced in Box 8.2. The higher the score, the higher the hazard for each factor. Details of the stand hazard index can be found in Brooks *et al.* (1985). For *C. occidentalis*, the key hazard attributes are presence of a high proportion of climax tree species within a dense stand, especially on warm, dry sites, all of which can be exacerbated by external stress factors such as drought, poor nutrition, overcrowding and root

Box 8.1 Stand hazard rating form for a range of biotic and abiotic hazards in central Idaho forests (from Steel *et al.* 1996). The specific hazard ratings for each component are given in the authors' original paper, except for western spruce budworm, *Choristoneura occidentalis,* **which is reproduced in Box 8.2.**

STAND HAZARD RATING FORM

Stand No.   _____   Worksheet for one stand

Location   _____   Habitat type   _____
T. _____ R. _____ S. _____ Date _____   Tree layer type   _____
Observer (s)   _____   Stand basal area   _____

		Hazard rating	Interaction

<u>   Line numbers from hazard ratings   </u>
<u>1</u> 2 <u>3</u> <u>4</u> <u>5</u>   <u>   6   </u>   <u>7</u>   <u>8</u>

A. Douglas fir beetle   _ + _ + _ = _ × _   --- =   ___   ___
B. Mountain pine beetle   _ + _ + _ = _ × _   × _ =   ___   ___
C. Spruce beetle   _ + _ + _ = _ × _   --- =   ___   ___
D. Western pine beetle   _ + _ + _ = _ × _   --- =   ___   ___
   (+MPB in PIPO)
E. Tussock moth   _ + _ + _ = _ × _   --- =   ___
F. Western spruce budworm   _ + _ + _ = _ × _   × a __ × 1 = __.
                                                     × b __ × 0.5 = __.
                                                     Total = __ × __ = __.

G. Dwarf mistletoes   _ + _ = _ × ,__ ---   --- =   ___
H. Annosus root disease   _ + _ = _ × a.__ × 1 = ___
                                        b.__ × 0.5 = ___
                                        c.__ × 0.25 = ___
                                 Total = __ × __ --- = __.
I. Armillaria root disease _ + _ = _   × ,__ × __ = __.
                                        × ,__ × __ = __.
                                        × ,__ × __ = __.
                                        × ,__ × __ = __.
                                 Total = __ × __ --- = __.
H. Schweinitzii root/butt rot _ + _ = _ × a.__ × 1 = __
                                         b.__ × 0.5 = __
                                         c.__ × 0.25 = __
                                 Total = __ × __ --- = __.
K. Wildfire   _   _   <u>Spp.</u>   *Bth* =
                                    <u>DBH</u>   *TLS* =
                                    <u>Ht.</u>   *Pm* × *10* =   4.__ 5.__ = ___ ___
                                    <u>C.R.</u>

Interaction effects (column 8)
If row B totals 9 or 10, add 1 to row K (if row K<10).
If row C totals 9 or 10, add 1 to row K (if row K <10).
If row D totals 9, add 1 to row K (if row K <10).
If row E totals 9, add 0.5 to row A (if row K>0), 0.5 to row I (if row I>0), and add 1 to row K (if row K<10).
If row F totals 6, add 0.5 to row A (if row A>0), 0.5 to row I (if row I>0).
If row G totals 6 or 8, add 1 to row K (if row K<10).
If row H totals>4, add 1 to row D (if row D>0).
If row I totals>3, add 1 to row A (if row A>0).
If row J totals>4, add 2 to row A (if row A>0).
If row K is 4 to 5, add 0.5 to row A (if row A>0), 0.5 to row D (if row D>0), and 0.5 to row J (if row J>0).

Caution: If any row exceeds 10 after interactions are included, reduce the value to 10.
Stand Hazard Rating: Combine the totals from columns 7 and 8.

Box 8.2 Hazard rating system for western spruce budworm, *Choristoneura occidentalis*, **in central Idaho forests on Douglas fir (DF), Engelmann spruce (ES), grand fir (GF), lodepole pine (LP), ponderosa pine (PP), western larch (WL) and subalpine fir (SF). See text for explanation of scoring system. (From Steele** *et al.,* **1996).**

1. Average age of host overstorey:
    If average age of combined host species in the overstory is:
    >80   Enter 1.5
    20–80   Enter 1.0
    <20   Enter 0.5   _____

2. Basal area (ft^2/acre) of stand (all species).
    If:
    >100   Enter 1.5
    80–100   Enter 1.0
    <80   Enter 0.5   _____

3. Site condition: if site index (SI$_{50}$) of the dominant species is:

DF	ES	GF	LP	PP	SF	WL	
<40	<49	<50	<47	<56	<46	<51	
					Enter 1.5		
40–60	49–68	50–60	47–59	56–68	46–65	51–63	
					Enter 1.0		
>60	>68	>60	>59	>68	>65	>63	
					Enter 0.5		_____

4. Stand structure:
    If host species create a stand that is:
    Multistoried   Enter 1.5
    Two-storied   Enter 1.0
    Single storied or patchy   Enter 0.5   _____

5. Total lines 1, 2, 3 and 4   _____

6. Percentage of host in stand
    Enter percent basal area, in decimal form, of:
    (a) Douglas fir+grand fir+subalpine fir   ____ × 1 =   _____
    (b) Engelmann spruce   ____ × 0.5 =   _____
    Total lines a and b   _____

7   Multiply total in line 6 by line 5
    Enter on the hazard rating form, Row F, Column 7   _____

---

diseases. From Box 8.1, it can be seen that if the spruce budworm rating reaches 6 (the highest likely score), then hazards from Douglas fir beetle (*D. pseudotsugae*) and *Armillaria* can be expected to increase. Similar studies of hazard rating by Wilson *et al.* (1998) showed that loss of overstorey lodgepole pine in mixed conifer forests in north central Washington increased the risks from *C. occidentalis* in the remaining trees. Additional factors, such as climate change effects, are also being incorporated into spruce budworm predictive models, both in relation to potential outbreaks and with respect to management for conservation of biodiversity (Fleming *et al.* 1998). Such complex interactions are not typical of the majority of intensively managed forest systems where tree species mixtures will tend to be more limited. However, they do serve to illustrate how natural ecosystems can suffer extensive damage from biotic and abiotic factors, but can be categorized according to recognizable risk ratings.

Using knowledge of the spruce budworm system to illustrate how management strategies can be developed to cope with infestations

The extensive knowledge that has been gathered on spruce budworms has helped to raise awareness of the reasons for outbreaks and suggest ways in which preventative as well as protective strategies can be developed. One of the aims of IFM is to develop multidisciplinary strategies that both prevent outbreaks and deal with them if they do arise. Spruce budworm dynamics suggest that outbreak frequencies are increasing over time, but not all researchers agree with this. Two explanations for the apparent outbreak cycles have been

advanced and offer different possibilities for long-term management. There is a considerable literature on these aspects (e.g. Mattson *et al.* 1988; Régnière & Lysyk 1995; Régnière *et al.* 1995) but, for brevity, the key points have been summarized in Fig. 8.4. This illustrates, schematically, that the observed outbreak frequencies of spruce budworm can be explained by two different hypotheses: the double equilibrium and the continuous oscillation theories. Management options differ considerably, depending on which theory is favoured.

It would appear that the continuous oscillation theory requires large-scale, regional changes in forest management and silviculture to prevent

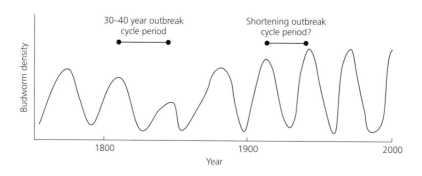

Alternative management options:

Double equilibrium theory: outbreaks commence in distinct epicentres. Management to prevent epicentre build-up and spread.

- Regional monitoring to detect early signs of epicentre development
- Spray populations to prevent build-up
- Long term: reduce rotation, change tree species mix, mixed forest structure

Continuous oscillation theory: outbreaks are synchronized over wide areas. Populations fluctuate around the average population level. Management must be regional and concentrate on damage control rather than on prevention of population build-up.

- Monitor regionally in relation to expected harvesting or other values
- Treat only those areas where tree cover needs to be retained
- Long term: reduce rotation, change tree species mix, mixed forest structure. Consider changes to spatial distribution of stands

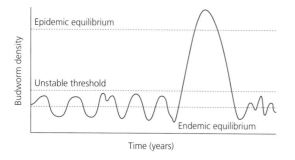

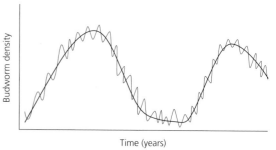

Fig. 8.4 Potential management options in relation to two theories of spruce budworm (*Choristoneura fumiferana* and *Choristoneura occidentalis*) dynamics (see text for sources of information).

outbreaks and, thus, is more likely to be based around protection of the forest resource than through selective use of insecticides. This would certainly be the case where access to sites is difficult and there is relatively little financial return in intervention purely to re-structure the forest. Where re-structuring has come about naturally, for example by widespread fire (Bergeron *et al.* 1998) or where the susceptible host tree species are isolated in islands of non-susceptible forests (Cappuccino *et al.* 1998), population densities and tree mortality arising from spruce budworm tends to be much lower than in highly susceptible sites.

It would appear therefore that management of spruce budworm will concentrate on foliage protection in the future, even though hazard-rating models suggest ways of re-structuring forests to reduce or prevent outbreaks. Such decisions have to be made on both financial and practicability grounds. Protection strategies, employing insecticides or microbial agents such as *Bacillus thuringiensis* (*Bt*) or baculovirus, have been used extensively in North America. In each case the decision to spray is based on likely infestation levels and desired foliage protection requirements (Fleming & Shoemaker 1992). Depending on the active ingredients employed for spray operations, there may be one or two applications during the vulnerable third instar stage feeding on newly flushed foliage that is the optimal target, thus giving a spray window of around 10 days only (Fleming & Van Frankenhuyzen 1995). Operational control of spruce budworms is carried out mainly using aerial application of *Bt*, usually based on a single application because of its high cost. A great deal of effort has gone into optimization of spray parameters and into formulation of *Bt* for both atomization and concentration of ai (active ingredient) per droplet (Van Frankenhuyzen 1993; Fleming & Van Frankenhuyzen 1995). Although there have also been encouraging results from use of the baculovirus against spruce (eastern and western) and pine budworms, all of which are susceptible to the nucleopolyhedrovirus originally isolated from *C. fumiferana*, there has been relatively little operational use of this microbial agent (Cunningham 1995). Operational management for foliage protection is there-fore increasingly dominated by use of *Bt*, with decisions being made at a regional scale (for example see Armstrong & Ives (1995), which includes several chapters on regional pest problems and pest control operations in Canada).

Although control measures applied during outbreaks can therefore provide adequate levels of foliage protection, they do not appear to be effective in breaking the cycle or in preventing the rise of the next outbreak of spruce budworms. Use of silvicultural management systems to avoid creating conditions conducive to the development of outbreaks has been considered as a longer-term alternative. Mattson *et al.* (1988) provided a useful summary of the factors leading to outbreaks and considered the potential for manipulation of those factors. In summary, they provided potential management options in relation to the various alternative scenarios that have been put forward to explain the regular and widescale appearance of budworm outbreaks.

8.3.3 Gypsy moth, *Lymantria dispar*, as an example of a non-indigenous pest that is still expanding its range

Earlier, the dangers of international movements of forest pests were illustrated by reference to PRAs and pathways of entry to new locations. In some cases pests may have established following deliberate introduction into a country. For example, the gypsy moth, *Lymantria dispar*, a native of Eurasia (Lechowicz & Mauffette 1986), was introduced into north-eastern USA in 1868 in an attempt to hybridize with the silkworm, *Bombyx mori* (Elkinton & Liebhold 1990). Barbosa and Schaefer (1997) have analysed the patterns of spread of this species in relation to other introduced lymantriids. They put forward the interesting theory that foliage quality for the moth is more suitable at the advancing front of its continuing dispersal in the USA, compared with areas that have been defoliated. Some support for this comes from studies of foliage quality before and after defoliation episodes. Kleiner *et al.* (1989) showed that foliage quality and potential stand susceptibility declined after gypsy moth defoliation episodes. These, and other factors, have been included in the major research efforts that have been put into understanding the dynamics of the moth in several countries.

This is particularly the case in the USA where *L. dispar* is still spreading and is subject to a campaign termed 'Slow-the-Spread' (STS) in the north-east of the country (Sharov & Liebhold 1998a). Part of the knowledge base used to develop such campaigns is information on the phenological development of gypsy moth in relation to local temperature conditions. A model, called GMPHEN, has been developed to enable reasonably accurate prediction of egg hatch and development in each larval instar on the basis of accumulated day-degrees above 8°C (Sheehan 1992). This model provides good prediction of gypsy moth field performance in the USA, where it was developed, but in recent applications of the model in Britain, it has required information on actual egg hatch dates before it accurately predicted larval development (J. Head, personal communication).

Further information on natural rates of spread (Sharov & Liebhold 1998b), including relationships with local climate conditions (Liebhold *et al.* 1992) and availability of suitable host plants, have been indicated in a bioeconomic analysis of the STS strategy (Sharov & Liebhold 1998a). The models concluded that slowing spread, but not stopping spread, was an economically viable strategy. Stopping spread was cost-effective only if natural barriers to population spread were present, thus supplementing the impacts of strategies to slow spread; these could include locations where the mean minimum January temperature is less than −13.9°C (negative effect on egg survival over winter) (Liebhold *et al.* 1992), terrain with low numbers of suitable host trees, etc.

Other aspects of control of *L. dispar* include assessment of the roles of natural enemies as components of the IFM strategy. Although most of the 90 or so parasitoid species that attack gypsy moth in its native range have been introduced into the USA during this century, very few have actually established successfully (Montgomery & Wallner 1988). However, several species persist following introduction and have been cited as successful regulatory factors in suppressing dam-

aging numbers of moths in established populations (Gould *et al.* 1990; Berryman 1991). Recent successful control arising from the fungus *Entomophaga maimaiga* has indicated that this agent, discovered in 1989 and possibly introduced from Japan, may offer long-term suppression of the moth as part of the STS and IFM approaches to managing *L. dispar* in the USA (Hajek *et al.* 1993). Epidemics of the disease have occurred naturally and after introduction of the fungus at the leading edge of the advancing gypsy moth population front (Hajek *et al.* 1996) or as a homeowner tactic to manage the pest locally (Hajek & Webb 1999).

8.3.4 *Dendroctonus micans* in Britain as an example of successful integrated forest management of an introduced forest pest

The great spruce bark beetle, *Dendroctonus micans*, is a pest of spruce across the entire natural and man-influenced range of spruce in Eurasia (Grégoire 1988). During the twentieth century the range of the beetle has expanded westwards to colonize the Republic of Georgia (Kobakhidze *et al.* 1976), Turkey (Benz 1985), France (Carle *et al.* 1979) and Great Britain (Evans & Fielding 1994). Unlike the majority of North American members of the genus *Dendroctonus* in which the individuals employ mass attack strategies to colonize and kill trees, solitary *D. micans* females are capable of successfully invading apparently healthy spruce trees (King & Fielding 1989). A relatively slow life cycle (each generation takes between 12 and 24 months) and the moderate tolerance to attack by its main natural host, Norway spruce, *Picea abies*, in Western Europe, contribute to the slow build-up to lethally damaging populations of *D. micans* (Grégoire 1988; Fielding *et al.* 1991a). Hazard rating in relation to both host preference and to likelihood of tree mortality has been carried out by Bejer (1985) and indicates that preferences within the genus *Picea* are ranked in the order *abies* > *pungens* > *sitchensis* = *glauca* > *omorika*, while likelihood of successful brood development and tree mortality is ranked *pungens* = *orientalis* > *sitchensis* = *glauca* > *abies* > *omorika*. Thus, although *D. micans* is an important forest pest, the extent of tree mortality tends to vary considerably in relation to the species of spruce being attacked and to the proportion of spruce within a given forest block, attacks generally being more frequent and successful in forests where there is a high proportion of preferred spruce species (Ruhm 1968).

Patterns of attack have been similar in all countries where *D. micans* has established. It would appear that new infestations establish quite readily but, because tree mortality is initially very low, tend to go unnoticed for several years while populations gradually build up. This phase is followed by a period of rapid attack and expansion of population density giving rise to serious damage and tree mortality, after which the population subsides to a significant, but generally non-epidemic, background level (Grégoire 1988). The reasons for this pattern of infestation differ between regions, but include tree stress arising from drought (Petersen 1952; Battisti 1984; Bejer 1988) or very low winter temperatures (Bejer 1985), the presence of a high proportion of susceptible host tree species such as *P. orientalis* in the Republic of Georgia (Shavliashvili & Zharkov 1985) and forest operations (thinning, partial felling) that damage trees and open up the forest canopy (Evans *et al.* 1985). Following the outbreak phase, populations have generally declined to levels below an EIL, although there may be resurgences as observed by Battisti *et al.* (1986) in Italy where severe outbreaks of *D. micans* were discovered in 1985 even though damaging populations of the beetle had not been noted since the end of the nineteenth century.

One of the key factors contributing to the decline of *D. micans* populations virtually throughout its range is the presence of the predatory beetle *Rhizophagus grandis* (Coleoptera: Rhizophagidae). This beetle is unusual among predators in being specific to *D. micans*, as a result of the very precise host-location behaviour of the adult stage (Wainhouse *et al.* 1991, 1992; Grégoire *et al.* 1992). *R. grandis* has been implicated in population reduction in most of the range of *D. micans* and is regarded as an effective natural regulatory factor, particularly during the post-outbreak phase of infestations (Kobakhidze *et al.* 1970; Grégoire *et al.* 1989).

Development of integrated forest management in relation to D. micans *in Britain*

The expansion of forest cover in Britain during the twentieth century, following the establishment of the Forestry Commission as the government's forestry department in 1919, has been dominated by the growing of exotic conifers, particularly Sitka spruce. Recognizing that the availability of large areas of spruce increased the potential for colonization by insect pests, measures were put in place to reduce the risks of importation of exotic organisms (Phillips 1980). Bark beetles were considered to be the main insect threats, a situation that remains today, and measures such as freedom from bark or kiln drying have been part of plant health legislation for a number of years, more recently under EU Plant Health Directives. *D. micans* was regarded as a serious threat and risk assessment was carried out to ensure that measures to prevent importation were in place and possible control measures evaluated (Brown & Bevan 1966). Despite these measures, *D. micans* has established in Britain, and was discovered in 1982 near Ludlow, Shropshire on the English–Welsh border (Bevan & King 1983). Intensive surveys revealed that the infestation was already widespread in Wales and the bordering areas of England and, through tree-ring analysis, had clearly been in the country since at least 1973 (Fielding *et al.* 1991a).

A full account of the development of the IFM strategy has been provided by Evans and Fielding (1994) and is summarized in Fig. 8.5. The initial response to discovery of the pest followed practice typically employed in managing *D. micans* outbreaks in Europe: surveys were accompanied by sanitation felling to reduce the bark beetle population in an attempt to contain the outbreak. However, this policy was dropped when survey results and retrospective tree-ring analysis revealed that the beetle had been in Britain since at least 1973 and that, although it had spread quickly to reach virtually the full extent of the *D. micans* control area (DMCA) shown in Fig. 8.5, the rate of development in Britain indicated a cycle of between 12 and 24 months and relatively little tree mortality (Evans & Fielding 1994). Extensive

pine beetle *D. frontalis* (Coulson *et al.* 1989). The rate of development of the latter is extremely rapid, reflecting its distribution in the Southern USA and, thus, hazard models have had to include a greater temporal element that the other bark beetle models. Coulson *et al.* (1996) have recently incorporated landscape heterogeneity into a hazard rating model for *D. frontalis*, based primarily on tree species, radial growth, tree height, diameter at breast height (d.b.h.), stand basal area, species mixture, degree of canopy closure and landform classification. Overlaid on this basic hazard map are data on lightning strikes (a predisposing factor for pioneer beetle infestations). The data sets are combined, using GIS models, to produce a functional heterogeneity map that incorporates beetle dispersal behaviour and host plant suitability as a visual prediction of high hazard areas. This level of refinement may offer ways of incorporating disparate data sets on host and pest biology with explicit data on dynamics at both the stand and the wider landscape scales.

8.4 FUTURE DEVELOPMENTS IN FOREST PEST MANAGEMENT

The fact that many of the major forest pests require repeated intervention indicates that further information is required to develop more sustainable long-term management options. The general principles described in section 8.2 and the three examples of pest management that were described in more detail serve to illustrate that long-term, sustainable management strategies are difficult to achieve. At the very least, they require repeated intervention in order to avoid unacceptable damage or tree mortality, especially when the knowledge base that could point to more sustainable strategies is inadequate or too difficult to obtain. Hazard rating systems and increased use of spatial data are likely directions for the future, allowing managers to consider regional strategies to prevent rather than control outbreaks. Within these strategies examples have been given of the use of silvicultural and other techniques that rely on improvements in the defensive qualities of trees as well as reducing their apparency to potential insect pests. Such an approach is at the core of IPM and of the more holistic IFM, both of which rely on multidisciplinary approaches to management (Kogan 1998).

8.4.1 Genetically modified trees

The approaches that currently form part of integrated management regimes are 'conventional' in that they rely on site management, tree species choice, use of pesticides, etc. to manage the pest threat. Other, more novel, approaches are also being developed, although they are not yet entering the main armoury of IPM/IFM strategies. Genetic modification of biopesticides and of the trees themselves are, perhaps, the most radical of the potential new approaches.

Although it has proved difficult to produce stable transformations of trees and to evaluate the potential expression of introduced genes, there has been considerable interest and progress on this aspect of pest management (Schuerman & Dandekar 1993). In common with transformation of a range of plants, it has been possible to use vectors, such as the crown gall bacterium *Agrobacterium*, to introduce genes into trees. Most progress has been made in the transformation of poplars (*Populus* spp.) (Klopfenstein *et al.* 1997). With respect to management of invertebrate pests, the tactics have been to introduce *Bacillus thuringiensis* (*Bt*) toxin genes (Kleiner *et al.* 1995) and proteinase inhibitor genes (Confalonieri *et al.* 1998). Field tests of transgenic poplars expressing *Bt* have confirmed that expression of the δ-endotoxin was sufficient to cause reduced feeding in two Lepidoptera (forest tent caterpillar, *Malacosoma disstria*, and gypsy moth, *L. dispar*) (Kleiner *et al.* 1995). Incorporation of proteinase inhibitors into poplars has also shown promise in reducing defoliation damage from various leaf feeders. Cottonwood leaf beetle, *Chrysomela scripta*, larvae ate significantly less foliage on transformed poplars than on untransformed controls (Kang *et al.* 1997). By contrast, no significant effects were observed in tests of poplar transformed to express proteinase inhibitor when assayed using the moths *L. dispar* and *Clostera anastomosis* (Confalonieri *et al.* 1998). Similar approaches are being used to transform other trees, especially elms, chestnuts and conifers, in relation to pest and disease control. For example incorporation of *Bt* into spruce using direct gene transfer particle acceleration methods has resulted in plants that gave sublethal expression of δ-endotoxin sufficient to reduce feeding and weight gain in eastern spruce budworm, *C. fumiferana* (Ellis *et al.* 1993). There is no doubt that further developments will be made as the technology and ecological significance of field use of transgenic trees are evaluated and refined. Issues of public concern and risk assessment must be addressed at the same rate as the science is developing, particularly in relation to the issues of gene transfer from transgenic trees and the possibilities of resistance developing as a result of continuous expression, for example, of *Bt* (James *et al.* 1998).

8.4.2 Improvements in monitoring and use of conventional pest management techniques

Although it could be argued that improvements to existing technologies do not constitute novelty, there is still considerable scope for further developments in most aspects of IPM and IFM. Monitoring of pest population change is central to decision support systems and there is a need for increased accuracy and cost-effectiveness in this area. Use of pheromone traps for many moths and beetles is a well-established technology, but there may be poor correspondence between trap catches and pest risk. This is being addressed through the use of GIS and spatial statistics in the case of *L. dispar* in North America (Liebhold *et al.* 1998). Modelling the links between pheromone trap captures, egg mass density estimates and distances from the advancing front of the gypsy moth infestation indicated the inadequacy of current methods and pointed to improved methods for the future, based around the predictive models.

Advances in detection and synthesis of biologically active compounds for a range of pests that have been difficult to monitor previously will offer further potential in monitoring. Thus the pheromone of European pine sawfly, *Neodiprion sertifer*, has been synthesized and used in field tests to establish links between trap catches and field populations (Simandl & Anderbrant 1995; Wedding *et al.* 1995). Such methods could replace costly surveys for other developmental stages such as pupae, eggs or larvae.

Clearly, monitoring systems must be as quantitative as possible and be linked to the EIL, thus giving a threshold for management action. Decision support systems may range from single responses to one or more of a complex of alternative measures for pest management. Spruce budworm, described in section 8.3.2, provides a good example of how complex the decision support process can be. Bark beetle hazard rating has also received wide attention and has resulted in various systems, including expert systems, that have aided management decisions. For instance, risk rating of factors affecting attack rates of spruce beetle (*Dendroctonus rufipennis*) in Alaska was developed after consultation with experts on both the pest and on the forests in which it lives (Reynolds & Holsten 1994). The expert system included nine factors—stand hazard, size and trend of spruce beetle populations in neighbouring stands, degree-days in the previous June, total rainfall in the previous summer, and availability of four types of breeding material—as the main determinants leading to outbreaks. A procedure called the analytic hierarchy process (AHP) was used for pairwise comparisons of the factors, ultimately leading to the conclusion that stand hazard and presence of windthrown trees were the key risk factors leading to *D. rufipennis* attack.

Once the risk of infestation and likelihood of exceeding the EIL is established, pest management options can be implemented, with specific reference to the purpose of forest management at the local or regional scale. All aspects of pest control, from use of pesticides (chemical and microbiological) through to silvicultural management and biological control, can then be evaluated against the desired reduction in pest population size.

As has been emphasized throughout this chapter, decision-support systems such as those described above can only be developed if detailed knowledge of pests and their interactions with their host trees and other biotic and abiotic factors is available. It is likely that, to be sustainable, future management will increasingly take this into account, especially as the multipurpose nature of forestry continues to dominate future management options.

REFERENCES

Armstrong, J.A. & Ives, W.G.H. (1995). *Forest Insect Pests in Canada*. Canadian Forest Service, Ottawa.

Barbosa, P. & Schaefer, P.W. (1997) Comparative analysis of patterns of invasion and spread of related lymantriids. In: Watt, A.D., Stork, N. & Hunter, M.D., eds *Forests and Insects*, pp. 153–75. Chapman & Hall, London.

Battisti, A. (1984) *Dendroctonus micans* (Kugelann) in Italy (Coleoptera Scolytidae). *Frustula Entomologica* 7, 631–7.

Battisti, A., Menardi, R. & Sala, G. (1986) Sulla presenza del coleottero scolitide *Dendroctonus micans* Kugelann in boschi di abete rosso del Veneto. *Italia Forestale E Montana* 41, 197–203.

Bejer, B. (1985) *Dendroctonus micans* in Denmark. In: Grégoire, J.C. & Pasteels, J.M., eds. *Biological Control of Bark Beetles (Dendroctonus micans)*, pp. 2–19. Commission of the European Communities, Brussels.

Bejer, B. (1988) Sitka spruce and *Dendroctonus micans*. *Dansk Skovforenings Tidsskrift* 73, 34–42.

Bentz, B.J., Amman, G.D. & Logan, J.A. (1993) A critical assessment of risk classification systems for the mountain pine beetle. *Forest Ecology and Management* 61, 349–66.

Benz, G. (1985) *Dendroctonus micans* in Turkey: The situation today. In: Grégoire, J.C. & Pasteels, J.M., eds. *Biological Control of Bark Beetles (Dendroctonus micans)*, pp. 43–7. Commission of the European Communities, Brussels.

Bergeron, Y., Leduc, A., Engelmark, O., Harvey, B., Morin, H. & Sirois, L. (1998) Relationships between change in fire frequency and mortality due to spruce budworm outbreak in the southeastern Canadian boreal forest. *Journal of Vegetation Science* 9, 493–500.

Berryman, A.A. (1972) Resistance of conifers to invasion by bark beetle fungus associations. *Bioscience* 22, 598–602.

Berryman, A.A. (1988) Towards a unified theory of plant defense. In: Mattson, W.J., Levieux, J. & Bernard-Dagan, C., eds. *Mechanisms of Woody Plant Defenses Against Insects*, pp. 39–55. Springer-Verlag, New York.

Berryman, A.A. (1991) The gypsy moth in North America: a case of successful biological control? *Trends in Ecology and Evolution* 6, 110–11.

Bevan, D. & King, C.J. (1983) *Dendroctonus micans* Kug—a new pest of spruce in UK. *Commonwealth Forestry Review* 62, 41–51.

Billany, D.J. (1978) *Gilpinia hercyniae* (Hertig): a pest of spruce. *Forestry Commission Forest Record* 117, 1–11.

Billany, D.J. & Brown, R.M. (1980) The web-spinning larch sawfly, *Cephalcia lariciphila* Wachtl. (Hymenoptera: Pamphiliidae). A new pest of *Larix* in England and Wales. *Forestry* 53, 71–80.

Billany, D.J., Winter, T.G. & Gauld, I.D. (1985) *Olesicampe monticola* (Hedwig) (Hymenoptera: Ichneumonidae) redescribed together with notes on its biology as a parasite of *Cephalcia lariciphila* (Wachtl) (Hymenoptera: Pamphiliidae). *Bulletin of Entomological Research* 75, 267–74.

Bird, F.T. & Elgee, D.E. (1957) A virus disease and introduced parasites as factors controlling the European spruce sawfly, *Diprion hercyniae* (Htg) in central New Brunswick. *Canadian Entomologist* 89, 371–8.

Birgersson, G. & Bergstrom, G. (1989) Volatiles released from individual spruce bark beetle entrance holes. Quantitative variations during the first week of attack. *Journal of Chemical Ecology* 15, 2465–83.

Blais, J.R. (1983) Trends in the frequency, extent and severity of spruce budworm outbreaks in eastern Canada. *Canadian Journal of Forest Research* 13, 539–47.

Blais, J.R. (1985) The ecology of the eastern spruce budworm: a review and discussion. In: Sanders, C.J., Stark, R.W., Mullins, E.J. & Murphy, J., eds. *Recent Advances in Spruce Budworm Research*, pp. 49–59. Canadian Forestry Service, Ottawa.

Brooks, M.H., Colbert, J.J. & Stark, R.W. (1985) Managing trees and stands susceptible to western spruce budworm. *USDA Technical Bulletin* 1695, 1–111.

Brown, J.M.B. & Bevan, D. (1966) The great spruce bark beetle, *Dendroctonus micans*, in north west Europe. *Forestry Commission Bulletin* 38, 1–41.

Cappuccino, N., Lavertu, D., Bergeron, Y. & Regniere, J. (1998) Spruce budworm impact, abundance and parasitism rate in a patchy landscape. *Oecologia* 114, 236–42.

Carle, P., Granet, A.-M. & Perrot, J.-P. (1979) Dispersal and destructiveness of *Dendroctonus micans* (Coleoptera, Scolytidae) in France. *Revue Forestiere Francaise* 31, 298–311.

Cavey, J.F., Hoebeke, E.R., Passoa, S. & Lingafelter, S.W. (1998) A new exotic threat to North American hardwood forests: an Asian longhorned beetle, *Anoplophora glabripennis* (Motschulsky) (Coleoptera: Cerambycidae). I. Larval description and diagnosis. *Proceedings of the Entomological Society of Washington* 100, 373–81.

Christiansen, E. (1985) *Ips/Ceratocystis*-infection of Norway spruce: what is a deadly dosage? *Zeitschrift fur Angewandte Entomologie* 99, 6–11.

Christiansen, E. & Bakke, A. (1988) The spruce bark beetle of Eurasia. In: Berryman, A.A., ed. *Dynamics of*

Forest Insect Populations: Patterns, Causes, Implications, pp. 479–503. Plenum Press, New York.

Christiansen, E., Waring, R.H. & Berryman, A.A. (1987) Resistance of conifers to bark beetle attack: searching for general relationships. *Forest Ecology and Management* **22**, 89–106.

Confalonieri, M., Allegro, G., Balestrazzi, A., Fogher, C. & Delledonne, M. (1998) Regeneration of *Populus nigra* transgenic plants expressing a Kunitz proteinase inhibitor (KTi3) gene. *Molecular Breeding* **4**, 137–45.

Coulson, R.N. (1979) Population dynamics of bark beetles. *Annual Review of Entomology* **24**, 417–47.

Coulson, R.N., Feldman, R.M., Sharpe, P.J.H., Pulley, P.E., Wagner, T.L. & Payne, T.L. (1989) An overview of the TAMBEETLE model of *Dendroctonus frontalis* population dynamics. *Holarctic Ecology* **12**, 445–50.

Coulson, R.N., Fitzgerald, J.W., McFadden, B.A., Pulley, P.E., Lovelady, C.N. & Giardino, J.R. (1996) Functional heterogeneity of forest landscapes: how host defenses influence epidemiology of the southern pine beetle. In: Mattson, W.J., Niemela, P. & Rousi, M., eds. *Dynamics of Forest Herbivory: Quest for Pattern and Principle*, pp. 272–86. USDA Forest Service, St. Paul, MN.

Cunningham, J.C. (1995) Baculoviruses as microbial insecticides. In: Reuveni, R., ed. *Novel Approaches to Integrated Pest Management*, pp. 261–92. Lewis Publishers, Boca Raton.

Dobesberger, E.J., Lim, K.P. & Raske, A.G. (1999) Spruce budworm moth flight from New Brunswick to Newfoundland. *Canadian Entomologist* **115**, 1641–5.

Elkinton, J.S. & Liebhold, A.M. (1990) Population dynamics of gypsy moth in North America. *Annual Review of Entomology* **35**, 571–96.

Ellis, D.D., McCabe, D.E., McInnis, S. *et al.* (1993) Stable transformation of *Picea glauca* by particle acceleration. *Bio/Technology* **11**, 84–9.

Entwistle, P.F., Adams, P.H.W., Evans, H.F. & Rivers, C.F. (1983) Epizootiology of a nuclear polyhedrosis virus (Baculoviridae) in European spruce sawfly (*Gilpinia hercyniae*): spread of disease from small epicentres in comparison with spread of baculovirus diseases in other hosts. *Journal of Applied Ecology* **20**, 473–87.

EPPO (1993) Guideline on pest risk analysis, no. 1. Check-list of information required for pest risk analysis. *EPPO Bulletin* **23**, 191–8.

Evans, H.F. & Fielding, N.J. (1994) Integrated management of *Dendroctonus micans* in Great Britain. *Forest Ecology and Management* **65**, 17–30.

Evans, H.F. & Fielding, N.J. (1996) Restoring the natural balance: biological control of *Dendroctonus micans* in Great Britain. In: Waage, J.K., ed. *Biological Control Introductions: Opportunities for Improved Crop Production*, pp. 45–57. BCPC, Farnham.

Evans, H.F. & King, C.J. (1988) *Dendroctonus micans:*

guidelines for forest managers. *Forestry Commission Research Information Note* **128**, 1–8.

Evans, H.F., King, C.J. & Wainhouse, D. (1985) *Dendroctonus micans* in the United Kingdom: the results of two years experience in survey and control. In: Grégoire, J.C. & Pasteels, J.M., eds. *Biological Control of Bark Beetles (Dendroctonus micans)*, pp. 20–34. Commission of the European Communities, Brussels.

Evans, H.F., Stoakley, J.T., Leather, S.R., Watt, A.D., Raske, A.G. & Wickman, B.E., eds (1991) Development of an integrated approach to control of pine beauty moth in Scotland. *Forest Ecology and Management* **39**, 19–28.

Evans, H.F., McNamara, D.G., Braasch, H., Chadoeuf, J. & Magnusson, C. (1996) Pest Risk Analysis (PRA) for the territories of the European Union (as PRA area) on *Bursaphelenchus xylophilus* and its vectors in the genus *Monochamus*. *EPPO Bulletin* **26**, 199–249.

Evertsen, J.A., Evans, H.F., Hall, G.A., Knaggs, G.R., Marzin, H. & McAree, D. (1991) *The Development of Treatment Schedules to Ensure Eradication in Timber of the Pinewood Nematode* (Bursaphelenchus xylophilus) *and its Vectors*. EOLAS, Dublin.

Fielding, N.J. (1992) *Rhizophagus grandis* as a means of biological control against *Dendroctonus micans* in Britain. *Forestry Commission Research Information Note* **224**, 1–3.

Fielding, N.J. & Evans, H.F. (1997) Biological control of *Dendroctonus micans* (Scolytidae) in Great Britain. *Biocontrol News and Information* **18**, 51–60.

Fielding, N.J., Evans, H.F., Williams, J.M. & Evans, B. (1991a) Distribution and spread of the great European spruce bark beetle, *Dendroctonus micans*, in Britain — 1982–89. *Forestry* **64**, 345–58.

Fielding, N.J., O'Keefe, T. & King, C.J. (1991b) Dispersal and host-finding capability of the predatory beetle *Rhizophagus grandis* Gyll. (Col., Rhizophagide). *Journal of Applied Entomology* **112**, 89–98.

Fielding, N.J., Evans, B., Burgess, R. & Evans, H.F. (1994) Protected zone surveys in Great Britain for *Ips typographus*, *I. amitinus*, *I. duplicatus* and *Dendroctonus micans*. *Forestry Commission Research Information Note* **253**, 1–6.

Flamm, R.O., Coulson, R.N. & Payne, T.L. (1988) The southern pine beetle. In: Berryman, A.A., ed. *Dynamics of Forest Insect Populations: Patterns, Causes, Implications*, pp. 531–553. Plenum Press, New York.

Fleming, R.A. & Shoemaker, C.A.S. (1992) Evaluating models for spruce budworm-forest management: comparing output with regional field data. *Ecological Applications* **2**, 460–77.

Fleming, R.A. & Van Frankenhuyzen, K. (1995) Population dynamics considerations in targeting double applications of *B.t.* to control the spruce budworm. In: Hain, F.P., Salom, S.M., Ravlin, W.F., Payne, T.L. & Raffa, K.F., eds. *Behavior, Population Dynamics and*

Control of Forest Insects, pp. 441–51. Ohio State University and USDA Forest Service, Maui, HI.

Fleming, R.A., Candau, J.N. & Munn, R.E. (1998) Influences of climatic change on some ecological processes of an insect outbreak system in Canada's boreal forests and the implications for biodiversity. *Environmental Monitoring and Assessment* **49**, 235–49.

Gao, R.T., Qin, X.F., Chen, D.Y. & Chen, W.P. (1993) A study on the damage caused by *Anoplophora glabripennis*. *Forest Research* **6**, 189–93.

Gould, J.R., Elkington, J.S. & Wallner, W.E. (1990) Density-dependent suppression of experimentally created gypsy moth, *Lymantria dispar* (Lepidoptera: Lymantriidae), populations by natural enemies. *Journal of Animal Ecology* **59**, 213–34.

Grégoire, J.C. (1988) The greater European spruce beetle. In: Berryman, A.A., ed. *Dynamics of Forest Insect Populations: Patterns, Causes, Implications*, pp. 455–78. Plenum Press, New York.

Grégoire, J.C., Baisier, M., Merlin, J. & Naccache, Y. (1989) Interactions between *Rhizophagus grandis* (Coleoptera: Rhizophagidae) and *Dendroctonus micans* (Coleoptera: Scolytidae) in the field and the laboratory: their application for the biological control of *D. micans* in France. In: Kulhavy, D.L. & Miller, M.C., eds. *Potential for Biological Control of Dendroctonus and Ips Bark Beetles*, pp. 95–108. Stephen F. Austin University, Nacogdoches, TX.

Grégoire, J.C., Couillien, D., Krebber, R., König, W.A., Meyer, H. & Francke, W. (1992) Orientation of *Rhizophagus grandis* (Coleoptera: Rhizophagidae) to oxygenated monoterpenes in a species-specific predator–prey relationship. *Chemoecology* **3**, 14–18.

Haack, R.A., Law, K.R., Mastro, V.C., Ossenbruggen, H.S. & Raimo, B.J. (1997) New York's battle with the Asian long-horned beetle. *Journal of Forestry* **95**, 11–15.

Hajek, A.E. & Webb, R.E. (1999) Inoculative augmentation of the fungal entomopathogen *Entomophaga maimaiga* as a homeowner tactic to control gypsy moth (Lepidoptera: Lymantriidae). *Biological Control* **14**, 11–18.

Hajek, A.E., Larkin, T.S., Carruthers, R.I. & Soper, R.S. (1993) Modeling the dynamics of *Entomophaga maimaiga* (Zygomycetes, Entomophthorales) epizootics in gypsy moth (Lepidoptera, Lymantriidae) populations. *Environmental Entomology* **22**, 1172–87.

Hajek, A.E., Elkinton, J.S. & Witcosky, J.J. (1996) Introduction and spread of the fungal pathogen *Entomophaga maimaiga* (zygomycetes: entomophthorales) along the leading edge of gypsy moth (lepidoptera: lymantriidae) spread. *Environmental Entomology* **25**, 1235–47.

James, R.R., Difazio, S.P., Brunner, A.M. & Strauss, S.H. (1998) Environmental effects of genetically engineered woody biomass crops. *Biomass and Bioenergy* **14**, 403–14.

Kang, H., Hall, R.B., Heuchelin, S.A. *et al.* (1997) Transgenic *Populus*: *in vitro* screening for resistance to cottonwood leaf beetle (Coleoptera: Chrysomelidae). *Canadian Journal of Forest Research* **27**, 943–4.

King, C.J. & Fielding, N.J. (1989) *Dendroctonus micans* in Britain—its Biology and Control. *Forestry Commission Bulletin* **85**, 1–11.

Kleiner, K.W., Ellis, D.D., McCown, B.H. & Raffa, K.F. (1995) Field evaluation of transgenic poplar expressing a *Bacillus thuringiensis* cry1A(a) d-endotoxin gene against forest tent caterpillar (Lepidoptera: Lasiocampidae) and gypsy moth (Lepidoptera: Lymantriidae) following winter dormancy. *Environmental Entomology* **24**, 1358–64.

Kleiner, K.W., Montgomery, M.E. & Schultz, J.C. (1989) Variation in leaf quality of two oak species: implications for stand susceptibility to gypsy moth defoliation. *Canadian Journal of Forest Research* **19**, 1445–50.

Klopfenstein, N.B., Chun, Y., Kim, M. *et al.* (1997) *Micropropagation, Genetic Engineering, and Molecular Biology of Populus*. General Technical Report RM-297. 1–326. Rocky Mountain Forest and Range Experiment Station, USDA Forest Service.

Kobakhidze, D.N., Tvaradze, M.S. & Kraveishvili, I.K. (1970) Preliminary results of introduction, study of bioecology, development of methods of artificial rearing and naturalisation of the effective entomophage, *Rhizophagus grandis* Kugel, in spruce plantations in Georgia (in Russian). *Bulletin of the Academy of Sciences of the Georgian SSR* **60**, 205–8.

Kobakhidze, D.N., Tvaradze, M.S. & Kraveishvili, I.K. (1976) Preliminary results of introduction, study of bioecology, development of methods of artificial rearing and naturalization of the effective entomophage, *Rhizophagus grandis* Gyll., against the European Spruce beetle, *Dendroctonus micans* Kugel, in spruce plantations in Georgia. *Translation, Environment Canada* no. OOENV TR-1030.

Kogan, M. (ed.) (1986) *Ecological Theory and Integrated Pest Management Practice*. John Wiley, New York.

Kogan, M. (1998) Integrated pest management: historical perspectives and contemporary developments. *Annual Review of Entomology* **43**, 243–70.

Kulhavy, D.L. & Miller, M.C. (1989) *Potential for Biological Control of Dendroctonus and Ips Bark Beetles*. Stephen F. Austin University Press, Nacogdoches, TX.

Lawrence, R.K., Mattson, W.J. & Haack, R.A. (1997) White spruce and the spruce budworm: defining the phenological window of susceptibility. *Canadian Entomologist* **129**, 291–318.

Leather, S.R. (1987) Pine monoterpenes stimulate oviposition in the pine beauty moth, *Panolis flammea*.

Entomologia Experimentalis et Applicata **43**, 295–303.

Leather, S.R. (1992) Forest management practice to minimise the impact of the pine beauty moth. *Research Information Note Forestry Commission Research Division* (217), 1–3.

Lechowicz, M.J. & Mauffette, Y. (1986) Host preferences of the gypsy moth in eastern North American versus European forests. *Revue d'Entomologie Du Quebec* **31**, 43–51.

Liebhold, A.M., Halverson, J.A. & Elmes, G.A. (1992) Gypsy moth invasion in North America: a quantitative analysis. *Journal of Biogeography* **19**, 513–20.

Liebhold, A., Luzader, E., Reardon, R. *et al.* (1998) Forecasting gypsy moth (Lepidoptera: Lymantriidae) defoliation with a geographical information system. *Journal of Economic Entomology* **91**, 464–72.

Mamiya, Y. (1988) History of pine wilt disease in Japan. *Journal of Nematology* **20**, 219–26.

Mattson, W.J., Simmons, G.A. & Witter, J.A. (1988) The spruce budworm in eastern North America. In: Berryman, A.A., ed., *Dynamics of Forest Insect Populations: Patterns, Causes, Implications* pp. 309–30. Plenum Press, New York.

Montgomery, M.E. & Wallner, W.E. (1988) The gypsy moth: a westward migrant. In: Berryman, A.A., ed. *Dynamics of Forest Insect Populations Patterns, Causes, Implications*, pp. 353–75. Plenum Press, New York.

Morris, R.F. & Miller, C.A. (1954) The development of life tables for the spruce budworm. *Canadian Journal of Zoology* **32**, 283–301.

Mulock, P. & Christiansen, E. (1986) The threshold of successful attack by *Ips typographus* on *Picea abies*: a field experiment. *Forest Ecology and Management* **14**, 125–32.

Nealis, V.G. & Lomic, P.V. (1994) Host-plant influence on the population ecology of the jack pine budworm, *Choristoneura pinus* (Lepidoptera: Tortricidae). *Ecological Entomology* **19**, 367–73.

Nebeker, T.E., Hodges, J.D., Blanche, C.A., Honea, C.R. & Tisdale, R.A. (1992) Variation in the constitutive defense system of loblolly pine in relation to bark beetle attack. *Forest Science* **38**, 457–66.

O'Neill, M. & Evans, H.F. (1999) Cost-effectiveness analysis of options within an Integrated Crop Management regime against great spruce bark beetle, *Dendroctonus micans* (Kug.) (Coleoptera: Scolytidae). *Agricultural and Forest Entomology* **1**, 151–6.

Paine, T.D., Raffa, K.F. & Harrington, T.C. (1997) Interactions among scolytid bark beetles, their associated fungi, and live host conifers. *Annual Review of Entomology* **42**, 179–206.

Petersen, B.B. (1952) *Hylesinus micans*, artens udbredelse og en oversigt over dens optraeden i Danmark. *Dansk Skovforenings Tidsskrift* **37**, 299–322.

Phillips, D.H. (1980) International plant health controls: conflicts, problems and cooperation. A European experience. *Research and Development Paper, Forestry Commission* **125**, 1–20.

Raffa, K.F. (1988) The mountain pine beetle in Western North America. In: Berryman, A.A., ed. *Dynamics of Forest Insect Populations: Patterns, Causes, Implications*, pp. 505–30. Plenum Press, New York.

Raffa, K.F. & Berryman, A.A. (1983) The role of host plant resistance in the colonization behavior and ecology of bark beetles (Coleoptera: Scolytidae). *Ecological Monographs* **53**, 27–49.

Raffa, K.F. & Smalley, E.B. (1995) Interaction of pre-attack and induced monoterpene concentrations in host conifer defense against bark beetle-fungal complexes. *Oecologia* **102**, 285–95.

Reeve, J.D., Rhodes, D.J. & Turchin, P. (1998) Scramble competition in the southern pine beetle, *Dendroctonus frontalis*. *Ecological Entomology* **23**, 433–43.

Régnière, J. & Lysyk, T.J. (1995) Population dynamics of the spruce budworm, *Choristoneura fumiferana*. In: Armstrong, J.A. & Ives, W.G.H., eds. *Forest Insect Pests in Canada*, pp. 95–105. Natural Resources Canada, Canadian Forest Service, Ottawa.

Régnière, J., Lavigne, D., Dickison, R. & Staples, A. (1995) *Performance Analysis of BioSIM, a Seasonal Pest Management Planning Tool, in New Brunswick in 1992 and 1993*. Information Report, Laurentian Forestry Centre Quebec Region Canadian Forest Service no. LAU-X-115, 1–28.

Reynolds, K.M. & Hard, J.S. (1991) Risk and hazard of spruce beetle attack in unmanaged stands on the Kenai Peninsula, Alaska, under epidemic conditions. *Forest Ecology and Management* **43**, 137–51.

Reynolds, K.M. & Holsten, E.H. (1994) Relative importance of risk factors for spruce beetle outbreaks. *Canadian Journal of Forest Research* **24**, 2089–95.

Reynolds, K.M. & Holsten, E.H. (1996) Classification of spruce beetle hazard in Lutz and Sitka spruce stands on the Kenai Peninsula, Alaska. *Forest Ecology and Management* **84**, 251–62.

Royama, T. (1984) Population dynamics of the spruce budworm *Choristoneura fumiferana*. *Ecological Monographs* **54**, 429–62.

Royama, T. (1997) Population dynamics of forest insects: are they governed by single or multiple factors? In: Watt, A.D., Stork, N. & Hunter, M.D., eds. *Forests and Insects*, pp. 37–48. Chapman & Hall, London.

Ruhm, W. (1968) Zur mechanisch-chemischen und oekologischen Bekampfung des Riesenbastkaefer *Dendroctonus micans* KUG. *Zeitschrift Fur Angewandte Entomologie* **43**, 286–325.

Schuerman, P.L. & Dandekar, A.M. (1993) Transformation of temperate woody crops—progress and potentials. *Science &. Horticulture, Amsterdam* **55**, 101–24.

Sharov, A.A. & Liebhold, A.M. (1998a) Bioeconomics of managing the spread of exotic pest species with barrier zones. *Ecological Applications* **8**, 833–45.

Sharov, A.A. & Liebhold, A.M. (1998b) Model of slowing the spread of gypsy moth (Lepidoptera: Lymantriidae) with a barrier zone. *Ecological Applications* **8**, 1170–9.

Shavliashvili, I.A. & Zharkov, D.G. (1985) Effects of ecological factors on the interactions between populations of *Dendroctonus micans* and *Ips typographus* (Coleoptera: Scolytidae). In: Safranyik, L., ed. *Proceedings of the IUFRO Conference on the Role of the Host Plant in the Population Dynamics of Forest Insects*, pp. 227–32. Canadian Forest Service, Banff, Canada.

Sheehan, K.A. (1992) GMPHEN: a gypsy moth phenology model. *USDA Forest Service General Technical Report, Northeastern Forest Experiment Station* NE-158.

Shepherd, R.F. (1985) A theory on the effects of diverse host-climate environments in British Columbia on the dynamics of western spruce budworm. In: Sanders, C.J., Stark, R.W., Mullins, E.J. & Murphy, J., eds. *Recent Advances in Spuce Budworms Research*, pp. 60–70. Canadian Forestry Service, Ottawa.

Simandl, J. & Anderbrant, O. (1995) Spatial distribution of flying *Neodiprion sertifer* (Hymenoptera, Diprionidae) males in a mature *Pinus sylvestris* stand as determined by pheromone trap catch. *Scandinavian Journal of Forest Research* **10**, 51–5.

Speight, M.R. & Wainhouse, D. (1989) *Ecology and Management of Forest Insects*. Oxford University Press, Oxford.

Steele, R., Williams, R.E., Weatherby, J.C., Reinhardt, E.D., Hoffman, J.T. & Thier, R.W. (1996) *Stand Hazard Rating for Central Idaho Forests*. General Technical Report INT-GTR-332, pp. 1–29. Intermountain Research Station, USDA Forest Service, Ogden, UT.

Stoakley, J.T. (1987) Pine beauty moth control: past, present and future. In: Leather, S.R., Stoakley, J.T. & Evans, H.F., eds. *Population Biology and Control of Pine Beauty Moth*. Bulletin 67, pp. 87–90. Forestry Commission, Edinburgh.

Straw, N.A. & Fielding, N.J. (1997) The impact of aphids on Sitka spruce. *Forestry Commission, Report on Forest Research* 40–4.

Talerico, R.L. (1983) *Summary of Life History and Hosts of Spruce Budworms*. General Technical Report NE-85, pp. 1–4, Northeastern Forest and Range Experiment Station, USDA Forest Service, Newhaven, CT.

Task Force on Pasteurization of Softwood Lumber

(1991) *The Use of Heat Treatment in the Eradication of the Pinewood Nematode and its Vectors in Softwood Lumber*. Forestry Canada, Ottawa.

Thompson, J.N. & Pellmyr, O. (1991) Evolution of oviposition behavior and host preference in Lepidoptera. *Annual Review of Entomology* **36**, 65–89.

Trewhella, K.E., Leather, S.R. & Day, K.R. (1997) The effect of constitutive resistance in lodgepole pine (*Pinus contorta*) and Scots pine (*P. sylvestris*) on oviposition by three pine feeding herbivores. *Bulletin of Entomological Research* **87**, 81–8.

Van Frankenhuyzen, K. (1993) The challenge of *Bacillus thuringiensis*. In: Entwistle, P.F., Cory, J.S., Bailey, M.J. & Higgs, S., eds. *Bacillus thuringiensis, An Environmental Biopesticide: Theory and Practice*, pp. 1–35. John Wiley, Chichester.

Wainhouse, D. & Beech-Garwood, P. (1994) Growth and survival of *Dendroctonus micans* larvae on six species of conifer. *Journal of Applied Entomology* **117**, 393–9.

Wainhouse, D., Cross, D.J. & Howell, R.S. (1990) The role of lignin as a defence against the spruce bark beetle *Dendroctonus micans*: effect on larvae and adults. *Oecologia* **85**, 257–65.

Wainhouse, D., Wyatt, T., Phillips, A. *et al.* (1991) Response of the predator *Rhizophagus grandis* to host plant derived chemicals in *Dendroctonus micans* larval frass in wind tunnel experiments (Coleoptera: Rhizophagidae, Scolytidae). *Chemoecology* **2**, 55–63.

Wainhouse, D., Beech-Garwood, P., Howell, R.S., Kelly, D.R. & Orozco, M.P. (1992) Field response of the predator *Rhizophagus grandis* to prey frass and synthetic attractants. *Journal of Chemical Ecology* **18**, 1693–705.

Wainhouse, D., Rose, D.R. & Peace, A.J. (1997) The influence of preformed defences on the dynamic wound response in spruce bark. *Functional Ecology* **11**, 564–72.

Warren, J.E. & Linit, M.J. (1992) Within-wood spatial dispersion of the pinewood nematode, *Bursaphelenchus xylophilus*. *Journal of Nematology* **24**, 489–94.

Wedding, R., Anderbrant, O. & Jonsson, P. (1995) Influence of wind conditions and intertrap spacing on pheromone trap catches of male European pine sawfly, *Neodiprion sertifer*. *Entomologia Experimentalis et Applicata* **77**, 223–32.

Wilson, J.S., Isaac, E.S. & Gara, R.I. (1998) Impacts of mountain pine beetle (*Dendroctonus ponderosae*) (Col., Scolytidae) infestation on future landscape susceptibility to the western spruce budworm (*Choristoneura occidentalis*) (Lep., Tortricidae) in north central Washington. *Journal of Applied Entomology* **122**, 239–45.

Witter, J.A., Lynch, A.M. & Montgomery, B.A. (1983) *Management Implications of Interactions Between the Spruce Budworm and Spruce–Fir Stands [in Eastern North America]*. General Technical Report no. NE-85, pp. 127–32. Northeastern Forest and Range Experiment Station, USDA Forest Service, Newhaven, CT.

Wood, D.L. (1982) The role of pheromones, kairomones, and allomones in the host selection and colonization behaviour of bark beetles. *Annual Review of Entomology* **27**, 411–46.

Zumr, V. & Stary, P. (1993) Baited pitfall and flight traps in monitoring *Hylobius abietis* (L.) (Col., Curculionidae). *Journal of Applied Entomology* **115**, 454–61.

9: Management of the Disease Burden

JOHN N. GIBBS

9.1 INTRODUCTION

Disease can be a major factor in determining the productivity and sustainability of forest ecosystems. This chapter begins with an analysis of the nature of forest diseases, because understanding the forces at work is a prerequisite for intelligent disease management. The issues of disease exclusion and eradication will then be covered, before considering the approaches which can be adopted towards established diseases. Before informed decisions on the institution of control measures can be taken, there is a requirement for accurate diagnosis and for the determination of current and future disease impact. If measures are deemed appropriate, they can be exercised through cultural means, through the introduction of disease resistance or through the use of chemical or biocontrol agents. These options are evaluated with examples taken from around the world.

9.2 THE CHARACTERISTICS OF TREE DISEASES

Before systems can be devised to minimize the disease burden, we need to analyse the processes influencing the relationships between agents of damage and forest trees.

9.2.1 Causes of disease

Chief among biotic agents are the fungi but serious tree diseases are also caused by certain plants, such as mistletoes, some nematodes, a variety of bacteria including the xylem-limited bacteria, viruses, viroids and phytoplasmas. Abiotic agents of disease include drought, flood-

ing, frost and winter cold, snow and ice, high temperatures, wind and harmful chemicals (Sinclair *et al.* 1987; Manion 1991; Butin 1995; Tainter & Baker 1996; Phillips and Burdekin 1992).

9.2.2 The nature of disease

It is important to recognize that the relationship between a damaging agent and the disease it causes can vary from the simple to the highly complex—and complexity is common. There are very few situations as straightforward as that which occurs when an unseasonal frost kills the shoots and foliage of a susceptible tree. With all the biotic agents of damage there are environmental factors to consider and these can operate on both host and pathogen to influence the amount of disease that occurs. These relationships are often illustrated in the form of a disease triangle (Fig. 9.1).

The so-called 'declines and diebacks' exhibit the greatest complexity. Here no single agent of damage can be identified; rather disease results from the combined effects of many factors. Various 'models' have been devised to describe this situation. Perhaps the simplest and most satisfactory is that developed by Houston (1992). It can be summarized as follows:

1 Healthy trees + stress → Altered trees (tissues) (dieback begins).

2 Altered trees + more stress → Trees (tissues) altered further (dieback continues).

3 Severely altered trees (tissues) + organisms of secondary action → Trees (tissues) invaded. (Trees lose ability to respond to improved conditions, decline and perhaps die.)

The stress factors can be either abiotic, for

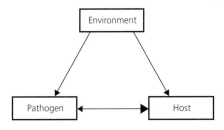

Fig. 9.1 The disease triangle. The environment influences both pathogen and host, while pathogen and host interact, each influencing the other.

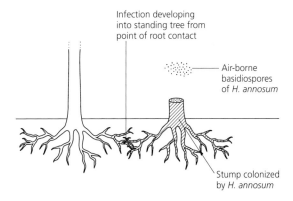

Infection developing into standing tree from point of root contact

Air-borne basidiospores of *H. annosum*

Stump colonized by *H. annosum*

Fig. 9.2 Diagram to illustrate the role of freshly cut thinning stumps in allowing *Heterobasidion annosum* to enter first rotation conifer plantations. Cross-hatching indicates wood colonized by the pathogen.

example drought, or biotic, for example defoliation by caterpillars. The organisms of secondary action are usually insects or fungi which cannot cause serious damage to a healthy tree but can be very damaging once the resistance of the host has been reduced.

9.2.3 Disease caused by indigenous pathogens

Indigenous pathogens can cause serious diseases on indigenous hosts if something occurs to upset the ecological balance that has evolved between the trees and their associated microorganisms. This can happen naturally or under the influence of man. A natural example is provided by a situation in which weather that is favourable for infection by a pathogen is prolonged over a number of growing seasons, thus leading to a progressive increase in the amount of disease. Some of the shoot and foliage diseases caused by Ascomycetes illustrate this point well: a succession of wet springs can lead to major damage. An example from southern Britain concerned *Venturia saliciperda* (asexual stage *Polaccia saliciperda*), the cause of scab on *Salix fragilis* and other willows. The spring weather in the years 1992–94 was unusually wet and, by the end of this period, constant loss of foliage shortly after flushing had resulted in the death of many branches. It was only with a dry spring in 1995 that the spiral of damage was broken.

Man can disturb the ecological balance in many ways, an important one being through woodland management operations. Thus in conifer crops the creation of fresh stumps through thinning or clearfelling can greatly increase the damage caused by the pathogenic wood-rotting basidiomycete *Heterobasidion annosum*. This is because, as shown in Fig. 9.2, the fungus is disseminated by air-borne basidiospores which colonize the fresh stump tissue and then spread by root contact to cause disease in neighbouring trees (Rishbeth 1950, 1951).

9.2.4 Disease caused by introduced pathogens

Many of the most serious disease problems have resulted from the movement of pathogens from one major forest region of the world to another. These organisms have not coevolved with their new hosts and the host may have no defensive systems to counteract them. A classic example is provided by *Cryphonectria parasitica*, the cause of chestnut blight on *Castanea* spp. This fungus is native in East Asia where it causes virtually no damage to the native *Castanea* species. However, it has been highly destructive when introduced to other temperate forest regions. Thus, within some 50 years of being detected in New York City, it had killed more than half the population of *C. dentata* through the extensive native range of that species (Fig. 9.3). In southern Europe, serious damage to the European chestnut, *C. sativa*, followed the introduction of the pathogen to Italy in *c.* 1938. For a more detailed analysis of

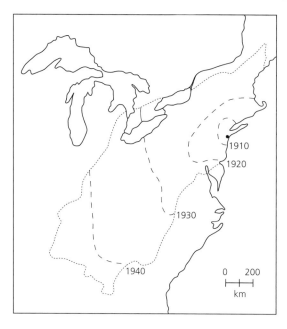

Fig. 9.3 Spread of chestnut blight through eastern North America. The dotted outline indicates the natural range of *Castanea dentata*. The dashed lines show, for a series of dates, the boundary of the area within which at least half the chestnut population had been killed. (From Gravatt 1949.)

this phenomenon in respect of both pests and pathogens, see Gibbs and Wainhouse (1986).

9.2.5 Diseases caused by pathogens that have undergone genetic change

Both indigenous and introduced pathogens are subject to genetic change, with the possibility that this will increase their potential for damage. These changes can involve both mutation and recombination during sexual reproduction. Much will depend on the nature of the host–pathogen relationship, on the durability of the disease-resistance mechanisms and on the selection pressure that is placed on the pathogen. The dangers are well illustrated by recent experience in Europe with poplar leaf rust caused by *Melampsora larici-populina* (Pinon & Frey 1997).

When related pathogens that have been previously isolated from each other are brought into contact, there is the possibility for hybridization

and the consequent emergence of new forms. An example of this is the newly recognized 'alder *Phytophthora*'. This fungus is causing mortality in riparian alder in a variety of places across Europe (Gibbs *et al.* 1999). Molecular studies have shown that it is a species hybrid between two *Phytophthora* species, *P. cambivora* and a fungus close to *P. fragariae*, neither of which has the capacity to attack alder (Brasier *et al.* 1999).

9.3 THE EXCLUSION AND ERADICATION OF TREE DISEASES

9.3.1 Exclusion

The first line of defence against non-native pathogens is to seek to exclude them through quarantine regulations. The stimulus for this approach was the discovery that much long-distance dissemination of tree pathogens was linked to the activities of man—most notably through trade in plants or plant products. Key examples are provided by chestnut blight, as mentioned above, and the probable introduction of white pine blister rust (*Cronartium ribicola*) from Europe to North America on pine seedlings.

Initially each country developed its own quarantine regulations, but with the formation of FAO (the Food and Agriculture Organization of the United Nations), the International Plant Protection Convention (IPPC) was drawn up in order to bring a global view to bear. The IPPC works through regional organizations like the European Plant Protection Organization (EPPO), which covers the whole of Europe and also some African countries which border the Mediterranean (Smith 1979). Over the years a framework of regulations has been devised, all designed to give quarantine benefits without causing unreasonable interference with trade. In recent years considerable emphasis has been placed on the identification of potential pathogens not present in the region in question. Then a Pest Risk Analysis (PRA) is conducted, in which the ability of the pathogen to cause damage is evaluated and the likely pathways of arrival are identified. Steps are then taken to devise suitable measures to prevent these pathways from being used.

It must be recognized that there are difficulties

with this approach when it comes to forestry. First, the great majority of forest pathogens are classified using morphological criteria which are inadequate for the recognition of differences that may be critical to their ability to cause disease (Brasier 1997). Therefore, the fact that a particular fungal species has been recorded in a forest region does not mean that there is nothing to fear from introductions of what is said to be the same species elsewhere. Here the classic example is Dutch elm disease. In Britain, this disease, caused by *Ophiostoma ulmi*, caused significant but not catastrophic damage in the period 1910–50. In the 1960s a highly destructive disease epidemic began which owed its origin to the introduction of another form of *O. ulmi* from North America (Brasier & Gibbs 1973). This new form was subsequently named *O. novo-ulmi*, but it should be noted that 'classical' morphological differences between it and *O. ulmi* are very small. Characterization depends much more on the differences in growth in culture, in genetic behaviour and in molecular constitution (Brasier 1991, 1997).

Another example of the complexity that can exist within a named species is provided by *Mycosphaerella dearnessii* (syn. *Scirrhia acicola*), the cause of the serious brown spot needle disease of pine. *M. dearnessii*, with a possible centre of origin in Central America, has now been recorded from both North and South America, Europe, Oceania and Asia. Recent molecular work (Huang *et al.* 1995) has confirmed earlier studies of physiological attributes to show that in the USA there are two distinct populations of the fungus, a northern and a southern group. So far in China, only southern group isolates have been found, and there are obviously good reasons for trying to ensure that the northern group isolates are not introduced. By contrast, isolates of the fungus in Europe, where it has recently been recorded, only seem to correspond to the northern type. Here therefore the objective must be to exclude isolates of the southern group.

The second weakness of the PRA approach is that it results in the development of highly specific quarantine measures which are likely to leave 'loop-holes' that unrecognized pathogens can use to move from one forest region to another. As indicated earlier in this chapter, many

organisms with a great potential for damage live in ecological balance in their own forest regions and therefore are never going to be identified and subjected to a PRA.

There is a now a growing feeling among forest pathologists that the international policies that relate to the exclusion of exotic pathogens need to be re-examined. In particular it is considered that certain types of imported material pose such a great risk that the requirement is for blanket 'non-specific' control measures. Thus during 1999, officers of the American Phytopathological Society wrote to US Vice President Al Gore asking for *all* unprocessed wood products (including logs, dunnage, etc.) to be heat treated before arriving in North America. The suggested treatment is one which results in the temperature at the centre of the material reaching 71°C for 75 min (Anonymous 1999).

The long-distance movement of living material, such as plants for planting, is clearly dangerous and plants of important forest genera are often prohibited entirely. The importation of 'germplasm' for tree improvement programmes demands special consideration. A relevant initiative in this respect is the publication in booklet form of guidelines on the safe movement of *Eucalyptus* germplasm (Ciesla *et al.* 1996). This combines a series of recommendations on how to move living *Eucalyptus* material, with brief well-illustrated accounts of the key quarantine diseases and pests.

Seeds are considered to offer fewer risks. However, these risks must not be disregarded. There is evidence that the leaf pathogen *Marssonina brunnea* reached New Zealand in poplar seed (Spiers & Wenham 1983), and recently it has been found that *Fusarium subglutinans* f. sp. *pini*, the cause of the dangerous pitch canker disease of pine, can be carried deep within the seed of *Pinus radiata* (Storer *et al.* 1998). In this case even the fungicidal treatments normally considered effective may not be able to kill the pathogen.

Quarantine regulations need to be supported by inspection procedures. The examination of large quantities of woody material is an exceedingly difficult task, particularly where microorganisms are concerned. Suspect material must be referred to a research specialist. In practice, it is often an

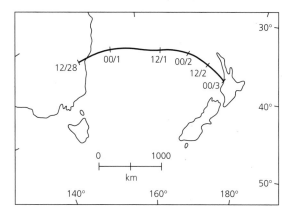

Fig. 9.4 Trajectory (recorded every 12 hours) of spores of poplar rust, *Melampsora*, from New South Wales to New Zealand, 28 February–3 March 1973, assuming drift on winds at a height of 3000 m (Pedgley 1982 after Wilkinson & Spiers 1976.)

entomologist who is the first to be consulted: insects are more conspicuous and more readily identifiable than most fungi and information on the types of insects present can often indicate whether or not dangerous pathogens are likely to be present. In this connection, it should be noted that many destructive pathogens are capable of only very limited survival in the tissues they have killed. Mould growth on bark and wood is often formed by 'secondary' microorganisms that carry no plant health threat.

Before leaving this subject, it should be recognized that a few pathogens are capable of long-distance dispersal by natural means. Spores of the rust fungi, for example, can survive the desiccation and UV exposure consequent upon being airborne to initiate new infections far from their place of origin. One of the best documented examples (Fig. 9.4) is the downwind drift of *Melampsora medusae* and *M. larici-populina* over a distance of 3000 km from Australia to New Zealand (Wilkinson & Spiers 1976). For a useful analysis of factors affecting windborne dispersal of pests and pathogens, see Pedgley (1982).

9.3.2 Eradication of an exotic pathogen

Eradication of a plant pathogen that is sufficiently well established to have caused symptoms is all but impossible. There is a record of success with European larch canker, caused by *Trichoscyphella willkommii*. This disease was introduced to Essex County, Massachusetts in the early years of the twentieth century on larch seedlings from Scotland. An eradication programme was instituted in the 1920s and over 5000 trees (European larch and Japanese larch) destroyed. Subsequent surveys showed progressively fewer diseased trees, and in 1965 it was declared that the disease had been eradicated. Unfortunately a separate outbreak, this time on the native *Larix laricina*, was discovered in the Canadian province of New Brunswick and subsequently in the State of Maine (Miller-Weeks *et al.* 1983). In Maine, examination of diseased material showed that it had been present for at least 11 years before detection. In this case eradication was not deemed to be feasible.

Where a specific threat has been identified and the pathogen has a number of distinctive features, as for example with *Ceratocystis fagacearum*, the cause of oak wilt in North America, it may be possible for countries at risk to produce well-illustrated literature that will increase the likelihood of early detection (e.g. Gibbs *et al.* 1984). However this is not normally possible. The best that can usually be achieved is to publicize the risk posed by exotic pathogens and to encourage forest staff to report all unusual occurrences of damage. An excellent example of an organized approach to the problem is provided by the New Zealand Forest Biology Survey which was established in 1956, as much to detect introduced pathogens (and pests) as to monitor the effects of native ones (Kershaw 1989). Plantations and high hazard areas around ports are monitored using a mixture of aerial survey, 'drive-through' road surveys and random point sampling (Carter 1989).

9.4 MANAGEMENT OF ESTABLISHED DISEASES

9.4.1 Diagnosis

Tree diseases are not easy to diagnose. There are many reasons for this. Symptoms are often not specific to a particular cause and may be complex in origin. If microorganisms are involved, they are often not easy to see and identify. As indicated above, the causal organisms often disappear quickly from the damaged tissue and are replaced by secondary organisms of no pathological significance. Awareness of these points has led to a lively debate among pathologists as to the extent to which diagnostic guides should be placed in the hands of forest staff. Some would argue that 'a little knowledge is a dangerous thing', and it cannot be denied that there is a danger that an important new disease may be misdiagnosed and in consequence disregarded. However, the balance of opinion around the world is that it is helpful to place diagnostic aids in the hands of forest staff. In the process, awareness of the hazards posed by diseases is raised and a partnership between forester and specialist can be created.

In this connection, it is interesting to see how the widespread public concern about 'Waldsterben' (death of the forest) or 'forest decline' that developed during the 1980s in Europe and North America led to the production of some useful guides aimed at helping those involved with trees to diagnose a range of insect and disease problems and thereby reduce the chance that *all* forest damage would uncritically be ascribed to air pollution. A good example from the USA is *Diagnosing Injury to Forest Trees* (Skelly *et al.* 1985) and from Germany the *Farbatlas Waldschäden* by Hartmann *et al.* (1988). Neither of these books requires the use of a microscope and both provide a wealth of coloured photographs.

Good illustrations are vital for field diagnosis and none is better or more comprehensive than the coloured photographs provided by Sinclair *et al.* (1984) in their classic work on the diseases of North American trees and shrubs. Splendid as this book is, however, it does have some weaknesses as a diagnostic guide, first because of its size and second because of its organization by 'disease type' rather than by host. For practical use, guides gain by being compact in size and by being organized, at least to some degree, by host species. An introduction to the practice of disease diagnosis can be very helpful and excellent examples have been provided in two British books, those by Strouts and Winter (1994) and Gregory and Redfern (1998). Where forest cover is dominated by only a few species, as in New Zealand

with *Pinus radiata*, it may be possible for a guide to be produced that is both compact and comprehensive (see Chapman 1998).

The role of computers is reflected in the production of guides on disk, and more recently, on the Internet. A good example here is provided by *Common Tree Diseases of British Columbia*, which is now available in book form (Allen *et al.* 1996), on CD ROM and on the World Wide Web (www.pfc.cfs.nrcan.gc.ca/health/td_web/). The last version incorporates an interactive 'expert key' for diagnosis. Another example available on the Internet is an expert system for diagnosing a wide range of tree problems in Sweden. This system, developed by staff of the Swedish University of Agricultural Sciences, is at present available only in Swedish but it is hoped that, eventually, it will be used on a Europe-wide basis. One of the major attractions of an Internet-based system is that its use can readily be monitored by the pathologists who have developed it. Not only can this allow for the collection of disease data but also it provides some safeguard from the problem of inappropriate interpretation (D R Rose, personnel communication).

Nothing that has been written here reduces the need for the forest manager to have direct access to suitably qualified pathologists to whom causes of damage can be reported and by whom detailed investigations can be made. This brings a two-fold benefit: first, the forester receives relevant first-hand advice, and second, the pathologist gains information on the behaviour of a particular disease under a particular set of conditions.

9.4.2 Determination of disease distribution and severity

Worldwide, many systems have been developed to provide information on disease distribution and severity. Some rely principally, if not entirely, on an analysis of data obtained through disease diagnostic services; others are more pro-active. These latter almost invariably involve assessment of both pest and pathogen problems, with the latter usually taking second place, this largely because of the difficulties of identifying diseases as discussed above. Thus the Forest Insect and Disease Survey (FIDS) that operated across

Canada for many years began its existence in 1935 purely as a forest insect survey; responsibility for diseases only being accepted in 1961. Six regional units covered Canada with rangers at field stations being supported by specialists at the regional headquarters. Access to many parts of the country is difficult and considerable use was made of aircraft to detect and map damaged areas of forest. In an evaluation of the work of FIDS in the Pacific and Yukon Region conducted in 1989, my colleague D. B. Redfern (unpublished report) commented that the system coped well with routine identification of recognized diseases and disorders but seemed to be less effective when it came to dealing with 'new' diseases and problems of complex cause. This is perhaps more or less inevitable!

A system which has developed impressively during the last decade is that operated by the Département de la Santé des Forêts (DSF) in France. Some 240 correspondents working for various forest management operations spend part of their time on tree health work. Standard forms have been devised for collecting information and this is entered into a national database by each observer using a telephone-linked computer system. Observations are evaluated by specialists based at five regional centres and these, in turn, have access to experts, for example in the identification of fungi. An excellent annual report is produced (e.g. Département de la Santé des Forêts 1997).

When it comes to obtaining quantitative information on disease impact, suitable sampling procedures are needed. Some 'generalized' techniques have been developed. Thus the DSF observers follow specific protocols based on the examination of clusters of 10 adjacent trees spaced 50 m apart along two parallel 200-m transacts (Flot 1998). In many cases, however, methods need to be tailored to the characteristics of the tree population and the disease. An example is provided by a survey established in 1994 to provide information on the development of the newly described Phytophthora disease of alder on riparian populations of *Alnus glutinosa* in southern Britain (Gibbs *et al.* 1999).

A major feature of the last two decades has been the development in Europe and North America of various plot-assessment procedures for determining forest health, these stemming from concerns, mentioned earlier in this chapter, about forest decline due to the effects of air pollution. The main emphasis is on the assessment of defoliation, a concept which works well enough for evergreen conifers but not for other types of tree: in the UK the emphasis is, more helpfully, on the 'density' or 'transparency' of the tree's crown. Other attributes of the tree may also be assessed, such as the degree of foliage discoloration and the amount of branch dieback. The procedures followed in Europe have been evaluated in some detail by Innes (1993). From a pathological viewpoint, a major drawback of much of this work is that investigation of the causes of any damage is often not integral to the process of health assessment. As a result, interpretation of the data is very difficult. An interesting perspective on this topic has, however, recently been offered by Dr E Boa of CABI Bioscience (personal communication). His view is that in much of the world there is a real need for the development of standardized damage assessment procedures which could be adopted by forest staff to build up quantitative information on the distribution and severity of a problem in advance of the involvement of a pathologist.

The value of an aerial view in obtaining information on the distribution and severity of a forest disease has long been obvious to forest scientists. With a foliage disease of a plantation crop, like Dothistroma needle blight on *Pinus radiata*, trained observers in helicopters or fixed-wing aircraft may even be able to obtain quantitative information by flying 100 m or so above the canopy (Kershaw *et al.* 1988). However, it is more normal to employ aerial photography and good use has been made of photographs—for example, for dating an episode of oak dieback in southwestern France (Riom 1985) or for determining the characteristics of *Armillaria* foci in various natural forest types (see Kile *et al.* 1991). Both black and white and colour film have been used effectively. False colour infrared film is sometimes preferred to other film on the basis that significant changes in leaf moisture content may occur before the changes in foliage coloration. However, Nandris *et al.* (1985), working in rubber plantations in Central Africa affected by root disease, found no benefits from the use of false colour. The resolution of satellite imagery is increasing, but has not yet reached a point where it has been found to be of practical value in forest disease assessment.

All the procedures described above depend on making use of forest staff with at least some professional or technical qualifications. The value of inviting the general public to provide information depends on the nature of the problem. The chief requirement is for the host tree to be readily identifiable and for the disease symptoms to be striking and highly characteristic. During the period 1993–94, good use was made in the UK of suitable television programmes to alert the public to the symptoms of the newly recognized Phytophthora disease of alder. Although some erroneous reports were received (elms with Dutch elm disease for example), knowledge of disease distribution was significantly and usefully extended.

Assessment of loss

Where commercial forest crops are concerned, it is important that relationships should be established between the amount of disease and the associated loss in productivity. When it comes to the effect of a foliage disease on tree growth, these are not necessarily closely related. Even mortality is not necessarily closely related to yield loss because surviving trees may make compensatory growth. Few forest diseases have been subject to such study in this respect as has *Heterobasidion annosum*, the cause of root and butt rot of conifers. In a recent publication (Woodward *et al.* 1998), data on killing and butt rot for various parts of Europe and for North America are presented in some detail. The information presented ranges from studies on decay in 'old-growth' forests of western hemlock on the west coast of North America to the mortality of young pine plantations in eastern England. For the European Union as a whole, an estimate for loss due to decay is put at 466 million ecu per annum.

9.4.3 Disease prognosis

Information on disease distribution and impact

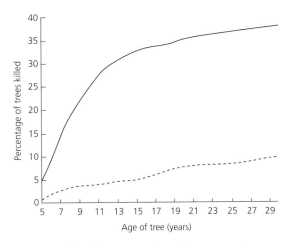

Fig. 9.5 Killing due to *Heterobasidion annosum* in a plantation of Corsican pine on a high-pH soil in eastern England after a previous crop of Scots pine. The solid line shows the situation where all the stumps of the previous crop remained, the dotted line shows the situation where the inoculum had been greatly reduced by stump removal. The first disease assessments were made when the trees were 5 years of age. (Data from British Forestry Commission files.)

should be combined with an evaluation of the likely course of events before any control measures are devised. An understanding of disease epidemiology is vital. This topic has been well reviewed for foresters by Schmidt (1996) in a paper which covers techniques for assessing the rate of disease development in time and space. In terms of temporal development, it is important to distinguish between 'simple interest' and 'compound interest' diseases. In the former, often exemplified by outbreaks of root disease soon after planting, it is the initial amount of inoculum provided by the stumps of the previous crop that is the key factor and there is little 'tree-to-tree' spread of disease. Figure 9.5 shows data on this kind of process in a stand of young Corsican pine (*Pinus nigra* ssp. *laricio*) affected by *Heterobasidion annosum*. In a compound interest disease, by contrast, diseased trees quickly produce inoculum for further infection. Such epidemics typically show an S-shaped curve—see that for Dutch elm disease in Fig. 9.6. For spatial disease development, a key element is the disease

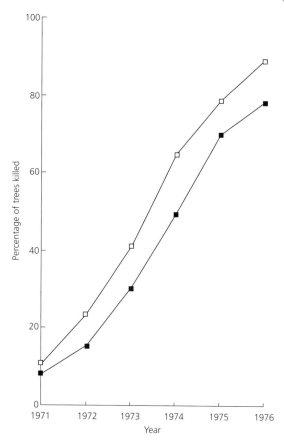

Fig. 9.6 Development of Dutch elm disease in two populations of English elm in southern Britain during the 1970s. Black squares, west and south-west England; white squares, south-east England. (From Gibbs 1978.)

gradient curve, the slope of which will depend upon the means of pathogen dissemination.

Good survey data are a prerequisite for epidemiological analysis. Ground-collected data are usually used, but a succession of aerial photographs can be used effectively. See, for example, the paper by Appel *et al.* (1989) on the expansion of oak wilt centres in stands of live oak, *Quercus fusiformis*, in Texas.

Development of models

As Pratt *et al.* (1998) have described in a recent paper on *Heterobasidion annosum*, there are two main types of model: analytical (based on observed responses) and simulation (derived from interpretation of fundamental concepts). In forestry, the development of disease modelling has tended to lag behind the development of models of tree growth. However, a well-known early example is that of Dickson and Hawksworth (1979) on the spread and intensification of the south-western dwarf mistletoe (*Arceuthobium vaginatum* ssp. *cryptopodum*) on ponderosa pine in western North America. More recently there has been much interest in the modelling of root rot diseases. The Western Root Disease Model (see Frankel 1998) was developed for *Armillaria* spp. and *Phellinus weirii* damage and allows for interactions with the effects of windblow and bark beetle attack. It is also linked into a stand growth and yield model to allow for 'forest vegetation simulation' (Teck *et al.* 1996).

9.4.4 Deciding on a course of action — or inaction

It should first be recognized that the most appropriate course of action is often to do nothing. This clearly applies if, by the time the disease has been diagnosed, it is apparent that its period of activity is over. It also applies if there is little likelihood of success in controlling the disease, perhaps because it would be impossible to carry out the operation with the necessary efficiency. A good example here might be the problem of controlling Dutch elm disease in mixed woodland on difficult terrain. Cost in relation to benefit may well be of crucial importance — no forester will need reminding that the value of wood is often not high enough to allow the use of potentially effective control measures. Quite sophisticated approaches may be appropriate here, such as the use of discounted revenues (see for example Greig 1984). It is, however, fair to say that general considerations, such as a commitment to the maintenance of a sustainable forest, sometimes justify measures which might not be used if they were judged solely on commercial criteria.

If it is decided that no action should be taken, it is worth remembering that there may be some beneficial consequences: dead wood is a valuable habitat for many organisms; glades created where trees have died may greatly increase biodiversity in a dense stand of trees.

9.4.5 Cultural control

Cultural control can be defined as that exercised by man through the application or adaptation of forest management practices.

Cultural control at planting

Experience has shown that some tree species should not be planted often over large geographical areas, because of their susceptibility to certain diseases. For example, no commercial use can be made of white pine in many parts of the world because of the likelihood that the trees will quickly fall prey to lethal attacks by the blister rust *Cronartium ribicola*. Similarly, Corsican pine, *Pinus nigra* ssp. *laricio* is an important plantation species in the south and east of Britain but it cannot be grown in the north and west where cool, wet conditions favour the development of shoot dieback caused by *Brunchorstia pinea*.

Alternative species can sometimes provide a management option when a disease is well established on a site. Thus broadleaf trees such as beech and oak are very much less susceptible than young conifers to attack by both the root pathogens *Armillaria* sp. and *H. annosum*.

Species mixtures can be used to provide a measure of insurance when the risk of disease is not clear. For example, in the UK wild cherry (*Prunus avium*) is currently not allowed to constitute more than 10% of a broadleaf plantation because of the threat of bacterial canker caused by *Pseudomonas syringae* pv. *morsprunorum*. In some situations mixtures can effectively delay the spread of disease. Thus underground spread of *Ceratocystis fagacearum*, the cause of oak wilt, depends upon the existence of functional root grafts between adjacent trees. Since grafting frequency depends, at least in part, on the genetic similarity between the two trees, it follows that the rate of disease development will be lower in mixtures of different oaks or in mixtures of oaks and other broadleaved species than in a pure stand of any one oak species (for relevant literature see Gibbs & French 1980). In considering mixtures of species, it must be remembered that there are often some disadvantages to their use: management may be more difficult and productivity lower.

In the context of disease control at planting, it is important to ensure that the planting stock is pathogen free. A number of pathogens, but perhaps most notably *Phytophthora* species, can readily be introduced to a site with the plants or associated soil.

Cultural control during crop development and exploitation

Some pathogens are more destructive on less vigorous trees than on more vigorous ones. The significance of this has long been known for certain pathogens—such as some *Armillaria* spp. which can invest root systems with their rhizomorphs and then quickly invade the tissues if tree vitality is reduced by adverse environmental circumstances. Much more recent is the recognition that the stem of a tree often contains pathogens living in an 'endophytic' or latent state. If the tree is stressed, for example by drought, these pathogens can quickly colonize the tissues, causing dysfunction of the wood and bark. Dieback and death of part or all of the tree can result. *Hypoxylon* spp. on various broadleaved trees commonly show this *modus operandi* (Chapela *et al.* 1988).

From the disease management viewpoint, the main approach to minimize the damaging effects of weak pathogens or stem endophytes is through the adoption of thinning or felling regimes to maintain stand vigour. In the longer term a more appropriate choice of species may be the best option.

Not all pathogens cause most damage on unthrifty trees. The rust fungi are an important group which, because of their high degree of adaptation to the host, often increase in severity as host vigour increases. Studies on fusiform rust of pine, caused by *Cronartium quercuum* f. sp. *fusiforme*, in the south-eastern USA have often shown that an improvement in growing conditions leads to more disease (Table 9.1). However it is often not clear whether this is due to greater susceptibility of an individual unit of tissue or to the fact that, with better conditions, there is more tissue available for infection.

Vigour is also of little benefit to the tree when it

Table 9.1 Incidence of fusiform rust (*Cronartium fusiforme*) in slash pine (*Pinus elliotii*) after 6 years. (After Burton *et al.* 1985.)

Treatment	Tree height (m)	% of trees with rust galls
Control	1.85	54
Weeded	2.8	72
Weeded and fertilized	4.4	81

comes to combating some exotic pathogens to which it has no evolved resistance mechanisms. This point is well illustrated by the response of both European and American elms to *Ophiostoma novo-ulmi*. Trees growing vigorously succumb more rapidly than trees growing poorly.

Control through sanitation can be looked upon as a form of cultural control, particularly if thinning and felling regimes are adjusted to allow for it. In the past considerable emphasis was placed on this approach. Thus in hardwood forests of East and North America, trees with cankers, such as those caused by *Nectria*, *Eutypella* and *Strumella*, were often felled and placed canker-face down to minimize spore dispersal. The value of such an approach can be questioned, not least in view of current knowledge of the endophytic existence of stem pathogens, as discussed above. By contrast, a full commitment to sanitation is absolutely vital if a pathogen such as *O. novoulmi* is to be controlled.

It is important to recognize that it is much easier to achieve benefits through sanitation with simple interest diseases than with compound interest ones. This is well exemplified by studies on 'de-stumping' after clear-felling in order to reduce the level of root disease in a subsequent crop (see for example the greatly reduced levels of killing by *Heterobasidion annosum* that can be achieved by such means (Fig. 9.5, p. 209)). Where the economics are favourable, de-stumping has been adopted as a practical control measure for *H annosum* (Greig 1984; Korhonen *et al.* 1998), *Armillaria* spp. (see Hagle & Shaw 1991) and *Phellinus weirii* (Morrison *et al.* 1988). It may not always be necessary physically to remove the stump from the ground. Various cultural practices

may be employed to reduce the period during which the pathogen survives within the tissues. Thus for the control of white root rot of rubber, caused by *Rigidoporus lignosus*, there has long been an interest in the role of a ground cover of creeping legumes in reducing the survival of the fungus within the stumps of the previous crop trees (see Fox 1965). The pathogen attacks the legume but in the process exhausts its food reserves in the stump without gaining equivalent ones in return.

9.4.6 Control through the use of disease resistance

World-wide there are many research programmes aimed at producing trees resistant to disease. Particularly well known examples include those for resistance to Dutch elm disease, to fusiform rust of southern pines and to various diseases of poplar, including leaf rust and bacterial canker. Where vegetative propagation of the host is easy, as with many broadleaves, the product of the programme is often a clone. Otherwise the aim is to produce populations of resistant seedlings. When considering the relative merits of clones and seedlings, much depends on whether every planted tree is expected to be equally productive, as is the case in crops of poplar at wide spacing, or whether less suitable trees can be progressively removed through selective thinning. Clones may well be favoured in the former situation, seedlings in the latter.

One of the main issues facing tree breeders is whether or not to make use of resistance based on single genes ('major gene' or 'specific' resistance). The history of agricultural crop breeding illustrates how short-lived this kind of resistance can be and similar problems have now shown up in trees, a relevant example being provided by the appearance of new races of the rust *Melampsora larici-populina* on previously immune poplar clones (see Pinon & Frey 1997). Some would wish to reject the use of major genes entirely, but it has been argued that they can find a useful place if they are embedded in protective 'general' resistance.

The deployment of resistant material remains a keenly debated issue. The dangers of large areas of monoculture are self-evident but there can often be penalties in using mixtures in terms of ease of management and the marketability of the final product. Moreover, Heybroek (1982), with principal reference to a variety of non-obligate pathogens (*Heterobasidion annosum, Nectria cinnabarina, Lophodermium pinastri*), has argued that the benefits of reduced disease on the more susceptible partners in a mixture may be offset by the 'disbenefits' of increased disease on the more resistant ones. His approach to spreading the risk was to suggest that a mosaic of pure stands of the different genotypes might be better than an intimate mixture. In a thoughtful review of the issue, Libby (1982) argued that the use of only two or three clones was often a poor option but that the use of modest numbers of clones (7–25) provided a robust and perhaps optimum strategy.

The genetic engineering of trees for disease resistance is in its infancy. In addition, the whole approach is subject to criticism from environmentalists. Research is currently under way on the engineering of the English elm, *Ulmus procera*, for resistance to Dutch elm disease; this seems a good candidate for such work as it is largely for environmental reasons that the public would wish to see this particular tree return to its former prominence (Brasier 1996).

9.4.7 Chemical control

There are economic and environmental constraints to the use of chemicals for the control of most forest diseases. These constraints increase year by year, most recently in relation to Forest Stewardship schemes aimed at ensuring that wood comes from sustainable sources, it being argued that the use of chemicals is not sustainable.

One of the very few situations in forestry in which fungicides play the kind of role that is common with agricultural crops is in respect of the copper-spraying of *Pinus radiata* in New Zealand to reduce Dothistroma needle blight (Kershaw *et al.* 1988). Costs of carrying out the operation are relatively low. This is because it is a foliage disease and symptoms can be assessed and spraying conducted from the air. Some forestry companies spray when average disease levels reach 15% of the tree crown; others spray at 25%.

One of the few other uses of chemical sprays is for the control of *Phytophthora cinnamomi* in the native heath/tree communities of western Australia. The material used in this case is a solution of potassium phosphite. This is of course not a conventional fungicide, and the application rates are so low that adverse environmental effects are considered minimal. Small trees are sprayed with the chemical and larger trees are treated by injection (G. Hardy, personal communication).

Much more common than the situations described above are those in which chemicals are used in a highly targeted way. Perhaps the best example is their use to treat trees just before, or just after felling, in order to prevent the build-up of wood-rotting root pathogens that would otherwise be highly destructive to the next rotation of trees. The original work in this field was done at the Rubber Research Institute in Malaysia where sodium arsenite and, subsequently, 2,4,5-T were employed (see Fox 1965).

A somewhat similar approach was adopted by Rishbeth (1951) in his work on the control of *Heterobasidion annosum* in the young pine plantations of eastern England. He showed that a chemical would prevent the colonization of freshly cut stump tissues by air-borne basidiospores of the pathogen. Creosote was employed initially but the material of choice is now either urea or a boron compound. Again, neither of these materials is a conventional fungicide, and this illustrates the point that the aim is not to exclude all fungi from the stump but to alter the fungal succession so the stump is colonized by harmless microorganisms rather than by the pathogen.

In the forest nursery, there is scope for the use of a rather wider spectrum of fungicides than in the forest. However, care should be taken that fungicides are not used in such a way that they suppress disease expression in the nursery only for the pathogen to be carried to the outplanting site where full host invasion and disease expression occurs. This often happens with diseases caused by *Phytophthora* species.

9.4.8 Biological control

Early in this chapter the point has been made that in natural forest systems, pathogens are kept in ecological balance. Natural biological control is a key component of this balanced situation, although it is not often that the mechanisms are understood. One partial exception is provided by the case of oak wilt, caused by the fungus *Ceratocystis fagacearum*, in West Virginia, where it has been estimated that the disease kills only about one tree per square mile of forest per year. Studies in the 1950s revealed something of the assemblage of microorganisms that play a part—most notably in limiting the saprotrophic survival of *C. fagacearum* in the trees it had killed. As a result the opportunities for disease dissemination are greatly reduced. *Hypoxylon* spp. are important in this process, as they rapidly colonize the sapwood of the diseased tree from endophytic infections. For a long time it was standard practice in West Virginia to girdle recently diseased trees with an axe; to a large extent this worked through enhancing and accelerating the competitive colonization of the tissues by *Hypoxylon* (see Gibbs & French 1982).

Competitive colonization of host tissue is also the key to the best-known example of 'artificial' or man-made biological control of a tree disease. This is the use of spores of the Basidiomycete *Phlebiopsis* (*Peniophora*) *gigantea* to treat the stumps of freshly felled pine to prevent establishment of the pathogen *Heterobasidion annosum*. *P. gigantea* achieves its effects by colonizing the stump tissues and thereby denying them to *H. annosum*. As a result, the latter is unable to develop into the root system and spread to neighbouring trees at points of root contact. *P. gigantea* has been used by the British Forestry Commission in one of its pine forests for well over 30 years. It has been used in other countries also, with most recent interest being directed to another product, Rot-Stop, which comprises a strain of *P. gigantea* capable of colonizing stumps of Norway spruce (*Picea abies*) as well as those of pine.

After years of use by the Forestry Commission, it became clear in 1995 that the use of *P. gigantea* would have to be registered under the UK Control of Pesticides Regulation 1986. In meeting this requirement it was evident that, while there was widespread recognition that *P. gigantea* is exactly the kind of environmentally friendly pesticide that should be used, current registration procedures take little account of the features of biocontrol agents—particularly those, such as *P. gigantea*, that achieve their effects through competition (Pratt *et al.* 1999).

In contrast to *P. gigantea*, most biocontrol agents for plant pests and diseases work through a direct attack by the agent on the pest organism. While there are many microorganisms that are capable of damaging tree pathogens, very few have been brought anywhere near the point of commercial exploitation. The most notable example is provided by the double-stranded (ds) RNA viruses that debilitate the chestnut blight pathogen *Cryphonectria parasitica*. The effects of these viruses were first noted in southern France in the 1970s, and since that time various systems for their use have been devised. The main limitation is that they can only spread within the fungus population via the fusion of infected and healthy mycelium and the fungus has an 'incompatibility system' that prevents many potential fusions from occurring. In an attempt to circumvent this problem, the current approach is to engineer the virus into the gene of the pathogen and to determine if it can be disseminated within the fungal population in that way. Similar ds RNA viruses have been found in the Dutch elm disease pathogens and are the subject of a rather similar research approach (Brasier 1996).

9.5 CONCLUSIONS

Some approaches to minimizing the adverse effects of forest diseases can only be made at the national (or even the international) level—the decision to conduct an eradication programme on a particular disease for example. Others are necessarily made entirely at the local level—for example the decision on whether or not to apply a control treatment to a particular forest stand. Whatever the circumstances, both national and local forest managers need access to a cadre of pathologists who have the expertise to provide authoritative information on the cause of any damage that may occur and on its likely future impact. If the lines of communication between forester and scientist are good, cost-effective and ecologically sound management can be devised.

ACKNOWLEDGEMENTS

In producing this chapter, I am indebted to colleagues around the world who promptly and generously responded to my Email requests for information on various topics. I would like to acknowledge the help of my colleagues in Forest Research, and in particular Dr Derek Redfern, for helpful discussion and critical reading of the manuscript.

REFERENCES

Allen, E.A., Morrison, D.J. & Wallis, G.W. (1996) *Common Tree Diseases of British Columbia*, 3rd edn. Canadian Forest Service.

Anonymous (1999) APS resolution on wood importation. *Phytopathology News* **33**, 165.

Appel, D.N., Maggio, R.C., Nelson, E.L. & Jeger, M.J. (1989) Measurement of expanding oak wilt centres in live oak. *Phytopathology* **79**, 1318–22.

Brasier, C.M. (1991) Ophiostoma novo-ulmi sp. nov., causative agent of the current Dutch elm disease pandemics. *Mycopathologia* **115**, 151–61.

Brasier, C.M. (1996) *New Horizons in Dutch Elm Disease Control. Report on Forest Research 1996*, pp. 20–8. Forestry Commission, Edinburgh.

Brasier, C.M. (1997) Fungal species in practice: identifying species units in fungi. In: Claridge, M.F., Dahwah, H.A. and Wilson, M.R., eds. *Species: the Units of Biodiversity*, pp. 135–70. Chapman & Hall, London.

Brasier, C.M. & Gibbs, J.N. (1973) Origin of the Dutch elm disease epidemic in Britain. *Nature, London* **242**, 607–9.

Brasier, C.M., Cooke, D.E.L. & Duncan, J.M. (1999) Origin of a new *Phytophthora* pathogen through interspecific hybridisation. *Proceedings of the National Academy of Sciences, USA* **96**, 5878–83.

Burton, J.D., Shoulders, E. & Snow, G.A. (1985) Incidence and impact of fusiform rust vary with silviculture in slash pine plantations. *Forest Science* **31**, 671–80.

Butin, H. (1995) *Tree Diseases and Disorders*. Oxford University Press, Oxford.

Carter, C.S. (1989) Risk assessment and pest detection surveys for exotic pests and diseases which threaten commercial forestry in New Zealand. *New Zealand Journal of Forestry Science* **19**, 353–74.

Chapela, E.H. & Boddy, L. (1988) Fungal colonisation of attached beech branches. II. Spatial and temporal organisation of communities arising from latent invaders in bark and functional sapwood under different moisture regimes. *New Phytologist* **110**, 47–57.

Chapman, S. (1998) *Field Guide to Common Pests, Diseases, and Other Disorders of Radiata Pine in New Zealand*. Forest Research Bulletin no. 207. New Zealand Forest Research Institute Limited, Rotorua.

Ciesla, W.M., Diekmann, M. & Putter, C.A.J. (1996) *FAO/IPGRI Technical Guidelines for the Safe Movement of Germplasm No. 17. Eucalyptus spp.* Food and Agriculture Organisation of the United Nations, Rome/International Plant Genetic Resources Institute.

Département de la Santé des Forêts (1997) La santé des forêts (France) en 1996. *Les Cahiers du DSF*, 1–1997, 87pp.

Dickson, G.E. & Hawksworth, F.G. (1979) A spread and intensification model for south-western dwarf mistletoe in Ponderosa pine. *Forest Science* **25**, 43–52.

Flot, J.-L. (1998) Organisation and activities of the French Forest Health Department. In: *Methodology of Forest Insect and Disease Survey in Central Europe. Proceedings from the IUFRO WP 7.03.10 Workshop Ustron—Jaszowiec (Poland)*, pp. 17–20.

Fox, R.A. (1965) The role of biological eradication in root-disease control in replantings of *Hevea brasiliensis*. In: Baker, K.F. & Snyder, W.C., eds. *Ecology of Soil-Borne Plant Pathogens*, pp. 348–62. University of California Press, Berkeley.

Frankel, S.J. (1998) *User's Guide to the Western Root Disease Model, Version 3.0*. General Technical Report PSW-GTR-165. Pacific South-West Research Station, USDA Forest Service, Albany, California.

Gibbs, J.N. (1978) Development of the Dutch elm disease epidemic in southern England: 1971–76. *Annals of Applied Biology* **88**, 219–28.

Gibbs, J.N. & French, D.W. (1980) The *Transmission of Oak Wilt*. USDA Forest Service Research Paper NC-185. St. Paul, MN.

Gibbs, J.N. & Wainhouse, D. (1986) Spread of forest pests and pathogens in the Northern Hemisphere. *Forestry* **59**, 142–53.

Gibbs, J.N., Liese, W. & Pinon, J. (1984) Oak wilt for Europe? *Outlook on Agriculture* **13**, 203–8.

Gibbs, J.N., Lipscombe, M.A. & Peace, A.J. (1999) The impact of phytophthora disease on riparian populations of common alder (*Alnus glutinosa*) in southern Britain. *European Journal of Forest Pathology* **29**, 39–50.

Gravatt, G.F. (1949) Chestnut blight in Asia and North America. *Unasylva* **3**, 1–7.

Gregory, S.C. & Redfern, D.B. (1998) *Diseases and Disorders of Forest Trees: A Guide to Identify Causes of*

Ill-health in Woodlands and Plantations. Forestry Commission Field Book 16. The Stationery Office, London.

Greig, B.J.W. (1984) Management of East England pine plantations affected by *Heterobasidion annosum* root rot. *European Journal of Forest Pathology* **14**, 392–7.

Hagle, S.K. & Shaw, C.G., III (1991) Avoiding and reducing losses from armillaria root disease. In: Shaw, C.G., III & Kile, G.A., eds. *Armillaria Root Disease*, pp. 157–73. USDA Forest Service, Washington, DC.

Hartmann, G., Nienhaus, F. & Butin, H. (1988) *Farbatlas Waldschäden: Diagnose von Baumkrankheiten.* Eugen Ulmer, Stuttgart.

Heybroek, H.M. (1982) Monoculture versus mixture: interactions between susceptible and resistant trees in a mixed stand. In: Heybrook, H.M., Stephan, B.R. & Von Weissenberg, D.K., eds. *Resistance to Diseases and Pests in Forest Trees*, pp. 326–41. Centre for Agricultural Publishing and Documentation (Pudoc), Wageningen.

Houston, D.R. (1992) A host-stress–saprogen model for forest dieback-decline diseases. In: Manion, P.D. & Lachance, D., eds. *Forest Decline Concepts*, pp. 3–25. APS Press, St Paul, MN.

Huang, Z.Y., Smalley, E.B. & Guries, R.P. (1995) Differentiation of *Mycosphaerella dearnessii* by cultural characters and RAPD analysis. *Phytopathology* **85**, 522–7.

Innes, J.L. (1993) *Forest Health: Its Assessment and Status.* CAB International, Wallingford, Oxon.

Kershaw, D.J. (1989) History of forest health surveillance in New Zealand. *New Zealand Journal of Forestry Science* **19**, 375–7.

Kile, G.A., McDonald, G.I., Byler, J.W. (1991) Ecology and disease in natural forests. In: *Armillaria Root Disease*, Shaw, C.G., III, Kile, G.A. eds. Service, 102–121, USDA Forest Service, Washington.

Korhonen, K., Delatour, C., Greig, B.J.W. & Schönhar, S. (1998) Silvicultural control In: Heterobasidion annosum *Biology, Ecology, Impact and Control.* CAB International, Wallingford, UK.

Libby, W.J. (1982) What is a safe number of clones per plantation? In: Heybrook, H.M., Stephan, B.R. & von Weissenberg, K., eds. *Resistance to Diseases and Pests in Forest Trees*, pp. 342–60. Centre for Agricultural Publishing and Documentation (Pudoc), Wageningen.

Manion, P.D. (1991) *Tree Disease Concepts*, 2nd edn. Prentice-Hall, Englewood Cliffs, NJ.

Miller-Weeks, M. & Stark, D. (1983) European larch canker in Maine. *Plant Disease* **67**, 448.

Nandris, D., van Canh, T., Geiger, J.-P., Omont, H. & Nicole, M. (1985) Remote sensing in plant diseases using infrared colour aerial photography: applications trials in the Ivory Coast to root diseases of *Hevea brasiliensis. European Journal Forestry Pathology* **15**, 11–21.

Pedgley, D.E. (1982) *Windborne Pests and Disease: Meteorology of Airborne Organisms.* Ellis Horwood, Chichester, Sussex.

Phillips, D.H. & Burdekin, D.A. (1992) *Diseases of Forest and Ornamental Trees*, 2nd edn. Macmillan Press, Basingstoke.

Pinon, J. & Frey, P. (1997) Structure of *Melampsora larici-populina* populations on wild and cultivated poplar. *European Journal of Plant Pathology* **103**, 159–73.

Pratt, J.E., Shaw, C.G., III & Vollbrecht, G. (1998) Modelling disease development in forest stands. In: Heterobasidion annosum *Biology, Ecology, Impact and Control*, pp. 213–23. CAB International, Wallingford, Oxon.

Pratt, J.E., Gibbs, J.N. & Webber, J.F. (1999) Registration of *Phlebiopsis gigantea* as a forest bio-control agent in the UK, recent experience. *Bio-control Science and Technology* **9**, 113–18.

Riom, J. (1985) Le deperissement du chene en France— apports de la teledetection. 1. Forets des Pyrenees atlantiques. 2. Foret de Troncais (Allier). *Colloques-de-l'INRA* **32**, 117–45.

Rishbeth, J. (1950) Observations on the biology of *Fomes annosus*, with particular reference to East Anglian pine plantations. I. The outbreaks of disease and ecological status of the fungus. *Annals of Botany, NS* **14**, 365–83.

Rishbeth, J. (1951) Observations on the biology of *Fomes annosus*, with particular reference to East Anglian pine plantations. (II) Spore production, stump infection, and saprophytic activity in stumps. *Annals of Botany NS* **15**, 1–21.

Schmidt, R.A. (1996) Epidemiology. In: Tainter, F.H. & Baker, F.A., eds. *Principles of Forest Pathology*, pp. 237–71. John Wiley, New York.

Sinclair, W.A., Lyon, H.H. & Johnson, W.T. (1987) *Diseases of Trees and Shrubs.* Cornell University Press, Cornell, NY.

Skelly, J.M., Davis, D.D., Merrill, W., Cameron, E.A., Brown, H.D., Drumond, D.B. & Dochinger, L.S. eds. (1985) *Diagnosing Injury to Eastern Forest Trees.* Pennsylvania State University.

Smith, I.M. (1979) EPPO: the work of a regional plant protection organization, with particular reference to phytosanitary regulations. In: Ebbels, D.L. & King, J.E., eds. *Plant Health, The Scientific Basis for Administrative Control of Plant Diseases and Pests*, pp. 13–22. Blackwell Scientific Publications, Oxford.

Spiers, A.G. & Wenham, H.T. (1983) Poplar seed transmission of *Marssonina brunnea. European Journal of Forest Pathology* **13**, 305–14.

Storer, A.J., Gordon, T.R. & Clark, S.L. (1998) Association of the pitch canker fungus, with Monterey pine seeds and seedlings in California. *Plant Pathology* **47**, 649–56.

Strouts, R.G. & Winter, T.G. (1994) *Diagnosis of Ill-Health in Trees*. HMSO, London.

Tainter, F.H. & Baker, F.A. (1996) *Principles of Forest Pathology*. John Wiley, New York.

Teck, R., Moeur, M. & Eav, B. (1996) Forecasting ecosystems with the Forest Vegetation Simulator. *Journal of Forestry* **94**, 7–10.

Wilkinson, A.G. & Spiers, A.G. (1976) Introduction of the poplar rust *Melampsora larici-populini* and *M. medusae* to New Zealand and their subsequent distribution. *New Zealand Journal of Science* **19**, 195–8.

Woodward, S., Stenlid, J., Karjalainen, R. & Hüttermann, A. (1998) Heterobasidion annosum *Biology, Ecology, Impact and Control*. CAB International, Wallingford, Oxon.

Part 3
Sustaining Social Values and Benefits

If written 30 years ago this book would not have included a section on social values and benefits. It is a reflection of today's better priorities that all 'stakeholders', as they are called, not only have an interest in how forests are managed but also can influence, at least to some extent, decisions affecting them.

Genuine sustainability must ultimately be people centred. This is spelt out clearly by Stephen Bass (Chapter 10), drawing on years of actual in-country experience, and his succinct account provides a universally applicable basis for involving all interested parties in forest development. Chapter 11, by N. C. Saxena, was originally to be a case study but has been included in this part because his purview of Indian experience, in particular the development of participatory forest management, is an exemplar of such develop-

ment. Also, the author is unsparing of the problems in evolving such holistic policies and sustainable management and shows how a genuine attempt to break out of the professional straightjacket is fraught but ultimately worth it.

Chapter 12 is welcome in recognizing that the world's increasingly urbanized population still benefits greatly from trees. Kjell Nilsson and colleagues distil many years' experience in summarizing the essentials of sound arboriculture from street (shade) trees to urban forests. For the vast majority of forests sustainability is largely about understanding how forests function and working with that knowledge, but in the massively altered and unnatural urban environment, it is more to do with understanding how the built environment affects urban forest, and especially street (shade) trees, and to modify management accordingly.

10: Working with Forest Stakeholders

STEPHEN BASS

10.1 ASSESSING SOCIAL VALUES AND PLANNING WITH STAKEHOLDER GROUPS

The importance of social values in sustainable forest management (SFM) was discussed in Chapter 14 in Volume 1 of this handbook. This chapter provides guidance on identifying the social values associated with key interest groups in particular forest circumstances, integrating these values in forest management, working with the interest groups, and keeping track of how values are both being met and are changing. Social values cover everything from firewood and forest-based foods for subsistence in poor communities, to landscape aesthetics and recreation for rich communities. But this chapter is *not* about the many particular forest management techniques for different forest products and services, which are covered in other chapters.

10.1.1 The importance of participation in assessing and realizing social values

It is very much a maxim that the lower the level of public trust in an institution (whether this is an authority or an enterprise), the higher the requirement for participation. Trust requires the development or acceptance of common language. This is being gained through the various participatory forest criteria and indicators processes—multiple forest values, and rights, have been reflected in these to varying degrees. Trust also requires the development of methods and channels for participation, consultation and communication. There has been rather less development here; and, indeed, quite a lot of confusion about where and how participation should be conducted. Exhortations for example to conduct all affairs 'with the maximum possible participation' (as in Agenda 21) are impracticable to respond to. While the notions of policy consultation, or a 'national forest forum', are now commonly advanced, there is little experience of how to operate these mechanisms. The same is even more evident at the forest management unit level: many managers of forest operations just have no experience even of talking to their neighbours.

SFM is about achieving a balance between economic, environmental and social objectives. Sometimes many of these objectives can be met; at other times, choices will have to be made between them. The choice is not merely an economic or technical one. A sound choice will reflect multiple values, many of which, for example biodiversity and aesthetics, are not covered by market decisions. Participation is therefore a special requirement where social values are concerned, in order to:

1 Ensure forestry standards, objectives and targets are credible. Every stakeholder group must know who was involved in developing them, and have confidence in the procedures used to negotiate them.

2 Make use of a broader range of ideas, skills and inputs. Many perspectives can be heard, so that social values are reflected. Participation can bring important information to the table that may otherwise have been unavailable.

3 Ensure practicality and focus of resulting standards, objectives and targets. Participation can achieve a best consensus on trade-offs given current conditions and knowledge in the area in question. Furthermore, it can modify that con-

sensus, from time to time, according to actual results—and hence get closer to SFM.

4 Build a stronger foundation of stakeholder trust and accountability. The process of participation strengthens partnerships and commitment to implementation (Higman *et al.* 1999).

The principle of participation in forestry activities is now widely accepted. But there is a new danger: a lack of quality in participation work. Participation was originally introduced in order to avoid the problems evident with standardized, 'top-down' approaches to forestry. Yet, ironically, there is now a tendency for participation itself to become a standardized approach, or to be inadequately implemented. Often full participation is claimed when only partial consultation has been practised, for example. A useful typology of participation is presented in Table 10.1. It is suggested that categories 4–6 are most appropriate for forestry that combines local social values and the resources of state or commercial enterprises.

However, 'participation' cannot be 'switched

on' as if it were a new machine. Considerable skills are required. For large forestry operations, specialist skills or consultancy are required, although all staff must also be familiar with what participation is required for performing their jobs. For smaller operations, short training courses are available and appropriate for site managers.

Furthermore, effective participation takes time. Relationships need to be formed first. But relationships cannot form until good communication systems are in place. And communication is more effective once 'stakeholders' have been identified.

10.1.2 The forest 'stakeholder'

Demands made on forests can come from many different groups of people. The idea of 'stakeholders' has emerged to describe all these people, but it implies that they have some legitimate claims to forest values.

The stakeholders in a forest are all the people

Table 10.1 A spectrum of degrees of participation. (Source: Cornwall 1996, with modifications.)

Mode of participation	Local stakeholder involvement	Relationship of forestry activity to stakeholders
1. Coercion	Compulsion of local groups Forced giving up of rights/powers	Done to people
2. Cooption	Token involvement Representative chosen by forest organization Locals have no real input or powers	Done to people
3. Compliance	Tasks are assigned Incentives to comply are given Outsiders decide agenda and direct process	For people
4. Consultation	Local opinions asked Outsiders analyse and decide on course of action	For and with people
5. Cooperation	Locals and outsiders work together to determine priorities Outsiders tend to direct process	With people
6. Co-learning and joint action	Locals and outsiders share knowledge, to create new understanding and work together to decide actions Outsiders facilitate	With and by people
7. Collective action	Locals set own agenda and mobilize to carry it out No outside initiators or facilitators	By people

and organizations who have a 'stake', or interest, in the forest and may be affected by any activity in it, or who may have an impact or influence on the forest.

Stakeholders in forests commonly include the following interest groups (Higman *et al.* 1999):

• *People who live in or near the forest.* For example, groups who have lived in the area for many years and whose culture, livelihood and economy is strongly linked to the forest. They may have useful knowledge about their environment and have evolved particular forest management knowledge and practices. With the exception of larger local forest owners, they often have less political power and money than other groups, and tend to be the main losers if commercial approaches to forest management are implemented without thought for their livelihoods.

• *People who live further away and who come to the forest.* These may vary from tourists seeking recreation to refugees and migrants coming to forests in times of stress, or at certain seasons, for example, to gather forest products or graze livestock. They may have an opportunistic outlook and varying levels of indigenous knowledge.

• *Settlers from elsewhere in the country, or from other countries.* They are typically poor and often unfamiliar with ways to thrive in forest environments, especially if they are more familiar with settled agricultural lifestyles. They may practise slash-and-burn farming and be expecting to move on to new areas relatively soon.

• *Forest workers* tend to be men seeking wages and other direct benefits, although women may work in the informal sector. Forest workers may be far from their families and other traditional sources of social support or control.

• *Small-scale entrepreneurs* have information and money which enable them to engage in the marketing and processing of forest products. They may also run small-scale logging operations or be contracted to do forest-related work for larger companies.

• *Managers of forest companies.* Often from urban areas and having power over other groups in the forest area. They may possess good forest knowledge for timber production, but may have poor knowledge of other forest goods and services and local social values. Logging companies tend

to have fewer long-term interests in the forest than forest management-based companies.

• *Environmentalists.* Usually urban-based, concerned with putting pressure on other groups, particularly the government and timber companies, to improve environmental sustainability. They may be influential through political lobbying, demonstrations, fund raising, obtaining media coverage, organizing local resistance and spreading information worldwide through their links with international organizations but may have poor knowledge of forestry operations.

• *Forestry officials* work for the government, often in bureaucratic systems with much paperwork. They may look down on other groups as being 'in the way' of their mandate to protect and administer the forest. Some may also use their position for extra personal gains.

• *Politicians* participate in decision making at local and national levels. Some politicians may be influenced by the preferences of people in their constituencies, others by national-level aims which may clash with local needs. For many, the allocation of land and forests is a basis of their power.

• *National citizens.* People from all over the country may have a voice in forest management—by voting or by writing letters they can influence politicians and government officials. While many are increasingly concerned with environmental issues and recreation, their interests may be highly varied.

• *Global citizens* represented through intergovernmental agreements and markets, for example international agreements on climate change, may affect national forest management.

• *Consumers* who buy forest products. They are usually most concerned about affordable supplies of products, but some—especially those in northwestern Europe and North America—want proof of sustainable forest management practices.

The term 'stakeholder' implies that a person or group has the power to make real inputs into decisions about what forest values are important, how forests should be managed, and who should bear the costs and benefits. However, some groups often have neither an effective voice, nor the means to pursue their interests. A certain amount of empowerment may be needed for them to become effective stakeholders. Without such (political) conditions being in place, it may be disingenuous to claim that stakeholder processes are being operated.

An understanding of two issues is crucial to improving interest groups' relations and the outcome of their negotiations:
• Who should be involved—especially whose values should be weighted most strongly.
• What local participatory mechanisms have worked in the past to bring interest groups together—in ways which redress major imbalances in their access to policy/decision-making.

These issues are considered in the next section.

10.1.3 Identifying the main stakeholders, their interests and their relations

Different stakeholders will have different objectives and interests, knowledge, skills, rights or influence in affected forests. To be sure of these, stakeholder identification, consultation and participation processes are now recognized as central aspects of SFM. However, it is important that consultation and participation is done sensitively and realistically (Bass *et al.* 1995; ODA 1995, 1996).

Identifying stakeholders:

To identify the broad scope of stakeholders some key questions can be asked concerning the likely degree of involvement of stakeholders with the forestry activity in question:
• Who is—or might be—affected, positively or negatively, by the forestry activity?
• Who can make forest management more effective through their participation, or less effective by their non-participation or opposition?
• Who can contribute useful resources and information?
• Who is likely to mobilize for, or against, forest management?

Following this scoping, which could be carried out internally by the forest organization or with key informants (below), a key question concerns 'who counts most', i.e. the issue of stakeholder weighting. Some stakeholders have real rights and strong dependence on forests, but perhaps little access to decision making; others merely have interests in forests, but (perhaps due to political,

financial or intellectual power) can more easily influence decisions. Colfer (1995) has developed an approach for attempting to redress imbalances among stakeholders in access to forestry decisions by ensuring that *local forest factors* are fully identified and 'weighted' against certain criteria. Building on this, we suggest stakeholders should be identified, and weight should be accorded to them, depending upon their:
• *proximity* to forests, woodlands or trees on farms;
• *dependence* on forests for their livelihoods (i.e. where there are few or no alternatives to forests for meeting basic needs);
• *cultural linkages* with forests and uses of forest resources;
• *knowledge* related to stewardship of forest assets;
• *pre-existing rights* to land and resources, under customary or common law;
• *organizational capacity* for effective rules and accountable decision-making about forest goods and services;
• *economically viable forest enterprise* that internalizes environmental and social costs, bringing equitable local benefits.

Colfer (*op.cit.*) also strongly suggests that an 'inverse' criterion should also be used, i.e. if a local group has a *power deficit* it should be weighted more (to make up for such a deficit).

There are several methodologies for identifying stakeholders (Borrini-Feyerabend 1997; Higman *et al.* 1999):
• *Identification by a forest operation's staff.* Those who have worked in the area for some time can identify groups and individuals whom they know to have interests in forest issues and to be well informed about them.
• *Identification by other knowledgeable individuals.* Land and agricultural agencies may be able to recommend relevant farmers and settlers; local government, religious and traditional authorities, forest agencies and other forest enterprises may all be able to identify key representatives of different forest interest groups.
• *Identification through written records, and population data.* Forestry operations often have useful records on employment, conflicting land claims, complaints of various kinds, people who have attended meetings, financial transactions,

etc. Forestry officials may have important histori-cal information on forest users, records of permit-holders, etc. Census and population data may provide useful information about numbers and locations of people by age, gender, religion, etc. Contacts with NGOs (non-governmental organizations) and academics may reveal relevant surveys and reports and knowledgeable or well-connected people.

• *Stakeholder self-selection.* Staff of the forest operation could make announcements in meetings and/or in newspapers, local radio or other local means of spreading information. Groups and individuals then come forward and ask to be involved. The approach works best for groups who already have good contacts and see it in their interests to communicate. Those who are in more remote areas, or are poor and less well educated, and those who may be hostile to the forest operation, may not come forward in this way. There is a risk that local elites, or others with inequitable or corrupt objectives, will put themselves forward.

• *Identification and verification by other stakeholders.* Early discussions with those stakeholders who are identified first can reveal their views on the other key stakeholders who matter to them. This will help the forest enterprise to better understand stakeholder interests and relations.

It is important that the individuals dealt with actually represent their constituencies. The dimensions of *representation* are:

• *Identity.* Does the representative share the views of the group/constituency in relation to forests? Or will the representative bring other/multiple identities to the process, for example tribal/class or political affinities? Where can such other identities help, and where might they hinder representation and forest management?

• *Accountability.* Was the representative chosen by a particular group/constituency? And/or does s/he consult with that group regularly? What kind of specificity and sanction has the group attached to the representative's accountability?

Once stakeholders or their representatives have been identified, it is important to assess:

• their capacities for participation in SFM;
• their interests in forests;
• their relationships (which often revolve around their forest interests).

These issues, described in detail by Dubois (1998), Carter (1996) and Higman *et al.* (1999), are discussed below.

Assessing stakeholder capacities, interests in forests and relations

Dubois (1998) introduces the '4Rs' approach for assessing stakeholders' Rights, Responsibilities, Rewards (or revenues or returns) and Relationships with other groups (Fig. 10.1). Establishing these 4Rs, and getting other interest groups to acknowledge and respect them, is essential for stakeholders to practise effective forest stewardship. As an example, Tables 10.2 and 10.3 illustrate use of the 4Rs approach by a Zambian group to analyse stakeholder situations and diagnose problems.

A range of tools can be used to assess stakeholder capacities, interests and relations.

Participatory learning and action. This comprises various means of obtaining information from local stakeholders, without introducing the bias of the researcher or planner on the one hand, or the leaders or narrow segments of stakeholders' groups on the other hand. They include Participatory Learning and Action (PLA) methodologies, of which there are several hundred worldwide. These methodologies have been developed in the last 15 years to foster people-first, sustainable development objectives. They have been tested in many participatory forestry projects, especially in developing countries, with the aim of helping stake-

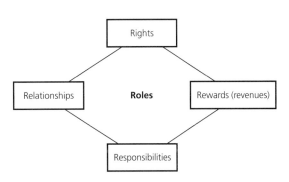

Fig. 10.1 The 4Rs approach to identifying stakeholder roles.

Table 10.2 A summary of rights, responsibilities and revenues in Lukolongo, Zambia.* (Source: Dubois 1998, adapted from Makano *et al.* 1997.)

Stakeholders	Responsibility	Rights	Revenues
Subsistence farmers	Custodians to land	Forest harvesting, cultivation of the land	Income from forest and agricultural products
Emergent farmers	Some land management	Land cultivation	As above
Charcoal producers	None	Wood harvesting	Income from forest products
Charcoal traders	None	Charcoal marketing	Income from trade
Curio-makers	None	Wood harvesting	Income from forest products
Fishermen	None	Fishing	Income from fishing
Forestry Department	Forest management, forest law enforcement	Collection of revenue from forest taxes	Revenue from forest taxes
ECAZ (an NGO)	Facilitator of development	To facilitate development	Indirectly, creation of employment

* This example suggests an imbalance between private operators' responsibilities and their rights and benefits.

Table 10.3 Stakeholders' relationships in Lukolongo, Zambia.* (Source: Dubois 1998, adapted from Makano *et al.* 1997.)

	Subsistence farmers	Emergent farmers	Charcoal producers	Curio-makers	Fishermen	Forestry Department	ECAZ
Subsistence farmers							
Emergent farmers	Good						
Charcoal producers	Good	Good					
Curio-makers	Good	Good	Fair				
Fishermen	Good	Good	Good	Good			
Forestry Department	Fairly good	Fairly good	Poor	Poor	Fair		
ECAZ	Good	Good	Good	Fair	Good	Good	

* This example highlights the poor relations between the State and other stakeholders.

holder groups to identify their forestry resources, problems and objectives of both the majority and minorities. A challenge for the next decade is to get these methodologies integrated into forestry management systems—as a better way of working with local stakeholders—and into national forest policy/information systems. The 4Rs approach described above suggests key categories of information that should be obtained through PLA, i.e. rights, responsibilities, revenues and relationships.

Carter (1996) provides an extensive discussion of the application of participatory approaches to forest resource assessment. IIED (1995) offers an extensive and practical trainers' guide. The following is just a summary of PLA methodologies:

• *Village/community meetings.* Existing village bodies, or community/interest groups, can be very effective if they are broadly representative, as the group already knows how to function internally. Otherwise, special meetings may have to be called to give out information on forestry and to obtain feedback. The latter has the advantage of being able to communicate the specific intentions of a forest operation—which is useful in the early stages of identifying stakeholder groups and possible impacts, but it can often get 'hijacked' by other issues.

• *Focus groups.* Special groups can be convened to discuss a particular topic, for example recreational interests requiring access to a forest, or farmers wanting land for seasonal use within the forest, or hunters and their practices.

• *Key informants.* It is often very rewarding to consult with acknowledged local experts, elders or knowledgeable individuals who can provide critical information.

• *Interviewing.* Semistructured interviews are often more effective than either rigid questionnaires on the one hand or loose discussion on the other, i.e. using a guide or checklist to pose questions, as well as to investigate topics as they arise.

• *Participatory mapping.* One way of dealing with location-specific or overlapping problems and opportunities is for stakeholders to pre-pare maps of resources/rights/responsibilities/problems/conflicts. This can be done on paper or blackboards, or can use local materials such as sticks, leaves and stones on the ground. One map can lead to others, as more and more people get involved. They can also lead to focus group discussions.

• *Time lines.* Groups can prepare histories of recollected events with approximate dates, and discussion of which changes have occurred and why (cause and effect).

• *Matrix scoring.* Matrices can be used to agree the ordering and structuring of information or values, and then for planning. Ranking criteria are agreed and presented as matrix rows, and relevant issues, for example forest values, are then assessed as matrix columns.

• *Participatory monitoring.* Conducted by stakeholders themselves, this involves processes for gathering, analysing and using the information which stakeholders agree is relevant. Stakeholders themselves both define the indicators of good forest management (in terms which are meaningful to them) and conduct the monitoring of these indicators (through means which are accepted). The results are then fed into the forestry operation's planning and review process or management system.

Social science approaches. The following can accommodate PLA, but are essentially more extractive methodologies, operated by professional social scientists:

• *Market research.* Where there are very large numbers of stakeholder groups, and/or inadequate representation of stakeholders, formal

market research techniques can be valuable. These have become quite commonly used by forest authorities in industrialized countries to assess stakeholders' changing values concerning forests, and changing attitudes to forest authorities. They may be undertaken through the telephone, mail or internet, or within the forest itself to survey attitudes to landscape and recreation.

• *Social impact assessment* involves collecting information about stakeholders and their interests, and then considering the actual, or possible, impacts of forest operations on these interests, in order to propose mitigation measures. It should generally be combined with environmental impact assessments, as an ESIA (environmental and social impact assessment), because stakeholders invariably raise environmental concerns, too. Social impact assessment should include:

(i) identification and evaluation of potentially significant social impacts;

(ii) consideration of options for mitigating negative impacts and optimizing positive impacts;

(iii) prioritization of activities required to address impacts.

Assessing stakeholder relations

Stakeholders have different sorts of relationships with each other. Some may not be aware of each other or may ignore each other, others may be in varying states of conflict or cooperation over different issues related to forest management. It is important to develop a general picture of these actual or possible areas of conflict or cooperation, so that the likely consequences can be incorporated into plans of the forest enterprise. Matrices describing relations between groups could be developed (see Table 10.3 for a Zambian example). Within such a matrix, four criteria could be used to describe relationships (Foteu *et al.* 1998):

• formal/informal;

• frequent/infrequent contacts;

• converging/diverging values and opinions;

• type of outputs, or performance of the linkages between stakeholders.

Note that conflict is not necessarily a bad thing. In many cultures, 'organized' conflict with (tacit) local rules for its conduct, is a normal way to achieve more or less equitable outcomes.

10.2 PARTNERSHIPS IN FOREST MANAGEMENT—CREATING, REALIZING AND SUSTAINING SOCIAL VALUES

A consideration of social values would usually appear to point towards the need for forests to be managed for multiple purposes. This would appear to be in contrast to the global trend towards the development of plantations and intensively managed forests of plantation-like characteristics, which tend to concentrate on timber (FAO 1997). However, there are two basic ways of ensuring multiple values—in essence:

• *A spatial mix at the landscape level*—zoning of specialized functions, taking account of local social needs and ecological conditions, such as ecological landscape planning as practised in Scandinavia.

• *A product mix at the stand level*—management for various products within and below the forest canopy, sometimes by encouraging a variety of species and changes over time, such as natural forest management, rotational shifting cultivation, agroforestry, etc.

It is not the intention of this chapter to provide technical guidelines on these forest management approaches—rather to cover one element which is key to both types of approach, i.e. managing the 'stakeholder mix'. Such mixes, or partnerships, should aim to include social values in forestry objectives (section 10.1) and to ensure their efficient production and distribution.

Just as there has been recent experience in participation in forestry planning, there have also been two main spheres of recent partnership experience in forest management:

• Collaborative management agreements between forest authorities and communities such as joint forest management schemes.

• Corporate/community partnerships, such as outgrower schemes.

10.2.1 Collaborative forest management (government approaches to social values)

Much of this experience has been in developing countries, in resource-poor environments where corporations may be unwilling to invest and where governments may have neither the resources nor the incentive to manage forests directly for multiple benefits. The approach has involved (re)asserting and recognizing local peoples' rights to forests; pooling skills and resources; developing robust and representative organizations to make decisions and to manage forests; and (re)introducing community forest management regimes and rules integrated with local livelihood activities such as farming.

Development assistance agencies have been major catalysts for such approaches in developing countries, acting on the observation that government-only forestry projects have inadequately helped to supply local communities' livelihood and development needs. Britain's Department for International Development (DFID) has supported over 200 such projects. A recent review of some of these found that:

• A management agreement that reflects stakeholder values is essential; this should not just cover very local communities—agreements are most effective if they involve a coalition of interested stakeholders.

• Effective partnerships do not necessarily need forest ownership to be transferred to communities, but securing rights is important; unless rights and boundaries are resolved, disputes can escalate

• An agreement is not a once-and-for-all affair, but needs to be kept under review, depending upon outcomes and changing needs.

• Misunderstandings about the level and type of participation have been common (Table 10.1); 'full' participation is not always necessary, however, matching participation to local demands for it, and ensuring full transparency, are more important.

• Collaborative approaches are time consuming—it is important not to move faster than stakeholders can consult with their constituencies—and they are costly to initiate, although ultimately appear to be sustainable.

• The process of collaboration can itself promote better stakeholder relations, improve local capacities, and help to revitalize stale policy processes and government authorities (although policy and institutional constraints may be so strong as to stymie partnerships in the first place).

• Transparent and efficient mechanisms for cost–benefit sharing, and for ensuring tangible benefit flows to local groups, are crucial (the agriculture sector often has more experience in a given locality, and this could be assessed). Agreed indicators of success, and participatory monitoring, appear to be important.

• However, experience so far also reveals some negative attributes, which are addressed in 10.2.4 (ODA 1996).

The experience of joint forest management (JFM) in India is discussed in more detail in Chapter 11. Although the Indian context is unique—most countries will not need to adopt such a bureaucratic approach, nor will many have such entrenched inequities—JFM reveals many general lessons, particularly concerning the need to establish strong community rights, and to reform government forest bodies so that they evolve as support structures for village groups—and put away their 'policemen' roles. This continues to be a rocky road.

In other countries, the approach of government has been less directly interventionist, relying more on informational instruments. An excellent example is the Swedish Forestry Board's campaigns to improve the democracy of information and knowledge about multiple forest values among all stakeholders. Annually, a subject such as biodiversity is taken, and course material/self-assessment guidelines are developed for private forest owners to work on individually and in small groups.

10.2.2 Corporate/community partnerships (private sector approaches to social values)

A growing trend amongst forestry corporations is to develop partnerships with local stakeholders. In 1996, the International Institute for Environment and Development (IIED) surveyed 18 of the largest corporations producing pulp wood in all continents. This revealed that in general:

• far more progress had been made on environmental management than on social issues (there being more legislative requirements and more immediate cost-savings associated with this).

However:

• most companies had some sort of access agreement with local people;

• 80% supported subsistence or non-commercial use of forests for public fishing or recreation;

• 60% either ran an outgrower scheme or provided extension services to small landowners who grow trees for them; and

• 25% produced wild meat, fish, oils, firewood and/or honey on a commercial basis (IIED 1996).

The *outgrower* approach provides the most significant corporate experience of partnerships. It involves companies entering into agreements with communities or resource-user groups, such as farmers, to share the costs, benefits and risks of growing forest products (usually trees). For example, often the company will provide plants and inputs, and will guarantee the purchase of resulting wood. IIED's study of outgrower schemes in Brazil, India and the Philippines suggests that companies benefit from such schemes because other wood sources may not be available, or may be expensive, and because social risk is reduced by involving local groups. Furthermore, company costs and financial risks can be reduced (30% cost reductions in certain Brazilian cases) (Roberts & Dubois 1996).

Farmers benefit from regular income and a guaranteed market, as well as from technical assistance. For middle-income farmers especially, tree crops can produce a higher return on labour than farming. But there are risks. For farmers to benefit from outgrower schemes, the following appear to be necessary:

• secure land tenure and clarity over rights to the trees being grown;

• access to financial support or sources of income while the trees mature;

• higher net returns from trees than from, for example, agriculture or livestock, or alternatively trees are a low-cost complement to farming;

• secure markets for the wood (but not being tied to one price-setting buyer);

• good means of participation with the company, and ability to appeal to third parties in disputes (Roberts & Dubois 1996).

Where these prerequisites are not present, this has led to inequalities in these terms of partnership, and reduced take-up of outgrower approaches. IIED's conclusions from outgrowing schemes show that the more beneficial schemes are not based solely on wood growing. There are, however, fewer examples of other forms of partnership involving different goods and services valued by communities. Partnerships between (large) South African forestry companies and local communities, for example, include other social values:

• land sharing (with community rules to protect the company's forest resource);

• access agreements for grazing and firewood;

• joint ventures in wood and fruit production;

• communities sharing equity within the company (Clarke & Foy 1997).

10.2.3 Preliminary guidance on partnerships that optimize social values

Building on recent reviews (Carter 1996; IIED 1996; ODA 1996), on principles established by the European Commission (1996), and on guidance by Higman *et al.* (1999), some *illustrative* goals, objectives and activities for corporate/community partnerships can be suggested. Something along these lines might be included in a forest organization's management plan, or formal management system:

Illustrative overall goal

To ensure commercially viable forestry operations which integrate social values and thereby contribute to local livelihoods and development

Illustrative objectives for partnerships

1 Assess local community values and needs, including those of the poorest and marginalized groups; consider forestry values in the context of broader livelihood and land-use patterns.

2 Provide support to those local decision-makers who can best address the non-market, forest-related needs of the majority.

3 Provide secure employment opportunities for members of local communities at all levels of the forest organization.

4 Provide training and skills development, recognizing and incorporating traditional knowledge and skills.

5 Agree, in advance, compensation for forest resources lost to the community, for example provide/subsidize building materials where access to timber trees has been stopped.

6 Establish benefit-sharing arrangements, for example a community's share of the timber harvested, of other forest products, or rights to recreation.

7 Develop income-generating opportunities, for example purchasing local goods and services, and outgrower schemes.

8 Develop profit-sharing or share-holding options in the forestry organization.

9 Provide infrastructure for local communities (e.g. roads, schools, hospitals and communications) based on needs identified by the community.

10 Stay up to date on local legislation and best practice concerning employment, health and safety, and partnerships, and incorporating changes as necessary.

Illustrative activities for forming a corporate/community agreement

1 Work with local groups to identify key actors and values:

(i) legal and traditional resource rights and tenure;

(ii) areas of cultural, historical, economic or religious significance to the community which may be excluded from forest management activities;

(iii) cultural practices which should be respected, such as taboo days, festivals, use of forest products in cultural events;

(iv) practical socioeconomic considerations which should be respected, such as roads and extraction route location, protection of drinking water sources, timing and planning of logging operations so as not to disrupt farming activities;

(v) existing community infrastructure and agricultural crops which should be respected, for example making agreements to maintain existing bridges used in forest operations, or compensation for damage to crops.

2 Negotiate an agreement with representatives of local and indigenous communities:

(i) agree which territory or set of resources, and associated boundaries, are applicable;

(ii) agree the range of functions and sustainable uses that are to be provided;

(iii) agree the relevant stakeholders and their representatives;

(iv) agree the 4Rs of each group—rights, responsibilities, rewards and relations;

(v) agree a management plan, with equitable cost- and benefit-sharing mechanisms;

(vi) agree that people with legal or customary tenure or use rights must be able to maintain these rights with respect to any forestry operations on the land in question; or else define where rights are ceded or delegated, based on free and informed consent;

(vii) agree compensation for the application of traditional knowledge—ensuring free and informed consent and full awareness of potential values;

(viii) agree conflict-resolution means, and sanctions for failure to comply with agreed actions, to ensure that all sides comply; these may be agreed and enforced between forest management and local groups or may involve external sanctions by government forest officers or others.

3 Sign an agreement with witnesses and make this public.

4 Institute a monitoring/learning process for all stakeholders:

(i) ensure a learning approach, especially as for most forest managers, management for social values is relatively new, and there will be uncertainties about what the priority social values really are, and about management methods to produce them;

(ii) as far as possible, integrate monitoring into the forest (environmental) management system, and use the participatory tools outlined above, especially participatory monitoring (section 10.1.3);

(iii) agree who can monitor operations and how; this may include monitoring by communities, the forestry organization, government officers or local NGOs;

(iv) institute a stakeholder liaison committee to monitor the agreement, and encourage stakeholder monitoring;

(v) establish useful and regular two-way means of communication;

(vi) document the whole process and results, including a record of who was involved, what values and issues were raised, what was agreed and responsibilities for compliance.

10.2.4 Remaining challenges for partnerships

Experience with forest management partnerships is still not strong. Some fundamental dilemmas are common and should be addressed in future. For example, DFID's review of collaborative forest management concerning government–community partnerships (section 10.2.1) revealed that:

1 It is too easy for forest project protagonists to assume that certain forestry roles are a priority to local groups, rather than to find out real local values. For example, food production within forests may be the most important local forest value, but 'imposed' partnerships for wood production may displace this.

2 Disadvantaged groups are rarely the immediate beneficiary of partnerships, as opposed to middle-income groups (because secure tenure to land, and some proven capacity and power, are often needed to join partnerships).

3 Once 'win-win' opportunities have been exhausted, conflicts among stakeholders become apparent; however, means to resolve them are not always available.

4 There are dangers that costs and risks are passed unduly from government or the forest organization to local stakeholders (ODA 1996).

Regarding corporate/community relations, current work at IIED suggests several further dilemmas which need addressing before partnerships can deliver social values:

5 Private sector companies' willingness to collaborate with community groups is the exception rather than the rule: the places where partnerships are most needed, in environmental and social terms, are still the places which are least

attractive to the private sector in economic terms. Market instability and management problems frequently undermine partnerships, no matter how well intended.

6 It is rare to find a true partnership of equals— the company usually dictates the pace, and its pursuit of quick profits may clash with community group aspirations. Local groups may get locked in to linkages with the private sector through debt. And partnerships which rely on producing for a single buyer may be temporary phenomena—lasting only as long as the company can hold on to a monopsony.

7 Governments are at the very early stages of knowing how to intervene to support partnerships. NGOs and other civil society groups may have more 'brokering' and mediating experience to offer.

An emphasis on partnerships will be key for taking forestry through the transition to a more sustainable future—a future in which policies, laws, markets, and ethics ensure that social and environmental concerns are given equal weight to commerce. We need to cultivate forests for multiple goods and services. In order to do this well, foresters will also have to start cultivating relationships, too—relationships between the 'sustainable development triad' of government, civil society and the market. This will take time. There is no 'quick fix'. A careful, well-monitored, step-by-step approach will be needed. Those who make a start now—even if it is a modest one— will be better placed to meet burgeoning legislative and market requirements.

REFERENCES

Bass, S., Dalal-Clayton, B. & Pretty, J. (1995) *Participation in Strategies for Sustainable Development.* Environmental Planning Issues no. 7. IIED, London.

Borrini-Feyerabend, G., ed. (1997) *Beyond Fences: Seeking Social Sustainability in Conservation.* IUCN, Gland, Switzerland.

Carter, J. (1996) *Recent Approaches to Participatory Forest Resource Assessment.* Rural Development Forestry Study Guide 2. Overseas Development Institute, London.

Clarke, J.M. & Foy, T.J. (1997) *The role of the forest industry in rural development and land reform in South Africa.* Paper prepared for the Fifteenth Commonwealth Forestry Conference, Zimbabwe, 12–17 May 1997.

Colfer, C.J.P. (1995) *Who Counts Most in Sustainable Forest Management?* CIFOR Working Paper no. 7. CIFOR, Bogor, Indonesia.

Cornwall, A. (1996) Towards participatory practice: participatory rural appraisal (PRA) and the participatory process. In: de Koning, K. & Martin, M., eds. *Participatory Research in Health: Issues and Experience.* Zed Books. London.

Dubois, O. (1998) *Capacity to Manage Role Changes in Forestry: Introducing the '4Rs' Framework.* IIED, London.

European Commission (1996) *Forests in Sustainable Development.* Office for Official Publications of the European Commission, Luxembourg.

FAO (1997) *State of the World's Forests 1997.* United Nations Food and Agriculture Organization, Rome.

Foteu, K.R., Essam, S., Mieugem, P., Nzokou, P., Tandjeu, J.-B. & Zang-Zang, U. (1998) *Les droits, relations, responsabilités et revenus de différents intervenants dans l'utilisation des ressources forestières à Oman, Cameroun.* Etude préparée pour l'IIED. IIED, London.

Higman, S., Bass, S., Judd, N., Mayers, J. & Nussbaum, R. (1999) *The Sustainable Forestry Handbook.* Earthscan, London.

IIED (1995) *Participatory Learning and Action: A Trainers' Guide.* IIED Participatory Methodology Series. IIED, London.

IIED (1996) *The Sustainable Paper Cycle.* IIED/WBCSD, London.

Makano, R.M., Sichinga, R.K. & Simwanda, L. (1997) *Understanding Stakeholders' Responsibilities, Relationships, Rights and Returns in Forest Resource Utilisation in Zambia: What Changes are Required to Achieve Sustainable Forest Management.* Final report prepared for IIED (International Institute for Environment and Development), London (unpublished).

ODA (1995) *A Guide to Social Analysis for Projects in Developing Countries.* Overseas Development Administration, London.

ODA (1996) *Sharing Forest Management: Key Factors, Best Practice and Ways Forward.* Overseas Development Administration, London.

Roberts, S. & Dubois, O. (1996) *The Role of Social/Farm Forestry Schemes in Supplying Fibre to the Pulp and Paper Industry. Towards a Sustainable Paper Cycle Sub-study Series,* No 4. IIED/WBCSD, London.

11: The New Forest Policy and Joint Forest Management in India

NARESH C. SAXENA

11.1 INTRODUCTION

International concern in forestry has generally been focused on resource degradation, declining biodiversity and the impact it has on global climate. The impact of diminishing forest cover on the local people has attracted less attention. Nationally, people have been given even lesser importance, as forest dwellers have generally no political voice, and green cover has been associated in the minds of planners with sustained production of timber for revenue generation and exports. In this context India's experiment since 1988 with a new people-orientated forest policy and with involving the local communities in forest management assumes great significance for the developing countries both for resource development and poverty alleviation.

Twenty-three per cent of India's geographical area equivalent to 76.5 million ha has been declared as forests, which is now mostly under government control. According to the Forest Survey of India (FSI 1998), in 1997, 48% of area notified as forests had a crown density of more than 40%, 34% between 10 and 40%, and the rest (18%) had less than 10% or no tree cover at all. Of the area under forests, 37% is tropical moist deciduous forest (where *Shorea robusta*, better known by its local name sal, is the main species), 29% is tropical dry forest (where teak, *Tectona grandis*, is the valued species), 8% is tropical wet evergreen forest, and the rest (26%) is subtropical, temperate, alpine and other forests.

The FSI estimated current productivity for the entire forests at 0.7 m^3 (cubic metres) of wood per hectare per year, which includes both recorded and unrecorded removals from forests. These levels are dramatically lower than the potential, which has been estimated at 2 m^{3}ha^{-1}year^{-1}. Achieving this potential, which is about three times the current productivity, would bring considerable improvement in the economic and environmental well-being of India's land and people.

Forests are not spread evenly in India, but are concentrated more in the poorer regions of low agricultural productivity and poor soils, where India's indigenous communities, called *adivasis* or tribals, live. India's forests have generally speaking not been uninhabited wildernesses. Even in the remote forests people have either been living traditionally or were brought by the forest department (FD) in the colonial period and settled there to ensure the availability of labour. Today, there are about 100 million forest dwellers in the country living in and around forestlands and another 275 million* for whom forests have continued to be an important source of their livelihoods and means of survival (Lynch 1992).

Changes in forest policy and management over time in India can best be understood in terms of the competing claims and relative influence of the various interest groups and stakeholders. Within an overall structure of state ownership and control, two groups seem to have had a major impact over the last hundred years: foresters and industrialists (Guha 1994). For almost a century and until 1988, forest management strategies were decided almost exclusively by the foresters

---

* There is no uniformity in the number of dependents as assessed by different authorities. One reason for this variation could be that dependence has not been rigorously defined in these studies.

and were markedly biased in favour of commercial and industrial exploitation, with little attention paid to sustainability or to social justice.

However, in the last decade and a half, as the forestry debate has intensified both locally and in the media, the state has increasingly responded to the claims of forest dwellers voiced by the activists and NGOs (non-government organizations). Their call for a decentralized and participatory system of forest management has finally been accepted, at least in theory, through the programme of joint forest management or JFM. This has attracted considerable donor support in the last 10 years, although the policy instruments of improving the effectiveness of communities in local management are still not very well understood. This chapter discusses the evolution of the current forest policy and the unresolved issues in its design and implementation, and offers some practical suggestions. In particular, it seeks to answer in sections 11.6 and 11.7 the question, 'what constrains success in participatory management and how does this relate to our understanding of social structure, ground economic realities and the science of growing trees?'

A glossary of terms commonly encountered in Indian forest management is supplied in Box 11.1.

11.2 FOREST POLICY BEFORE 1988

At the beginning of the nineteenth century more than two-thirds of the land mass in India was lying uncultivated (Singh 1986). As lands close to villages were enough to satisfy the subsistence needs of the people, forests remote from villages were generally underexploited. Often these virgin forests were concentrated in infertile highlands, with a heavy tribal concentration who were forced in the medieval period to seek refuge in forests, being driven from fertile lands by the more aggressive warrior communities.

The British presence from the late eighteenth century onwards started making a difference to land and forest usage in India. Guided by commercial interests the British viewed forests as crown lands, limiting private property rights only to continuously cultivated lands. Uncultivated land remote from habitation was declared by the government as forests and managed under the new forest regulations for timber production. This orientation did not change even after Independence, as the Forest Policy 1952 declared that village communities should in no event be permitted to use forests at the cost of 'national interest', which was identified with defence, communications and vital industries.

The next policy declaration contained in the report of the National Commission on Agriculture (NCA) in 1976 too put its stamp of approval on this commercial approach in the following terms (GOI 1976: 32):

> Production of industrial wood would have to

Box 11.1 Glossary of terms commonly used

Adivasis: Also legally termed as scheduled tribes, people who until a few decades ago lived by hunting and gathering of forest products, or practised shifting cultivation

Bamboo: *Bambusa arundinacea* and *Dendrocalamus strictus* are the two most common species, wanted by both the paper industry and the poor

Bidi: Indian cigarettes

Casuarina: *Casuarina equisetifolia*; widely grown in coastal areas for poles and fuelwood

CHIPKO: Name of the movement for protection of environment in the central Himalayan region

Jharkhand: Name of the movement to create a new state in eastern India, literal meaning 'land of trees'

Naxalites: Members of a militant ultra-left movement

Panchayat: Village council; the lowest form of local government; consists of elected members headed by a chairman

Sal: *Shorea robusta*, a common but slow-growing large tree in Indian forests; yields both timber and important minor forest products like seeds and leaves

Tendu: *Diospyros melanoxylon*; used as wrappers of tobacco to produce *bidi*, Indian cigarettes be the *raison d'être* for the existence of forests. There should be a change over from the conservation-orientated forestry to more dynamic programme of production forestry. The future production programme should concentrate on clearfelling of valuable mixed forests, mixed quality forest and inaccessible hard wood forests and planting these areas with suitable fast growing species yielding higher returns per unit area.

In tune with the above recommendation the states stepped up the programme of industrial forestry. For instance, the 6th Five Year Plan (1980–85) of Madhya Pradesh, the central Indian state with highest area under forests, stated,

> To produce 25 million m³ of industrial wood it would be necessary to subject 5.5 million ha of production forest lands to the intensive management, that is to clear-felling and planting . . . with the massive plantation programme being launched in the state, there would be extensive monocrops of teak in the forests . . . we should clear-fell and plant roughly one lakh hectare annually if we want production of industrial wood to keep pace with demand in future.
> [10 lakh = 1 million.]

As regards efforts made to meet tribal demands for fruit, medicinal herbs, etc. from forestlands, the same Plan document admitted, 'no special programme was taken, which could directly contribute to the upliftment of the tribal economy. The programmes executed were essentially the forest development programmes which benefited the tribals only indirectly, . . . (through) wage earning opportunities.'

Between 1952 and 1980 over 3 million ha of plantations were established in India, the major proportion of which were used to fulfil industrial needs (CSE 1982). Out of the 670 million Rs (1 US$ =42.5 Indian rupees in 1998) spent on afforestation during 1966–74, roughly 560 million Rs* was on production forestry alone (GOI 1981, p. 45). The use of bamboo for paper manufacture accelerated from a low of 58 000 tonnes at the end of the Second World War to over 5 million tonnes by 1987 (Hobley 1996). Plantations have usually

been of single species, equally entailing loss of diversity and access, and often on a large scale, and in practice hardly pursuing an objective of benefiting the local people, beyond wages (even wage employment becomes insignificant after the first year of plantation). This was recognized by the Inspector General of Forests, Mr Dalvi, who while addressing the 1981 International Conference on tropical forest management at Dehradun illustrated the inherent conflict arising out of forest plantations in the following terms:

> Let us consider another example of a natural forest predominantly of sal. This forest represents to poor forest-fringe-dwellers a source of livelihood yielding seeds for sale, branches and leaves for fuel and manure. The decision to convert this sal forest to industrially more valuable species like teak may satisfy the needs for higher revenues which may or may not be used for the welfare of these same people, but would certainly deprive them of an output from the forest which they were enjoying.

Other writers have been less charitable about the intentions of government. An ex-Forest Secretary of MP (Madhya Pradesh) writes:

> This (the policy of giving priority to industries and subsidizing industrial raw material) is clearly discriminatory. The rights of a huge section of society cannot be wiped out in order to benefit a few industrialists. For instance, the Orient Paper Mills was promised a lakh ton of bamboo per year from four districts of the state. This eliminated all bamboo from Rewa, Panna, Satna and Shahdol. When such a situation arises the Forest Department tells the villagers to fend for themselves because there is nothing in the forests for them (NCHSE 1987, p. iv).

Turning a complex forest into a genetically simplified plantation may help produce industrial raw material, but ignores the basic needs of forest dwellers—mostly tribals—and is therefore of questionable sustainability. Conventional forestry based on clearfelling disrupts the annual cycling of nutrients, and increases soil erosion (Spurr & Barnes 1980, p. 240). Monoculture plantation forestry is also prone to pest attack. Thus

*Both figures are not adjusted for inflation.

eucalyptus plantations in Kerala raised after clear-felling dense green and deciduous forests have been devastated by fungal diseases, and the consequent low productivity defeated the very purpose for which they had been raised (Nair 1985). The experience of teak plantations in north Bastar, MP was similar (Anderson & Huber 1988, p. 63). However, mixed forests draw and give nutrients to the soil at different stages of their growth, and hence are ecologically far more beneficial than plantations. One of the main outputs from forests should be water, which is possible only when forests are considered more in the context of local rather than 'national' needs.

Tropical forests support complex ecological chains while playing an essential and salutary role in the earth's climate and atmosphere. They can return as much as 75% of the moisture they receive to the atmosphere. Thus they have a profound effect on rainfall. Yet these vast natural forests, surrounded by poor populations, were rapidly diminished as they were being converted into plantations to meet market demands.

Thus the entire thrust of forestry during the first four decades after Independence was towards the production of a uniform industrial cropping system, created after clearfelling and ruthless cutting back of all growth, except of the species chosen for dominance. Three sets of factors have been at work in shaping such policies. First, development until the mid-1970s was associated in the minds of planners with creating surplus from rural areas and its utilization for value addition through industry. Hence output from forestlands was heavily subsidized to be used as raw material for industries. The impact of such policies on the environment, on the existing natural forests or forest dwellers was not considered to be serious, as the resource was thought to be inexhaustible. Second, tribals and forest dwellers, with little voice or means to communicate, were remote from decision making, and politically their interests were not articulated. Third, foresters were trained to raise trees for timber. Other intermediate and non-wood forest products were not valued, indicated by their usual description as 'minor products', leading to the adoption of technologies which discouraged their production. The combination of these forces led to

perpetuation of a timber- and revenue-orientated policy that harmed both environment and the people, but was argued to be meeting the goals of the nation-state.

11.2.1 Social forestry

The NCA in the 1976 policy did, however, concede that peoples' basic subsistence needs should be met from non-forestlands in order to check the process of deforestation and save forests. This was then sought to be achieved through social forestry, a programme of fuelwood and fodder plantations on village and private lands. It is significant that social forestry was not tried on forestlands, except in small measures, as such lands were, as in the past, used for producing timber. Thus social forestry was seen as a programme which would release industrial forestry from social pressures. The core objective for forestlands remained the production of commercial timber. But in order to keep people out it was necessary to make them produce what they consumed free of charge using community and private lands to draw off the pressure on forestlands.

The programme of social forestry had two main components: planting trees on private lands, called farm forestry, and afforestation of village lands. In terms of sheer production of trees the programme of farm forestry had been immensely successful, especially in north-west India, Gujarat and Karnataka, leading even to a glut of eucalyptus wood (Arnold & Stewart 1991). However, two serious problems have surfaced lately with this programme. First, its geographical spread has remained confined to a few agriculturally prosperous regions only, and second, even in these regions the initial enthusiasm for farm forestry could not be sustained after 1990. The craze for planting eucalyptus, which was the main farm forestry species, declined as fast as it had risen during the 1980s (Saxena 1994).

The sustainability of the social forestry component on village lands could not be ensured because, despite the rhetoric, the programme remained essentially a departmental one where local people were not involved; village councils merely transferred common lands to the forest department and played no role in decision making or choice of species. There was no continuity in the management and control of thousands of scattered pieces of planted village lands, creating enormous problems of protection. Projects were designed around the ultimate felling of the planted trees, degradation often set in after the trees were harvested. The area available as village lands was also far less than anticipated at the project stage. Projects failed to define, establish and publicize the rights to the trees and the procedures for marketing and allocating benefits. The shares which would go to the individuals, village, panchayat (the village council for self-government) and the forest department were not clearly laid down. Insecurity about benefits led to indifference on behalf of the people. Poor soil conditions and tenurial insecurity led to a low survival rate of plants, especially in deforested regions and hillsides prone to soil erosion, where trees were needed most (Barnes & Olivares 1988).

In addition, market-orientated trees were planted which did little to improve consumption within the village. Fodder trees were generally ignored. Close spacing to accommodate more trees affected grass production. As projects did little to meet the demands of the poor for fuelwood and fodder, pressure on forestland continued unabated. Finally, as state funds became locked to meet the matching contributions required for external assistance for social forestry projects, forestlands became starved of funds, with several adverse effects. The neglect of forestlands hurt forest dwellers and tribals. It reduced timber supplies to the markets, resulting in price escalation, which further increased smuggling from forestlands.

11.3 THE 1988 FOREST POLICY

In a mixed economy, where both government and private sectors work, it is generally the government sector that looks after the infrastructural or welfare needs of the people, whereas market needs are met by the private sector. Thus, health, education, roads, etc. which are non-commercial programmes, come under the domain of the government in India, whereas the private sector has been primarily motivated by profit orientation. It was strange that in forestry this distribution of

responsibility was not being followed, and the reverse was being attempted. Forestlands were to meet the commercial needs of the economy and farmlands were to produce 'fuelwood and fodder'. This conceptual weakness was perhaps one of the main reasons for the failure of the two programmes: industrial forestry in 1950s–1970s, and social forestry in the 1980s.

The correction came in 1988 with a new forest policy which was radically different from the two previous policies. According to this, forests are not to be commercially exploited for industries, but they are to conserve soil and the environment and meet the subsistence requirements of the local people. The policy gives higher priority to environmental stability than to earning revenue. Derivation of direct economic benefit from forests has been subordinated to the objective of ensuring environmental stability and maintenance of ecological balance. It discour-ages monocultures and prefers mixed forests. The focus has shifted from 'commerce' and 'investment' to ecology and satisfying minimum needs of the people, providing fuelwood and fodder, and strengthening the tribal–forest linkages. It declared that the domestic requirements of fuelwood, fodder, minor forest produce and construction timber of tribals should be the first charge on forest produce. It advises industry to establish direct relationship with farmers who can grow the raw material if supported by industry with inputs including credit, technical advice and transport services. As these linkages may take time, in the interim it suggested that the import of wood and wood products should be liberalized, but the practice of supply of forest produce to industry at concessional prices should cease.

11.3.1 Political factors behind change

The striking policy reversal in 1988 suggests that there were no strong political constraints in effecting a radical shift in forest policies. This may surprise political economists and therefore requires an explanation. In a democratic country with 70% rural population, rural interests cannot be ignored for long. A subsistence-orientated forest policy does not hit the rural elite at all; it in fact reduces the control of the centralized bureaucracy, besides curtailing the outflow of forest products to industries. Hence such a policy should not attract political impediments, which are inherent in distributive programmes such as land reforms.

From the early 1970s intellectuals and activists have also picked up the longstanding grievances of forest-dependent communities. Consequently, in the last two decades the working of the forest department has come under close and critical scrutiny. It has been demonstrated that state policies by promoting commercial forestry have contributed significantly to the decimation of biological diversity and to an increase in soil erosion and floods (CSE 1985; Gadgil & Guha 1992a).

The battles on behalf of forest dwellers have not only been fought in the press on an intellectual plane, but have actually been carried out in the countryside. There have been both armed struggles and political movements to regain control over what they perceived to be their lands. Several heavily forested districts in Andhra, Maharashtra and Madhya Pradesh are witnessing armed rebellion, called the Naxalite Movement, directed against the state. One of the main demands of the Naxalites is better community control over forest resources. In Bihar, in 1978, local people protested in what is called the 'Tree War' against the replacement of natural forests by teak plantations (CSE 1982). Even today the Jharkhand (meaning land of trees) movement calls for the creation of a separate state in the central region of India.

There have also been peaceful and non-political forest movements in the country. In many places during the 1970s people on their own initiative started protecting forests, of which CHIPKO is a well-known example. It started in 1973 when a local village group was denied access to forests for making agricultural implements, whereas the same coupe was allotted to a sports goods company. This favouritism provoked the villagers who prevented the company from felling trees by hugging them (Gadgil & Guha 1992a, p. 223). It spread throughout the hills in north India, and led the government to impose a ban in 1979 on all commercial felling in the hills above 1000 m, which continues to this date.

When different interest groups have claims on a resource, policy and politics do not always favour the same group. In the years preceding 1988, while many decisions were taken (Chambers *et al.* 1989) to make forest policy acceptable to the environmentalists, the state government of Karnataka decided to lease out 30000 ha of common lands in 1986 to a paper company on a lease rent of 12% of the produce. When agitation and petitions to government did not yield any result, a public interest writ petition was filed by several NGOs before the Supreme Court. The Government of India in this litigation decided to oppose the decision of the state government and side with the NGOs, with the result that the state government was forced to cancel the lease in 1991 (Hiremath *et al.* 1994). This was celebrated as a big victory against the lobby of forest-based industry.

In addition to the industry getting discredited in the popular press, two other factors must have weighed upon the minds of policy planners in suggesting a diminished role for forest industries on forestlands in the new Forest Policy. First, the popularity of eucalyptus among farmers increased the availability of pulpwood at a cheap price for the paper industry. In some states, such as Gujarat, large farmers even took to teak plantations, a timber crop which takes 30–60 years to mature. Second, liberalized imports of timber and pulpwood were permitted which eased the supply for the industry. With new sources of supply, it was no longer considered crucial for the industry to depend on forests.

11.4 THE JUNE 1990 GUIDELINES AND JOINT FOREST MANAGEMENT

Although the 1988 Policy redefined the objectives of forest management, it did not envisage any direct role for the people in the day-to-day management of forests. It implicitly believed that government alone should control forests, albeit with changed objectives.

The case for public management of forests hinges on a number of factors (Commander 1986, p. 9). First, forest management is associated with a wide range of externalities, as these provide external benefits to the rest of the ecosystem. Second, forest department operatives have often argued that the management of forests requires a level of professional training and scientific competence that lies outside the capacities of peasants and forest users (Shyam Sundar & Parameshwarappa 1987). Third, the time horizons for forest management would favour public ownership and public investment. And last, it will allow for major economies of scale and a longer-term planning framework.

The strong case for exclusive government management is weakened because government is not in a position today to enforce its property rights. Forests are subject to intense pressure from humans, livestock and urban markets. Overexploitation by the people, which increased in the last three decades, is caused by several factors. First, increasing marginalization of small landowners has forced them to seek new avenues of income, like headloading.* A study (Agarwal 1987, p. 181) estimated that at least 3–4 million people were involved in headloading, making it India's biggest source of employment in the energy sector. Second, as village commons deteriorated, villagers turned to government forests for succour. And third, government policies raising commercial plantations further alienated the people from the resource. More than the official revenues which such policies brought to the government exchequer, it nurtured a new culture of rent seeking by those in power. The indiscriminate tree felling by the contractor–official–politician nexus has had a corrupting influence on the forest dwellers, who also wish 'to make hay while the sun shines'.

Unlike many developing countries, extension for agriculture and shifting cultivation, the two familiar causes of deforestation, have not been the main cause of deforestation on forestlands in India at least in the last 40 years. The alienation of forestlands from the people who need it for satisfying their needs, and consequently forests turning into open access lands, has been one of the main causes of degradation as well as for increasing misery of the people. Until the mid-1980s the response of the government to this crisis of deforestation was to bring more land under the reserved category (which increased from 26 million ha in 1951 to 46 million ha in 1988), and plant non-browsable and market-orientated single-product timber trees in order to reduce pressure from the local population and increase state revenues. This strategy became counterproductive and hastened the very degradation process it was designed to prevent.

Given the ease of access to forests, indiscipline and sociopolitical culture it has been impossible, in practical terms, for the forest department to enforce its property rights, which required that people should fear from interfering with the state-owned property and that a symbolic presence of forest staff representing the authority of law would be sufficient to caution against law breakers. Such conditions unfortunately do not exist in India today. These weaknesses in the enforcement of access to government property have led to forest areas being exploited as an open access resource by those who have no stake in its health, where all basic decisions are guided in terms of current income flows rather than capturing of delayed returns arising out of protection and long-term management.

Realizing these realities, the Government of India introduced participation of the people in managing forests by issuing a Joint Forest Management (JFM) resolution in June 1990, making it possible for the forest departments to involve people in the management of forests. The resolution breaks a new path as for the first time it specifies the rights of the protecting communities over forestlands. Those protecting the forests are to be given usufructs like grasses, lops and tops of branches, non-timber forest produce, and a portion of the proceeds from the sale of trees when they mature. This varies from 20 to 60% of the timber sold. The order exhorts the state forest departments to take full advantage of the expertise of committed voluntary agencies for building up meaningful people's participation in protection and development of degraded forestlands.

* Fuelwood collection by the poor from public lands and carrying it on their heads to the nearest market. In Orissa for instance, a headloader would earn Rs 25–30 a day for a shoulder load and Rs 50–60 for a cycle load in 1992. This was against the agricultural wage rate of Rs 25 per day fixed by government, and actual payment of about Rs 16–17 a day (Jonsonn 1994). However, agricultural work is available in monocropped areas barely for 4 months in a year.

Thus in the previous policies people and the environment were seen, all too often, as antagonistic. The forest–people interaction was conceptualized as a zero-sum game, in which both parties could not win. According to the JFM philosophy, the conflict model is neither necessary nor useful. On the contrary, ways can be sought in which the interests of people and of long-term sustainability are harmonized in a mutually supporting manner.

11.4.1 Progress of JFM in the states

Participative policies have taken different forms in different states today. In some, forests have been leased to local communities or village forest communities share the responsibility for protection with forest department. In others, villagers on their own had been protect-ing forests without any goading from the forest department. By 1997, 18 state governments had issued enabling resolutions (GRs) permitting partnerships with local people. These states have 80% of the country's forestland and 92% of the country's tribal population. The JFM programme has now become the central point of future forest development projects funded by the Government of India and the donor agencies. According to government data, it is estimated that 35 000 village communities are protecting about 3.5 million ha of forests, which comprise 5.5% of the total forest cover in India. With more imaginative policies and innovative silviculture (discussed in sections 11.6 and 11.7), this area could be increased to 15 million ha in about a decade, thus covering about 23% of the total forest cover.

In the eastern states of Andhra Pradesh, Orissa, West Bengal and Bihar there are a large number of groups protecting forests either on their own or initiated by the forest department, as Table 11.1 indicates.

Although the quality of participation is not uniformly good in all the JFM villages, the results of involving people in forest management have been quite encouraging in some places at least. For instance, in a DFID (British Government, the Department for International Development) assisted project in Karnataka the villagers' earlier fear of the forest department had declined (Saxena & Sarin 1998). Indirectly, this has resulted in the VFC (Village Forest Committee) members' improved access to forests for meeting their *bona*

Table 11.1 Actual forest cover and area under JFM in certain states. (SPWD 1998.)

State/UT	Forest cover (km²)	Area under JFM (km²)	JFM area as % of forest cover	External donor, if any
Andhra Pradesh	43 290	10 540	24.35	World Bank
Bihar	26 524	7 192	27.12	None
Gujarat	12 578	692	5.50	OECF
Haryana	604	630*	104.30	European Union
Himachal Pradesh	12 521	60	0.48	DFID
Jammu and Kashmir	20 440	141	0.69	None
Karnataka	32 403	814	2.51	OECF and DFID (British) Government
Kerala	10 334	20	0.19	OECF
Madhya Pradesh	131 195	3 500	2.67	World Bank
Orissa	46 941	4 360	9.29	SIDA up to 1994
Rajasthan	13 353	1 857	13.91	OECF
West Bengal	8 349	4 493	53.81	World Bank up to 1997
Total (including other states)	633 397	34 837	5.50	

* Includes non-forest commons too.

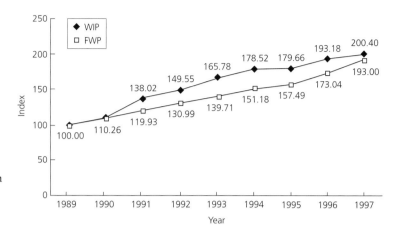

Fig. 11.1 Comparative position between average firewood price (FWP) and wholesale prices (WPI) during 1989–97. (Based on data from Labour Bureau in Shimla collected by the author.)

fide requirements of firewood, grazing and fodder. In some VFCs irresponsible cutting by the local people in forests has been reduced due to improved relations. The consumption of fuelwood had also reduced in some places because of the use of fuel-saving devices, sup-plied by the forest department. There is substantial goodwill towards the project among NGOs, who are playing an important role in organizing the people, and in helping them through non-forestry activities. People's participation is far higher in those VFCs where NGOs are active.

11.4.2 Impact of the new forest policy and JFM on deforestation

There is some evidence to show that the rate of forest degradation in India has declined after 1989, and that timber and fuelwood prices have stabilized since 1987. The rise in price in fuelwood during the period 1989–97 has been slightly less than the rise in wholesale prices, as Fig. 11.1 shows.

Recent remote sensing data about changes in forest cover is shown in Table 11.2.

One explanation of the improved scenario could be the general ban on green felling which many Indian states have enacted since the late 1980s. The relative contribution of forests to state revenues, similar to land revenue, has been falling dramatically since Independence, because of expansion of economic activity outside land. This has enabled the states to forego incomes from

Table 11.2 Changes in forest cover in India annually as per remote sensing data. (FSI 1998.)

Period	Area gained/lost annually (ha)
1981–87	(–) 147 000
1987–89	(–) 47 500
1989–91	(+) 28 000
1991–93	(+) 2 200
1993–95	(–) 25 000

logging.* Therefore it may be premature to suggest that the improvement in forest cover or the fall in wood prices is due to the change in forest policies only; other factors, such as a ban on felling, imports, the spread of prosopis (*Prosopis juliflora*)† shrubs, and the success of farm forestry, may be better explanations. However, with the success of JFM in certain areas, there are examples of a local glut of wood (Poffenberger & McGean 1996), and such instances are likely to multiply as the protected forests mature for harvesting. If the trend of improvement in forest

---

* In an important judgment in January 1997 the Supreme Court of India has reinforced this ban, which led to the closure of several sawmills, especially in the north-east.
† An excellent coppicer with high calorific value; in the prosopis-abundant districts sale of its twigs has emerged as a cottage industry for the poor, especially for women and children.

cover continues for a longer period, it could well be due to participatory policies.

JFM projects have been in operation for more than 5 years in many states. The interim evaluation studies highlight two kinds of constraints in scaling up the participatory management of forests; factors which are endogenous to the protecting group, and factors which are in the domain of public policy. Theoretical work done on the subject supports this, as why some groups succeed in collective action and others do not depends upon both internal and external factors (Ostrom 1990, 1994). The internal variables include: total number of decision-makers, interdependence among the participants, the discount rate or risk perceptions of the group, similarities of interest, leadership, information about expected benefits and costs, and shared norms and opportunities. The external variables refer to the policies of the implementing agency, political climate and government. Locally selected systems of norms, rules and property rights that are not recognized by external authorities may collapse if their legitimacy is challenged or if government does not have faith in the capacity of local communities. Blueprint thinking, i.e. the imposition of uniform solutions to a wide variety of local problems, centralization of authority, overdependence on external sources of help, and corruption or other forms of opportunistic behaviour on the part of external agencies are likely to cause threats to sustainable community governance of common property resources. Besides, government may still continue the old timber-orientated technical package while paying lip service to participative policies.

The implication for community involvement of these three sets of issues—factors which are internal to the community, government policies and the changes required in silvicultural practices—is discussed in the next sections.

11.5 INTERNAL FACTORS IN COMMUNITY MOBILIZATION

The management of forests before JFM was such that there was a dichotomy between resource users (including both industry and the people) and those responsible for resource regeneration, the user having no responsibility for plantation establishment and maintenance. Granting the management role exclusively to users, however, would succeed only if the village groups have the necessary inclination and management capability.

Whether the local population has been conservation orientated and managed commons in the past is debatable. While Gadgil (1989) has studied people-orientated management in a few villages of South India that survived the colonial period, other scholars (Nadkarni *et al.* 1989; Locke 1995; Buchy 1996) studying the same region have doubted whether such practices were widespread.

Nadkarni concludes that historically people's institutions existed only in a few villages which managed local commons; elsewhere forests survived because of limited biotic pressure. Buchy rejects a history of community management on the basis of her archive research and questions the extent to which precolonial prohibitions over certain types of resource use can be said to constitute a system of communal management. Locke also concludes that it seems likely that although reduced rates of degradation in precolonial South India in some cases may have been due to culturally determined patterns of behaviour, for the most part they arose from levels of demand too low to cause depletion. Thus the romantic assumption often made by environmentalists that the local population has been conservation orientated in the past does not seem to be true.

Exclusive village management of natural resources will require socially cohesive villages, about which there are polarized views. One social science view of an Indian village is 'an atomised mass, composed of individuals who are not in any organized fold except the family and extended kin-groups which form the subcaste' (Gaikwad 1981, p. 331). According to this view, rigid stratification of village society inhibits the development of institutions representing a common will. Grossly unequal land tenure and access to markets ensure that only a powerful minority gains in the name of the community (Eckholm 1979).

Bandyopadhyay *et al.* (1983), however, disputes that social and economic inequalities have hindered the possibility of community ownership, participation and control in India. The management of village commons has been an historical reality for two reasons. First, whereas private resources in India were governed by individualistic and class-dominated norms, there have been communally shared norms when it comes to community resources. Second, the self-sufficient nature of the traditional village economy guided the exploitation of common resources through a system of self-control. He therefore concludes that there are no structural barriers to achieving community participation in forestry projects. Many environmentalists support this viewpoint (Agarwal & Narain 1990). It is difficult to settle this controversy.

Why does collective action succeed in some cases and not in others? Empirical evidence discussed in this section, although from limited ecological regions of India, suggests several clues. First, most effective local institutions develop in those small communities where people know each other. Collaboration is easier among small and homogeneous groups (Attwood 1988).

Second, the topography of the upland villages makes their common lands visible from most of the dwellings, so that any unauthorized felling cannot escape notice. If a village is located inside the forest or at the bottom of a slope where the forest is located, individual households can easily provide protection services from their households, thus cutting down on protection costs. By contrast, the area of a flat village in Central–South India may well be spread over 5–10 km in one direction, which makes it easy for 'free riders' to escape undetected.

Third, hill and upland villages usually have better land resources. This attracts better management from the people, as its protection is more vital for their survival. On the other hand, once degradation sets in, people may become indifferent to protection. In district Kheda in Gujarat, grazing lands are so degraded that the dependence of an average milk producer on the grazing lands — and hence his or her stake in their preservation — is low (Shah 1987). There are thus two syndromes: a valuable resource well managed because it is worth managing well; and a degraded resource neglected because it is degraded.

Fourth, remoteness from roads and markets further helps in retaining mutual obligations,

and discourages poaching by outsiders. Fifth, in remote villages, fear of reprisals from village elders deters too frequent abuse of common resources. By contrast, the old system of authority in the modernized villages has been undermined without being replaced effectively by a new one, resulting in a hiatus of confidence (Wade 1987). Sixth, upland settlements are more homogeneous in caste, with one caste or tribe usually dominating both in land and number, whereas other villages tend to be multicaste, which makes social control more difficult.

Finally, collaboration is likely to succeed if all families including the rich are highly dependent on forests for their survival needs of fodder and fuelwood. As the productivity of the commons declines, the rich shift to privately grown fodder on their own land, while the poor still use the commons for their sheep and goats. The rich then lose interest in the upkeep of commons, while the poor lack the power and organization to manage the commons themselves. This may be one reason why in Gujarat, the Indian state best known for community endeavour, only 60 out of 18 000 villages have come forward to start community fodder farms whereas many times more villages have opted for income-generating eucalyptus plantations on grazing lands (Shah 1987).

The personal interests of village elites in the management of the commons appears a crucial element. Even a market-dominated village may develop collective action, if it is in the interest of all, including the powerful people of the village (Wade 1985, 1988). A study of fodder farms on common lands in Gujarat (Shah 1987) found that some panchayat leaders took a great deal of personal interest in setting up and running such farms. They seemed to work for non-economic rewards, such as power, reputation and social status. Once established, the successful farms were run along business lines, with the interaction between a farmer and the fodder farm similar to that between a buyer and a private seller. An average farmer did not show any interest in running the farm. In fact its success was due more to the initiative of the leader than to community spirit. This makes such enterprises precariously dependent upon the quality and integrity of leaders willing to work in the group interest.

It can be concluded, then, that community control and management can work in three circumstances: first, in villages which are small, homogeneous, remote from markets and dependent upon produce from commons; second, where gains from organization are high, for both the village elite and the commoners; and third, where a leader is willing to run the show for non-monetary gains. Analysing the success of sugar cooperatives in Maharashtra, Attwood (1988, pp. 69–87) reached similar conclusions: cooperatives succeed not because there are no classes in the village society, but because an alliance between the rich and poor farmers for the successful running of the cooperative sugar factories is in the economic interest of both the classes.

Local collective action has been undermined in the last 30 years by a number of political and economic processes (Bardhan 1993). Village societies have become heterogeneous, and market forces have commercialized the erstwhile subsistence economies, integrating them with urban and national economies. Possibilities for migration and mobility tend to work against cooperation. Intragroup heterogeneity in payoffs may thus adversely affect the sustainability of cooperative agreements. Cooperation works best in small groups with similar needs and clear boundaries, and shared norms and patterns of reciprocity. Swallow and Bromley (1994) suggest that a group agreement is more likely to collapse where there are more than 30–40 members, whereas the average size of villages in India is about 100–200 households. Participatory politics erodes the traditional authority structures, and modernization improves the options of both exit and voice for the common people. As the old authority structure crumbles, appeals to government for conflict resolution and arbitration become more common, and dependence on government for local resource management increases. Many rural communities in developing countries are now in this difficult transition period, with traditional institutions on the decline, while new self-governing institutions are yet to be born.

The overall result has often been to forestall local action. One way out of this hiatus is 'comanagement—cooperative management

arrangements between state and local organizations in which states assign group rights to specific resources, establish overall guidelines for inter-group interactions, and help to create more positive environments for the operation of local organizations. The latter can then mobilize local participation in resource management and advise the state on the desirability of proposed management practices' (Swallow & Bromley 1994, p. 5).

The precise distribution of control and management between the state and the local group should depend on a number of situation-specific factors, such as the ease with which groups can be formed and can retain cohesiveness. The process of sharing decision making and management of forestlands will then proceed at different paces in different conditions.

To sum up, the emergence of community cohesiveness and participation cannot always be taken for granted, even when the people face a distressing situation. In Orissa, where success stories can be located in every district, the total area under protection by indigenous groups is much less than 10% of the total forest area of the state. This could be enhanced substantially if the problems faced by the groups, and especially the ones which groups cannot resolve on their own, such as intervillage disputes, demarcation, relations with external institutions and markets, are taken care of by the government. This has often been the case in many success stories of joint forest management, in which the forest department helps the group in removing encroachments, providing funds and technical help, mediating in intervillage and intravillage conflicts, and taking legal action against 'free riders'. In other words, the capability of a village group to manage commons may be enhanced if the group is supported in its efforts by government.

11.6 CONSTRAINTS OF GOVERNMENT POLICY AND ACTION

New policies bring new issues. What is important is to develop policy mechanisms to address these issues. Rather than trying to locate barriers to community action in the structural and sociological factors discussed in the previous section, greater attention needs to be paid to governmental policy, which has often hampered such initiatives. This section discusses the problems which have been noticed in the implementation of JFM, and the policy changes which are necessary to redress them.

11.6.1 Rights of non-protecting people

The legal and organizational framework for joint management remains weak and controversial. First, the old rights and privileges of the people (usually established in the colonial period) have continued in most degraded forests, and often such rights include free access to expensive timber. Privileges without corresponding responsibility is counterproductive. Second, often more than one village have their rights in the same forest, with the result that it becomes difficult to promote village protection committees. Third, a large number of new settlers in a village (they may be the poorest) have no traditional rights in forests, as their ancestors did not live in the village at the time of forest settlement. They get deprived of benefits, and are compelled to obtain these illegally. Fourth, sometimes people living several kilometres away from forest have customary rights in forests. With no possibility of getting involved in forest management, they have been customarily using these lands as an open-access resource without any restriction, for grazing and collection of fuelwood and non-timber forest products (NTFPs). Often the forest officials, while recognizing the VFC (village forest committee) formed in a village with respect to a particular forest tract, give permission for collection of firewood from the same forest area to right holders from other villages, who do not contribute to protection. Migratory tribes from other states also send their cattle for grazing, and their rights have been upheld by the Supreme Court (S. Vira, unpublished).

Thus, a forest patch does not have a well-defined and recognized user-group, and may admit the rights of the entire population of that region or the entire forest area (Singh *et al.* 1996). This type of 'right-regime', which makes forests open-access lands, is not conducive to successful protection, as rights of contiguous villages protecting forests may come into conflict with those of distant villages, not protecting but still having rights to enjoy usufruct.

Therefore, at least in JFM areas, use rights should be reviewed in order to put them in harmony with the 'care and share philosophy' which is the basis of JFM. Even in unclassed forests, where no previous settlement has been done, the task is not simple due to the practice of use by a large class of stakeholders. Elsewhere, old settlement rights may have to be modified with a view to make these amenable to the formation of viable VFCs. This is easier said than done, as changing customary or legal rights would be perceived as an unpopular step and may face political hurdles. Such a policy can be made acceptable if it is accompanied by other pro-people changes in technology, nature of species, secure rights over produce, etc.

11.6.2 Intervillage disputes

Depriving communities far from the resource but having traditional rights is a ticklish question. Some close communities have solved this by charging fees from distant villages on the grounds that they do not have to protect the resource. In West Bengal, some VFCs negotiated with neighbouring communities to clarify rights and territorial responsibilities when they began to initiate protection activities. As the user groups have a strong incentive to avoid conflicts, they have often demonstrated that they can conduct much of the negotiation on their own or with the help of the panchayat (elected village councils) leaders. However the forest department holds ultimate responsibility for seeing that management groups do not create conflicts over pre-existing usufruct.

Confusion over forest boundaries is a recurring problem for the VFCs. In one case, members from Chandmura village in West Bengal thought that they were also protecting the Arabari forests. Only when the forest was harvested for timber did they realize that they were not part of the programme. The village took the government to court, thus delaying harvest benefits to others. The problem could have been avoided had there been maps and constant dialogue between the participating villages (Roy 1993; Chatterji 1996).

The Rajasthan GR provides for the formation of one VFC per one revenue village, which may consist of several hamlets removed from each other. This makes the smooth functioning of the VFC very difficult (VIKSAT 1995a). Where multi-hamlet forest protection committees have been formed, field experience shows that the component communities keep their independent identity within the large group, maintaining clear boundaries of their area and retaining exclusive control over harvests in their territory. Often such groups surrounding a large tract of forest form an apex committee to coordinate their activities and represent themselves to the forest department. Although the larger group may facilitate joint protection and dispute resolution, informal partitioning of the resource has no validity in law and may not be sustained over a long period.

Most VFCs want their forest tract boundaries to be formally demarcated. Rough agreements between villages over these boundaries may be sufficient when the resource is degraded, but once valuable products are regenerated, conflicts will ensue in the absence of formal notification. Often forest maps are not available which delays the formalization of boundaries. This is not a simple exercise, as natural, administrative and customary boundaries do not coincide. In practice, under existing customary use, different boundaries apply to different products, for example grazing and fuelwood. Boundary disputes between neighbouring VFCs are likely to increase as harvesting approaches.

11.6.3 VFCs and panchayats

Another legal problem concerns the status of VFCs versus the village panchayats, which may cover a larger area than controlled by VFCs. The state government resolutions recommend VFCs as functional groups. However, these committees have no legal and statutory basis, and it may be difficult for them to manage resources on a long-term basis. Their relationships with the statutory village panchayats will need to be sharply defined (Poffenberger & Singh 1993).

The 1989 West Bengal GR stated that the local panchayat land management committee shall select beneficiaries to constitute the VFC. This indicated that the panchayat, which is outside the user group, would determine who could and who could not participate. Although in 1990, the West Bengal Government allowed every member in the village to be a member of the management group, the hold of the panchayat remained strong. The Orissa order prescribes that the lady Deputy Chief of the local panchayat will be the head of the VFC, but the panchayats are not working well and her stewardship is not seen as legitimate by the indigenous VFCs.

Experience over the last 20 years from Indian social forestry programmes indicates that in many cases panchayats had difficulties in effectively managing community woodlots due to their inherent political nature and often diverse constituencies. Panchayats are political organizations based on the electoral system, whereas conflict can be quite harmful for the effective functioning of VFCs. Protection can work only if there is near unanimity and consensus among the user group.

Unlike panchayats, powers to the VFC are not given under any law, which may affect their powers to check 'free-riding' in the longer run. Thus, most successful VFCs charge fees for the collection of forest produce, although this practice is technically against the Forest Act. The illegality can be removed if the allotment of forestland to the VFCs is carried out under section 28 of the Forest Act. At present it is allotted administratively.

Due to the increasing importance of panchayats in decision making in India many field activists feel that community forest management must take place at the smallest possible level of those who actually use the resource. This would require statutory changes in the current panchayat laws.

There is also some concern that if JFM groups were absorbed by village panchayats, vested interests might exert control over decision making. As small user communities may consist of less powerful groups, they may lose authority to the elite if the management becomes a direct adjunct of the panchayat. VFCs are recognized only by the forest department, all other government departments recognize panchayats making them much more powerful than the VFCs. On the whole, there is a need to clarify the relationship of local forest management groups to panchayats; simply subsuming them as part of the panchayat would almost certainly threaten their effectiveness.

11.6.4 Marketing of NTFPs

The JFM programme is no doubt orientated towards the subsistence needs of local communities, but once the produce of forests increases through proper protection there is every likelihood of production increasing beyond what can be consumed within the village itself, and hence the importance of marketing. Moreover many NTFPs have traditionally been used by the gatherers to generate cash incomes (Shepherd 1989). However, old restrictions imposed in the past on their processing and sale are still in place—the poor have no right to process these items and sell them freely in the market.

As the commercial importance of NTFPs increased in the past, the state governments nationalized during the 1960s and 1970s almost all important NTFPs; that is, these can be sold only to government agencies or to agencies so nominated by the government. In theory, this right was acquired ostensibly to protect the interest of the poor against exploitation by private traders and middlemen. In practice, such rights were sublet to private traders and industry. Thus, a hierarchy of objectives developed: industry and other large end-users had the first charge on the product at low and subsidized rates; revenue was maximized subject to the first objective which implied that there was no consistent policy to encourage value addition at lower levels; tribal interests and the interest of the poor was relegated to the last level, or completely ignored (Saxena 1995b).

Studies (Chambers *et al.* 1989) indicate that while collectors of NTFPs are usually some of the lowest income groups in India, they often receive only 5–20% of the retail value of their goods. Various governments run marketing and cooperative schemes and have established parastatals for this purpose, but these have frequently failed to result in major improvements in prices.

The state marketing agencies have reached a stage where they are unable to play the roles for which they were intended. The policy framework, wherein a state monopoly was considered necessary to counteract severe market imperfections, has also become counterproductive and is encouraging market monopolies. If the poor are to enjoy the fruits of their labour (and of the forests which they are supposed to protect), a drastic overhaul of the policy as well as the supportive institutional framework is necessary so that it is consistent with the 1988 Policy objectives.

Owing to several practical constraints government is incapable of effectively administering complete control and do buying and selling of NTFPs itself. It is better for government to facilitate private trade, and to act as a watchdog rather than try to eliminate it. Monopoly purchase by government requires sustained political support and excellent bureaucratic machinery. It is difficult to ensure these over a long period and hence nationalization has often increased exploitation of the poor. Experience shows that open markets may give producers the best chance of gaining a competitive price for their products. For marketing NTFPs, government should not have a monopoly, nor create such a monopoly for traders and mills. The solution is to denationalize NTFPs gradually so as to encourage healthy competition. Government should set up promotional marketing boards, as distinct from commercial corporations (which are inefficient, and hence demand nationalization), with responsibility for the dissemination of information about markets and prices to the gatherers. The boards would help in bridging the gap between what the consumers pay and what gatherers get. Free purchase by all and sundry would also be in tune with the current liberalization and open market climate.

Encouraging the setting up of processing units within the tribal areas is also to be recommended. In other cases, NGO-run programmes to develop NTFPs, which make processing more efficient and improve market access, can enhance the income of forest communities. In West Bengal, the presence of an NGO that provided improved sal plate processing and marketing support allowed village producers to improve incomes to Rs 11.50 for an 8-hour day equivalent versus

Rs 5–6 for other communities dependent on middlemen.

To sum up, with a view to promoting people's participation, state governments permit the collection of NTFPs from JFM areas. However, this hardly acts as an incentive for two reasons. First, even from non-JFM areas people have *de facto* or *de jure* collection rights. Second, the marketing environment for realizing the full value from NTFPs is constrained by exploitative governmental regulations restricting sale, processing and transport. At least in JFM areas, markets must be freed from unnecessary government controls so that gatherers are able to optimize returns on their labour.

11.6.5 Insensitivity to gender issues

The protection of a degraded area under JFM often increases women's drudgery as they have to travel a greater distance to collect their daily requirements of fuelwood and fodder (Sarin 1998). Despite the good intentions of forest protection, community forest management has often burdened women with additional hardships, or concentrated it on the shoulders of younger women. They have also had to switch over to inferior fuels like leaves, husk, weeds and bushes.

Obviously, merely shifting the protection role from the forest departments to the community does not provide any immediate relief to women. Further, the gender-differentiated impact is not restricted to firewood—it applies equally to other forest produce. For example, protecting sal (*Shorea robusta*) trees with the existing technology of multiple shoot cutting results in the leaves getting out of reach. This affects the making of sal leaf plates, which is a common source of income primarily for poor women in many parts of West Bengal, Orissa and Bihar.

Another problem is providing adequate representation for women in JFM committees. In this respect, women's rights and entitlements have been almost totally overlooked in the JFM Rules. For instance, Bihar, Karnataka, Madhya Pradesh and Tripura provide for the membership of only one representative per household; Gujarat, Rajasthan and Maharashtra have left the matter open; Punjab has no provision for a general body

at all; and in Jammu and Kashmir, it is unclear whether both a man and a woman or either can represent a household. In cases where one person represents a household, it invariably ends up being a man (except in the case of widows with no adult sons).

Membership of committees is not synonymous with a share in rights or of benefits. One needs to ensure both, and not one or the other. Forced inclusion of women through legislation has not led to genuine participation. Participation of women is token, i.e. cosmetic and symbolic; it is however better where NGOs or a gender-sensitive forest official were active (Oxfam 1998). Often meetings are scheduled in the evenings to suit men, but at times when women tend to be cooking. When attending meetings, women feel marginalized, unlistened to and shy to talk. It is considered against Indian culture to talk in the presence of men, much less to question their ideas. Often they become busy in arranging tea and snacks for the male members and thus are unable to concentrate on the deliberations of the Committee. As a result, there is a bias in favour of those forest products of interest primarily to men. In the village Kilmora in a van panchayat in Uttar Pradesh (Britt-Kapoor 1994) the female member who has the added advantage of living at the house where meetings are held rarely stayed longer than was necessary to sign the register. In another village, Katuul (Britt-Kapoor 1994), a female member said, 'I went to three or four meetings. My suggestions never got implemented. No one ever listened. I marked my signature in the register. I am illiterate, so I couldn't tell what was written in the meeting minutes. I was told that my recommendations would be considered, but first that the register had to be signed. They were uninterested.'

Given the sex-segregated and hierarchical nature of Indian society, separate women's organizations and staff are needed to work among women, to instil confidence in them, so that they can fight for their rights. Therefore, whenever there is recruitment, more women need to be recruited in the forest department. The village level committees should have adequate representation of women. Forestry staff should be sensitive to gender issues through orientation programmes. As women in many societies still feel inhibited in expressing themselves in mixed gatherings, each committee should have a separate women's cell for raising their consciousness and for improving their skills. The quality of women's participation and the control they exercise over decision-making processes is more important than the sheer number of women present in such bodies.

11.6.6 Balance of power between forest departments and communities

A frequent complaint against forest departments is that they have still not given up the 'police' image, and are not sufficiently sensitized to undertake extension work. To this charge the forest officials' response is that in the initial stages the community is at a low level of formation, and therefore they cannot dilute their basic responsibility of protection. The forest department also has powers to cancel or dissolve the VFCs, which itself constrains the confidence and autonomy of the village committee. While forest departments will require some statement in the resolution to dissolve the management agreement if their community partners fail to uphold their responsibilities under the JFM programme, it is also important that the identity of the user group is respected. In Rajasthan and Haryana, where the GR requires that the user group become registered societies, these would have greater independence, and will continue even if their relations with the forest department are severed. In Gujarat, VFCs are registered as cooperative societies, which in addition to being legal entities provide functional autonomy. Once the user group has a separate legal status this can be used for several purposes. For instance, in Haryana 14 groups met together to request the Haryana Forest Department to modify the terms of the grass lease pricing and payment system. The need for autonomy and democratic process at the community level are currently not reflected in the state resolutions, but should be given careful consideration when these documents are revised.

Often literate villagers are chosen by the forest department as leaders. These are generally younger people and therefore their authority lacks legitimacy. Excessive dependence of the village communities on the forest department has frustrated village autonomy.

Another important element is the response of forest department staff to VFCs' grievances. In the initial stages, VFCs look forward to getting support from the forest department in booking offenders, negotiating with other villages/departments, etc. VFCs also need flexibility, and field staff should not 'throw rule books at them'. Many protecting villages have complained that the forest department unilaterally overrules their decisions without explaining the reasons. In some places protection is preceded by clean-ing, weeding, etc. for which the forest department pays wages to the labour (Femconsult 1995). This must change and the forest department must hand over the responsibility of handling funds to the VFCs. This would make financial dealing more transparent and foolproof against the misuse of funds. VFCs will also develop a greater sense of accountability in the process.

For instance, in Gujarat the committees raided houses and confiscated illegally poached wood. When they wanted to conduct an auction of the seized material, not only did the forest department not permit them to utilize the money so received, but it objected to the site of the auction too, pointing to the rules that an auction must be conducted at the range forest depot only.

With the exception of clauses in the National and West Bengal resolutions, most state guidelines do not address the long-term rights of participating communities. Clear tenure security enhances the authority of community management groups to carry out protection activities, especially when under pressure from neighbouring villages and private interest groups. It is necessary that the time-frame for such agreements is clear, as well as the basis for extensions. It may be appropriate for the time period of the agreement to correspond to the production cycle (rotation) of the primary products.

11.6.7 JFM and the poorest

JFM has often failed to give fair attention to the poorest forest-dependent communities, such as artisans, headloaders and podu (shifting) cultivators. Although the forest department does not evict podu cultivators, it includes old podu lands within the scope of JFM, which increases the fears of the tribals that after wage-earning programmes are withdrawn, their food security would be jeopardized (Oxfam 1998).

Not all social groups are hit equally by the decision to protect and keep livestock out of the proposed area to be protected in JFM. For low income rural families to participate, it is important that benefits start flowing as early as possible, either in the form of gatherable biomass or new opportunities of employment. This will require policy shifts and afforestation programmes on other categories of land—farmlands, village lands and forest department lands remote from village habitation. Otherwise, mere protection of a degraded area may simply transfer biotic pressure to somewhere else. Therefore production of biomass through quick-growing shrubs, bushes and grasses must be undertaken on more remote lands, so that peoples' demands are met in a sustained manner from these bushes and shrubs, while people protect forestlands in anticipation of more valuable NTFPs and forest products. The issue of how to meet the economic needs of the people for the first few years, during which they have to reduce their dependence on the protected land, must be faced squarely. It is impractical to expect that people will give up grazing or reduce their consumption in 'the national interest' without expecting any tangible gains in return. Simultaneous development of all categories of land in the same region will also provide short-term benefits to the most needy.

One practical implication of including programmes on different types of land in the same project would be that these would have to be implemented in the same range or group of villages, rather than being implemented, as happens currently, in different ranges, which loses complementarity.

Coordinating JFM with other departments

Afforestation may be the first priority of the forest departments, but communities needs may be drinking water, irrigation or waged employment during the slack months. JFM projects should therefore either include these as entry point programmes, or coordinate the JFM-related efforts with the activities of other departments, such as irrigation, animal husbandry and cottage industries. It may therefore be desirable if such activities are taken up in the same area as the JFM for better results and multidimensional development of these villages. This will improve relations between people and the forest staff.

Need to develop strategies for conflict resolution

Capitalizing on this potential of joint management on a large scale will require significant shifts in investments and strategy, and some of the constraints have already been discussed above. Where regenerating forests are already beginning to increase in value, conflicts will increase between contending resource users such as adjoining villagers, migrating herders, or more distant and periodic forest users. As a more lucrative range of non-timber products begins to mature, and the sharing of timber harvests becomes regularized, questions of equity and the distribution of benefits will create new management challenges and conflict resolution skills. Strategies to deal with these problems are yet to be evolved.

11.7 CHANGES IN SILVICULTURE

Some conservative field officials understand JFM as an arrangement in which wages are paid in kind (100% of NTFPs and a share of final harvest) in place of cash. Others define it as a new management regime in which protection is to be carried out by the people and technology is to be controlled by the department. These narrow perspectives assume that the objectives of forest management need not be redefined, and could continue as before to be timber orientated. However, with a new Forest Policy in favour of local needs and usufructs, silvicultural practices and management options should also be radically altered to meet these new objectives. Foresters will have to accept a reduction in yield of timber from stem or bole of trees, and settle for a diverse menu of biomass-based products (Campbell *et al.* 1996).

For instance, local people often prefer the production of grasses to wood. In the case of a pastoral tribe, Bashir Khan was persuaded to reduce his stock in order to allow regeneration on the forest patch allotted to him in the alpine pastures (Rizvi 1994). He found that although tree density increased due to control of grazing, the output of natural grasses and carissa bush, which he used to feed to goats and sheep, had gone down. He wanted the coupe to be thinned in order to get more grasses, but unfortunately the forest department is not geared to the silvopastoral system of using a forest compartment to produce the kind of biomass which is useful to the herdsmen. Thus there is a danger of his becoming alienated from the department and reverting to the old unsustainable practice of uncontrolled grazing. The onus is now on the forest department to shift to a new silvicultural practice of maximizing biomass and NTFPs rather than timber.

Similarly, the forest department's present management of sal seems to be for timber, and hence only one shoot is allowed to grow. Because sal is an excellent coppicer, degraded forests and hills close to a village should be managed under a coppice or a coppice with standards system for fuelwood and sal leaves, as is being done in South-west Bengal, where JFM has been highly successful.

As forests have been traditionally looked upon as a source of revenue, the entire thrust of forestry has been towards the high forest system which calls for clearfelling and ruthless cutting back of all growth, except of the species chosen for dominance. This has the major defect of creating a bias in favour of coppice origin plantations which, in the long run, are more amenable to biotic and climatic factors, and second, it results in the removal of all the material which could serve gathering needs. The high forest system, which neglects the understorey so vital for the prevention of run-off, has resulted in pure forests being created, but with gathering falling a casualty in the process. It is in this context that a major policy change is required.

While some distant forests may continue to produce high value timber as one but not the only output (provided these could be saved from smugglers), most forest department lands should be used for mixtures and multiple use with timber a byproduct. Silvicultural practices should change from 'conversion to uniform' to 'improvement felling and protection'. Only over-mature, malformed, dead or dying trees should be removed, with no particular reservation by species. Ground flora and the understorey should be largely left undisturbed, except for the improvement of hygiene of the forest flora through the removal of noxious weeds (Buch 1992). Canopy manipulation, tending, and thinning, etc. should be so adjusted as to optimize gatherable produce. The crop would be representative of all age groups because no attempt would be made to achieve a uniform crop in terms of variety or age. In those areas where teak and sal are the naturally dominant species, they would continue to predominate even without silvicultural intervention to achieve a uniform crop. However, because of age and species mix the forests would be able to maintain a continuous supply of miscellaneous small timber and fuelwood for use in gathering. Commercial working would taper off because clearfelling by blocks would be totally abandoned, but there would be some production of timber from the overmature trees that would be felled.

From the people's point of view, crown-based trees are important for usufruct, but forests still remain largely stem based. Timber is a product of the dead tree, whereas NTFPs come from living trees allowing the stem to perform its various environmental functions. Moreover, gathering is more labour intensive than mechanized clearfelling. Local people living in the forests possess the necessary knowledge and skills for sustainable harvesting. Lastly, NTFPs generate recurrent and seasonal as opposed to one-time incomes, making its extraction more attractive to the poor. Thus if access to NTFPs can be assured, standing trees can generate more income and employment than the same areas cleared for timber, and also maintain the land's natural biodiversity.

Norms for silvicultural practices were developed in times prior to the current scenario of high human and cattle pressures, and must now be adjusted accordingly. If the national objectives have changed to prioritize people's needs there must be an accompanying change in silvicultural practices and technology. Besides, forests should have more of such produce as are useful to farmers, such as trees producing fodder, green manure and wood for agricultural implements and fuel (Nadkarni *et al.* 1989; Nadkarni & Pasha 1991).

To ensure that growing space is maximized it is essential that all levels of forest architecture is utilized. This includes shade-tolerant shrubs and herb layers; introduction of herbaceous medicinal plants; management of forest floor to enrich soil and encourage natural regeneration and the production of natural tubers (Campbell *et al.* 1996). Often advance closure and deferring the planting activity for 2–3 years on barren lands, while the closed area is treated with soil and moisture conservation inputs, allows the regeneration of grasses and root stock. Similarly, plant manipulation methods such as pruning, lopping, pollarding, ratooning and weeding can all be used to increasing the production of gatherable NTFPs on a periodic basis while reducing and delaying the production of timber. However, such changes would require budget to be made available for innovative silvicultural practices.

Thus, as the objectives of managing forestlands change, so should the technology, nature of species, spacing, etc. Keeping technology and silviculture the same as before 1988 will amount to sabotaging the changed objectives, as it is technology, besides power of organization, which would influence who will benefit and who will lose. Although after the advent of the new forest policy in 1988 there has been some effort to involve forest communities in management, little thought has been given to make necessary changes in the technology which will be suitable to achieve the changed objectives. Multiple objectives to maximize outputs from many products will require innovative and experimental silviculture, which must focus more on the management of shrub and herb layers, and on forest floor management to enrich the soil and encourage natural regeneration. Unfortunately there is not much evidence that state governments have appreciated the need for change in silvicultural practices as a tool to promote JFM.

One must make a distinction between the objectives of forest management and the means through which these objectives are to be realized. The objectives are set out in the 1988 Forest Policy, whereas people's participation is a means, and not necessarily an essential means, to achieve the defined ends. It is important to keep this distinction in mind, especially for distant and well-stocked reserve forests, the management of which may continue with the forest department, but the objectives should change as spelled out in the 1988 Policy.

Increased production of NTFPs must, however, be accompanied with greater discipline in its use, as new opportunities for livelihood promotion may also lead to serious threats of unsustainable and irresponsible NTFP harvesting. Such restraint is almost impossible to achieve without consultation with the people. For instance, the widespread shift to use of forest sweepings* to meet domestic fuel needs has a negative effect on regeneration and nutrient recycling essential for maintaining soil productivity. When this issue of ecological effect of sweeping leaves from the forest floor was discussed with the VFC members in West Bengal, they candidly admitted to the adverse effect, but requested for alternative energy devices (Campbell *et al.* 1996). This would need provision by the project of solar cookers and gas plants based on cowdung which do not require cash inputs to run them. The challenge for the forest department is to devise policies that strike the correct balance between livelihoods of collectors and sustainability of NTFP harvesting.

Silvicultural practices for bamboo

As an example of what technology can do to both raise productivity and increase incomes to the poor, the case of bamboo is discussed below.

Bamboo is an important ground crop in central and eastern India. Yet, the productivity and quality of the bamboo has been far below its potential due to the dense build-up of dead leaves and other organic material. The abundance of litter within the clump has suppressed the growth of new shoots and poses additional fire hazards during the dry season. If the stands were routinely cleaned and thinned, the danger of fire would be

---

* This practice is more common in eastern India where leaves are an important source of fuel, especially for parboiling rice.

reduced, productivity would increase several fold, and a regular flow of bamboo stands will be ensured to the bamboo artisans. However, budgets for cleaning bamboo clumps and introduction of its protection by the bamboo artisans must be simultaneous, as bamboo is a highly browsable crop.

Thus traditionally the department's bamboo-harvesting policy systematically maximizes dry bamboo output for paper mills rather than green bamboo output for artisans. Whether by design,* or because of lack of funds, the present practice of leaving bamboo clumps unworked helps industry, reduces productivity and is antiartisan. In fact, if bamboo forests are carefully worked and green bamboos regularly harvested, total bamboo output of an average clump would increase. Artisans who are living close to forests can be involved in the management of bamboo forests, so that they extract bamboo themselves without damaging the clump.

The artisans require young and green bamboo, which is not produced by the forest department, in fact the present silvicultural practices ban felling of green bamboo. At least in some coupes this restriction needs to be relaxed. If tendu (*Diospyros melanoxylon*; used as wrappers of tobacco to produce *bidi*, Indian cigarettes) can be maintained only at the bush level and not allowed to grow into a mature tree because of revenue interests of the state, similar concessions should be available to the poor artisans.

Issues of technology are interlinked with issues of management. Along with changing the technology that will maximize the production of green bamboo, one would also have to streamline the procedure for making this available directly to

---

* There is a practical reason why production of green bamboo desired by artisans is not preferred by the forest department. Harvesting green bamboos from the outer rings of the clumps would leave completely unprotected new shoots sprouting on the inside which are a delicacy for man and beast alike. Since protecting bamboo clumps is well nigh impossible, survival rates from the tender shoots would be negligible once green bamboos from the outer ring are harvested. However, this can be avoided if bamboos from the interior of a clump are removed.

the artisans, give them better designs, and link their products with up-markets.

11.7.1 Regeneration versus planting on degraded lands

The success of JFM is generally seen as proving the superiority of natural regeneration over plantation as a technique for improving productivity and biodiversity in forests. While natural regeneration is cheaper, there are several circumstances where planting cannot be avoided. Three such situations could be: (i) creating a fuelwood reserve before beginning protection by the community; (ii) planting on lands incapable of regeneration; and (iii) where indigenous species that emerge as a result of protection are not considered of value by the community. These are discussed below.

Mere protection of a not-so-degraded area may transfer human and cattle pressure to some other area, as people have to meet their daily requirement of fuelwood somehow. Therefore production of biomass through quick-growing trees, shrubs, bushes and grasses must be undertaken on degraded and barren lands before the beginning of community protection, so that peoples' demands are met in a sustained manner from these bushes and shrubs, while people protect forestlands in anticipation of more valuable NTFPs and forest products. The issue of how to meet the economic needs of the people for the first few years, during which they have to reduce their dependence on the protected land, must be faced squarely. Although the success of many JFM experiments is generally credited to leadership or people's efforts, it is seen that in almost all such cases there was an alternative source of fuel available to them. In South-west Bengal, the task of peoples' protection of degraded forestlands became easier because the farm forestry programme in that area had been highly successful, increasing fuelwood supplies and incomes even for the poor. In Eklingpura, Udaipur, where community protection has been highly successful, plenty of prosopis shrubs in and around the village provide fuelwood to everyone almost at zero opportunity cost. On the other hand, in another village of the same district, Shyampura, which had no prosopis in its vicinity, a local

NGO was struggling to promote protection, but was finding it difficult to prevent unauthorized removals from the area (Choudhury 1993). These examples illustrate the importance of creating a fuelwood reserve before expecting people to start protection.

The other situation warranting planting is where land is so degraded that regeneration is slow, or root stock is absent. There may be other barriers to natural recovery such as the presence of weeds, unfavourable soil and climatic factors, a low presence of fertile trees and a lack of symbiotic microbial associations necessary for seedling establishment (Perera *et al.* 1995). In extreme situations where soil erosion has reached conditions characterized by gullies and ravines with little or no vegetation, natural regeneration alone may have very little or no impact on improving vegetative cover. In such a situation, there may not be sufficient incentive for the people to give their time and labour for protection in lieu of the intermediate and final products, which may be available after an inordinate amount of waiting. Intensive soil working is required in such cases. Natural regeneration can also be enhanced or accelerated by soil and water conservation measures like contour trenching, vegetative bunding and small check dams. In denuded areas where severe overexploitation has reduced possibilities for rapid natural regeneration, nurseries and plantations will be needed to provide employment, and so would be turned into fodder and fuelwood in the quickest possible time. The whole idea is to consider how a continuous flow of forest products can be ensured to the communities.

The third situation where planting may be necessary is where, due to protection, species which emerge do not coppice well. As peoples' demands cannot be curbed for a long period, some amount of harvesting becomes unavoidable after a few years of patient waiting. In cases where the species do not coppice well, harvesting leads to a non-sustainable situation, and land may become denuded again. A similar situation will be when the root stock is already quite degraded, and species likely to emerge are not valuable in the perception of the people, and therefore they may be reluctant to contribute their labour for protection. In such cases the strategy of natural regeneration alone may not be enough to enforce the necessary discipline. If plantations of species desired by the people are established, the perception of the tract's value may increase, and everyone may cooperate. Enrichment planting could also increase the supply of raw material for the local craft- or artisan-based activity. An active strategy of forest restoration through management is likely to be more successful than simply abandoning lands and hoping that the regenerating forest will survive the numerous threats to it. Such a strategy would include accelerating growth by simple silvicultural methods, such as trenching, ploughing, weeding and thinning. However planting should be done in such a way that existing rootstocks or advanced regeneration are encouraged to grow as part of the mixed stand.

Choice of species

Having thus established the need for artificial planting in many situations, the question arises 'which species should be given priority'? We suggest that even on barren forestlands, where plantations are required, one should consult the people and their choice may often be in favour of planting of usufruct-based trees. These should be supplemented with grasses, legumes, shrubs and bushes to yield fuelwood and fodder in the shortest possible time. An immediate identification of quick-growing shrubs with high calorific value, with their retention in the forest to serve fuel requirements, the development of pastures, and the development of massive fuelwood plantations around centres of high consumption and encouragement of silviculturally sensible exploitation of fuelwood species would also be important components of the new policy. This would strengthen tribals' access to forests, and therefore benefits would be directly appropriated by them.

Foresters and foreign experts who advise the government and the donor agencies, because of their training and experience, have looked upon trees as timber to be obtained after felling. Therefore, even in the social forestry programmes market-orientated species were planted. The traditional Indian way of looking at trees has, however, been different. As opposed to trees for timber, for centuries Indian villagers have depended on trees for their sustenance. There has been little felling. Instead, trees have been valued for the intermediate products they provide, which sustain and secure the livelihoods of the people. The rich fell trees; the poor pollard them!

Given the inefficiency of administration and 'soft' character of the political system, one could generalize that out of a tree on public lands the stem goes to the rich and the towns, whereas branches, leaves and twigs belong to the poor. Therefore the strategy should be to opt for species which have high proportions of branches and twigs relative to stem wood.

The difference can be understood by comparing how fuelwood species are viewed in the two perspectives. As per received wisdom, fuelwood is obtained by felling trees having a high calorific value, or as a byproduct from lops and tops of timber trees. Casuarina and eucalyptus therefore seem perfectly justified on public lands. But the poor tribals obtain fuelwood from twigs and branches of living trees, and not by felling trees, and often get little from the felling of so-called fuelwood trees. Casuarina and eucalyptus may be justified on farmlands, if they improve farm incomes on a sustainable basis. But these hardly serve the poor, when raised on public lands. To sum up, several changes are needed in forest management which are summarized below (Table 11.3).

11.8 SUMMING UP

The last two sections referred to some of the shortcomings of government policy in the way JFM is being implemented in the field. A short summary of recommendations is given in Table 11.4.

Despite evidence from several states highlighting the problems discussed in this paper, state governments have hardly taken any ameliorative action in removing the constraints of policy, and initiating measures on the lines listed in Table 11.4. Such indifference could be because of many reasons. First, state governments treat JFM as another programme, which they think can be implemented without making any changes in other sectoral programmes. JFM, however, requires a paradigm shift and will be successful only when radical changes are introduced in rights and privileges over forests, policies and laws pertaining to NTFPs, Working Plans, and silvicultural arrangements, etc. Second, field officials are willing to entrust protection to the communities, but hesitate in involving them in management and control of government forests, thus reducing JFM to 'I manage, you participate'. Unless serious efforts are made to trust the

Table 11.3 Technical and policy options in forest management.

	Old practices which may still be continuing	New practices mandated by the new forest policy, but not properly understood or introduced
Objective	Reduce people's dependence on forest lands	Increase supply of goods desired by people
'Look' of the forest	Stem based	Crown based
Species	Exotics and commercial	Usufruct giving trees, grasses, bushes, shrubs and NTFPs
Client	Market and industry	Forest dwellers and local people
Main product	One-time timber from felled, i.e. dead, trees	Recurrent non-timber produce from living trees with timber as byproduct
Silviculture	Monoculture and conversion to uniform conditions	Polyculture and improvement felling and protection
Production through	Planting	Mainly regeneration, supplemented by planting
Usage through	Harvesting	Gathering

Table 11.4 Analysis of goverment orders and suggested changes. (Saxena 1997.)

Existing	Suggested changes
No thinking on tenurial issues	Protecting communities should have clearly defined property rights over forests in comparison to distant villages
VFC is a creation of forest department	Should be an independent and spontaneous entity
VFCs kept independent of panchayats	VFCs as a parallel organization to panchayats may not be sustainable, hence linkages are necessary
VFC has been given mainly protection role, for which wages are delayed and in kind, control and authority is with FD	VFC should manage and control all natural resources within their domain. There should be genuine partnership
NTFP markets distorted by government regulations	Markets should be freed from overcontrol
No thinking on changing silviculture	New silvicultural practices required
Focus is only on forests	Focus should also be on sustained benefits to the people, and on empowerment
No deviation from norms, membership determined by government order	Flexibility and decentralization should leave many decisions to the judgement of the village
No specific provision about the short-term suffering of disadvantaged groups, such as women and headloaders	Sensitivity on gender and poverty issues is needed

communities with control functions, peoples' efforts in protection may not be sustained for long. Third, government resolutions tend to over-prescribe what communities may or may not do, leaving little flexibility for them to adjust to the local situation. Fourth, the main support for JFM has come so far from environmentalists, academicians, NGOs, and the Ford Foundation in India. While their support is crucial in documenting the dynamics of community behaviour in different ecological conditions and the emergence of policy issues, the hold of this class of people on instruments of policy formulation is rather weak. In addition to forest bureaucracy, which is often indifferent to the idea of empowering the people, politicians too have not put JFM high on their agenda. They see greater political advantage in espousing schemes which bring individual benefits.

To sum up, though internal factors behind the success of collective action such as cohesiveness and size of the community are no doubt important, often these are like 'givens' of a situation, not amenable to outside intervention, at least in the short run. On the other hand, external factors of government policies and the way government

functionaries interact with the local communities may have greater influence on why communal action is sustained over time, because these variables are in the realm of public policy, and are of greater practical significance in influencing collective behaviour. A good policy process also requires careful planning and skilled staff, together with mechanisms to learn from mistakes.

Despite these problems, exciting beginnings have been made in a number of states. Local VFCs are proliferating, some spontaneously and others with the encouragement and assistance of forest department field staff and NGOs. Old attitudes are changing. After over a century of unilateral custodial control over forests, the forest service is gradually shifting to a collaborative form of management (Poffenberger & McGean 1996). NGOs and forest departments, once distrustful of one another, are now working cooperatively in a number of regions. Badly abused forests are making a come-back, new stems are sprouting out of ancient stumps. Villagers who have depended on forests for generations are now participating legally in their protection and management, in partnership with their old adversaries, the forest departments. The potential returns from such efforts are immense because natural regeneration is a low-cost option when compared with that of longer gestation plantations. Collecting NTFPs generates a lot of self-employment and potentially reduces conflicts between forest departments and rural communities. Much, however, is still to be learnt before it is known that the emerging patterns are sustainable.

A development concept faces very different constraints and opportunities when it is new, unproven and unaccepted, compared with when it is long established and widely accepted—and the role of those who are in charge of its promotion must vary accordingly (Dove 1992). For example, much of the effort of the 'sympathizers' of JFM to date has concentrated on promoting the principles of JFM to the government, NGOs, and local communities. Such promotion may be valuable in the early phases of a programme, but there are potential problems in sustaining it for too long. The nature of promotion results in too much emphasis on the positive aspects of the programme and too little critical analysis. At the outset it is important to be able to persuade key actors of the merits of JFM, but it eventually becomes important to temper this with critical appraisal, long-term strategies, and the building of capacity to implement such policies. Care should be taken to ensure that JFM does not just become the next development bandwagon.

REFERENCES

Agarwal, A. (1987) Between need and greed—the wasting of India; the greening of India. In: Agarwal, A., D'Monte, D. & Samarth, U., eds. *The Fight For Survival.* Centre for Science & Environment, New Delhi.

Agarwal, A. & Narain, S. (1990) *Strategies for the Involvement of the Landless and Women in Afforestation: Five Case Studies from India—A Technical Cooperation Report.* World Employment Programme, International Labour Office, Geneva.

Anderson, R. & Huber, W. (1988) *The Hour of the Fox: Tropical Forests, the World Bank and Indigenous People in Central India*, pp.169–95. Vistaar Publications, New Delhi.

Arnold, J.E.M. & Stewart, W.C. (1991) *Common Property Resource Management.* Tropical Forestry Paper no. 24. Oxford Forestry Institute, University of Oxford, Oxford.

Attwood, D.W. (1988) Social and political preconditions for successful cooperatives. The cooperative sugar factories of Western India. In: Attwood, D.W. & Baviskar, B.S. eds. *Who Shares? Cooperatives and Rural Development*, pp. 69–90. Oxford University Press, New Delhi.

Bandyopadhyay, J., Shira, Vandana & Sharatchandra, H.C. (1983) The challenge of social forestry. In: Walter, F. & Sharad, K., eds. *Towards a New Policy*, pp. 48–72. Indian Social Institute, New Delhi.

Bardhan, P. (1993) Managing the village commons. *Journal of Economic Perspectives* 7, 87–92.

Barnes, D.F. & Olivares, J. (1988) *Sustainable Resource Management in Agriculture and Rural Development Projects; Policies, Procedures and Results.* Environment Department Working Paper no. 5, Policy Planning and Research Staff, World Bank, June. Washington D.C.

Britt-Kapoor, C. (1994) *A Tale of Two Committees: Village Perspectives on Local Institutions, Forest Management and Resource Use in Two Central Himalayan Indian Villages.* ODI Rural Development Forestry Network Paper 17a. ODI, London.

Buch, M.N. (1992) *Forests of Madhya Pradesh*, Government of Madhya Pradesh Bhopal.

Buchy, M. (1996) *Teak and Arecanut: Colonial State, Forest and People in the Western Ghats 1800–1947.* Indira Gandhi National Centre for the Arts, Pondicherry.

Campbell, J.Y., Rathore, B.M.S. & Branney, P. (1996) The New Silviculture—India and Nepal. In: Hobley, M., ed. *Participatory Forestry: The Process of Change in India and Nepal*, pp. 175–210. ODI, London.

Chambers, R., Saxena, N.C. & Tushaar Shah (1989) *To the Hands of the Poor: Water and Tree.* Oxford and IBH, New Delhi/Intermediate Technology, London.

Chatterjee Mitali (1996) *Women in Joint Forest Management: A Case Study from West Bengal*, Technical Paper—4, IBRAD, Calcutta.

Choudhury, S.D. (1993) A challenging task in Shyampura. *Wastelands News*, **8**, 67–8 February–April.

Commander, S. (1986) Managing Indian forests: a case for the reform of property rights. *Development Policy Review*, **4**. London.

CSE (1982) *The State of India's Environment 1982: The First Citizen's Report.* Centre for Science and Environment, New Delhi.

CSE (1985) *The State of India's Environment 1984–85: The Second Citizen's Report.* Centre for Science and Environment, New Delhi.

Dove, M.R. (1992) *Joint forest management in India.* Report to Ford Foundation, New Delhi.

Eckholm, E. (1979) *Planting for the Future: Forestry for Human Needs.* Worldwatch Institute, Washington.

Femconsult (1995) *Study of the incentives for joint forest management, May 1995.* Report to the World Bank, New Delhi.

FSI (1998) *The State of Forest Report.* Forest Survey of India, Dehradun.

Gadgil, M. (1989) Deforestation. Problems and prospects. *Indian Journal of Public Administration* **7**, 752–801.

Gadgil, M. & Guha, R. (1992) *Ecological History of India.* OUP, New Delhi.

Gaikwad, V. (1981) Community development in India. In: Dore, R. & Mars, Z., eds, *Community Development: Comparative Case Studies in India, the Republics of Korea. Mexico and Tanzania*, pp. 245–331. Croom-Helm, London.

GOI (1976) Report of the National Commission on Agriculture, Part IX, Forestry, Department of Agriculture and Cooperation, Government of India, New Delhi.

GOI (1981) Report on Development of Tribal Areas, National Committee on the Development of Backward Areas, Planning Commission, Government of India, New Delhi.

Guha, R. (1994) Forestry debate and draft forest act: who wins, who loses? *Economic and Political Weekly*, 20 August **xxix** (3), 2192–6.

Hiremath, S.R., Kanwalli, S. & Kulkarni, S. (1994). *All About Draft Forest Bill and Forest Land.* Samaj Parivartan Samudaya, Dharwad.

Hobley, M. (1996) *Participatory Forestry: The Process of Change in India and Nepal.* Overseas Development Institute, London.

Locke, C. (1995) *Planning for the participation of vulnerable groups in communal management of forest resources: the case of the Western Ghats Forestry Project.* PhD thesis, Centre of Development Studies, University of Wales, submitted.

Lynch, O.J. (1992) *Securing Community-based Tenurial Rights in the Tropical Forests of Asia, Issues in Development.* WRI, Washington.

Nadkarni, M.V. & Pasha, S.A. (1991) Developing uncultivated lands: some issues from Karnataka's experience in social forestry. *Indian Journal of Agricultural Economics* **46**, 543–54.

Nadkarni, M.V., Pasha, S.A. & Prabhakar, L.S. (1989) *The Political Economy of Forest Use and Management.* Sage Publications, New Delhi.

Nair, C.T.S. (1985) Crisis in forest resource management. In: Bandyopadhyaya, J., Jayal, N.D., Schoettli, U. & Singh, C., eds, *India's Environment: Crisis and Responses*, pp. 7–25. Natraj Publishers, Dehradun.

NCHSE (1987) *Documentation on Forest and Rights*, Vol. 1. National Centre for Human Settlements and Environment, New Delhi.

Ostrom, E. (1990) *Governing the Commons: The Evolution of Institutions for Collective Action.* Cambridge University Press, Cambridge.

Ostrom, E. (1994) *Neither Markets nor State: Governance of Common Pool Resources in the 21st Century.* IFPRI, Washington.

Oxfam (1998) *Joint Forest Management in Andhra Pradesh.* OXFAM (India) Trust, Hyderabad.

Perera, D., Brown, N. & Burslem, D. (1995) Restoring the degraded dry zone woodlands of Sri Lanka. *Tropical Forest Update* **5** (3), 5, 8–10.

Poffenberger, M. & McGean, B., eds. (1996) *Village Voices, Forest Choices.* Oxford University Press, New Delhi.

Poffenberger, M. & Singh, C. (1993) The legal framework for Joint Forest Management of forest lands in India. In: Fox, J., ed. *Legal Frameworks for Forest Management in Asia: Case Studies of Community/State Relations.* East West Center, Honolulu.

Rizvi, S.S. (1994) Managing the forests—herdsman's way. *Wastelands News* May–July.

Roy, S.B. (1993) Forest protection committees in West Bengal, India. In: Fox, J. ed. *Legal Frameworks for Forest Management in Asia: Case Studies of Community/State Relations.* East West Center, Honolulu.

Sarin, Madhu (1998) Grassroots Initiatives vs. Official Responses: The Dilemmas Facing Community Forest Management in India. In: Victor, M., Lang, C. & Bornemeier, J., eds. *Community Forestry at a Crossroads: Reflections and Future Directions in the Development of Community Forestry*, RECOFTC, Bangkok.

Saxena, N.C. (1994) *India's Eucalyptus Craze: The God that Failed*, Sage, New Delhi.

Saxena, N.C. (1995) *Forests, People and Profit.* Natraj, Dehradun, India.

Saxena, N.C. (1997) *The Saga of Participatory Forest Management in India.* CIFOR, Indonesia.

Saxena, N.C. & Sarin, M. (1998) Western Ghats Forestry Project—a preliminary assessment. In: Jeffery, R. & Sundar, N. eds. *A New Moral Economy for Indian Forests*, pp. 181–215. Sage, New Delhi.

Shah, T. (1987) *Gains from Social Forestry: Lessons from West Bengal.* ODI Social Forestry Network Paper 5e. Overseas Development Institute, London.

Shepherd, G. (1989) *Putting Trees into the Farming Systems: Land Adjudication and Agroforestry on the Lower Slopes of Mount Kenya.* ODI Social Forestry Network Paper 8a. Overseas Development Institute, London.

Shyam Sundar, S. & Parameshwarappa, S. (1987) Forestry in India. The forester's view. *AMBIO* **26** (6).

Singh, C. (1986) *Common Property and Common Poverty: India's Forests. Forest Dwellers and the Law.* Oxford University Press, New Delhi.

Singh, Neera M. (1996) *Communities and Forest Management in Orissa.* Orissa Forest Department, Bubaneshwar.

Spurr, S.H. & Burton Barnes, V. (1980) *Forest Ecology*, 3rd edn. John Wiley, New York.

Swallow, B.M. & Bromley, D.W. (1994) Co-management or no management: the prospects for internal governance of common property regimes through dynamic contracts. *Oxford Agrarian Studies* **22**, 3–16.

VIKSAT (1995) *Research Methodologies in JFM*. A Training Report. Vikram Sarabhai Centre for Development Interaction, Ahmedabad.

Wade, R. (1985) *Common Property Resource Management in South Indian Villages*. IBRD, Washington.

Wade, R. (1987) The management of common property resources: collective action as an alternative to privatisation or state regulation. *Cambridge Journal of Economics* **11**, 95–106.

Wade, R. (1988) Why Some Indian villages cooperate. *Economic and Political Weekly*. April 16, 773–6.

12: Trees in the Urban Environment: Some Practical Considerations

KJELL NILSSON, THOMAS B. RANDRUP AND BARBARA L. M. WANDALL

12.1 INTRODUCTION

In the urban landscape, growing conditions of plants differ from those in the rural landscape. Paving and buildings characterize the city, in which wind speed is decreased, temperatures and precipitation are raised, the humidity is lowered and shading is common in many street canyons. Hence, urban growing conditions differ significantly from those in the rural landscape, and may also be difficult as a result of recreational uses, disturbed soil conditions and pollution of soil and air.

This chapter covers the more practical application of the scientific challenges related to urban forestry, as described in Chapter 13 in Volume 1 of this handbook.

12.2 DESIGN ASPECTS

Designing with respect to trees as living organisms demands a thorough knowledge about the potentials and limitations of the different species of trees for urban use. The way trees are used in urban areas often implies problems with both growth and function. Craul (1992) stresses, that in many cases, little thought is given to the interrelationship of design considerations, for example between the soil physical properties. Due to the heavy increase in pressure on urban trees and green areas from traffic and buildings, there has therefore been a great effort to develop new establishment techniques for urban planting. However, designing with trees involves knowledge of more than the establishing techniques.

According to Scully (1991) there are two diametrically opposite ways of relating to nature. One of them, in which man-made monumental structures imitate the shapes of nature, seems to have been characterizing the way most humans originally have thought of the earth–man relationship. This approach seems to have been worldwide. In modern architecture it can be found in the late work of Frank Lloyd Wright. In Greek architecture, the opposite approach was used when building. In this the man-made structures monumentally contrast with the natural shapes, and have been dominating the European tradition ever since. A modern reference would be the works of Le Corbusier.

These different design traditions have developed from ancient cultural traditions. Some of the basic principles of planning and designing are based upon distinct cultural symbolism, and are results of the evolution of ways to solve certain basic problems. In environmental design complexity is one of the great problems. A range of different theories have been developed in order to find methods of solving these problems. The different methods of working leads to different solutions to problems and, thus, to different kinds of design (Turner 1996).

Based on a simplification of people's perception of a city's structure, Lynch (1960) developed a language for describing the physical surroundings. He names these recurring structural contents of our cities as paths, edges, districts, nodes and landmarks, each of them representing different physical elements. Using these elements, Lynch argues that through an understanding of how people perceive these elements, and by designing so that they are significant, it is possible for designers to build more imaginative and psychologically satisfying cities.

Another approach to solving the complex matters of planning is Christopher Alexander's pattern language. These patterns are based on archetypal designs of solutions to recurrent problems that humans have developed through time (Alexander *et al.* 1977). The patterns are grouped into different parts of the 'pattern language', working its way through the process of planning, designing and building. These patterns are inter-related, describing first the global patterns of the large-scale structures like towns, on which individuals have no influence as such. Second, the patterns are describing individual buildings and the space between buildings—patterns that are under the control of individuals—and finally the patterns are describing the detailed construction of these. When using the pattern language the design will be based on patterns of human behaviour, developed throughout time.

Trees have not always been a part of the city. In medieval times, there were practically no trees in the cities, but the cities were often placed near to forest settings. In most cities, the parks are remnants from the land or gardens of the nobility or the rich landowners (Bradshaw *et al.* 1995). The European tradition of garden design has contributed to the overall history of garden design, with four different eras. The first was the Spanish garden, a stylistic depicted irrigation system of which the Alhambra in the city of Granada is an extraordinary example. This kind of garden design was introduced by the Moors, who ruled Spain during the years 700–1400. The Italian Renaissance garden from the 1500s can, in a very simplified way, be described as a stylistic mountain stream. One of the clearest examples of the Renaissance style is the Villa d'Este in the old city of Tivoli just outside Rome. The third era is constituted by the French baroque garden, most distinctly represented by the gardens of the castle of Versailles, which were commenced in the 1660s by the French garden designer André le Notre. The fourth contribution of European tradition to the history of garden design is the romantic English landscape garden, is represented for instance by Stourhead, in Wiltshire, England, which was established in the later half of the 1700s.

During the eighteenth and nineteenth cen-turies, the development of urban trees increased steadily. Napoleon III added to Paris its radial street pattern with the tree-lined boulevards as late as the 1850s (Miller 1989). Some time later, as a response to the needs and demands of a growing middle class, to be able to enjoy open green spaces within the city, British city developers planned the city squares (Bradshaw *et al.* 1995). Later, from the early 1900s, the concept of using trees and woodlands as a setting for urban development was adapted to be the fundamental aspect of the development of the British New Towns, of which the first generation comprises towns that were developed as early as 1903 (Letchworth) and 1908 (Welwyn Garden City) (Simson 2000).

12.3 SPECIES SELECTION FOR URBAN USES

The selection of urban trees is linked to two main problems. The first is that urban trees are common cultivars selected for 'landscape' planting. This is a problem because landscape planting is not always the same as 'urban' planting, because growing conditions can vary widely between rural and urban landscapes. Secondly, growing conditions vary between regions. This can, for instance, cause problems for trees originating in the northern or eastern parts of America when they are planted in South America.

Species selection plays an important role in the selection of cultivars for use in urban situations, to determine the main biological processes that control the growth patterns and the structure of trees at any point in their life. By focusing on both stem and root systems, the morphological features that reflect a tree species habit, form and physiological attributes can be determined. From this, diagnostic criteria for tree vitality can be established and the processes of regeneration and growth of young trees planted in urban settings can be better understood.

There is a need for an integrated focus on the identification and selection of cultivars used for urban greening. This selection must be coordinated by simultaneous testing, in different locations, as the selection of plants is always a two-part process. Site conditions must first be characterized, then matched with the requirements and tolerances of appropriate plants.

In Mexico City, 72% of the total tree population is represented by only nine species (Chacalo *et al.* 1994), and in Hong Kong two-thirds of the tree stock is dominated by 14 of 149 species (Jim 1995). The limited number of species used as urban trees often is a result of the prolonged use of cultivars; this has shown which species are the most hardy, aesthetically pleasing and easy to propagate. Also, the plants used as rootstock for these grafted cultivars are often simply those that are most available locally and grow to usable size the fastest. This, however, is not always equal to the selection of trees of good survivability.

The conservation of the genetic resources of trees and shrubs is of great importance. The immediate objective of gene resource conservation should be to secure the ability of the species to adapt to environmental change and to maintain the opportunity for future improvement work. In general, genetic resources should be conserved in conservation stands, which form a network that covers the spectrum of presumed genecological variability. The required number of stands in the network should be estimated on the basis of genecological zoning and the biology and distribution of the species.

Provenance selection (the collection of propagation material from different geographical locations) has been a mainstream idea in forestry since the beginning of the nineteenth century. Provenance selection of genetically superior tree species with a high value for use in urban and roadside planting, however, remains relatively uninvestigated. Both planted and naturally occurring trees in coastal regions that are continuously exposed to salt spray and high winds may, for instance, offer an abundance of largely untapped genetic resources.

New environmental constraints, such as de-icing salts and ozone, have made the selection of urban tree cultivars more restricted than previous selections. More than 700000 trees die annually from the damage caused by de-icing salt applied to roads and roadsides in Western Europe (Dobson 1991). Given a conservative value of US$ 35/Euro 30 per tree, the replacement costs

would be approximately US$ 25 million/Euro 21 million. Replacement costs caused by other factors, such as drought, atmospheric pollution, low soil moisture and low soil oxygen, could cost in excess of US$ 35 million/Euro 30 million.

Owing to Dutch elm disease (*Ophiostoma novo-ulmi*), one of the most important urban trees (*Ulmus* spp.) has practically disappeared from many regions of America and Europe since the 1970s. A satisfactory alternative to the elm has yet to be found for urban settings in northern Europe. Similar concern can be recorded in the south of Europe for plane trees (*Platanus* spp.), attacked by *Ceratocystis fimbriata* f. sp. *platani*, and for cypress (*Cupressus sempervirens*), affected by *Seiridium cardinale*. A satisfactory alternative to the plane tree and cypress has yet to be found in southern Europe.

The most obvious approach to environmental degradation is to control it. Santamour (1990) describes the use of a model for urban plantings, to secure species diversity and to guard against large-scale insect or disease pests. For maximum protection against such pests it is propounded that the urban forest should contain no more than 10% of any single tree species, no more than 20% of species in any tree genus and no more than 30% of species in any tree family.

In general, many attempts to grow trees under most inhospitable conditions are carried out. Therefore, planners tend to choose potentially large trees, because of the constricted growing conditions. The assumption seems to be that, by choosing a potentially large tree for a difficult location, planting will be more successful even if the tree only grows for a limited number of years, whereas a small tree in a difficult location might not grow at all. The wrong choice of species though, placed in an inappropriate location, has to do with a lack of planning, rather than with species selection.

12.4 ESTABLISHMENT TECHNIQUES FOR URBAN TREES

Planting of urban trees is often carried out with little attention to the character and quality of the growing conditions beneath the surface. The negative influences of the urban environment may be minimized by new establishment techniques and improved planning.

Because of the harsh growing conditions, the development in establishment techniques is of great importance, as the success of establishment is dependent on the development of the new root system (Craul 1992; Watson & Himelick 1997).

12.4.1 Planting holes

Along streets, the average volume of a planting hole in Denmark is 3.4 m^3 for planting trees with a diameter at breast height (d.b.h.) of 4.5–6 cm. Moll (1989) showed that the average planting hole in an American downtown area is 2.7 m^3. An increase in soil volumes will increase the growth rate (e.g. Urban 1989; Bradshaw *et al.* 1995; Watson & Himelick 1997). Even though the average planting hole has become larger in recent years, there is still a long way to go if optimum growing space is to be achieved in urban situations.

Lindsey and Bassuk (1992) showed that the daily water requirements of an urban tree could be calculated by estimating evapotranspiration and relating it to soil type and local climatic factors. With a 10-day water supply interval, the amount of soil needed for a single tree varied from 3 to 4.3 m^3 in different British locations, and from 5.9 to 14.2 m^3 in different North American locations. This gives an estimated soil volume/crown projection ratio (m^3/m^2) that varies from 0.10 to 0.15 in Britain and from 0.21 to 0.50 in the USA.

In the urban environment, a stable site is often required for traffic and buildings. Paved areas are therefore heavily compacted or sealed with concrete, asphalt, etc. A new technique has been introduced in order to obtain both stability and to create rooting space, in one and the same location (Garborsky & Bassuk 1996; Kristoffersen 1998). Uniformly graded gravel or stones are mixed with top soil. The stone/soil mix is compacted to optimum densities, leaving uncompacted soil in the voids between the stones. This mixture has been successfully tested in several countries in northern Europe, as well as in the USA. Another variation of an artificial soil mix, allowing both site stability and root growth, is the Amsterdam Tree Soil (ATS). This soil mixture consists of medium coarse sand containing 4–5% (w/w)

organic matter and between 2% and 4% (w/w) clay. It is made by mixing soils rich in well-decomposed organic matter and low in clay content with medium coarse sand using an industrial mixing device. The ATS is compacted until the soil has a penetration resistance between 1.5 and 2 MPa due to the known optimum for root growth—see this volume, Chapters 5 and 7. This is checked during compaction using a penetrometer (Couenberg 1994). The stone/soil mixes and ATS can be used under pavements that normally bear light loads, especially pavements and bicycle roads.

The general increase in planting hole size reflects the common approach of today. It is not so much a question of whether we are going to have urban greening, but how we can achieve the best urban greening. Apart from the selection of new and better plant material, new and better establishment techniques are continuously needed.

12.4.2 Securing adequate tree-water supply

One of the major causes of the death of newly planted trees is drought (Bradshaw *et al.* 1995). If a tree suffers drought stress on transplanting, it becomes trapped in a vicious circle: it is unable to develop the root system with which it would be able to absorb the water that would relieve the stress (Walmsley *et al.* 1991).

In overcoming problems of tree-water supply, the main difficulties are planting trees with sufficiently large root systems to increase the soil volume available for supplying water, and providing the necessary irrigation to restore soil water content before it reaches dangerously low levels. In many countries, lack of water may not be the main problem, but watering in the establishment period must be regarded as being vital.

Urban plantings can be irrigated in a number of ways, for example by forming a basin, using a furrow or sprinkler, to mention a few. The method used will depend on the type of planting; the amount, quality, and source of water; the terrain; available funding, and sources of labour. Regardless of the methods chosen, the main delivery system to individual plants or beds should be planned and installed together with the other utilities at the site, before paving and planting takes place.

For street trees, drainage is just as important as irrigation. A poorly drained soil is most often caused by a compacted soil, as is the case in many urban areas (Craul 1992). If the downward movement of water is impeded by compacted soil, rock or another hardpan, an internal drainage system may provide the only suitable rooting zone (Harris 1992). The desirable depth and spacing of drain lines depend on the soil texture and structure, the depth of impervious soil or rock, the depth of drainage necessary for the plants and the amount of water from rain, irrigation, seepage or ground water. The drain line should be located above any impervious layer. It is in general very difficult to relieve an impervious layer such as a compacted soil, but excess water can be drained by deep tillage operations as described by Rolf (1991). In an urban planting hole, the base and the sides should be sufficiently loosened to allow excess water to percolate away. Where drainage is likely to be a permanent problem, a proper drainage system must be installed for each planting hole (Bradshaw *et al.* 1995).

12.4.3 Staking

Trees growing in natural situations do not require artificial support. However, this may not be the case in many urban situations. Larger trees are needed to provide an instant visual impact and to deter their removal by vandals. Bradshaw *et al.* (1995) describes the problems caused by supporting trees as often outnumbering the benefits. Harris (1992) explains this dilemma by stating that the extent of staking for trees will depend on tree strength and conformation, expected wind conditions, the amount of vehicular and foot traffic, the type of landscape planting and the level of follow-up maintenance. Harris (1992) lists several consequences of staking which should be taken into consideration when deciding whether or not to stake: a staked tree will grow taller, grow less in trunk thickness near the ground but more near the top support tie, produce a decreased or even reversed trunk taper, develop a smaller root system, be more subject to rubbing and girdling from stakes and ties and may be unable to stay upright when untied.

There are a number of different ways of staking a tree. Protective staking is primarily used to keep mowing equipment, vehicles and vandals away from the trees. Support staking is needed for trees whose trunks are not strong enough to stand without support or to return upright after a wind. Bradshaw *et al.* (1995) state that support should be minimized because of the costs, the need for the trees to sway and for limiting the need for the maintenance required to prevent damage to the stem. Whenever support is required this should be kept as low as possible on the stem. No staking is necessary for most shrubs and for many conifers and other trees with limbs close to the ground.

12.4.4 Protection during construction

Root injuries during construction often causes the death of trees. Protecting the roots located under the tree branches will give the tree a reasonable chance of survival, though root loss could still be substantial. Major parts of the water- and nutrient-absorbing root system, which proliferates near the soil surface, may be removed when excavating soil around trees. Protecting a larger area will give the tree a better chance. Coder (1995) emphasizes that trees standing within 20 m of a building need special protection during the construction period.

If digging near the base of a tree, for example for a utility trench, nearly half the roots will be cut off. This will both severely damage the tree's ability to support the canopy, and also destabilize the tree. Trunk damage during construction may be avoided by using wooden boards which are strapped to the trunk. The boards should be separated from the trunk by rubber bands, thick rope or drain pipes, which have been wrapped around the trunk previously. Damage from cars and lawn mowers is a much too common problem; protection guards may be used. To avoid lawn mower damage, a lawn-free area around the trunk is the best 'protection'. Also, it benefits root establishment not to be competing with the lawn for water, nutrients, etc.

12.5 CARE AND MAINTENANCE OF URBAN TREES AND FOREST STANDS

There is a need for a range of care and maintenance practices related to the many urban tree species, situations and purposes. Many practices are used around the world. They vary according to geographical region and are dictated by local and national traditions.

In the care and maintenance of trees in the urban environment, special attention should be paid to the detection of tree vitality and hazardous trees. This is related to the compartmentalization of tree wounds (wood biological methods, computer tomography), the influence of the time of wounding on wound reactions, decay (pathology), wound treatment, wound dressings, root damage, protection, development and pruning.

12.5.1 Pruning

There are a number of reasons for pruning urban trees: training of young plants, maintenance of health and appearance, reduction of a hazard, control of plant size, influence on flowering, fruiting and vigour, and compensation for root loss. Pruning as part of the training of young trees can ensure structurally strong trees, which will be safer and require less corrective pruning when mature (Harris 1992). Pruning trees in the nursery, to select a healthy leader and form a well-structured crown, is undoubtedly of importance when trees are to be planted in urban areas. The overall aim must be to produce stems that are structurally sound and that bear well-supported branches (Bradshaw *et al.* 1995).

Most tree species have evolved for life as part of a stand in a forest habitat. Although selection has been carried out for the purpose of urban greening, the growth patterns of most trees remain similar to those of their wild ancestors. If left unpruned, few species would remain with a single stem as the central leader and a well-developed, evenly balanced crown. Therefore, the pruning schedule started at the nursery must be maintained after a tree has been planted. The restoration of the natural root:crown ratio that existed before transplanting may be enhanced by reduction of the size of the crown at planting. However, this problem may have been overcome in recent years by proper pruning in the nursery. Proper thinning of the crown may also reduce wind resistance, which can create deformities or even uproot a tree. Dieback is often seen in newly established urban trees and can be regarded as a self-pruning

mechanism on the part of the tree itself, to restore the balance between the shoot and the root systems.

Crown lifting is almost always required, even within a few years of planting. This increases the clearance between the lowest branches of the tree and the ground, leaving room for vehicles and making vandalism of the canopy less likely. Pruning can also reduce shade and interference with utility lines and can prevent obstruction of the view and traffic.

Pruning and thinning techniques and schedules that have been developed over the last 30 years now appear well established and used worldwide. A further understanding of tree growth and natural tree response to pruning has been presented by Shigo (1991) and Lonsdale (1993). Pruning techniques and the time of pruning have now changed from flush cuts and stub pruning to 'target' pruning, in which understanding tree anatomy indicates where, when and how the tree should be pruned.

In the USA, where urban trees are often referred to as 'shade trees', heavy pruning or even pollarding is seldom seen. In other parts of the world (e.g. Europe), many street trees are heavily pruned annually or biannually, to maintain light in nearby apartments. Thus, street trees are definitely not regarded as 'shade trees' in northern Europe in the way they are in the USA.

12.5.2 Wound treatment

Rot in trunks, branches or roots is usually due to wood-decaying fungi infecting a wound mostly caused by physical damage to the tree. The damage may occur from car injuries, mowing, soil excavation or from building activities. Wrong pruning methods and even the timing of pruning may also affect the degree of wounding. Treating wounds is an issue which has been subjected to considerable attention. Wound closure depends on the wound's location, the season of injury, and which tissues are exposed. Most commonly, callus develops primarily from the sides of a wound, much less from the top, and least from the bottom (Harris 1992). There are no data to show that wound dressings stop decay (Shigo 1991). This being so, the best approach is to treat all wounds in a way that will maximize natural healing and minimize the opportunity for spore germination (Bradshaw *et al.* 1995).

12.5.3 Hazardous trees including detection of decay

Trees are hazardous when the failure of one or more of their parts results in property damage and/or personal injury (Matheny & Clark 1994). There are a number of instruments available for the inspection of decay in trees. Some of the most well known are described by, for example, Shigo (1991), Matheny and Clark (1994) and Mattheck and Breloer (1998). Some methods use transmission of sound or electricity, others use mechanical resistance, often in combination with visual inspection. All methods cause some physical damage to the tree. Therefore, it is usually recommended to use intrusions into the tree that are as small and sharp as possible (Shigo 1991).

12.5.4 Management of urban forest stands

This section is adapted from Konijnendijk (1999b).

Urban forest management has an interesting relationship with forest management at large. It has often acted as a 'test area' for bringing urban dwellers in contact with forestry. Krott (1998) sees urban forests as hot spots for the forestry profession. Konijnendijk (1997, 1999a) agrees and sees urban forestry as a type of forestry where new and innovative new concepts and techniques are elaborated.

The call for more natural processes and more 'nature' in urban forests, however contradictory it may be in many cases, has led to 'closer-to-nature forestry' becoming well established in European urban forestry. In the Amsterdamse Bos, for example, more emphasis is placed on nature and natural processes. Instead of a more uniform thinning, groups of trees are now cut to simulate the event of a heavy storm. Grazing is applied to increase the variation in the vegetation (e.g. Koop 1994). In the Dutch Bentwoud, the planting of native broadleaved species is favoured (Stuurgroep Bentwoud 1996). Both open and closed as well as wet and dry areas are created.

Management is small scale and includes grazing. The German city of Lübeck manages 4500 ha of forest. Since 1994, the city has applied the concept of 'Naturnahen Waldnutzung' (close-to-nature forestry), partly as a reaction to quarrels with the environmental movement. The concept includes, for example, the principle of least interference. Non-interference areas of at least 20 ha (10% of the total area) have been allocated. Dead wood, the use of site-specific species, natural regeneration and selective cutting are characteristics, while hunting only takes place for purposes of regulation. There are no clearcuts nor monocultures or exotic species, and working during 'sensitive' times is avoided (Fähser 1995). Closer-to-nature forest management is also practised in Berlin, according to its special 'Urban Forest Management Concept'. Management is aimed at forest transformation (more natural species, more broadleaves), sanitation of polluted areas, a greater flexibility and so forth (Kilz & Zietz 1996). In the urban forests of Warsaw, exotic or non-forest species such as *Quercus rubra* and *Robinia pseudoacacia* are removed (H. Seremet, personal communication, see Konijnendijk 1999b). However, in some cases exotic species may be used, for instance for increasing diversity in forest stands, mainly for the purpose of recreation (Konijnendijk 1999b). Fitting in with the trend towards closer-to-nature management and enhancement of diversity is the increase in grazing in urban forests as a form of management (Krott & Nilsson 1998; Konijnendijk 1999a). Redl (personal communication, see Konijnendijk 1999b) mentions the use of horses for skidding in the Viennese forests. The use of horses also has a 'public relations purpose', as the public likes to see the animals in action.

In general, forest management is adapted to urban settings (e.g. Hodge 1995; Grey 1996; Miller 1997). For example, to an increasing extent municipalities in Finland adapt their management of recreation forests in terms of less heavy thinnings, lighter methods of timber extraction, longer rotations, more natural regeneration and small regeneration areas, less fertilization and so forth (Löfström 1990). The development of more appropriate techniques and adaptations for urban forest management, including the selection of suitable species, is also mentioned by Jones (1996). Rydberg (1998) suggests the use of precommercial thinning and possibly selective coppicing to make urban forests—especially young, dense ones—more attractive by enhancing the structural diversity of the forests. During the establishment phase of urban forests, it has been suggested to start with pioneer species to create rapidly a forest environment, and to increase species diversity over time, by mixing with slower-growing species (e.g. Jones 1996; Simson 1998).

12.6 TREE INVENTORIES

For most managers, a tree inventory is a useful day-to-day tool for scheduling maintenance and planting activities. The inventory can be used to generate listings of trees in need of immediate attention or priority maintenance prescriptions (Laverne 1994). An inventory system gives an overview of the stock of trees in the city. But it is also a planning tool with which urban trees can be maintained properly and in line with the resources available and the optimum techniques. Tree health, planning of working schedules and preparation of tree-care specifications can be improved by tree inventory systems.

In many cases, a tree inventory may be more complicated to introduce than ordinary inventories. Urban trees occur in a wide variety of species, sizes and locations. Furthermore, a tree population will never be static. Trees grow and need to be maintained by pruning and watering. Over the years, trees will change from one type of maintenance practice to another and, eventually, they will need to be removed. Even trees of the same species and seed origin will require different maintenance techniques and routines, according to their location and purpose.

The acquisition and processing of data are therefore important. Clear specification of data required will determine the success of the inventory. The qualifications of the personnel who collect the data are also important. In its simplest form, the collection of data can involve manual recording. Nevertheless, a Geographical Information System (GIS) is a powerful tool in this respect, consisting of digitized maps able to contain a wealth of information on a given

location and in different layers. This can be applied to all aspects of the planning and management of urban green areas: the identification of site-adapted tree taxa; optimum localization of new plantations; the timing and routing of maintenance; and the design of management plans. Airborne videography has also been used for these procedures. These measures are reasonably accurate and useful, especially in estimating overall percentage of tree cover (Sacamano *et al.* 1995).

Most communities are or can be subdivided into smaller management units for planning tree and forestry work. In some cities, the use of streets, railway lines and park or school districts provide obvious dividing lines for the management areas. Management units can also be used to set priorities on the intensity of management activities, which can be divided into single streets, parks, etc. Tree numbers are the last item of location information and they identify each tree according to its position.

Although tree type can be noted simply as evergreen, conifer or deciduous, most inventories seek to identify trees by their genus and or species. It may also be of value to record information on potential planting sites. These are frequently classified by the size of the available planting space (small, medium or large), and the presence or absence of overhead utility lines. Tree size may be described by the diameter of the main stem or trunk. The tree condition rating assists in assessing overall tree health and in evaluating species performance. Often it is simply recorded in a few, defined categories as excellent, very good, good, fair, poor, critical or dead. Tree maintenance needs are related to tree condition. The combination of structural integrity and health of the tree and potential 'targets', such as people and traffic that could be harmed should the tree fail, influence the priority of the maintenance prescription. A well-designed inventory should contain maintenance categories that identify and assign priority to hazardous conditions, including removal and pruning, as well as routine maintenance for non-hazardous trees.

Two examples of municipal tree inventories show how the information obtained can be used as a tool for future planting and management. In Hong Kong, a 5-year plan for landscape improvement has been designed. Some 1094 potential planting sites (PPS) with capacity for 12 063 trees have been identified. Most PPSs are small and spatially scattered, and opportunities for additional planting will have to come from new development areas (Jim 1995). In New Orleans, Louisiana, tree condition was related to the type of land cover in the root growth zone under the tree canopy, the presence of cables in the tree canopy and the associated land use. The inventory showed that the New Orleans urban forest was a mature forest, which suggested special care in order both to protect and then to replenish it. This information was proposed for use as a planning and management tool for the New Orleans urban forest (Talarchek 1987).

12.7 CONCLUSIONS

An urban area may be described as a place where many people have decided to live close together. The presence of many people in a limited area creates a significant wear on the premises. In order to limit intensive wear, hard and lasting materials are needed—and used. This, in itself, conflicts with growing trees in the urban environment.

The scientists and professionals working in relation to trees in the urban environment originate from many different backgrounds. Researchers within biology, geography, sociology, forestry, political sciences, urban planning, horticulture, arboriculture and landscape architecture may all be involved in dealing with urban forests and urban trees. Very often the researchers will concentrate on aspects related to their own discipline. Furthermore, most of the research within 'urban forestry' is applied research. This means that the results are very often directed towards site-specific, local or regional problems. Common problems need to be highlighted through initiatives which will cover local, national and international problems related to urban forests and trees.

In recent years a number of new coordination efforts have come to light within urban forestry. These initiatives may be grouped into three main areas: research co-ordination, overview studies and proceedings based on seminars, symposium,

etc. In the following some key publications are mentioned for each area:

12.7.1 Research and professional coordination

• Randrup and Nilsson (1998): The Tree ROUTE Network (Research on Urban Trees in Europe).
• Randrup and Nilsson (1998): COST Action E12, Urban Forests and Trees.
• The Metropolitan Tree Improvement Alliance (METRIA) gives an overview of the state of the art within urban forestry in the USA. A number of proceedings on tree selection, breeding, etc. for urban use have been published since the mid 1970s.
• Krott and Nilsson (1998): Multiple use of town forest in international comparison.
• Kuchelmeister (1997): The urban forestry research agenda—tropical forest research. Reference to the TREE City Initiative, a collaborative effort to promote urban forestry in developing countries.

12.7.2 Overview studies

• Albrecht (1995): An urban forestry bibliography.
• Braatz (1994): Urban forestry in developing countries: status and issues.
• Cobo (1998): Urban forestry development in Latin America.
• Johnston (1997a): The development of urban forestry in the Republic of Ireland.
• Johnston (1997b): The early development of urban forestry in Britain.
• Johnston (1998): The development of urban forestry in Northern Ireland.
• Konijnendijk (1997): Urban forestry: overview and analysis of European forest policies.
• Konijnendijk (1999a): Urban forestry: comparative analysis of policies and concepts in Europe.
• Kuchelmeister (1998): Urban forestry in the Asia-Pacific region: status and prospects.

12.7.3 Proceedings

In the following a number of national or regional overviews are presented. Most are proceedings from seminars and symposia:

- Arensberg (1997): Urban greening in Latin America and the Caribbean.
- Chambers and Sangster (1994): Urban and community forestry in Britain.
- Collins (1996): Urban forestry in Ireland.
- Decembrini (1997): Arboriculture, worldwide.
- Kollin (1998): American urban forest research.
- Randrup and Nilsson (1996): Urban forestry in the Nordic countries.
- Sander and Randrup (1998): Urban forestry in the Nordic and the Baltic Countries.

These initiatives are important in order to keep the focus on problems related to urban trees and urban forests. The whole subject of urban forestry is fortunate, because the large diversity in involvement may be seen as a resource. However, at the same time, the large diversity is also an obstacle which has to be overcome, as coordination of different references, traditions and methods takes time and requires open-minded and creative participants in the process.

During recent years the focus has been on creating overviews and exchanging information on a regional, national and international basis. This has been needed. However, now it is time for new, innovative research and for carrying out experiments. Furthermore, the time for implementing existing results into real life seems to be near. In the Third World countries there is a need for *in situ* experiments, because there is a desperate need for results and there is no time to waste. In the developed world, the continuous fight for resources makes the success rate of urban forestry related initiatives, and thus the visibility of urban green structures, ever more important. In both cases increasing urbanization in itself is a sufficiently strong argument for the implementation of research and knowledge related to trees in the urban environment.

ACKNOWLEDGEMENTS

The authors sincerely thank Mr Cecil C. Konijnendijk for his valuable contribution to section 12.5.4.

The following experts are gratefully acknowledged for reviewing and commenting on the manuscript or having contributed to the preparation of this chapter in any other way: Mr Rune Bengtsson, Swedish Agricultural University, Sweden; Mrs Alicia Chacalo, Universidad Autónoma Metropolitana–Azcapotzalco, Mexico; Kevin D. Collins, Tree Council of Ireland, Ireland; Dr Larry Costello, University of California, USA; Dir. Clive Davies, Cleveland Community Forest, UK; Dr Dirk Dujesiefken, Institut für Baumpflege, Germany; Mr Cecil C. Konijnendijk, European Forest Institute, Finland; Dr Palle Kristoffersen, Danish Forest and Landscape Research Institute, Denmark; Dr Frank S. Santamour, US National Arboretum, USA; Mr Allan Simson, Leeds Metropolitan University, UK; Dr Jozef van Slycken, Institut voor Bosbouw en Wildbeheer, Belgium; Dr Frank Søndergaard Jensen, Danish Forest and Landscape Research Institute, Denmark; Dr Liisa Tyrvainen, University of Joensuu, Finland; Dr Gary Watson, Morton Arboretum, USA.

REFERENCES

Albrecht, J. (1995) *Urban Forestry: A Bibliography*. Department of Forest Resources. Staff Paper Series no. 101. University of Minnesota, Minnesota.

Alexander, C., Ishikawa, S., Silverstein, M., Jacobson, M., Fiksdahl-King, I. & Angel, S. (1977) *A Pattern Language*. Oxford University Press, New York.

Arensberg (1997) *Urban Greening in Latin America and the Caribbean*. Inter-American Development Bank, Social Programs and Sustainable Development, Environmental Division, Washington DC.

Braatz, S.M. (1994) Urban forestry in developing countries: Status and issues. In: Kollin, C., Mahon, J. & Frame, L., eds. *Growing Greening Communities: Proceedings of the Sixth National Urban Forest Conference, Minneapolis, MN, September 14–18, 1993*, pp. 85–8. American Forests, Washington, DC.

Bradshaw, A., Hunt, B. & Walmsley, T. (1995) *Trees in the Urban Landscape. Principles and Practice*. E. & F.N. Spon, London.

Chacalo, A., Aldama, A. & Grabinsky, J. (1994) Street tree inventory in Mexico City. *Journal of Arboriculture* **20**, 222–6.

Chambers, K. & Sangster, M. (1994) Urban and community forestry in Britain. In: *Proceedings of the Third International Conference on Urban and Community Forestry, Manchester Town Hall, August 31–Sepetember 2, 1993*. Forestry Commission, Edinburgh.

Cobo, W. (1998) Urban forestry development in Latin America. In: Palo, M. & Uusivuori, J., eds. *Forest, Society and Environment. World Forests*, Vol. 1. Kluwer Academic Publishers, Amsterdam.

Coder, K.D. (1995) Tree quality BMP's for developing

wooded areas and protecting residual trees. In: Watson, G.W. & Neely, D., eds. *Trees and Building Sites. Proceedings of an International Workshop on Trees and Building*, pp. 111–24. International Society of Arboriculture, Savoy, IL.

Collins, K.D., ed. (1996) *Proceedings of the Second National Conference on Urban Forestry. Limerick City, Ireland, March 27–29, 1996.* The Tree Council of Ireland, Dublin.

Couenberg, E.A.M. (1994) Amsterdam tree soil. In: Watson, G.W. & Neely, D., eds. *The Landscape Below Ground. Proceedings of an International Workshop on Tree Root Development in Urban Soils. September 30–October 1, 1993*, pp. 24–33. International Society of Arboriculture, Morton Arboretum.

Craul, P.J. (1992) *Urban Soil in Landscape Design.* John Wiley, Chichester, Sussex.

Decembrini, F., ed. (1997) *Proceedings from the III European Arboriculture Congress, Merano, Italy, May 14–16, 1997.* International Society of Arboriculture. Savoy, IL.

Dobson, M.C. (1991) *De-icing Salt Damage to Trees and Shrubs.* Forestry Commission Bulletin Number 101. Forestry Commission, HMSO, London.

Fähser, L. (1995) Das Konzept der Naturnahen Waldnutzung im Stadtforstamt Lübeck. [The concept of close-to-nature forestry in the City Forest of Lübeck.] *Der Dauerwald* 2–6.

Garborsky, J. & Bassuk, N. (1996) A new urban tree soil to safely increase rooting volumes under sidewalks. *Journal of Arboriculture* **21**, 187–201.

Grey, G.W. (1996) *The Urban Forest: Comprehensive Management.* Wiley & Sons, New York.

Harris, R.W. (1992) *Arboriculture: Integrated Management of Landscape Trees, Shrubs and Vines*, 2nd edn. Prentice Hall, New Jersey.

Hodge, S.J. (1995) *Creating and Managing Woodlands around Towns.* Forestry Commission Handbook 11. HMSO, London.

Jim, C.Y. (1995) Comprehensive street tree census and planting plan for urban Hong Kong. In: *Caring for the Forest: Research in a Changing World. Abstracts of Invited Papers. IUFRO XX World Congress, 6–12 August 1995, Tampera, Finland*, pp. 475–6. International Union of Forest Research Organizations, Vienna.

Johnston, M. (1997a) The development of urban forestry in the Republic of Ireland. *Irish Forestry* **54**, 14–32.

Johnston, M. (1997b) The early development of urban forestry in Britain: part I. *Arboricultural Journal* **21**, 107–26.

Johnston, M. (1998) The development of urban forestry in Northern Ireland. *Irish Forestry* **55**, 37–58.

Jones, N. (1996) A practical approach to urban forestry. In: Collins, K.D., ed. *Trees and Woodlands for Towns and Cities. Proceedings of the Second National Con-*

ference on Urban Forestry. Limerick City, Ireland, 27–29 March 1996*, pp. 81–4. The Tree Council of Ireland, Dublin.

Kilz, E. & Zietz, M. (1996) Berliner Stadtwald-Behandlungskonzept: Stadtentwicklung und Wald. [Berlin Forest Management Concept: urban development and the forest.] *AFZ/der Wald* **51**, 710–13.

Kollin, C., ed. (1998) *Proceedings of the 8th National Urban Forest Conference, Atlanta, GA, September 17–20, 1997.* American Forests, Washington, DC.

Konijnendijk, C.C. (1997) *Urban Forestry: Overview and Analysis of European Forest Policies. Part 1: Conceptual Framework and European Urban Forestry History.* EFI Working Paper 12. European Forest Institute, Joensuu.

Konijnendijk, C.C. (1999a) *Urban Forestry: Comparative Analysis of Policies and Concepts in Europe — Contemporary Urban Forest Policy-making in Selected Cities and Countries of Europe.* European Forest Institute, Joensuu.

Konijnendijk, C.C. (1999b) *Urban Forestry: Comparative Analysis of Policies and Concepts in Europe — Needs and Developments.* EFI Working Paper. European Forest Institute, Joensuu. [Currently under review.]

Koop, H. (1994) Kansen voor natuurlijker bos. [Opportunities for more natural forest.] *Stad en groen* **2** (13), 14–17.

Kristoffersen, P. (1998) Designing urban pavement subbases to support trees. *Journal of Arboriculture* **24**, 121–6.

Krott, M. (1998) Urban forestry: management within the focus of people and trees. In: Krott, M. & Nilsson, K., eds. *Multiple-Use of Town Forests in International Comparison. Proceedings of the First European Forum on Urban Forestry, 5–7 May 1998, Wuppertal*, pp. 9–19. IUFRO Working Group S.6.14.00. Wuppertal.

Krott, M. & Nilsson, K., eds. (1998) Urban forestry. In: *Multiple-Use of Town Forests in International Comparison. Proceedings of the First European Forum on Urban Forestry, 5–7 May 1998, Wuppertal* IUFRO Working Group S.6.14.00. Wuppertal.

Kuchelmeister, G. (1997) The urban forestry research agenda — one missing link in setting research priorities relevant to tropical forest research. *ETFRN* (European Tropical Forest Research Network) *News* **22**, 19–22.

Kuchelmeister, G. (1998) *Urban Forestry in the Asia-Pacific Region: Status and Prospects.* Regional Office for Asia and the Pacific, Asia-Pacific Forestry Sector Outlook Study Working Paper Series, no. APFSOS/WP/44. Bangkok.

Laverne, R.J. (1994) *Suggested Data Structures for Tree Inventories.* ACRT, OH.

Lindsey, P. & Bassuk, N. (1992) Redesigning the urban

forest from the ground below. A new approach to specifying adequate soil volumes for street trees. *Arboricultural Journal* **16**, 25–39.

Löfström, I. (1990) Kaupunkien ja kuntien metsien hoito. [Management of urban and municipal forests.] *Selvitys* **87**, 1990. [In Finnish, with English summary.]

Lonsdale, D. (1993) A comparison of 'target' pruning versus flush cuts and stub pruning. *Arboriculture Research Note* **116**, 93.

Lynch, K. (1960) *The Image of the City*. MIT Press, Cambridge, MA.

Matheny, N.P. & Clark, J.R. (1994) *A Photographic Guide to the Evaluation of Hazard Trees in Urban Areas*. Second Edition. International Society of Arboricuture, Savoy, IL.

Mattheck, C. & Breloer, H. (1998) *The Body Language of Trees. A Handbook for Failure Analysis*. HMSO, London.

Miller, R.W. (1989) The history of trees in the city. In: Moll, G. & Ebenreck, S., eds. *Shading Our Cities*, pp. 32–5. Island Press. American Forestry Association.

Miller, R.W. (1997) *Urban Forestry. Planning and Managing Urban Greenspaces*, 2nd edn. Prentice Hall, Upper Saddle River, New Jersey.

Moll, G. (1989) The state of our urban forest. *American Forests* November/December, **95** (11), 61–4.

Randrup, T.B. & Nilsson, K., eds. (1996) Urban Forestry in the Nordic countries. In: *Proceedings of a Nordic Workshop on Urban Forestry, Reykjavik, Iceland September 21–24, 1996*. Danish Forest and Landscape Research Institute, Hørsholm.

Randrup, T.B. & Nilsson, K. (1998) Research note: coordination of research on urban trees in Europe. *Arboricultural Journal* **22**, 173–7.

Rolf, K. (1991) Soil improvement and increased growth response from subsoil cultivation. *Journal of Arboriculture* **17**, 200–4.

Rydberg, D. (1998) *Urban forestry in Sweden—silvicultural aspects focusing on young forests*. PhD thesis. Acta Universitatis Agriculturae Sueciae—Silvestria 73. Swedish University of Agricultural Sciences, Umeå.

Sacamano, P.L., McPherson, E.G., Myhre, R.J., Stankovich, M. & Weih, R.C. (1995) Describing urban forest cover, an evaluation of airborne videography. *Journal of Forestry* May, **93**, 43–8.

Sander, H. & Randrup, T.B. (1998) Urban forestry in the Nordic and Baltic countries. In: *Proceedings of a Nordic Workshop on Urban Forestry, Tallinn, Estonia December 1–3, 1997*. Danish Forest and Landscape Research Institute & Estonian Agricultural University, Forests Research Institute, Tallinn, Copenhagen.

Santamour, F. (1990) Trees for urban planting: diversity, uniformity and common sense. Proceedings of the Metro Tree Improvement Alliance. *METRIA* **7**, 57–65.

Scully, V. (1991) Architecture: the natural and the manmade. In: Wrede, S. & Adams, W.H., eds. *Denatured Visions, Landscape and Culture in the Twentieth Century*, pp. 7–18. The Museum of Modern Art, New York.

Shigo, A. (1991) *Modern Arboriculture: A Systems Approach to the Care of Trees and their Associates*. Shigo and Trees Associates, Durham, USA.

Simson, A. (2000) Urban forestry in the UK new towns. In: Randrup, T.B., Konijnendijk, C.C. & Nilsson, K., eds. *Proceedings of COST Action E12, Urban Forests and Trees. Invited Papers*. Danish Centre for Forest, Landscape and Planning, Hoersholm, Denmark (in preparation).

Stuurgroep Bentwoud (1996) *Bentwoud: kleurbehoud voor het Groene Hart*. [Bentwoud: maintaining colour for the Green Heart.] Stuurgroep Bentwoud, Den Haag.

Talarchek, G.M. (1987) Indicators of urban forest condition in New Orleans. *Journal of Arboriculture* **13**, 217–24.

Turner, T. (1996) *City as Landscape—A Post-Postmodern View of Design and Planning*. E. & F.N. Spon. Chapman & Hall.

Urban, J. (1989) Evaluation of tree planting practices in the urban landscape. In: *Proceedings of the Fourth National Urban Forestry Conference, St Louis, Missouri, October 15–19*, pp. 119–27. The American Forestry Association.

Walmsley, T.J., Hunt, B. & Bradshaw, A.D. (1991) Root growth, water stress and tree establishment. In: Hodge, S.J., ed. *Research for Practical Arboriculture*, pp. 38–44. Forestry Commission, Alice Holt Lodge, Farnham, Surrey.

Watson, G.W. & Himelick, E.B. (1997) *Principles and Practice of Planting Trees and Shrubs*. International Society of Arboriculture, Savoy, IL.

Part 4
Cases Studies of Sustainable Management

The six case studies assembled show the importance of several elements: the value of ecologically based management (mainly Chapters 13, 16 and 17); the importance of long-term experimentation and monitoring (Chapters 17 and 18); flexibility and adaptability to enhance an unpromising resource (Chapter 15); the role of silvicultural investigations and trials (Chapters 13, 16 and 17); the resilience of forest through changing history and politics (Chapter 14); and the recognition and embracing of today's environmental imperatives (Chapters 15, 16 and 18). All these elements help add up to an understanding of sustainable forest management.

The case studies were also deliberately invited to illustrate experience from both temperate and tropical forests, from natural (old growth) forest and plantations, and from developed and developing countries. The reader can happily add Chapter 11 to the case studies, and even Chapters 2 and 3, to broaden further the scope of how genuine forest management issues have been tackled both over time and in the last decade or so. Here will be found the solid evidence of progress towards sustainable forest management.

13: The Structure, Functioning and Management of Old-growth Cedar–Hemlock–Fir Forests on Vancouver Island, British Columbia

GORDON WEETMAN AND CINDY PRESCOTT

13.1 INTRODUCTION

The coastal conifer forests of northern Vancouver Island, British Columbia, have developed over the last 10 000 years, since glaciation. The climate is wet and mild year-round and influenced by storms from the Pacific Ocean. Agriculture is not possible. At low elevation the dominant species are western red cedar (*Thuja plicata* Donn. Ex. D. Don), western hemlock (*Tsuga heterophylla* (Raf.) Sarg), amabilis fir (*Abies amabilis* (Dougl.) and Sitka spruce (*Picea sitchensis* (Bong.) Carr.). These forests grow to great ages and sizes due to the absence of fire or major insect or disease attacks. Windthrow is the primary disturbance agent. These forests are classified as Natural Disturbance Type 1 (NDT1), i.e. ecosystems with rare stand initiating events with gap regeneration and a disturbance interval of 250 years (British Columbia Ministry of Forests 1995, 1999; Voller & Harrison 1997). These are public forests owned by the Province of British Columbia. Large areas are managed by major forest companies who have rights to cut the timber with management obligations under long-term licences. These forests were not extensively cut until the mid 1900s; thus all stands cut today are naturally regenerated old stands. Today, most stands are clearcut in small irregularly shaped cut blocks. Clearcutting is used because of the great size of trees and problems of rot and windthrow in selectively harvested stands.

The forests are in the coastal western hemlock biogeoclimatic zone (CWH) as classified by Green and Klinka (1994). This system of classification, used throughout British Columbia for forestry purposes, uses a grid of soil moisture and nutrient regimes to identify individual forest sites. Silviculture and regeneration actions and productivity estimates are customized to each site. Licenses are required that specify regeneration of all cut blocks to prescribed stocking densities and species within a specified number of years following harvest.

Viewed from the air, the forests are in an obvious intermingled mosaic of two types:
1 very old, irregularly structured stands composed of cedar and hemlock (cedar–hemlock) that have not been blown down;
2 much younger, but mature even-aged stands composed of hemlock and Pacific silver fir (hemlock–amabilis) originating from a major windthrow event in 1906.

Regeneration problems on the cedar–hemlock cutovers became apparent in the 1970s when extensive areas of Sitka spruce plantations almost ceased growing and turned golden yellow in colour as salal (*Gaultheria shallon* Pursh), an ericaceous shrub, invaded the cutovers. Sitka spruce is not planted today due to attacks by spruce weevil (*Pissodes strobi*) which kills tree leaders.

These mature forests have a high commercial value. Western red cedar is especially valuable because it is decay resistant. Stands are logged by cable systems; the logs are trucked to the ocean where they are towed in rafts to sawmills for conversion into structural forest products. The chips from sawmills are barged to pulp mills. The rate of harvest is fixed by the provincial government (free of politics) on a sustainable basis, with due regard given for sustainable landscapes, landscape, stand, species and genetic biodiversity, and wildlife habitat. Under present management plans prepared by the licensees, and approved by the government, these natural, old stands will be cut for many more decades. Harvest rates are aligned to cutover productivity estimates, thus nutritional problems of regeneration are of concern.

Severe regeneration problems occur on cutovers of coastal old-growth cedar–hemlock forests in the very wet, maritime portions of the coastal western hemlock zone (CWHvm1). The problem appears as poor growth and chlorosis (yellowing) of regenerating western red cedar, western hemlock, amabilis fir and Sitka spruce (Fig. 13.1). These symptoms appear 5–8 years after clearcutting and slash burning, coincident with the expansions of the ericaceous shrub, salal, on the cutovers. This also coincides with the end of a period of high nutrient availability that follows harvesting on these sites. Some cutovers are nearly 50 years old and have still not achieved crown closure. The problem does not occur on adjacent cutovers that were formerly occupied by second-growth forests of hemlock and amabilis fir (Fig. 13.2), originating following a windstorm in 1906. Mean annual volume increment on hemlock–amabilis sites is 10–14 m^3 ha^{-1} year^{-1} compared to 4–6 m^3 ha^{-1} year^{-1} on cedar–hemlock sites. Lewis (1982) could not distinguish between the two forest types on the basis of topography or mineral soil characteristics, and included them

Fig. 13.1 Cedar–hemlock (CH) cutovers are characterized by dense salal cover and slow-growing conifers with chlorotic (yellow) foliage. These symptoms appear 5–8 years after clearcutting and result from low availability of nitrogen.

Fig. 13.2 Adjacent hemlock–amabilis fir (HA) cutovers (left) have abundant regeneration and rapid growth of hemlock compared with CH sites (right).

both in a 'salal–moss' ecosystem association. He further hypothesized that amabilis fir forests were a seral stage of cedar–hemlock forests, and that the superior regeneration and site growth conditions typical of hemlock–amabilis sites might be achieved on cedar–hemlock sites through silvicultural treatments. In the biogeoclimatic ecosystem classification scheme of Green and Klinka (1994), cedar–hemlock sites are classified as 10 '*HwBa—blueberry*' or 10 '*HwBa—deer fern*', with a modifier 's' that signifies the nutrient-poor to very poor 'salal phase' that occurs in subdued terrain of the west coast and north end of Vancouver Island.

Research has been conducted for a decade to determine the reasons for the poor nutrition of trees and the best silvicultural practices for alleviating the cedar–hemlock growth problem (summarized by Prescott & Weetman 1994 and Prescott 1996). The low availability of nitrogen (N) and phosphorus (P) in cedar–hemlock cutovers originates in forest floors of the old-growth forests prior to clearcutting. Nutrient availability is low in all layers of the forest floor in these forests. This is a result of three factors. First, cedar litter contains less N and more decay-resis-

tant material than other species, and produces forest floors with low rates of N mineralization. Second, the forest floors in cedar–hemlock forests are wetter and have less soil than in forest floors of amabilis fir forests, leading to incomplete decomposition and mineralization of N. Third, the salal understorey in cedar–hemlock forests interferes with mineralization of N through the production of tannins. Processes involved in the development of low nutrient conditions on these sites are illustrated in Fig. 13.3.

13.2 CHARACTERISTICS OF FORESTS ON VANCOUVER ISLAND

Characteristics of these two forest types have been reviewed by Keenan (1993) and Fraser (1993). The natural landscape along the west coast of North America, from northern California to the top of the Alaskan panhandle, is dominated by coniferous forests of unsurpassed form and stature (as illustrated on the cover of this volume!). Most of the genera composing these forests differentiated from earlier gymnosperms during the Jurassic period from 140 to 160 million years ago (Scagel *et al.* 1965). Therefore, conifers

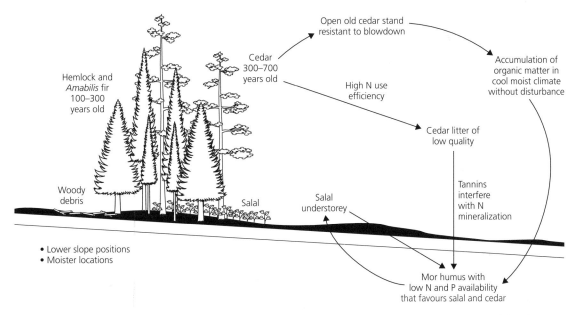

Fig. 13.3 In the absence of a major disturbance, nitrogen in cedar–hemlock forests is continually immobilized in humus, leading to poor N supply in old-growth forests.

have had a considerable period in which to adapt to a variety of environmental conditions and now occur in a wide range of climatic environments. However, Pacific coastal forests are thought to be remnants of vegetation types that once dominated the landmasses of the Northern Hemisphere. Their current range is now restricted to areas with a temperate, wet winter climate and mild to warm summers (Waring & Franklin 1979).

Northern Vancouver Island is located in the centre of this band of Pacific coastal forest, and its temperate climate has mild winters and cool wet summers. The distribution and forest vegetation across this area varies with topography, geological substrate, and the type and frequency of natural disturbance. On well-drained to somewhat imperfectly drained middle or upper slope situations, the forests form two distinct types: (i) old-growth type dominated by western red cedar with a smaller component of the western hemlock type, and (ii) a second-growth type dominated by western hemlock and amabilis fir, that appears even-aged and to have originated following a widespread windstorm in 1906 (the hemlock–amabilis type) (Fig. 13.3). The forest floors of both types are deep mor humus, generally of greater depth in the cedar–hemlock than the amabilis fir (Germain 1985). Mineral soils are duric or orthic Humo-Ferric Podzols.

Old growth is a difficult term to define, but for these sites the oldest trees are over 500 years old and very large in diameter. These stands also have the diversity of height, diameter, age and understorey structure considered characteristic of old-growth stands of the Pacific North-west by Franklin and Waving (1981).

In classifying the ecosystems of this area, Lewis (1982) could not distinguish between the two types on the basis of topography or mineral soil characteristics, and included them in the same ecosystem association. He further hypothesized that they were different stages (or phases) of a successional sequence.

Since the 1960s, these forests have been extensively harvested for timber. Following clearcutting and slash burning, major differences in the productivity of planted and naturally regenerated seedlings have been observed in the two forest types (Weetman *et al.* 1989a,b). Seedlings regenerated after cutting in the cedar–hemlock type grow slowly and exhibit chlorosis which is symptomatic of nutrient (particularly N) deficiency. The regeneration on the amabilis fir type grows relatively rapidly and exhibits no signs of nutrient deficiencies. The slower seedling growth on the cedar–hemlock type is considered to be partly due to lower forest floor nutrient availability, and partly to competition for nutrients from the ericaceous shrub salal which re-sprouts rapidly from rhizomes following clearcutting and slash burning (Weetman *et al.* 1990; Messier 1991; Messier & Kimmins 1991). The differences between these types are summarized in Table 13.1.

13.3 THE HYPOTHESIS: EXPLANTATIONS WHY HEMLOCK– FIR AND CEDAR–HEMLOCK SITES DIFFER

To attempt to explain the differences four hypotheses were developed:

13.3.1 The 'disturbance hypothesis'

It was argued that areas frequently disturbed by catastrophic windstorms will regenerate to western hemlock and Pacific silver fir, as is observed on amabilis fir sites. Windstorms cause tree falls resulting in soil mixing which promotes a well-drained and aerated soil with active organic matter decomposition and nutrient cycling. Improved soils increase the growth rate of trees and help to produce dense stands which exclude salal by shading. In areas that are not affected by the windstorms, such as cedar–hemlock sites, western hemlock and Pacific silver fir stands thin, allowing the regeneration of western red cedar. This improves the stand's ability to withstand windstorms because wind can pass through rather than strike against the forest stand.

13.3.2 The 'salal hypothesis'

It was argued that salal, an ericaceous plant, is a major competitor with conifer seedlings. Two theories are relevant. One theory proposes that salal suppresses the growth of conifer seedlings through an allelopathic agent which inhibits either mycorrhizal development, root development, or both (deMontigny & Weetman 1990; deMontigny 1992). This theory is related to obser-

Table 13.1 Comparison of old-growth cedar–hemlock (CH) and second-growth hemlock–fir (HA) stands growing side by side on the same soils and topography on northern Vancouver Island, BC.

	CH (cedar–hemlock)	HA (hemlock–fir)
Before		
Dominant conifers	Western red cedar and western hemlock	Western hemlock and Pacific silver fir
Wind disturbance	Absent	Repeated catastrophic disturbance
Age of stand	'Old-growth' >200 years old	Young <100 years old
Productivity of conifers	Low	High
Canopy	Open	Dense
Salal abundance	Plentiful	Rare
After		
Natural regeneration of conifers	Slow	Prompt, dense, fast growing
Performance of planted conifers	Poor	Good
Non-crop vegetation regeneration	Plentiful salal, little fireweed	Little salal, plentiful fireweed

vations of other ericaceous species in eastern Canada (Meades 1983) and the heathlands of Europe (Malcolm 1975) which have been attributed to growth problems of planted conifer seedlings after deforestation. The second theory suggests that salal is simply a better competitor than the conifer seedlings for soil nutrients in the cedar–hemlock clearcuts. In either case, the reason for high productivity in amabilis fir sites following a major disturbance is that the dense regeneration of hemlock and fir excludes salal, thus eliminating either allelopathy, nutrient competition, or both. Because the cedar–hemlock stands are not dense, salal can maintain a dense understorey, and inhibit most conifer seedling growth. Therefore, little ecological succession occurs at those sites and the 'old-growth' cedar forests remain relatively stable

13.3.3 The 'western red cedar hypothesis'

Western red cedar, unlike western hemlock and Pacific silver fir, is highly resistant to decay due to the presence of a fungitoxic chemical (thujaplicin) and a chemical (thujic acid) which repels a variety of insects (Swan *et al.* 1987). Consequently, a forest floor dominated by decomposing western red cedar will have a low mineralization potential and will immobilize nitrogen in the decomposer community, thus lowering the rate of nutrient cycling and nutrient availability. It has also been postulated that western red cedar is not a climax species, but a long-lived pioneer species that requires exposed mineral soil, or decaying cedar logs, and moderate levels of light to regenerate. Therefore, the conditions in windthrown hemlock–amabilis sites are suitable for western hemlock and Pacific silver fir. Because western red cedar is better able to regenerate on decaying cedar logs than western hemlock and Pacific silver fir, it can regenerate slowly in the cedar–hemlock sites.

13.3.4 The 'site-difference hypothesis'

This hypothesis proposes that hemlock–amabilis and cedar–hemlock sites are not different seral stages in the same succession, but rather two different plant associations determined by topography. Hemlock–amabilis sites are situated on knolls and upper slopes and are therefore more exposed to wind and better drained than the cedar–hemlock sites which are situated in lower areas.

These hypotheses were examined in a series of studies on cedar–hemlock and hemlock–amabilis ecosystems.

13.4 STAND STRUCTURE

Age–diameter relationships, diameter and height class structure, spatial pattern and seedling establishment substrate were studied by Keenan (1993) in old-growth cedar–hemlock stands, and in windstorm-derived, second-growth amabilis fir stands. There was a strong positive relationship between diameter and age for both species in the cedar–hemlock stands. Cedar ranged in age up to

1000 years, while the maximum longevity for hemlock was 300–400 years. Hemlock dominated the cedar–hemlock stands in terms of numbers, but these were largely in the lower size and height classes and cedar made up 73–85% of the basal area. The second-growth amabilis fir stands exhibited a range in ages despite their catastrophic origin but the size class distributions were unimodal. Hemlock exhibited differential regeneration strategies depending on the disturbance regime. The reverse-J diameter distribution indicated continuous establishment and recruitment to larger size classes in cedar–hemlock stands, but the past rapid response to disturbance indicated by unimodal distributions in hemlock–amabilis stands were more characteristic of an early seral species. The diameter–class distributions for cedar were generally flat, with many gaps in the size class distribution. However, its presence in most size classes at each site and relatively high numbers of seedlings and saplings suggested that, because of its longevity, cedar does form stable populations despite relatively low levels of recruitment. Establishment of elevated surfaces, such as decaying stumps and boles, above the dense understorey of the shrub *Gaultheria shallon* appeared critical for the regenerative success of both species in the cedar–hemlock stands.

The two stand types predominate on middle- or upper-slope situations on northern Vancouver Island. Distribution of the two types was found to have no obvious relationship with geology, topography or mineral soil. However, seedlings regenerated following cutting on the two types exhibit large differences in growth. Seedlings on cutovers in the cedar–hemlock type grow slowly and have symptoms of nutrient deficiency; those in the amabilis fir type grow relatively rapidly with no sign of nutrient deficiency. This difference in productivity is partly due to lower nutrient availability in the forest floors of the cedar–hemlock type.

The diameter–class structure of the cedar–hemlock type suggested it was a self-replacing, climax community. The diameter–distribution of western hemlock indicated continuous recruitment, while that of western red cedar suggested more periodic recruitment, at a slower rate than hemlock. Although the diameter–class distribu-

tion of the amabilis fir type was unimodal, suggesting an even-aged stand, a sample of tree ages indicated that many trees established some time before, or after, the 1906 windstorm.

13.5 NUTRIENT CYCLING

Measurements of nutrient availability in cedar–hemlock and amabilis fir forest floors demonstrated that lower N and P availability in cedar–hemlock forest floors existed prior to clearcutting (Prescott *et al.* 1993). All layers of cedar–hemlock forest floors had lower concentrations of total and extractable N and mineralized less N during 40-day aerobic incubations in the laboratory (Fig. 13.4). Total and extractable P was lower only in the litter layer of cedar–hemlock forest floors. Seedlings of cedar, Sitka spruce, hemlock and amabilis fir grown from seed in forest floor material from cedar–hemlock forests grew more slowly and took up less N and P than did seedlings grown in hemlock–amabilis forest floor material during a 1-year greenhouse experiment. Analysis of P forms by ^{31}P solution nuclear mag-

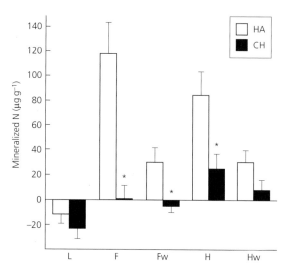

Fig. 13.4 Amounts of N mineralized during a 40-day aerobic incubation of each forest floor layer of adjacent hemlock–amabilis fir (HA) and cedar–hemlock (CH) forests. Each value represents the mean (+ 1 SE) of 15 samples; asterisks indicate significant differences ($P < 0.05$) between the two forest types based on two-factor ANOVA. L, litter; F, fermentation; H, humus; w, woody.

netic resonance (NMR) spectroscopy indicated that concentrations of total and available P were lower in cedar–hemlock forest floors, and there was less polyphosphate and more phosphate in cedar–hemlock forest floors (Cade-Menum 1995).

There was little evidence to support the hypothesis that more N and P were immobilized in detritus in cedar–hemlock forests, because the total amounts of N and P in coarse woody debris and forest floors were similar in the two forest types: 2.18 mg N ha^{-1} in cedar–hemlock, 2.05 in hemlock–amabilis; 142 kg P ha^{-1} in cedar–hemlock, 118.5 in hemlock–amabilis, respectively (Keenan *et al.* 1993). However, there was relatively more N in the humus (H) layer in cedar–hemlock forests and less in the F (fermentation) layer, so the N may be less available in cedar–hemlock forests. The greater mass of humus in cedar–hemlock forests may be the result of their greater age, or less complete decomposition, as discussed below.

Less N was returned in above-ground litter in cedar–hemlock forests (14.2 kg ha^{-1}) than in hemlock–amabilis forests (35.9 kg ha^{-1}), as a result of lower mass and N concentrations in foliar litter. There was greater internal recycling within the trees, which resulted from very efficient use of N by cedar, and from more efficient use of N by hemlock growing in cedar–hemlock forests. This is probably a response to low N availability and could also create a positive feedback that would exacerbate the low N availability in cedar–hemlock forest floors.

Decomposition rates of standard litter substrates were similar in the two forest types, and rates of CO_2 evolution from each forest floor layer during laboratory incubations were similar to or greater than those in hemlock–amabilis forests (Prescott *et al.* 1995a). These findings suggest that the decomposition potential of the two sites is similar. The lower N availability despite similar rates of litter decay may be attributable to the tannins, which bind proteins and immobilize N. There was evidence for tannins in the cedar–hemlock forest floor ^{13}C-NMR spectrum, as indicated by a peak at 145 p.p.m. (deMontigny *et al.* 1993). The dipolar-dephased cedar–hemlock spectrum also had higher intensity at 108 p.p.m., another feature diagnostic of tannins. This tannin

may be associated with salal; ^{13}C-NMR spectra of salal components indicated high levels of tannin.

The presence of cedar in cedar–hemlock forests could also contribute to low N availability. In a modelling study (Keenan *et al.* 1995), there was lower N availability in forest floor in simulated cedar forests than in hemlock forests. In trials at the UBC Research Forest (Prescott & Preston 1994), and in Ireland (Prescott *et al.* 1995b), low rates of N mineralization were measured in forest floors in cedar plantations, compared with adjacent plantations of other species including hemlock and firs. The relatively low concentrations of N and high concentrations of waxes and lipids in cedar litter may be responsible for slower N mineralization in cedar forests.

There is also evidence that moisture levels in cedar–hemlock humus and soil are greater than in hemlock–amabilis forests, and this may result in conditions that inhibit decomposition and nutrient cycling in cedar–hemlock forests. For instance, in the study of Prescott and Preston (1994), moisture contents of samples of humus and mineral soils were usually greater in cedar–hemlock forests, and sensors placed in the humus recorded consistently higher moisture levels in cedar–hemlock forests. Mineral soil in many of the cedar–hemlock forests studied was more compacted and cemented layers were continuous and shallower than in hemlock–amabilis forests (deMontigny 1992). This could lead to poorer drainage of humus, as indicated by the occurrence of hydromors in cedar–hemlock forests. Lower biomass of fauna, and greater representation by aquatic animals such as copepods and brachiopods in cedar–hemlock humus, is also indicative of wetter conditions (Battigelli *et al.* 1994). High moisture content and lower faunal biomass could result in less complete decomposition in cedar–hemlock humus. This was suggested by the higher concentrations of lipids and carbohydrates in ^{13}C-NMR spectra of cedar–hemlock forest floors and the poorer lignin biodegradation (acid–aldehyde ratio) in cedar–hemlock forest floors (Prescott *et al.* 1995a). The poorer drainage in cedar–hemlock humus may be attributed to the lack of soil disturbance by windthrow, or the tendency for the cedar–hemlock forests to be on lower topographic positions.

Together, these studies indicated that the growth check of conifer regeneration observed in cutovers in cedar–hemlock forests is a consequence of inadequate supplies of N and P. The low nutrient supply originates in the forest floor of the old-growth cedar–hemlock forests, prior to cutting (Fig. 13.1). Several factors appear to contribute to low nutrient availability in cedar–hemlock forest floors, and their relative importance is not clear. Cedar litter, tannins associated with salal, and greater soil water lead to incomplete decomposition and low N availability in cedar–hemlock forest floors. Under these conditions, there is more efficient use of nutrients by trees, and less nutrients recycled in litter, which further reduces nutrient availability in the forest floor. These conditions develop over several centuries without severe disturbance, and may be most prevalent on lower slope positions.

After clearcutting there is an assart effect causing a temporary improvement in N and P availability for growth of regeneration (Fig. 13.2). During this period, salal re-sprouts from rhizomes, immobilizing nutrients in biomass and causing growth check in conifers through N and P competition, mycorrhizal antagonisms and release of tannins. The growth check of the conifers can be relieved by fertilization with additions of 250–300 kg N ha^{-1} and 100 kg P ha^{-1} fish silage or sewage sludge. With this treatment, it is hoped that crown closure will be reached after a few years, shading the salal and leading to sustained improvement in growth rates.

13.6 MANAGEMENT IMPLICATIONS

13.6.1 Site preparation

Cedar–salal sites can be broadcast burned to reduce slash and salal cover, thereby creating suitable seedbed and plantable spots. In addition to temporarily disrupting the salal, burning also temporarily increases nutrient availability, leading to higher foliar nutrient concentrations and improved early growth rates of conifers. Intense burns will increase the amount of N lost, so low-impact spring burns are recommended. Burning costs in 1996 ranged from US$350 ha^{-1} to US$650 ha^{-1}, depending on the guarding costs for

the site. Burning may not be feasible in small blocks with convoluted edges.

Scarification and cultivation have been shown to increase early growth of western hemlock and western red cedar on cedar–hemlock sites. Responses are even greater when scarification is combined with fertilization (Fig. 13.5). The response to scarification appears to be largely due to the resulting disruption of salal, which reduces competition during the establishment of conifers. Cultivation or mixing of soil does not appear to increase nutrient availability on cedar–hemlock sites. Mounding may be successful on wetter sites such as cedar–spruce–skunk cabbage ecosystems and in the mid-coast. Mechanical site preparation can be used on sites that cannot be burned, such as those in the vicinity of riparian areas and wildlife corridors. Mechanical mulchers have caused unacceptable puddling; backhoes have been more successful on cedar–hemlock sites. Care must be taken with any equipment in wetter areas, to avoid interrupting water flow and creating unproductive wet depressions. The cost of mechanical site preparation on cedar–hemlock sites in 1996 is about US$1.00 per planting spot.

13.6.2 Regeneration

Cedar–hemlock sites should be planted as soon as possible after burning or site preparation so that conifer seedlings can take advantage of the nutrient flush and disruption of salal. Large planting plug stock and genetically improved stock will improve the competitive ability of conifers on cedar–hemlock sites. Survival rates for conifers planted on cedar–hemlock sites are over 90% for cedar and 85% for hemlock. Direct seeding has proven unsuccessful on burned sites, due to seed predation by rodents. Planting at fairly high densities (1400–1800 stems ha^{-1}) will hasten crown closure, but broadcast burning may be necessary to create this many plantable spots. In a trial with western red cedar and western hemlock at 500, 1500 and 2500 stems ha^{-1}, growth of individual trees during the first 7 years was greatest at 1500 stems ha^{-1}. At 2500 stems ha^{-1} individual tree growth of both species was reduced. Stand volume of western hemlock was greatest at 2500 stems ha^{-1}, but western red cedar stand

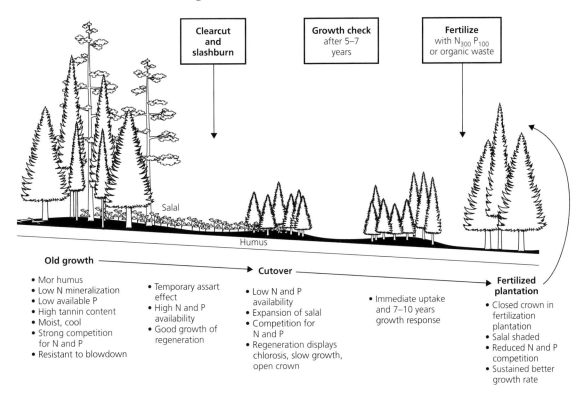

Clearcut and slashburn

Growth check after 5–7 years

Fertilize with $N_{300} P_{100}$ or organic waste

Salal

Humus

Old growth ────────→ **Cutover** ────────→ **Fertilized plantation**

- Mor humus
- Low N mineralization
- Low available P
- High tannin content
- Moist, cool
- Strong competition for N and P
- Resistant to blowdown

- Temporary assart effect
- High N and P availability
- Good growth of regeneration

- Low N and P availability
- Expansion of salal
- Competition for N and P
- Regeneration displays chlorosis, slow growth, open crown

- Immediate uptake and 7–10 years growth response

- Closed crown in fertilization plantation
- Salal shaded
- Reduced N and P competition
- Sustained better growth rate

Fig. 13.5 Strategy for improving conifer growth on cedar–hemlock cutovers.

volume was greatest at 1500 stems ha⁻¹. Competition between trees is unlikely to be significant at this early stage of development. The reduced growth at higher densities is more likely related to the limited number of suitable microsites for seedlings on cedar–hemlock sites, which are characterized by uneven terrain and a prevalence of air pockets in humus.

Western red cedar is the preferred species on cedar–hemlock sites, because it will grow at acceptable rates without fertilization. However, the mean annual increment of unfertilized cedar is low. Yellow cedar (*Chamaecyparis nootkatensis* (D. Don) Spach) may be planted on cedar–hemlock sites because it grows as well as western red cedar and has a higher market value. However, some mortality due to *Armillaria* infection has been observed in yellow cedar plantations on hemlock–amabilis sites, so a maximum of 10–20% yellow cedar is recommended. Western hemlock grows poorly on cedar–hemlock sites, but responds well to fertilization, so it is a pre-

ferred species on sites for which fertilization is planned. Douglas fir grows well on drier and warmer microsites, but poorly on others. Because the cedar–hemlock ecosystem is off-site for Douglas fir, it often has poor form and low wood quality, although it can develop a dense canopy and large piece sizes. Sitka spruce and *Amabilis* fir grow poorly on cedar–hemlock sites, unless they are fertilized. Fertilization of Sitka spruce increases its susceptibility to terminal weevil (*Pissodes strobi*). The incidence of weevil is low in hypermaritime areas, but there is some evidence that it is increasing, particularly on the west coast of Vancouver Island. Both Sitka spruce and amabilis fir are better suited to sites richer than cedar–hemlock. Western white pine (*Pinus monticola* D. Don) grows well on cedar–hemlock sites, and outgrows other species on some sites. However, the prevalence of blister rust limits the use of this species, unless a scheduled pruning regime is followed to reduce the incidence of rust. Lodgepole pine (*Pinus contorta* var. *contorta*

Dougl. Ex Loud) is common on cedar–hemlock sites particularly in wetter locations, and may be useful as a nurse species for other conifers.

In summary, if the site is not to be fertilized, cedar (western red cedar and yellow cedar) should be planted, as these species will grow at acceptable rates under the nutrient-poor conditions. If fertilization is planned, hemlock could also be planted, and growth of naturally regenerated western hemlock will also be enhanced. Repeated fertilization may be necessary to maintain adequate hemlock growth, which may result in western hemlock overtopping the planted western red cedar.

13.6.3 Fertilization

Conifer growth on cedar–hemlock sites is limited by the extremely low availability of N, so fertilization is critical for improving productivity. Western hemlock, amabilis fir and Sitka spruce all respond well to N addition (Fig. 13.6); western red cedar is less responsive. Greater responses are achieved when P is added in addition to N. The response period is 5–10 years. One application significantly advances the stand towards crown closure. The recommended rate of addition is 250–300 kg N ha^{-1} and 100 kg P ha^{-1}. The cost of fertilization with N and P in 1996 was (Can.) \$550 to (Can.) \$650 ha^{-1}. Fertilizer blends in which the N and P granules are the same size ensures the even distribution of both nutrients across the cutover block. For this reason, forest-grade urea

with diammonium phosphate is the preferred blend. Sites should not be broadcast fertilized until the trees are large enough to capture a significant portion of the nutrients added, i.e. at least 5 years after planting. Fertilization is most successful in immature stands on salal-invaded site series (Green & Klinka 1994) that have not reached crown closure (Weetman *et al.* 1989a,b).

Individual-tree fertilization at planting has been tested in two trials on cedar–hemlock sites. Spot application of granular NPK increased growth of newly planted cedar and western hemlock on cedar–hemlock sites. Hemlock response declined after 5 years, at which time fertilizer was reapplied in a broadcast application. Responses have also been observed in western red cedar and western hemlock seedlings that received either one Gromax teabag (24-4-7) or two Woodlace briquettes (14-3-3) at the time of planting. Additional trials are under way to compare broadcast fertilization at planting with spot application of granular fertilizer or teabags. Spot fertilization at planting may be particularly beneficial on sites which have not been burned, and may make it possible to use smaller seedlings. Some nurseries are experimenting with adding slow-release fertilizer to large-diameter plug-grown planting stock, which may provide an alternative to fertilization at planting.

Re-fertilization may be necessary after 5–10 years to maintain the growth responses of conifers on cedar–hemlock sites. It is not known at this time if fertilization will be required after crown

Fig. 13.6 Fertilization with 250 kg N ha^{-1} and 100 kg P ha^{-1} results in denser crowns, greener foliage and doubling or tripling of height growth. Hw, Cw, Ss and Ba are all responsive to fertilization of CH sites. Pictured is a plot of Hw on a CH site 5 years after N + P fertilization.

closure. Studies of the fate of fertilizer N have shown that less than 10–15% of the N applied is taken up by the trees (Tamm 1991). Most of the added N becomes immobilized in the soil and in the salal and is not available to trees after the first year, necessitating re-fertilization with N. By contrast, higher concentrations of P have been found in foliage and humus 10 years after fertilization with triple superphosphate at 100 kg P ha^{-1}. The sustained improvement in P availability after fertilization may negate the need for re-fertilization with P. Such long-term response to P fertilization has been shown for many sites (Allen *et al.* 1990). When re-fertilization is being considered, concentrations of N and P in foliage should be measured and compared with levels considered to be adequate for each species: 1.45% N and 0.35% P in hemlock, and 1.65% N and 0.16% P in cedar (Ballard & Carter 1986).

Trials with a variety of organic fertilizers have shown them to be effective for increasing conifer growth on cedar–hemlock cutovers. Response to these fertilizers applied at a rate of 500 kg N ha^{-1} are generally similar in magnitude and duration to those achieved with chemical N + P fertilizer at 225 kg N ha^{-1}. The major drawback to using organic residues is the high application cost due to low nutrient content per unit weight. Municipal sewage sludge and fish silage both increased the height and diameter of cedar trees. Mixing with pulp sludge reduced the response to sewage sludge but increased the response to fish silage (McDonald *et al.* 1994). Wood ash alone suppressed tree growth. Composted fish and wood waste and wheat straw both increased height increments and foliar nutrient concentrations of cedar during the first year after application. During the second year, height increments in the plots treated with straw increased even more, and foliar concentrations of N, P, K and S remained high. This suggests that there may be a long-term improvement in nutrient supply following the addition of fresh residues such as straw.

Sewage sludge was tested experimentally because the forest area is close to the ocean for barge transport from Vancouver and the area is relatively flat and roaded. These are factors which would facilitate the type of sludge-disposal methods already in long-time use by Seattle in the USA (Bledsoe 1981). Fish silage is available from numerous salmon farms on the coast. Fish guts and dead fish cannot be dumped in the ocean. In a greenhouse bioassay, lime applied at a rate equivalent to 5000 kg ha^{-1} did not affect growth of seedlings of cedar, hemlock or Sitka spruce.

13.6.4 Vegetation control

Several studies have demonstrated that a reduction in salal competition will increase growth of conifers on cedar–hemlock sites. Eradication of salal through manual removal followed by a single application of triclopyr ester (Release) was effective at reducing salal cover for at least 9 years. Salal removal increased growth of cedar in particular, whereas Sitka spruce and cedar responded more to fertilization. However, the cost of manual salal removal is prohibitive (US$5000 ha^{-1}), and herbicides alone are not highly effective at controlling salal. Triclopyr ester is somewhat effective (Biring *et al.* 1996), especially foliar applications in the fall with a diesel carrier (D'Anjou 1990). However, conifers can also be killed when sprayed with triclopyr ester and diesel. Burning and mechanical treatments that temporarily disrupt salal appear to be the most viable means of vegetation control on cedar–hemlock sites.

There is some evidence that repeated fertilization with N alone, totalling more than 600 kg N ha^{-1}, will cause a reduction in salal cover. Nearly total kill of salal has been observed in field trials on northern and southern Vancouver Island following the heavy application of ammonium nitrate or urea. More research is needed on the effects of high N additions on conifer growth and drainage waters to determine if this may be a viable option for controlling salal and enhancing conifer growth on cedar–hemlock sites.

13.6.5 Recommendations for regenerating cedar–hemlock sites

• Clearcutting is the most viable silvicultural system because of the need for site preparation.
• Mechanical site preparation or slash burning are necessary to disrupt the salal prior to planting conifers.
• Sites should be promptly planted with large, genetically improved stock, at 1400–1800 stems ha^{-1}.
• Western red cedar should be planted; western

hemlock may also be planted if the site is to be fertilized.

• Fertilization with 250 kg N ha^{-1} and 100 kg P ha^{-1} will increase the growth of conifers for 5–10 years. Individual tree fertilization at planting is also effective, but refertilization with N 5 years later is likely to be required to reach crown closure.

• Organic materials such as fish silage, fishwood compost, sewage sludge or fresh residues are effective fertilizers for cedar–hemlock sites.

• Manual removal of salal is effective but the cost is prohibitive; herbicides are not highly effective at controlling salal.

• A combination of salal removal and fertilization will yield greater responses than either treatment applied separately.

ACKNOWLEDGEMENTS

The research described in this chapter was conducted by an interdisciplinary team of scientists from the University of British Columbia, Canadian Forest Service, BC Ministry of Forests, and Western Forest Products Ltd, including J. Barker, J. Battigelli, S. Berch, S. Brown, B. Cade-Menum, X. Chang, L. deMontigny, R. Fournier, L. Fraser, A. Germain, B. Hawkins, L. Husted, R. Keenan, J.P. Kimmins, T. Lewis, M. McDonald, C. Messier, C. Prescott, B. Thompson, G. Weetman and G. Xiao.

Funding for the studies was provided by an NSERC University–Industry Grant, with Western Forest Products Ltd, MacMillan Bloedel Ltd and Timberwest Ltd, an NSERC Tripartite Grant, with Canadian Forest Service and Western Forest Products Ltd, and grants from the Science Council of British Columbia, South Moresby Replacement Fund, and the Canada–British Columbia Agreement on Forest Resource Development (FRDA II).

Western Forest Products Ltd made land available for experiments, and provided accommodation, facilities and logistical assistance and advice to all researchers. Chemical analyses were provided by McMillan Bloedel Ltd. The manuscript was typed by P. Quay and improved by the comments of two anonymous reviewers.

REFERENCES

Allen, H.L., Dougherty, P.M. & Campbell, R.G. (1990) Manipulation of water and nutrients—practice and opportunity in Southern U.S. pine forests. *Forest Ecology and Management* **30**, 437–53.

Ballard, T.M. & Carter, R.E. (1986) *Evaluating Forest and Stand Nutrient Status*. British Columbia Ministry of Forests, Victoria, BC.

Battigelli, J.P., Berch, S.M. & Marshall, V.G. (1994) Soil fauna communities in two distinct but adjacent forest types on northern Vancouver Island, British Columbia. *Canadian Journal of Forest Research* **24**, 1557–77.

Biring, B.S., Comeau, P.G. & Boateng, J.P. (1996) *Effectiveness of Forest Vegetation Control Methods in BC*. FRDA Handbook 011. British Columbia Ministry of Forests, Victoria, BC.

Bledsoe, C.S., ed. (1981) *Municipal Sludge Application to Pacific Northwest Lands*, Contribution no. 41. College Forest Resources, University of Washington, Seattle.

British Columbia Ministry of Forests (1995) *Biodiversity Guidebook*. British Columbia Ministry of Forests, Victoria, BC.

British Columbia Ministry of Forests (1999) *Landscape Unit Planning Guide*. British Columbia Ministry of Forests, Victoria, BC.

Cade-Menum, B. (1995) *Phosphorus forms in podzols on northern Vancouver Island*. PhD Thesis, Department. of Soil Science, University of British Columbia, Vancouver, BC.

D'Anjou, B. (1990) *Impact of Chemical and Manual Treatment on Salal and Conifer Development*. FRDA Red. Memorandum no. 15. British Colombia Ministry of Forests, Victoria, BC.

Franklin, J.F. & Waving, P.H. (1981) *Distinctive Features of the Northwestern Forest: Development, Structure and Function in Forests: Fresh Perspectives from Ecosystem Analysis*. Oregon State University of Press, Corvallis.

Fraser, L. (1993) *The influence of salal on planted hemlock, and cedar saplings on northern Vancouver Island*. MSc Thesis, University of British Columbia, Department of Botany, Vancouver, BC.

Germain, A.Y. (1985) *Fertilization of stagnated Sitka spruce plantations on northern Vancouver Island*. MF thesis, Department of Forest Sciences, University of British Columbia, Vancouver, BC.

Green, R.N. & Klinka, K. (1994) *Field Guide to Site Identification and Interpretation for the Vancouver Forest Region*. British Colombia Ministry of Forests, Victoria, BC.

Keenan, R.J. (1993) *Structure and function of western red cedar and western hemlock forests on northern Vancouver Island*. PhD thesis, Department of Forestry, University of British Columbia, Vancouver, BC.

Keenan, R.J., Prescott, C.E. & Kimmins, J.P. (1993) Mass and nutrient content of the forest floor and woody debris in western red cedar and western hemlock forests on northern Vancouver Island. *Canadian Journal of Forest Research* **23**, 1052–9.

Keenan, R.J., Kimmins, J.P. & Pastor, J. (1995) Modelling carbon and nitrogen dynamics in western red cedar and western hemlock forests. In: Kelly, J.M. & McFee, W.W., eds. *Carbon Forms and Functions in Forest Soils*, pp. 547–65. Soil Science Society of America, Madison, WI.

Lewis, T. (1982) *Ecosystems of Block 4, TFL 25*. Internal Report. Available from Western Forest Products Ltd, Vancouver, BC.

McDonald, M.A., Hawkins, A., Prescott, C.E. & Kimmins, J.P. (1994) Growth and foliar nutrition of western red cedar fertilized with sewage sludge, pulp sludge, ash silage and wood ash on northern Vancouver Island. *Canadian Journal of Forest Research* **24**, 297–301.

Malcolm, D.C. (1975) The influence of heather on silviculture practice: an appraisal. *Scottish Forestry* **29**, 14–24.

Meades, W.J. (1983) The origin and successional status of anthropologenic dwarf shrub heath in Newfoundland. *Advances in Space Research* **2** (8), 97–101.

Messier, C. (1991) *Factors limiting early growth of* Thuja plicata, Tsuga heterophylla *and* Picea sitchensis *seedlings on* Gaultheria shallon *dominated cutovers in coastal British Columbia*. PhD thesis, Department of Forest Sciences, University of British Colombia, Vancouver, BC.

Messier, C. & Kimmins, J.P. (1991) Above- and below-ground vegetation recovery in recently clearcut and burned sites dominated by *Gaultheria shallon* in coastal British Columbia. *Forest Ecology and Management* **46**, 275–94.

deMontigny, L.E. (1992) *Evidence of allelopathy by salal* (Gaultheria shallon *Pursh*) *in conifer plantations on northern Vancouver Island*. PhD thesis, Faculty of orestry, University of British Columbia, Vancouver, BC.

deMontigny, L.E. & Weetman, G.F. (1990) The effects of ericaceous plants on forest productivity. In: Titus, B.D., Lavigne, M.B., Newton, P.F. & Meades, W.J., eds. *The Silvics and Ecology of Boreal Spruces*. Forestry Canada Information Report N-X-271, St. Johns, New Foundland pp. 83–90.

deMontigny, L.E., Preston, C.M., Hatcher, P.G. & Kogel-Knaber, I. (1993) I. Comparisons of humus horizons from two ecosystem phases on northern Vancouver Island using C-13 CPMAS NMR spectroscopy and CuO oxidation. *Canadian Journal of Soil Science* **73**, 9–25.

Prescott, C.E. (1996) *Salal Cedar Hemlock Integrated Research Program Research Update*. Faculty of Forestry, University of BC, Vancouver.

Prescott, C.E. & Weetman, G.F. (1994) *Salal Cedar Hemlock Integrated Research Program: A Synthesis*. Faculty of Forestry, University of British Colombia, Vancouver, BC.

Prescott, C.E., Coward, C.P., Weetman, G.F. & Gessell, S.P. (1993) Effects of repeated nitrogen fertilization on the ericaceous shrub salal (*Gaultheria shallon*) in two coastal Douglas-fir forests. *Forest Ecology and Management* **61**, 45–60.

Prescott, C.E., deMontigny, L.E., Preston, C.M., Keenan, R.J. & Weetman, G.F. (1995a) Carbon chemistry and nutrient supply in cedar–hemlock and hemlock–amabilis fir forest floors. In: Kelly, J.M. & McFee, W.W., eds. *Carbon Forms and Functions in Forest Soils*, pp. 377–96. Soil Science Society of America, Madison, WI.

Prescott, C.E., Thomas, K.D. & Weetman, G.F. (1995b) The influence of tree species on nitrogen in the forest floor; lessons from three retrospective studies. In: Mean, D.J. & Comforth, I.S., eds. *Proceedings of the Trees and Soil Workshop*, Special Publication no. 10, pp. 59–68. Lincoln University Press, Agronomy Society of New Zealand, Canterbury.

Scagel, R.R., Bandoni, R.J., Rouse, G.E. *et al.* (1965) *An Evolutionary Survey of the Plant Kingdom*. Wadsworth Publishing Co., Belmont, Ca.

Swan, E.P., Kellogg, R.M. & Smith, R.S. (1987) Properties of western red cedar. In: Smith, N.J., ed. *Western Red Cedar—Does it Have a Future?*, pp. 147–60. Faculty of Forestry, University of British Columbia, Vancouver, BC.

Tamm, C.O. (1991) *Nitrogen in Terrestrial Ecosystems*, Ecological Studies, Vol. 18. Springer-Verlag, Berlin.

Voller, J. & Harrison, S., eds. (1997) *Conservation Biology Principles for Forested Landscapes*. University of British Columbia Press, Vancouver, BC.

Waring, R.H. & Franklin, J.F. (1979) Evergreen coniferous forests of the Pacific Northwest. *Science* **204**, 1380–6.

Weetman, G.F., Fournier, R., Barker, J., Schnorbus-Panozzo, E. & Germain, A. (1989a) Foliar analysis and response of fertilized chlorotic western hemlock and western red cedar reproduction on salal dominated cedar–hemlock cutovers on Vancouver Island. *Canadian Journal of Forest Research* **19**, 1501–11.

Weetman, G.F., Fournier, R., Barker, J., Schnorbus-Panozzo, E. & Germain, A. (1989b) Foliar analysis and response of fertilized chlorotic western hemlock and western red cedar reproduction on salal dominated cedar–hemlock cutovers on Vancouver Island. *Canadian Journal of Forest Research* **19**, 1512–20.

Weetman, G.F., Fournier, R., Barker, J., Schnorbus-Panozzo, E. & Germain, A. (1990) Post-burn nitrogen and phosphorus availability of deep humus soils in coastal British Columbia cedar/hemlock forests and the use of fertilization and salal eradication to restore productivity. In: Gessel, S.P. *et al.*, eds. *Sustained Productivity of Forest Soils*, pp. 451–9. Faculty of Forestry, University of British Colombia, Vancouver, BC.

14: The Beech Forests of Haute-Normandie, France

DAVID ROSE

14.1 HISTORICAL ASPECTS OF THE FORESTS OF HAUTE-NORMANDIE

The beech (*Fagus sylvatica*) forests of northern France are among the most magnificent in Europe with a long tradition of management based on sustainability. Those found in Normandy, mostly in the Region of Haute-Normandie (Seine-Maritime and Eure), are particularly fine examples. In the beech forests of Haute-Normandie the majestic trunks of the mature trees form the columns of 'living cathedrals' (Fig. 14.1), which rival the abbeys and cathedrals of stone that proliferate in the valley of the River Seine.

The total area occupied by beech forests in Normandy today is around 58000 ha, of which 47000 ha are to be found in Haute-Normandie and a greater part of this (37000 ha) in the Department of Seine-Maritime (Fig. 14.2).

The forests of Normandy are managed for high-quality sawlogs, with the high forest regime being applied to 90% of all forest areas, but when looking at Haute-Normandie on its own the figure rises to 93%. The pattern of Normandy's forests was set shortly after the last glacial period when temperate boreal forests moved northwards behind the retreating ice sheets. Until then much of the land to the north of what is now Paris was tundra, devoid of any tree cover. By the time the first agricultural tribes reached the area the main cover was deciduous woodland consisting of oak (*Quercus robur* and *Q. petraea*), beech, lime (*Tilia* spp.), elm (*Ulmus* spp.) and ash (*Fraxinus excelsior*) as the main species. On many sites beech was the climax woodland overtopping oak and other species. Early settlers had a profound effect on this forest and by the time of the Roman occu-

pation some types of woodland had become very limited (Peters 1997). The first species to suffer were lime and elm, which were cut for animal fodder and to open up clearings in the forest. Later on the clearances gathered momentum and much of the woodlands on the plains and plateaux were cleared. Elsewhere the woods were still dense and almost impenetrable, providing sanctuary for the Gauls that still resisted the Romans. The Romans were unused to fighting in woodlands and fared badly when they entered them and encountered those that they called *sylvatici* (from which we get the French *sauvages* and the English 'savages'). As a result most of the Roman roads and settlements were to be found on the plains or plateaux or clinging close to the banks of the River Seine (Reed 1954; Foubert 1985). In the more settled Gallo-Roman period that followed there was a renewed interest in the forest as a source of timber and it could be said that the first silvicultural exploitation began. Indeed, there are references to agents charged with managing the forests for continued production of timber (Reed 1954). In times of peace the forests provided grazing and timber and in times of war they provided shelter. Many settlements were established by clearing the forests close to the major rivers (e.g. the Seine), receiving shelter from the forests behind and easy access to travel from the river. The arrival of the Viking raiders ('Northmen' later to be known as Normans) saw widespread pillaging of the countryside and disruption of the population and as a consequence cultivation regressed. Many Gallo-Roman settlements were abandoned at this time and the land reclaimed by the forests (Reed 1954). Haute-Normandie contains a number of such sites which, owing to the rise of

Fig. 14.1 The living cathedral of Normandy (Forêt de Lyons).

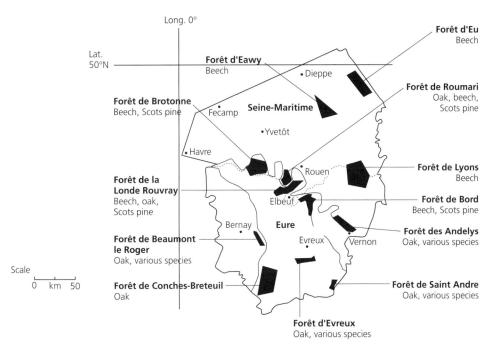

Fig. 14.2 The main forests of Normandy.

new settlements elsewhere, remained as forest when times became more settled. The forests of Brotonne, La Londe Rouvray and Bord are all examples of abandoned Gallo-Roman settlements. Only the Forêt de Lyons, with its central settlement of Lyons-la-Forêt, remains as an example of the Gallo-Roman forest settlement.

This unsettling effect of the early Normans was responsible for a resurgence in the fortunes of the forests which lasted only to the end of the first millennium. The period of the eleventh, twelfth and thirteenth centuries was the most destructive for the forests of Normandy yet, perversely, it also saw the introduction of a wide range of forest laws and rights. The abbeys, enjoying the support of the Normans, proliferated along the rivers close to the forests. The Benedictine rules required the cultivation of land for crops so the forest was cleared again for agriculture, a process known as *essartage*, and these ancient sites of forest clearance are revealed today by the many small hamlets in or near today's forests which bear the name *Les Essarts*. At the same time there were numerous gifts of forest land to the monasteries and abbeys by the Dukes of Normandy and their kinsmen and these were exploited heavily for their timber. During a long period of relative peace the population grew and more land was cleared to provide settlements. These settlements also needed timber for building and heating so the demands on the forests increased. This was a period of marked decline for the forests and there had yet to be a consistent approach to their protection and management. The nature of the demands, however, were continuing to shape the nature of the forests, with beech becoming more and more dominant in some areas by selective pressures on other species. The first forest laws to be put in place were not concerned with regulating felling directly but with generating tax income for the title-holders, mainly the church and royalty. There was a system known as *Tiers et Danger*, a cumulative tax that resulted in 43% of the selling price of forest timber being paid to the title-holder. The tax was based on thirds and tenths and for a typical sale was calculated as follows: $1/3 + 1/10 = 10/30 + 3/30 = 13/30$ of the product sold (Foubert 1985).

Fortunately for the forests, the advent of more troubled times, and the need for timber for such purposes as shipbuilding, led to the development of forest laws and the instigation of forest wardens in 1291. These were the first in a long line of laws and *ordonnances* for the management of forests (Box 14.1). The main system of management that was practised from this time was known as *tire et aire* where a certain section of forest, mostly on the outside, was felled each year. A few standards were left to ensure regeneration and the coupes were adjacent to each other or *proche en proche*. This type of management continued largely unchanged up to the time of the Revolution.

Box 14.1 Major dates in the history of French forestry (Husson 1995)

1291: Phillipe IV le Bel creates the corps of Masters of Water and Forests

1346: Phillipe VI de Valois draws up the first code of ordonnances for the Royal forests

1376: Ordonnances of Charles V le Sage, based on the management of water and forests

1561: Prohibition of cutting plantations aged less than 10 years. An obligation to leave one-third of all trees in timber-producing forests to continue to grow

1669: Grand Ordonnance of Colbert to bring all broadleaved forest under regulated felling

1789: The lands of the Church placed at the disposition of the Nation. Proposals for a multiplicity of forest projects

1824: Creation of the Forestry School at Nancy

1827: Drawing up of the Forestry Code

1854: Law restricting the rights of forest use

1859: Law on the regulation of clearing and restocking particular forests

1913: Law bringing all private forests under a forestry regime

1940: The Vichy regime increases normal levels of production by 50%

1946: Creation of the National Forestry Fund

1954: Constitution of forestry groups for reforestation and management

1958: Creation of permanent inventory of forest resources

1963: Creation of the Regional Centres for Forest Owners (CRPF)

1964: Creation of the National Forest Office (ONF)

1988: Creation of the Department of Forest Health (DSF)

The system worked fairly well under normal circumstances when the coupes did not vary much in size from year to year (*coupes ordinaire*). In spite of these laws there were periods of intense exploitation, most notably as a result of industrialization—iron-working, potteries and glassworks—or the need for ships for the navy.

Under these circumstances the coupes were much larger (*coupes extrordinaire*) and not always adjacent or confined to the outside of the forest (Reed 1954). France lacked the extensive coal reserves that fuelled the Industrial Revolution in Britain and relied heavily on its forests to fuel the furnaces. This made its operations less competitive than its close neighbour, leading to the extreme poverty of the industrial workers in France at the beginning of the nineteenth century. The problems this brought about were famously depicted by Victor Hugo in his novel *Les Miserables* and it is interesting to note that he lived the later part of his life in the shadow of the great beech forest of Brotonne. Coupled with this industrialization was the huge increase in population of the major cities and towns in Normandy, most notably Rouen. Such demands had a major impact on species diversity and as a result some forests changed dramatically. The forests of Bord, Lyons and Brotonne became predominantly beech as the oak was used in great quantities for house and boat building. Although records were kept quite assiduously of the volumes of timber produced in these forests, the turmoil of the Revolution resulted in the destruction or loss of many of these ledgers, so it is only possible to judge the level of exploitation from contemporary writings of historians. In spite of this, all the intentions of the management of these forests were to secure lasting—sustainable—supply of timber to meet all current and future needs. At the time of the Revolution the church was said to hold 800000 ha of forest land, much of it in the hands of monastic orders, mainly the Benedictines. The proposal by Talleyrand, Bishop of Autun, to confiscate church property had a profound effect on the future of many forests, especially in Haute-Normandie where Benedictine Abbeys were particularly numerous. These confiscated forests were passed to the State which then passed them to the various domains and communes as will be seen later under the sections dealing with the individual forests. The management thus moved from an exploitative system which frequently paid only passing attention to management principles to a generally more sympathetic form of management using dedicated forest managers.

14.2 CURRENT SILVICULTURAL PRACTICE

From the Grand Ordonnances of Colbert in 1669 to the drawing up of the Forestry Code in 1827 (Box 14.1), the aims became clear and the role of the forest manager grew in importance.

The Forestry Code of 1827 brought with it the introduction of a new system of forest management from Germany through the action of Bernard Lorentz, the first Director of the new Forestry School at Nancy. Well versed in the German systems, he introduced them to France with the aim of producing homogeneous growth over the period of rotation with natural regeneration as the means of replenishment at the end of the rotation. This is what we now call the shelterwood system. Intriguingly the German silviculturists gradually abandoned this system, over much of Germany, for one where replanting at the end of the rotation was the preferred method of restocking (Reed 1954). The rotation age of beech in Normandy is 120–150 years and it is managed today on the uniform shelterwood system (Matthews 1989). During the first 15 years beech should regenerate freely under the opened canopy of the old stand. Fifteen years are necessary to obtain complete regeneration, although it may be necessary to plant trees where, in exceptional circumstances, this has not been achieved. The remaining old trees are removed after 25 years with juvenile thinning and clearing continuing until the 30th year, after which commercial thinning starts. Around that time 80–100 promising trees per hectare are selected as final crop trees and thinnings favour these trees up to the 80th year. After 100 years thinnings are aimed at increasing the light to the forest floor to aid natural regeneration. In recent years the timing of the regeneration felling has been strictly controlled for each region to protect the seeds, seedlings and young trees. In the case of the public forests (*forêts dominales* and *forêts communales*) the restrictions are set out in *Clauses Générales* and *Clauses Communes* established by the Office National des Forêts (ONF). From these, the forest manager can draw up *Clauses Particulaires* and *Clauses Restrictives* for each forest. For private

Fig. 14.3 Regeneration in progress, Forêt de Brotonne.

forests the restrictions are set out in *Clauses Générales* drawn up by the Compagnie Nationale des Ingenieurs et Experts Forestiers et Experts en Bois (CNIEFEB) and the Federation Nationale du Bois. In Normandy the *Clauses Générales* prohibit felling and clearing between 1 April and 31 July, a time when newly emerged seedlings and young plants are most vulnerable to damage, while the *Clauses Particulaires* specify that the treatment of lop and top is by burning (Rotaru 1988).

Though initially forests managed under the uniform shelterwood system can have trees differing by as much as 15 years, by the time the first commercial thinnings take place this difference is much reduced and barely noticeable. Beech in Normandy usually produce a good mast year every 3–5 years and regeneration under this system is excellent (Fig. 14.3). In fact, a recent survey showed that nearly 70% of all beech regeneration in French forests is achieved in this way (Rotaru 1988) but the figure for Haute-Normandie is much higher at around 85% (ONF, personal communication). The forests of Haute-Normandie account for nearly 5% of all beech production in France and management success since 1950 has seen a steady rise in productivity (Fig. 14.4). The growing stock of beech in the region is estimated at around 4 million m^3 with

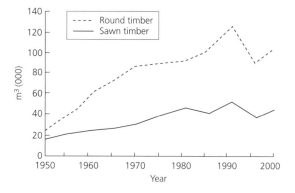

Fig. 14.4 Production of beech timber in Haute-Normandie (1950–95 actual, 2000 estimated).

an annual increment of around 0.15 million m^3 (CRPF 1992).

Recent years have seen a rise in the fortunes of beech in France and beech timber now commands a price on a par with oak at around 485 francs m^{-3} (ONF 1998; Rérat 1999). Reasons for this may be the dramatic increase in exports to China, rising from almost nothing in 1995 to 70 000 t in 1998, and the boycott of tropical hardwoods, particularly in northern Europe (Rérat 1999). Beech is eminently suited to replace many tropical hardwoods and now features widely in furnishings which are highly competitive in price with those that were made from tropical hardwoods.

14.3 FORÊT DE BROTONNE

This is one of the forests that took over an old Gallo-Roman settlement that was abandoned during the Dark Ages. Originally known as the forest of Arelaune, it took its present name from Saint Condéde the Breton who had a hermitage in part of the forest. It has had a colourful past, sheltering kings and princes in the period before the Norman occupation. In more settled times, it became the property of the Dukes of Normandy before large parts of it were granted to the church, mainly the abbeys of Jumièges and Fécamp. It has also been a witness to some draconian forest laws. In 1123 one of the Counts of Meulan, Galeran, cut off the feet of some peasants who had felled some trees without permission in the forest. In 1365 a census showed there to be 10 forest guards each responsible for a subdivision or *metier*. In 1407, faced with an abundance of brigands, the Count of Pont-Audemer organized a general sweep of the forest, rounding up large numbers who were then summarily punished (Foubert 1985). After the Revolution, the forest was appropriated by the State and, not wishing to keep its name with its religious origins, was renamed 'the Forest of National Unity'. Needless to say, this name was never popular and was dropped from usage almost as soon as it was coined. At this time, the town of Le Havre was Brotonne's biggest customer, taking each year around 24000 cords (a cord was variable, depending on the region, but was generally around 3.6 m^3 stacked), mostly of beech; this reminds us of the value of firewood in the days before the widespread use of coal. It is this historical background that explains the nature of the forest as we see it today. Today, the forest is one of the most compact in Normandy and covers 7000 ha. Beech represents 61%, Scots pine 21%, hornbeam 8.5% and oak 3% of the species, with other less important conifers and broadleaves making up the balance (6.5%). The very low figure for oak is due in part to its selective removal in the past. Although the forest is now an important regional natural park, it still manages to produce very high-quality timber. The high quality of timber from Brotonne assures it of a good market and most of it is processed in Normandy at a sawmill specializing in beech furniture. The forest, like many in Normandy, contains a number of large, venerable trees of considerable age. A spectacular beech with 22 m to the first branch and a height of 40 m known as 'Le Hêtre Chandelle' dwarfs the surrounding trees, which are a mere 150 years old compared to this fine beech's 230 years (Fig. 14.5).

There is also a remarkable oak known as 'Le Chêne Cuve' with four massive stems arising from a short main stem; it may have been the product of pollarding. Its circumference is 6.6 m and its height is 34 m with its age between 200 and 250 years. There used to be five large stems, but one was cut off in an act of vengeance

Fig. 14.5 Le Hêtre Chandelle, Forêt de Brotonne.

in 1830 by marauders who were unhappy with the vigilance of a forest guard named Letailleur!

14.4 FORÊT DE LYONS

This forest is widely regarded as one of the finest beechwoods in France and, indeed, in all of Europe. It covers an area of 10 600 ha and is composed of 70% beech, 20% hornbeam (*Carpinus betulus*) and 10% oak. Conifers, which were only introduced following the Second World War, cover a mere 70 ha. As with the Forêt de Brotonne, it is managed on a 150-year rotation under the uniform shelterwood system. It covers a plateau area dissected by both river valleys and ancient dry valleys and has an altitude ranging from 45 to 200 m. The major soil type is clay with flints though there are many limestone beds throughout the forest. At its centre is the ancient town of Lyons-la-Forêt, one of the few Gallo-Roman settlements that was not abandoned and reclaimed by forest during the Dark Ages.

From earliest times, the forest had an importance as a hunting forest for kings and dukes. Many fine hunting lodges were built in the area and continued to be built right up to the time of the Revolution. Initially, the rights of usage (grazing, pannage, fuelwood and timber) were dispensed locally but always subject to the requirements of hunting. Following the Revolution, the forest passed into public ownership and there then began a period of unrestrained exploitation. The forest guards were so worried by this that they sought to reason with the people. When reason failed, they brought in the cavalry of the Republic to restore order! However, this 'order' was little better. The forest was the main provider for firewood for Paris and Rouen and vast quantities were floated upstream and downstream on the Seine to these cities. The devastation of the forest was perhaps not all bad. Just prior to the Revolution, the forest was described as having a vast number of very ancient beech and oak which were decaying from old age and whose huge spreading branches prevented the development of coppice and, incidentally, of any natural regeneration (Reed 1954). It is the regeneration of beech that followed the depredations of the Revolution that created the magnificent beech woods of the mid-twentieth century. In addition to firewood, the forest also provided oak for Napoleon's navy and fuelwood for the iron and glass industries that flourished in the area in the seventeenth and eighteenth centuries. Another industry often overlooked, but nonetheless extensive, was the manufacture of sabots or clogs. In the Forêt de Lyons in the nineteenth century, there were 435 tree fellers and 240 clog makers producing 422 000 pairs of clogs each year. In total, towards 1880, around a thousand people worked in the forest, but by 1910 this was already halved (Foubert 1985). In the early 1880s, the foresters were again establishing control over the forest and were working towards establishing stability and sustainability.

The system we now call the shelterwood system was introduced to advance the success of natural regeneration and the concept of rotation age was established. From this time, the fortunes of the forest have improved but the oak (though never numerous at the best of times) did not recover from the demands placed on it for house- and ship-building timber. Nonetheless, the forest still holds a number of ancient, venerable beech and oak trees. Where Brotonne has three or four, Lyons has over 20 roughly equally divided between beech and oak.

14.5 FORÊT DE BORD AND FORÊT DE LOUVIERS

The Forêt de Bord and its immediate neighbour, Forêt de Louviers, cover some 4600 ha, with around 50% beech, 40% oak and 5% Scots pine. The forests are remarkable in that they have continued in more or less the same size and composition since at least 1014. After having constituted a seventh part of the ancient grand forest of Aulerques-Eburovices that covered much of Haute-Normandie and remained largely intact through the Roman occupation, they passed into the hands of the Dukes of Normandy around the start of the second millennium. On 16 October 1197, Richard the Lionheart offered de Louviers and other areas to the Archbishop of Rouen, Gautier de Coutance, in compensation for his appropriation of the town of Andelys and the construction of Chateau Gaillard. A clause in the agreement prohibited any form of clearance for cultivation within the forest area. So, in contrast to many of the church-owned forests in Normandy, which suffered extensive deforestation, the forests enjoyed an existence of strictly limited exploitation. In one interesting episode, the new proprietors sought to acquire, from the forest, the wood they needed for the renovation of the Bishop's Palace in Rouen. Thus, in 1228, the Archbishop, Thibaut, went ahead with some major felling in his domain. But, for some obscure reason, the Bailiff of Vaudreuil opposed the movement of the timber to Rouen and seized the timber for his own use. This interference by an officer representing the authority of the King in the affairs of the Church did not please the Archbishop, who promptly ex-communicated the Bailiff. The King then reminded the Archbishop that, in his view, he could only remove wood from the forest for the use in the manor of Louviers and nowhere else. With this remark, the King had only succeeded in fanning the flames of the confrontation and the affair was brought before the Pope. The Pope confirmed the sentence of ex-communication, and the final result was that the Bailiff was made to cart the wood to Rouen to lift the sentence. This was the last challenge to the right of the Archbishops of Rouen to manage the forest, a right that continued for nearly six centuries. Even when industrialization arrived, the forest remained largely unscathed as there were no glassworks or forges close-by. The only industrial activity consisted of lime kilns and tileworks.

With the Revolution, the forest was declared a national forest and promptly laid to waste, particularly during the winter of 1793 when violations abounded—thefts of timber and falsification of felling records by officials. Gradually, order was restored and by 1870 the State had extinguished the last of the rights of usage and brought the whole forest under the new management system.

Where the Forêt de Louviers passed into the hands of the church and remained so for nearly six centuries, the Forêt de Bord remained a Royal forest for the same period of time. Though areas were rented out, the whole forest was under almost continuous stewardship of various Forest Guards. One of the first guards, Etienne Louvel,

was appointed in 1418 by the then King of England, Henry V, as he passed through the forest on his way to Pont-de-l'Arche. In 1715, the writer, Hector Bernard Bonnet, visited the forest and described it as well managed and subdivided under six forest guards. As with Louviers, Bord suffered after the Revolution and it too was eventually brought under full State control. However, during the Franco-Prussian war, the forest suffered badly at the hands of local inhabitants. The depredations became so frequent and severe that the forest administration called in the gendarmerie. In December 1870, during the space of 15 days, men with saws and axes destroyed 25 ha of 70–80-year-old high forest in one canton. Thousands of people went into the forest seeking easy pickings of timber and fuelwood during that winter. In this way, a total of 263 ha of Bord and 115 ha of Louviers were destroyed. Damage was put at 160 000 francs and 451 people were taken to court; 266 were found guilty with penalties totalling over 80 000 francs and over 1000 days in prison. Even in these difficult times of war and insurrection, there was still the will to protect and preserve the forest.

14.6 FORÊT D'EAWY

This beech forest ranks alongside Lyons, Brotonne and Bord for the magnificence of its trees. Covering over 6550 ha, the forest consists of 92% beech with oak limited to a mere 5%. The name is derived from a Celtic word meaning wet, marshy, humid place and its spelling has varied through the centuries as 'd'Alwy', 'Eauy' then 'Eavi' before arriving at its present form. The nature of the soils and climate make this forest one where the process of *crochetage* or scarifying is necessary before regeneration will take place. The forest was completely cleared and settled during the Gallo-Roman period and remains of settlements abound in the forest. As with other areas of Gallo-Roman settlement, it fell into disuse during the Dark Ages and the forest began to take back its former territory. At the beginning of the second millennium, the forest passed into the hands of the Dukes of Normandy and became a hunting forest. The rights of the hunt were not always retained by the dukes but were leased out to local lords. The lords allowed the local monastery much freedom in the forest and, as a result of their clearances, large openings were created. A charter by William the Conqueror granted rights to the neighbouring monasteries which continued up until the Revolution. Not only did the monasteries benefit from the forest but also the local people. The small timber and branchwood was granted to local artisans such as butchers, for their skewers and meat hooks, and blacksmiths, for the handles of their hammers and bases of their anvils. With so much exploitation, the forest was soon degraded.

Two activities then have marked the life of this forest—industry and hunting. The forest was the main source of firewood and building timber for the town of Dieppe between the years 1650 and 1660, with large areas of the forest being cleared. However, it was the glassworks, of which there were four, that took large quantities of wood to fuel their furnaces from around 1670 up until 1812 when the last ceased operating. In spite of this extensive exploitation, the forest continued to provide excellent hunting, particularly of roe deer. Today, the forest is, like all the others, managed for the production of high-quality sawlogs but it still retains a reputation as a forest where deer and game abound.

14.7 FORÊT D'EU

This large beech forest covers nearly 9400 ha, of which 82% is beech, 9% oak, 7% conifers and with other broadleaves accounting for the balance of 2%. The forest has an interesting history in terms of ownership. Originally the property of the dukes of Normandy, it became part of the lands which passed to the Houses of Lusignan, de Brienne, d'Artois, de Bourgogne, de Cleeves, de Guise, etc. This form of inheritance continued for nearly seven centuries until 1660. The estate was heavily in debt and the forest was sold to Mademoiselle de Montpensier but, when she died in 1693, the Duke du Maine inherited the forest. The forest then passed through a series of inheritances to Louis Bourbon. However, after the Revolution, the rights were never sequestered by the State so, on Louis' death in 1793, the forest was simply taken into management by the State. In 1814, the House of Orleans reclaimed the forest and it became the personal property of King Louis-Phillipe. Following further resumptions, the forest was once again back to its former size and firmly under the control of the House of Orleans in 1872. At the start of the twentieth century, the Duke of Orleans sold the forest to a syndicate consisting of his friends. By this means, he sought to carry out massive exploitation of the forest. The Government, discovering this, made a tentative offer to buy the forest. When this was refused, they then decided to invoke the procedure of expropriation, paying the Duke one million gold francs in compensation. The forest was thus preserved from near destruction and passed into the control of the State.

Today, Forêt d'Eu is owned jointly between the State and the Department of Seine-Maritime. Exploitation in the past has followed that of the Forêt d'Eawy, except that it did not suffer from so much clearance for agriculture. It did, however, furnish wood for local glassworks. Today, it provides quality timber for solid beech furniture and for veneer.

14.8 CONCLUSIONS

The history of the beech forests of Normandy covers so much of the history of France that their preservation on historical grounds alone would be valid! They have a longevity that is truly remarkable even though many have passed through periods of abuse and overexploitation. Running through the whole story has been an underlying acceptance that the forests are important and that their products are valuable. There must be few other places with such a continuity of existence, utilization and management. These forests are now of such national importance that the management practices used in them are actually part of the Civil Code of France.

The difference in approach to management by the different owners is quite revealing. The early feudal lords allowed full use of the forests by the local population but brought in controls and forest guards to prevent overuse and exploitation. Subsequent division of forestlands saw the creation of many new forest owners who, in general, followed the same principles. The involvement of the Church in the forests was less positive, especially the Benedictine abbeys. They increased clearance within the forest and at times heavily exploited the timber reserves.

After the Revolution forests passed into public ownership and the role of the professional forest manager increased in importance. Though there were still times of overexploitation the overriding aims of sustainability were now firmly in place. Current systems of management, in use successfully since 1827, are likely to continue, providing high-quality timber from productive forests of great historical and social importance.

Normandy's beech forests have survived largely intact for more than a millennium. Actual historical records demonstrate this beyond doubt. But the historical records also reveal a continuous use and enjoyment of the forests, even if at times to the point of outright exploitation, which appear the key to their survival. As a case study focusing on forest history it shows that forests survive when they are useful and demonstrably seen to be so: they are more valuable to retain than to clear away. Such an observation has important implica-tions for forests under threat, notably those in the tropics, and for sustainable forest management generally (see also this volume, Chapters 10 and 11).

REFERENCES

CRPF (1992) *Centres Régionaux de la Propriété Forestière 1967–92.* ANCRPF, Paris.

Foubert, J.-M. (1985) *Bois et Forêts de Normandie.* Corlet, Condé sur Noireau.

Husson, J.-P. (1995) *Les Forêts Françaises.* Presses Universitaires de Nancy, Nancy.

Matthews, J.D. (1989) *Silvicultural Systems.* Clarendon Press, Oxford.

ONF (1998) *Rapport Annuel 1998.* ONF, Paris.

Peters, R. (1997) *Geobotany 24: Beech Forests.* Kluwer Academic Publishers, Dordrecht.

Reed, J.L. (1954) *Forests of France.* Faber & Faber, London.

Rérat, B. (1999) Une 'nouvelle essence'. *Forêt-Enterprise* **129**, 12–16.

Rotaru, C. (1988) *Exploitation forestière et dispositions relatives à la protection de la régeneration naturelle.* CTBA, Paris.

15: Restructuring of Plantation Forest : Kielder, United Kingdom *

GRAHAM GILL AND BOB McINTOSH

15.1 INTRODUCTION

Kielder Forest in Northumberland is the largest state-owned forest in Britain and one of the oldest. It is managed by Forest Enterprise, an agency of the British Forestry Commission. Afforestation began in 1926 and proceeded quickly, especially during the period 1945–60. By the late 1970s some 50 000 ha of previously treeless land had been planted with a relatively narrow range of coniferous tree species. In the interval between planting and the beginning of large-scale timber harvesting in the early 1980s, perceptions of the role and function of forestry in Britain underwent significant changes. The original, rather narrow, objectives of the state's involvement in forestry were to provide a strategic reserve of timber against the possibility of future wars and as a source of employment in rural areas. These were gradually replaced by a broader vision of the benefits that forests can provide, initially by the inclusion of secondary objectives such as the provision of public recreation and a duty to seek a balance between the needs of timber production and the environment. These desires have developed further into today's aim that state forests be managed sustainably to produce environmental, financial and social benefits, the environmental quality and productive potential being maintained or enhanced. This case study describes the afforestation history of Kielder and the current management strategy which is converting a single-purpose, rather uniform plantation into a diverse structured forest capable of the sustainable delivery of today's multipurpose objectives.

15.2 GEOGRAPHY, EARLY HISTORY AND LAND USE

15.2.1 Location

Kielder Forest is located in the north of England adjacent to the Scottish border. Situated in a relatively compact block it straddles the boundary between the counties of Cumbria and Northumberland and occupies a large part of the upper North Tyne valley. Some 58% (36 000 ha) lies within the area administered by Northumberland County Council, 20% (12 500 ha) lies within the area administered by the Northumberland National Park Authority and the remaining 22% (13 500 ha) is within the area administered by Cumbria County Council.

About 5% of the area is below 200 m elevation, 35% above 300 m, leaving the bulk of the forest between 200 and 300 m. The highest ground occurs in the north of the forest, where Peel Fell rises to 602 m, but exposure at this elevation is limiting to tree growth, and afforestation has not generally extended uphill beyond 400 m.

15.2.2 Geology and soils

Virtually the whole of the forest is underlain by rocks of the Carboniferous limestone series. The three main groups of rocks are the Cementstones, the Fell Sandstones and the Scremerston Coal Group. The Cementstones consist mostly of shales and form the low, relatively smooth and featureless land in the upper reaches of the North

* This case study is based on a revision of a paper by R. McIntosh (1995) published in *Forest Ecology and Management* **79**, 1–11.

Tyne valley and the Upper Rede valley. The Fell Sandstones occur to the north of Kielder village and give rise to the high hills such as Peel Fell and Deadwater Fell. The harder nature of these rocks has resulted in steeper slopes and more topographic shelter than is generally encountered in the rest of the forest, the bulk of which is underlain by the softer rocks of the Scremerston Coal Group. This group consists of successive layers of shales, limestones and coal seams and gives rise to more subdued, rolling topography, interspersed with peaty flats and dissected by occasional relatively shallow water courses. Occasional igneous dykes have intruded through the softer rocks and are quarried for road-building material. The area has a long history of coal mining. Apart from localized rock outcrops, the bulk of the area is covered by a layer of boulder clay or, on the upper slopes and in hollows, by peat deposits. The boulder clay is locally derived and, though it coarsens in nature on the steeper slopes over the Fell Sandstones, it is generally dense and fine grained.

Although the soils of Kielder have not been extensively mapped, sufficient pilot surveys have been carried out in different topographic zones to give an indication of the distribution of soil types within the forest (Table 15.1). Valley bottom sites have small areas of Brown Earth soils, and small areas of Podsol and Ironpan soils occur over the Fell Sandstones. In general, such freely draining soils are very poorly represented and the vast majority of the forest is on poorly drained, poorly aerated subsoils with a variable peat layer, the depth of which increases with increasing elevation. The preponderance of gleyed soils reflects the nature of the glacial till and has a significant impact on the productivity of the sites and the ability of trees to root deeply.

15.2.3 Exposure and windthrow risk

Mean annual rainfall recorded at Kielder Castle (elevation 250 m) is around 1300 mm. The driest months are April–July and the wettest October–January.

The combination of relatively high elevation and the lack of significant topographic shelter leads to relatively high exposure levels. These,

Table 15.1 Soil type distribution (%) within Kielder Forest.

Soil type	Topographic zone		
	Valley bottoms	Lower/mid slopes	Upper slopes
Brown Earth Ironpan, Podsol	13	2	5
Surface water gley	40	35	2
Peaty gley <45 cm peat	42	52	60
Deep peat >45 cm peat	3	7	29
Other	2	4	4
Sample size (ha)	1998	6161	1099

coupled with the poorly draining soils, with restricted rooting depth, leads to a high risk of trees suffering windthrow as a result of the normal winter gales. A system of classifying upland site types in Britain according to degree of windthrow risk has been described by Miller (1985) and Quine & White (1993) and is based on an assessment of the degree of windiness, the degree of exposure as assessed by Topex readings (Wilson 1984), the elevation and a soil scoring system. Most of Kielder Forest lies within Windthrow Hazard Classes 4 and 5, which effectively means that stands of trees are likely to suffer high levels (over 40%) of windthrow when they reach a mean height of about 20 m in unthinned stands and 17 m in thinned stands. The influence of wind on management practices in forests like Kielder is therefore profound and often limits rotation lengths for the predominant species to 45–50 years.

15.2.4 Early history

Fluctuations in the extent of native woodlands in Britain following the retreat of the ice sheets some 14 000 years ago have been well documented and are mirrored in northern England, where pollen analysis from areas of deep peat (e.g. Pearson 1960; Chapman 1964) indicates the former presence of predominantly broadleaved woodland, consisting of mainly oak (*Quercus* spp.) and alder (*Alnus* spp.) on the lower slopes, with an upper fringe of birch (*Betula* spp.) and possibly Scots pine (*Pinus sylvestris*) on sandy soils on the high sandstone outcrops. Hedley (1950) suggests that a substantial degree of forest cover survived into the sixteenth century but thereafter was gradually removed by man. By the early part of the twentieth century, the landscape of the Kielder area was dominated by open, tree-less hills supporting a relatively extensive form of agriculture dominated by sheep husbandry, with occasional small areas of late nineteenth/early twentieth century tree planting and small remnants of native woodland.

In 1910 a young Australian, Roy L. Robinson, was commissioned by the Board of Agriculture to investigate the possibilities of timber growth and the prospects for afforestation in Great Britain. In the Kielder area he found a large tract of poor agricultural land, the climate and soil of which he described as 'wet and difficult'; nevertheless, he considered that it had the potential for the creation of a large spruce forest. Planting began in 1926 and the acquisition of Kielder Castle and 19000 ha of land from the Duke of Northumberland in 1930 paved the way for further acquisitions of land throughout the period 1930–70.

The problems and difficulties involved in establishing the early plantations have been well documented (Forestry Commission, unpublished). The first trees were notched directly into the undisturbed ground surface but this was not successful and was quickly replaced, initially by the technique of planting on top of an upturned turf, and later (in the early 1940s) by the introduction of tractors and ploughs. Initially the tractors ploughed at 4.5-m intervals with the plough ridge cut and spread by hand in the intervening space at 1.5-m intervals, but by the late 1940s this method had been replaced by ploughing at 1.5-m intervals with planting on the upturned ridge.

It was recognized at an early stage that spruce should be the main species at Kielder, and both Norway spruce (*Picea abies*) and Sitka spruce (*Picea sitchensis*) were used. Scots pine featured in the early planting but suffered heavily from damage by black grouse (*Tetrao tetrix*), until the number of these birds declined across the region for reasons unknown. Lodgepole pine (*Pinus contorta*) and Corsican pine (*Pinus nigra*) were also tried, as were Douglas fir (*Pseudotsuga menziesii*) and various larches (*Larix* spp.) and firs (*Abies* spp.). A pattern of species choice in relation to vegetation type soon emerged (Table 15.2), as did the value of applying basic slag (phosphate fertilizer) around each tree at the time of planting on less fertile sites. By 1950 the species distribution on the planted area was 50% Sitka spruce, 24% Norway spruce, 11% Scots and lodgepole pines, 8% Japanese larch, 2% Douglas fir, 3% other conifers and 2% broadleaved species. The superior growth rate of Sitka spruce on most types of site became obvious, and was increasingly used, until by the mid 1980s it comprised 72% of the forest. The remainder consisted of Norway spruce (12%), lodgepole pine (9%), Scots pine (2%) and broadleaves (1%). During the 1960s and 1970s the interval between plough ridges was widened and the initial stocking density of trees was reduced from around 4000 stems ha^{-1} to 2500 stems ha^{-1}.

In 1974 the decision was taken to build Kielder

Table 15.2 Initial species choice in relation to vegetation type.

Dominant vegetation/soil type	Species choice
Molinia	Norway spruce below 180 m, Sitka spruce over 180 m and on exposed sites
Mixed grassland on mineral soils	Norway spruce or Douglas fir
Molinia/Calluna peat	Sitka spruce in mixture with Scots or lodgepole pine
Calluna on hard knolls	Scots pine, lodgepole pine or larch
Calluna/Tricophorum/Erica	Sitka spruce in mixture with lodgepole pine
Eriophorum/Calluna	Sitka spruce pure, or in mixture with Scots or lodgepole in pine
Pteridium	Norway spruce or Sitka spruce depending on elevation
Raised bogs	Mostly unplanted but small areas planted with Sitka spruce in mixture with lodgepole pine, pure lodgepole pine

Water, a reservoir of 1000 ha, and during the next 6 years some 425 ha of forest were cleared for its construction. The creation of this resource plus the associated road improvements changed the forest significantly and provided an impetus to the recreational use of the forest. Kielder Water supplies water to Newcastle and to the North-east Region. It can provide up to 909 million litres per day, nearly as much as all the other water resources in the region added together. The maintenance of water quantity and quality from the forested catchment is a significant consideration in forest management operations.

15.2.5 Land use and tree productivity

Current land use is shown in Table 15.3, along with an indication of the productivity of the main tree species used, expressed in terms of mean general yield class (GYC) (Hamilton & Christie 1971). GYC is a measure of the maximum mean annual volume increment of a stand of trees in $m^3\ ha^{-1}\ year^{-1}$. The superior productivity of Sitka spruce is clear, and contrasts with the poor growth rates of broadleaved trees on these site types. The major unplanted areas consist of high-elevation heather (*Calluna vulgaris*) moorland, not planted because of high exposure, areas of raised bog originally considered too wet to be planted, and areas still in agricultural use which

have been left unplanted to produce a more diverse land-use pattern.

15.3 MANAGEMENT OBJECTIVES

The stated aims of Forest Enterprise, the management agency of the Forestry Commission, are:

to produce the environmental, financial, social and other outputs sought by Ministers and the Forestry Commissioners in a way which meets the Government's objectives and international commitments and sustains both the environmental quality and the productive potential of the forest estate; and to offer an efficient service (Forestry Commission 1996).

Management at Kielder aims to create and sustain a forest which is attractive as well as productive, a pleasant place for people to visit, a forest rich in wildlife, both plant and animal, and where the natural and cultural heritage is safely conserved. It also aims to generate sufficient income to pay for these activities and provide a financial return on investment.

15.4 FOREST DESIGN PLANNING

The end of the first rotation has provided the opportunity to redesign Kielder Forest, transforming it from a vast even-aged plantation to a diverse

Table 15.3 Current land use in Kielder Forest.

Land use	Area	Mean GYC*
Sitka spruce	72%	12
Norway spruce	12%	9
Scots pine	2%	8
Lodgepole pine	9%	6
Larch	1%	8
Other conifers	3%	8–10
Broadleaves	1%	0–2
Total planted area	42 500 ha	
Open space intimately linked with plantations (roads, rides, deer glades)	7 500 ha	
Discrete areas of open space—raised bogs, agricultural areas, unplanted heather moorland	12 000 ha	
Total area	62 000 ha	

* General yield class.

structured forest better able to deliver the full range of benefits demanded by modern objectives.

To ensure effective translation of stated objectives into a multipurpose management plan, a structured approach to forest design planning has been developed (Hibberd 1985) and is being applied in Kielder and to all other Forestry Commission forests (Bell 1998). The steps involved are described below.

15.4.1 Design plan areas

The forest is divided into a number of landscape units or forest design plan areas ranging in size from 1000 to 10000 ha. Areas are chosen to have a degree of landscape integrity, for example where valleys are separated by well-defined ridges, the valley or water catchment is appropriate. In more rolling topography, the area between two main water courses may be a more natural definition. In either case, there is a need to ensure that separate plans are integrated across plan area boundaries. Plan areas should be as large as can be managed practicably to allow a weighting of objectives at an appropriate spatial scale. Too small a scale risks an excessively fine-grained approach, with suboptimization of multiple outputs at forest level.

15.4.2 Basic survey and appraisal

For each design plan area a series of maps at 1:10000 or 1:20000 scale is prepared showing the underlying landform, the current species and open space distribution, the predicted optimum felling date (age of optimal financial return) for each stand of trees and the location of environmental and recreational features such as water courses, picnic sites, rights of way and sites of archaeological or nature conservation interest.

15.4.3 Design plan concept

The next stage is to produce a concept map for the design plan area. Different objectives carry different priorities in different parts of the design plan area. Thus recreation may be a priority objective on the lower ground or close to public roads. In other places, due to rarity or diversity, wildlife conservation may be the priority. Usually, priority does not necessitate exclusivity, so for example timber production can still be accommodated in recreation areas. The concept map is a means of allocating weightings to multiple objectives in a spatially coherent manner.

15.4.4 Felling plans

This plan identifies the location, shape and size of felling coupes where clearfelling is the preferred system, and areas to be managed under a continuous cover regime (Plate 1). The plan also shows the 5-year period when each coupe will be felled. There are four principles involved in producing the felling plan. First, the felling coupe boundaries should be windfirm, so felling needs to stop at a definite break in the crop where adjacent trees have had the opportunity to develop a stable edge. Second, the shape and size of any coupe should reflect the underlying landform, and take account of the landscape scale. In practice, this means that on the lower slopes and valley bottoms, around public roads and recreation areas, felling coupes are generally small (5–15 ha). On the middle slopes, where the view of the forest is mainly in the middle to long distance rather than from within, coupes are generally 25–50 ha (as can be seen in inset photograph on back cover of this volume). On the upper slopes and high-level plateaux, less visible from public roads and recreation areas, coupe size is generally 50–100 ha. Coupe size is driven primarily by visual consideration. However, the range of coupe sizes is also believed, from what limited information is available, to be appropriate to the requirements of a range of wildlife species. Forest biodiversity, in principle, is expected to be enhanced by ensuring wide variation in coupe size. Third, the timing of fellings should be arranged so that the interval between felling of adjacent coupes is, whenever possible, 7 years and preferably more. This means advancing or delaying felling from the financial optimum. On wind-hazard class 5, Sitka spruce of GYC 12 reaches terminal height (the height at which 40% of the crop is expected to have blown down) at age 47, only 1 year beyond the financially optimum rotation length. Opportunities for delaying felling are therefore very limited at

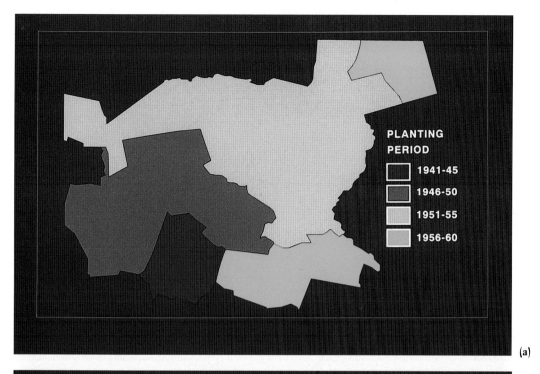

(a)

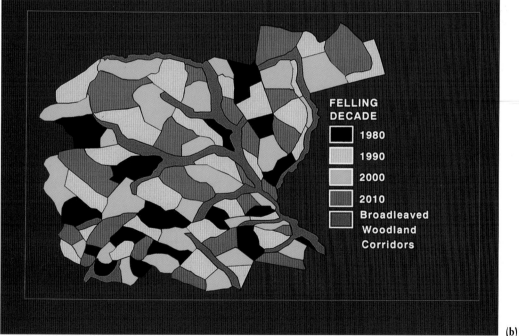

(b)

Plate 1 (a) A typical 1000-h section of Kielder Forest. The colours represent the planting years by 5-year bands which would be closely reflected by financially optimum felling years. (b) The preferred felling pattern, breaking up the even-aged areas of forest into coupes which in scale and shape are more in keeping with the underlying land form. Felling is spread over a period of 20 years. Riparian zones along main water courses become a permanent feature in the restructured forest.

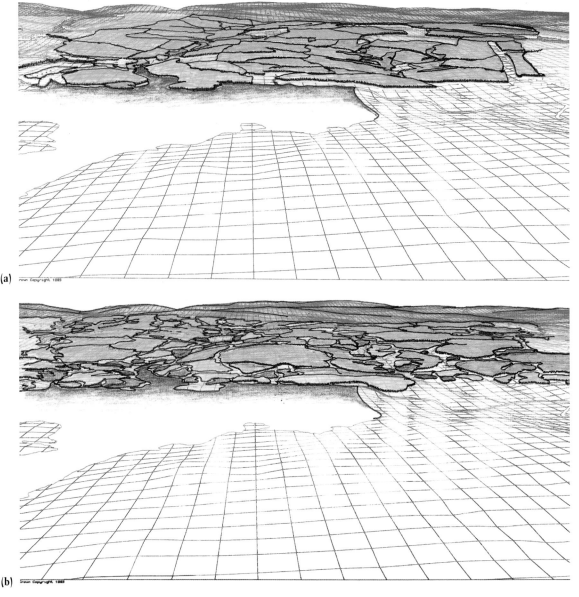

Plate 2 Computer-generated views of part of the forest north of Kielder Water. (a) The forest as it was in 1990. (b) The forest as it should appear in 2035, once the design plan has been implemented. The objective is to create a more natural forest in harmony with its surroundings. Computer-aided drawings are extremely useful in forest design as they allow the forest to be 'viewed' at any point in time from any location and postion (the pictures shown here are of elevated views from a popular viewing point).

Kielder, and age-class diversity has to be achieved by premature felling. This means that an element of revenue is forgone and the timber yield, both in total and in the more valuable sawlog content, is reduced. By making use of what variation there is in first rotation growth, and by careful juggling of felling dates, experience at Kielder is that the level of revenue forgone to achieve restructuring can usually be limited to 10% of the revenue which could have been achieved by felling all coupes at the financial optimum (carpet roll-up). The fourth requirement is that all coupes must have frontage to an existing or planned forest road: ground conditions are such that forwarders must travel on constructed roads or on mats of fresh branches. A computer-aided design package is used to simulate the visual impact through time of the planned pattern of felling and replanting. A separate production forecasting package is used to test that timber production from the proposed plan is compatible with marketing commitments.

15.4.5 Restocking plans

This plan shows the future distribution of tree species, along with the location and extent of open spaces (Plate 2). Major water courses are identified as the focus for most of the planting of broadleaved trees. The standard procedure is to create wide riparian zones along major water courses defined by the topography of the valley, with mainly native broadleaves interspersed with open space. Additional open space is built into the remainder of the forest in the form of deer glades and irregularly shaped perimeter rides around each compartment.

15.5 TREE SPECIES CHOICE

The choice of tree species at the time of restocking is an important issue which affects the future productivity of the forest, the marketability of the produce, the visual impact of the forest and wildlife and amenity value. Following an appraisal of the options available on the site types found in Kielder Forest, a species choice rationale has been drawn up with the aim of achieving the greatest productivity while meeting environmental objectives. The key issues are described below.

15.5.1 Productivity and marketability

Of all the conifer and broadleaved tree species capable of growing in Kielder Forest, Sitka spruce has the fastest growth rate and produces readily marketable timber. The mean general yield class (GYC) of current Sitka spruce stands is 12 (Table 15.3), and the use of selectively bred planting stock should enable GYC14 to be achieved. The GYC of other conifers ranges from 6 to 8 for Scots and lodgepole pine to 10 for Norway spruce. The marketability and unit value of pine timber is rather less than for spruces. Few broadleaved species are capable of achieving growth rates of $2–4\,m^3\,ha^{-1}\,year^{-1}$, even on the most fertile sites. The net effect is that any departure from the planting of Sitka spruce results in a loss of revenue through lower productivity unit values. Table 15.4 shows the profitability of the different tree species planted in Kielder Forest, based on a comparison of the discounted (at 5%) net revenue generated over one full rotation (unthinned), relative to the value of 100 for Sitka spruce.

15.5.2 Visual impact

The contrast in colour and shape among the individual tree species provides a basis for creating visual diversity to enhance both the within-forest experience and the view of the forest from afar. To provide a meaningful contrast when viewed from a distance the species chosen must have distinctly different appearances in terms of shape, colour and texture. Only larches and broadleaved trees are capable of providing such contrast with spruce when viewed from the middle to long

Table 15.4 Relative profitability of species planted in Kielder Forest.

Species	Relative profitability
Sitka spruce	100
Norway spruce	40
Scots pine	27
Lodgepole pine	34
Japanese larch	41
Douglas fir	71
Broadleaves	0–5

distance, although pines are distinguishable from spruces and firs at short to middle distances. The relatively small differences between Sitka spruce, Norway spruce and Douglas fir are unlikely to be noticed other than by close contact. Thus, where the emphasis is on the appearance of the forest when viewed from distant points, only larches and broadleaved trees provide the contrast with spruce, while in situations where the intimate, within-forest experience is the more important, a wider variety of species will provide visual diversity. In many instances, the value of the forest for informal recreation is related to its 'attractiveness' when viewed from within. Attractiveness is related to the degree of structural and species diversity. Because of the relatively restricted species choice available on infertile and exposed sites, much of the diversity will be provided by manipulating the structure of the forest rather than the species mix.

15.5.3 Silvicultural considerations

Browsing damage

The tree species discussed in this chapter vary considerably in their susceptibility to damage by roe deer (*Capreolus capreolus*), and this is a major factor affecting the cost of establishment. Current roe deer management aims to allow the establishment of Sitka spruce without fencing. By contrast, fencing or individual tree protection is required to establish broadleaved trees and most conifers other than Sitka spruce. The cost of deer fencing adds some £150–400 ha^{-1} to the cost of establishing alternative conifer species, and there may be a further additional cost associated with planting broadleaved trees because of the need to give individual protection in tree shelters.

Planting costs

Given similar attention to plant quality and plant handling as with Sitka spruce, there is no reason for the cost of planting and beating up (replacement of losses) to be substantially different for conifer species, but broadleaved trees are generally more expensive to buy and plant.

Natural regeneration

Examples of natural regeneration of Scots pine, lodgepole pine and Norway spruce can be found occasionally, but generally the only two species which regenerate profusely in Kielder are Sitka spruce and birch (and occasionally *Tsuga heterophylla*). In both cases this can be considered as an asset or a liability, depending on which site it appears, and on the extent and density of the regeneration.

15.5.4 Wildlife conservation potential

The enhancement of the wildlife conservation value of the forest involves both an increase in the general level of habitat diversity and the adoption of specific measures for the conservation of particular species. While these aims can be partly achieved by manipulating the structural diversity of the forest, tree species choice is also important The marked differences between conifers and broadleaves provide an opportunity to create distinctly different habitats, while the main value of conifer species for wildlife is due to the scale and periodicity of seed production and the size of seeds produced. Conifer seed is a crucial food for red squirrels (*Sciurus vulgaris*), crossbills (*Loxia curvirostra*) and siskins (*Carduelis spinus*). Evidence is accumulating that the relatively large seeds of Norway spruce, larches and pines make them more valuable as food than seed from Sitka spruce, particularly for red squirrels. Differences between conifer species in the periodicity of seed years indicates that maintaining proportions of pines, larches and Norway spruce will add significantly to the wildlife value of the forest and may be essential if a viable red squirrel population is to survive. However, recent research suggests that the proportion of these conifers should not be too high, otherwise it can make the forest too attractive to grey squirrel (P. Lurz, personal communication). The importance of large-seeded broadleaved trees (e.g. oaks and beech) as a food source for grey squirrels (*Sciurus carolinensis*) may become an important issue in future, since this species is gradually expanding its range towards Kielder from both north and south. If maintaining a red squirrel population were the only conservation consideration, a case could be made for excluding all or some broadleaved species from the forest. However, the situation does not warrant such a policy given an overall aim of multipurpose management and the many benefits that broadleaves bring to other plants and animals. The concentration of broadleaves in valley bottoms and riparian zones still leaves the potential to create large areas of predominantly broadleaf-free forest where invasion by grey squirrels may be less likely. However, to ensure that this policy does not prejudice the future of red squirrels in Kielder, the Spadeadam section (5000 ha) has been designated as a 'red squirrel reserve'. Here the intention is to exclude completely large-seeded species of broadleaves. Coupled with a policy of maintaining a higher than average proportion of the larger-seeded conifers this should provide habitat that is more attractive to red squirrels than grey squirrels.

Carefully planned and sited open space can greatly add to structural diversity and increase the proportion of edge within the forest. The proportion of open space (15%) incorporated at the time of initial planting is acceptable in terms of total area, but the distribution was influenced by the need then for a geometric pattern of rides and firebreaks. A redistribution of that open space is required at the end of the first rotation to ensure more effectively visual and wildlife diversity, and enhanced deer control opportunity.

The protection of freshwater habitats from physical and/or chemical pollution is also an important objective of management. The geology of Kielder is such that forest streams have a high buffering capacity and there is no evidence of reduced pH in streams with afforested catchments in comparison with those draining moorlands. A policy of creating broadleaved woodlands and open spaces in the major riparian zones will ensure that any potential for damage through the effects of siltation, overshading or acidification is minimized. Indeed, the relative absence of grazing with the resultant overhanging bankside vegetation should provide a considerably improved habitat for fish.

15.5.5 Prescription for the choice of tree species

Clearly Sitka spruce is relatively cheap and easy to establish and provides the option of using silvicultural systems based to some extent on natural regeneration. This species is the highest volume producer and its timber is preferred by a wide range of markets. The financial return from planting Sitka spruce is considerably higher than from any other species, and inevitably it will remain the dominant tree. Any departure from using Sitka spruce must be justified largely on grounds of visual amenity or wildlife conservation value. In areas heavily used by the general public, almost all of the species capable of growing in Kielder have the potential to add to the amenity value, while only larches and broadleaves (and to some extent pines) provide a significant visual contrast with Sitka spruce when viewed at the landscape scale. The use of Norway spruce, Scots pine, lodgepole pine and broadleaved tree and shrub species is likely to increase significantly the wildlife conservation value of the forest.

Based on these conclusions it is possible to describe a series of species choice prescriptions to achieve an acceptable level of balance between the conflicting demands of the forester, the public and wildlife. Table 15.5 gives these for the forest by site type and for the forest as a whole and Table 15.6 compares the net present values (NPV) generated by one rotation of four different species choice options. Option 1 represents the completely commercial option which would in practice be unrealistic because it would fail to meet the minimum environmental standards as laid down by the Forestry Commission (1998). Option 2 is the species choice strategy which is considered to be necessary to meet the standards laid down in the UK Forestry Standard and the UK Woodland Assurance Scheme. Option 3 is the planned species mix and Option 4 represents a further move away from Sitka spruce to broadleaves and other conifers. Net present value is directly related to the proportion of Sitka spruce and drops significantly as this proportion decreases. Taking Option 2 as baseline, Option 3 results in a drop in NPV from £1019 ha^{-1} to £857 ha^{-1}. In annual equivalent terms this

Table 15.5 Species choice proposals for second rotation stands in Kielder Forest.

| Site type | Approximate area (ha) | % of area | | | | Main species of other conifers |
		Open space	Broadleaves	Sitka spruce	Other conifers	
Areas of high visual amenity and recreation importance. Generally valley bottom sites	4 000	12	18	40	30	Douglas fir Norway spruce Hybrid larch Scots pine Lodgepole pine
Lower slopes with moderate levels of exposure. Gererally surface water gley soils of flushed peaty gleys with <30 cm peat	23 000	15	7	65	13	Norway spruce Hybrid larch Scots pine
Upper slopes and flat areas with relatively high exposure levels. Soils generally deep peats or peaty gleys with >30 cm peat	18 000	15	5	71	9	Lodgepole pine
Spadeadam red squirrel reserve	5 000	15	5	60	20	Norway spruce Scots pine Lodgepole pine
Whole forest	50 000	15	7	65	13	

Table 15.6 Kielder Forest species choice appraisals.

Option	% gross plantation area	% net plantation area	NPV (3%) (£ ha^{-1})	NPV loss cf. option 2 (£ ha^{-1})	Comments
1	90%SS 10%OS	100%SS	1599		Completely commercial, outside guidelines, unrealistic
2	70%SS 10%OC 5%BL 15%OS	82%SS 12%OC 6%BL	1019		Interpretation of minimum standard under UK Forestry Standard and UK Woodland Assurance Scheme
3	65%SS 13%OC 7%BL 15%OS	77%SS 15%OC 8%BL	857	162	Current proposal
4	60%SS 15%OC 10%BL 15%OS	70%SS 18%OC 12%BL	691	328	Higher proportion of other conifers and broadleaves

BL, broadleaves; NPV, net present value; OC, other conifers; SS, Sitka spruce.

represents a loss of about £7ha^{-1}year^{-1} or £300000 year^{-1} over the whole forest. The decision to opt for a species choice mix which is poorer, in NPV terms, than the minimum standards is justified partly by the fact that we are managing the Forest Park for multiple outputs, not just for timber. Thus, an increase in the proportion of broadleaves and other conifers is considered essential to achieve Forest Park objectives of landscape, recreation and wildlife conservation, including the needs of the red squirrel.

15.6 FOREST OPERATIONS

It is within the context of the forest design planning procedures previously described that forest operations take place and a brief description of the main activities is given below.

15.6.1 Harvesting of timber

The harvesting programme for Kielder Forest has more or less stabilized at the sustained yield level of some 400000 m^3 year^{-1}. Approximately 45% of the timber is sold standing, 25% is harvested by contractors and the remainder (30%) is harvested by Forest Enterprise. Virtually all the harvesting operations are mechanized, and most of the timber is produced from clearfelling rather than thinning operations. About half of the produce is in the form of sawlogs used for the production of construction timber, fencing material and pallet manufacture, while the remainder is small roundwood used for the production of paper and panel products such as chipboard, medium-density fibreboard and orientated strand board. A database containing environmental information is maintained, and all areas due for harvesting are subjected to a preharvesting appraisal to identify areas and issues of environmental importance, ensuring the production of a harvesting plan which identifies how environmental constraints and opportunities will be addressed.

15.6.2 Replanting (restocking)

The current harvesting programme results in the felling of around 1000 ha^{-1} year^{-1} and this generates an equivalent annual restocking programme.

There is currently no programme for the afforestation of additional land. As in the case of harvesting operations, all restocking operations are preceded by a full appraisal of environmental constraints and opportunities. Restocking is carried out with minimal ground preparation, and usually with bare-rooted transplants which have been dipped in permethrin solution to protect them from damage by the pine weevil (*Hylobius abietis*) (Bevan 1987). Weed control is generally only necessary on fertile mineral soils in the valley bottoms and use of herbicides is therefore minimal. While inorganic fertilizers were applied to many stands in the afforestation stage, this has not proved to be necessary on second rotation sites. The unpredictable occurrence of natural regeneration by Sitka spruce is an increasing problem which can result in expensive manual respacing programmes. The protection of plantations from deer damage involves both fencing and a culling programme that removes some 1200 roe deer annually.

15.6.3 Wildlife conservation

Enhancement of the wildlife conservation value of the forest arises largely from increased habitat diversity resulting from forest design planning and species choice strategies employed at Kielder. A wide range of specific conservation projects provides the means for protecting and enhancing particular habitats and species. Management of non-forest habitats includes working with other statutory and voluntary conservation agencies to protect and enhance the Border Mires, a series of raised bogs scattered throughout the southern part of the forest, and also Kielderhead Moor, a significant area of high-elevation heather moorland. Management plans for the Border Mires, agreed with English Nature, identified some 600 ha in total of original bog surface which had been planted with trees, where bog surfaces will be restored by a programme of tree clearance and ditch blocking. The aim is to restore the hydrological integrity of each of the mires, and to incorporate these within the forest design plans.

In addition to the Border Mires and Kielderhead Moor, there are a further five Sites of Special Scientific Interest within the forest. Forest-based projects include monitoring and protection of several important raptor species including goshawk (*Accipiter gentilis*), tawny owl (*Strix aluco*), merlin (*Falco columbarius*) and peregrine falcon (*Falco peregrinus*), bird box schemes to encourage and monitor song bird populations, pond creation and management, a bat-box scheme, special management of open spaces and an ongoing programme of survey work aimed at identifying areas of particular botanical and entomological interest. Key findings from these projects continue to be fed into the design process.

15.6.4 Recreation

An open-access policy is maintained for visitors on foot throughout the whole forest area, except where legal or safety considerations prevent this. In addition, a wide range of facilities and visitor services are provided, of which Kielder Castle Forest Park Centre is the focal point attracting some 80000 visitors a year. Other facilities include a 19-km long forest drive, several car parks and picnic sites, forest trails, long-distance walking, riding and cycling routes, viewpoints and a series of informal back-packing campsites. Overnight accommodation is available in a 70-pitch caravan/camping site and full-time recreation staff provide a range of visitor services including guided walks, four-wheel drive safaris and an educational package. Specialist activities include car rallying, husky dog sled racing, orienteering and mountain biking. It is estimated that the forest area together with Kielder Water receives some 400000 day visitors a year, but the size of the area, coupled with a planned approach to the integration of recreation activities with other forest operations, ensures that few problems are encountered, and that capacity exists to expand the range of facilities available, and to absorb an increasing number of visitors.

15.7 SUSTAINABILITY OF YIELDS

Multipurpose forestry yields a range of outputs, and a central management objective is that these outputs must be sustainable. With the longest established parts of the forest only halfway through the second rotation, it cannot yet be demonstrated whether yields are sustained through successive rotations. It is necessary to make predictions on the basis of processes which can be observed currently. Landscape values will be sustained, we believe, by the process of restructuring and forest design. Computer-generated predictions of future forest landscapes support this view. Similarly, forest design is expected to sustain, and indeed enhance, the recreational potential of the forest. In terms of biodiversity, restructuring has already had an impact in increasing both the variety and abundance of species present in the forest: some species, for example the cyclically very abundant short-tailed field vole (*Microtus agrestis*), have decreased in numbers while the less abundant, non-cyclic bank vole (*Clethrionomys glareolus*) and wood mouse (*Apodemus sylvaticus*) have increased (Petty 1999).

Sustainability of timber yields depends primarily on sustaining the productive value of the soil. Overall, second rotation Sitka spruce appears to be growing as well as or better than first rotation, and with no requirement to re-apply the phosphate fertilizer which was essential for first rotation establishment. While site-specific first and second rotation yield comparisons are not generally available, the recorded average yield class for 1500 ha of second rotation Sitka spruce at Kielder is $12.8 \, m^3 ha^{-1} year^{-1}$ compared with $11.4 \, m^3 ha^{-1} year^{-1}$ for the remaining 24000 ha of first rotation Sitka spruce. Caution needs to be applied to the interpretation of these data: although restructuring means that second rotation crops are well distributed through the whole forest, the comparison is not between identical site types. Also, some of the second rotation crops comprise genotypes which have been selectively bred for improved yield.

In a review of the influence of forest operations on the sustainable management of forest soils, Worrell and Hampson (1997) conclude that the impacts of some of the more intensive forest management practices may be quite closely balanced with the recovery capacity of some of the more susceptible sites. Compliance with current environmental guidelines for site preparation and harvesting can prevent soil erosion problems on most sites, and should restrict instances of soil compaction exceeding the capacity of the sites to recover, although this is less well researched. Research has indicated that nutrient levels may be quite finely balanced on less fertile sites: for example, inappropriate use of whole tree harvesting on nutrient-poor sites could reduce second rotation growth rates compared to conventional stem-only removals (Proe *et al.* 1999), although it is unlikely to cause a problem on more fertile sites. Clearly, the adoption of best operational practice is essential to sustainable management.

15.8 CONCLUSIONS

The last 20 years have witnessed a substantial move towards the concept of multiple-benefit forestry and the changing structure of forests like Kielder is increasingly reflecting this underlying philosophy. It will take time for the impact of these policies to be fully reflected in the appearance of the forest, but to date it is clear that as the forest moves away from the plantation phase towards a forest which is diverse in structure and species composition, the variety of wildlife species that can adapt to and be accommodated by this emerging habitat is considerable. It means that Kielder Forest was well placed to achieve the independent certification which has been awarded to all Forest Enterprise-managed forests. Our forest management is assessed against the requirements of the UK Woodland Assurance Scheme, a standard recognized by a wide range of industry and environmental bodies. It provides our wood customers with the independent assurance of sound forest management which they and their customers are seeking.

The process of restructuring through forest design has transformed Kielder, and it is a process which is now being applied to large, even-aged forests throughout upland Britain. Indeed, it is a process which would lend itself to any large-scale, even-aged plantation where multiple benefit outputs are now being sought. The process, as described at Kielder, has had to contend with an upland forest with high windthrow risk; in situations with lower windthrow risk, the options are

wider but the principles of restructuring are the same. It is impossible to predict what new pressures forests and forestry policy will be subject to in the future, but it seems likely that the non-market benefits provided by forests like Kielder will become increasingly valued. This does not and need not conflict with the ability of the forest to remain productive and to produce a sustainable yield of valuable timber.

REFERENCES

Bell, S. (1998) *Forest Design Planning.* Forestry Commission, Edinburgh.

Bevan, D. (1987) *Forest Insects.* HMSO, London.

Chapman, S.D. (1964) The ecology of Coom Rigg Moss, Northumberland. *Journal of Ecology.* **52**, 299–313.

Forestry Commission (1996) *Forest Enterprise Framework Document.* Forestry Commission, Edinburgh.

Forestry Commission (1998) *The UK Forestry Standard.* Forestry Commission, Edinburgh.

Hamilton, G.J. & Christie, J.M. (1971) *Forest Management Tables (Metric).* HMSO, London.

Hedley, P. (1950) The medieval forests of Northumberland. *Archaeologia Aeliana, 4th Series* **27**, 96–204.

Hibberd, B.G. (1985) Restructuring of plantations in Kielder Forest District. *Forestry* **58**, 119–29.

Miller, K.F. (1985) *Windthrow Hazard Classification.* Forestry Commission Leaflet 85. HMSO, London.

Pearson, M.C. (1960) Muckle Moss, Northumberland. I. Historical. *Journal of Ecology* **48**, 647–66.

Petty, S.J. (1999) Diet of tawny owls (*Strix aluco*) in relation to field vole (*Microtus agrestis*). Abundance in a conifer forest in Northern England. *Journal of Zoology, London* **248**, 451–65.

Proe, M.F., Craig, J., Dutch, J. & Griffiths, J. (1999) Use of vector analysis to determine the effects of harvest residues on early growth of second-rotation Sitka spruce. *Forest Ecology and Management* **122**, 87–105.

Quine, C.P. & White, I.M.S. (1993) *Revised Windiness Scores for the Winthrow Hazard Classification: The Revised Scoring Method.* Forestry Commission Research Information Note 230. Forestry Commission, Edinburgh.

Wilson, J.D. (1984) Determining a Topex Score. *Scottish Forestry* **38**, 251–6.

Worrell, R. & Hampson, A. (1997) The influence of some forest operations on the sustainable management of forest soils—a review. *Forestry* **70**, 61–85.

16: Sustainable Management of the Mountain Ash (*Eucalyptus regnans* F. Muell.) Forests in the Central Highlands, Victoria, Australia

PETER M. ATTIWILL AND JANE M. FEWINGS

16.1 INTRODUCTION

Mountain ash (*Eucalyptus regnans* F. Muell., Fig. 16.1) grows in cool-temperate parts of south-eastern Australia and is arguably the world's tallest flowering plant. It is mostly restricted to the mountains of the eastern half of Victoria at altitudes of 150–1100 m (there are small occurrences in the western half), and in Tasmania it occurs from sea-level to 600 m in the north-east, south-east and in the valleys of the Derwent and Huon Rivers (Boland *et al.* 1992). The species name (from *regnare*—to rule) is appropriate to the immense height, girth and dominance of the trees. The common name 'mountain ash' is most widely recognized, but 'swamp gum' and 'stringy gum' are used in parts of Tasmania, where it occurs in lower landscape positions. The common name mountain ash refers both to its mountain habitat in Victoria and to the similarity of its timber to that of ash (*Fraxinus*) in the Northern Hemisphere (Boland *et al.* 1992), although the timber is also marketed in Tasmania as 'Tasmanian oak'.

In Victoria, the ash-type eucalypts *E. regnans* and *E. delegatensis* R.T. Baker (alpine ash) grow in essentially pure stands. These ash-type eucalypts, together with more restricted areas of *E. nitens* (Deane and Maiden) Maiden, *E. obliqua* L'Hér. (messmate) and *E. viminalis* Labill. (manna gum), form forests that are classified as 'wet sclerophyll forests' (Beadle & Costin 1952) or 'tall open-forest' (Specht 1970). In Tasmania, *E. regnans*, *E. delegatensis* and *E. obliqua* form 'wet eucalypt forests', which have been termed 'mixed forest' where there is a dense understorey of cool-temperate rainforest species, and 'wet sclerophyll forest' where there is an understorey of broadleaved shrubs (Forestry Commission of Tasmania 1994; Wilkinson 1994). The ash-type eucalypts make up 8.4% of the native forest area in Victoria and 27.7% of the native forest area in Tasmania (Table 16.1).

While Australia's forest industries have an annual turnover of AUS$11 billion per year (=US$6.5 billion; this and other conversions based on a mid-2000 rate of AUS$ approximately equal to US$0.59) which contributes around 2.5% to the gross domestic product, Australia's trade deficit in forest products in 1996–97 was AUS$1.4 billion (US$0.826 billion). Australia's lack of manufacturing capacity products has been attributed, in part, to uncertainty about future access to forest resources and to high environmental standards for the establishment of pulpmills. This lack of manufacturing capacity has resulted in the exportation of low-value unprocessed wood, principally in the form of woodchips, and the importation of high-value, processed, paper products.

The management of forests in Australia has been the subject of numerous and extensive reviews and inquiries, culminating in a decision to implement a system of Regional Forest Agreements between the Commonwealth and state governments, as we explain later. As at June 1999, only four Regional Forest Agreements had been signed. The second Regional Forest Agreement to be finalized was for the Central Highlands of Victoria, a region that contains extensive forests of mountain ash. The Central Highlands includes 181 000 ha of ash-type eucalypt forest, about 52% of which is available for timber production, with the remainder (85 000 ha) in reserves or in otherwise protected areas (Department of Natural

Fig. 16.1 Mountain ash forest near Toolangi, Victoria. The forest is even-aged regrowth following the 1939 bushfires. The trees are 60-year-old; mean diameter at breast height (over-bark) is 80 cm, and mean dominant height is 69 m. Regrowth forest from the 1939 bushfire is the major source of mountain ash for the timber industry.

Resources and Environment Victoria 1998). The timber-based industries of the Central Highlands employ 1537 people, pay an annual royalty of AUS$13 million (US$7.7 million) for an annual log consumption of 860 000 m³, and have an estimated gross value of turnover of AUS$323 million (US$191 million—data for 1995–96; Commonwealth and Victorian Regional Forest Agreement (RFA) Steering Committee 1997). These forests are highly valued as water catchments for the city of Melbourne, for fauna and flora conservation, for recreation and for timber and fibre production. It is the sustainable management of these forests for all their values that forms the case study presented in this chapter.

Table 16.1 Forest areas in Victoria and Tasmania (Attiwill *et al.* 1996, adapted from Department of Conservation and Natural Resources Victoria 1991, and Forestry Commission of Tasmania 1994).

Forest type	Area (10³ ha)
Victoria	
Native forest	
Snow gum forests	184.5
Forests dominated by ash-type eucalypts	629.9
Rainforest	
Cool temperate	6.9
Warm temperate	8.9
Mixed forests of stringybarks, peppermints and gums	
Dominated by *E. obliqua*	1019.9
Dominated by *E. radiata* and *E. dives*	1362.7
Dominated by *E. sieberi*	562.1
Other mixed species forests	946.9
Red gum forests	213.6
Box—ironbark forests	969.6
Mallee forests	1573.5
Other forests and woodlands	43.9
Total for native forest	7522
Plantations	
Hardwood (mainly *E. regnans, E. nitens, E. globulus*)	26.2
Softwood (mainly *Pinus radiata*)	208.4
Total for plantations	234
Total forest area in Victoria	7756
Tasmania	
Native forest	
Wet eucalypt forest	885
Dry eucalypt forest	1594
Rainforest	565
Other	152
Total for native forest	3196
Plantations	
Hardwood (mainly *E. nitens, E. globulus, E. regnans*)	48
Softwood (mainly *Pinus radiata*)	76
Total for plantations	124
Total forest area in Tasmania	3320

It will be impossible to cover all aspects of sustainable forest management here. This chapter is presented in three parts. In the first part, an Australia-wide perspective is taken in a historical review of the social and political environment for sustainability. Issues discussed are state ownership of public forests, the rise of environmental concerns over forest management, and state and Commonwealth interactions and agreements from which the structures for sustainable forest management and the Regional Forest Agreement for the Central Highlands derive. In the second part, some aspects of the associated Forest Management Plan are summarized, again using mountain ash as the basis. In the third part, the ecology of the mountain ash forest as the basis for a discussion of silvicultural systems is reviewed. Thus this chapter has an emphasis on the development of the political framework for sustainability, and on the biological aspects of sustainability.

The first part—a review of the social and political environment for sustainability of forest management in Australia—is the largest part, and to that extent the chapter might be viewed as unbalanced. In defence, it is proposed that long-term planning for sustainability is not possible until social and political issues are resolved and appropriate structures are in place (see also Chapter 2, this volume).

16.2 THE SOCIAL AND POLITICAL ENVIRONMENT FOR SUSTAINABILITY IN AUSTRALIA

16.2.1 Ownership of public forests in Australia

By contrast to other developed countries, Australian forests are predominantly publicly owned. Government actions and policies are therefore central to forest management and to the controversies surrounding their use (Young 1996). Australia is a federation with three spheres of government—the Commonwealth, the states and territories, and local government. The responsibilities of each of these spheres of government for forest management are not always clearly delineated, and this adds another dimension to the role of government policy to the sustainability debate.

At federation in 1901, the six former British colonies which became the States (the two internal territories were formed at a later date from land surrendered by the states and accepted by the Commonwealth) retained responsibility for the land and its accompanying resources. State government powers cover crown land, land-use deci-

sion making, systems of land classification and resource-related industries. Local governments have responsibilities for local land-use planning and rating systems which affect public and private forest management and use. In addition to the three levels of government, private owners have responsibility for management of private forests.

Legislation for forest management varies from state to state and region to region. However, most states and territories have enacted legislation covering environmental impacts, national parks and protection of endangered species. Forest management and planning is the responsibility of a single, integrated agency in some states and in others there are separate agencies for multiple-use forests and for forest reserves. Codes of practice used to guide field operations are based in legislation in some states and are enforced through contracts for the sale of forest products in others.

16.2.2 Interactions between the states and the Commonwealth

Although the states have constitutional responsibility for all land-use matters, the Commonwealth has been able to exert influence and a large degree of control through use of its powers (under Section 51 of the Australian Constitution) with respect to trade and commerce with other countries and between the states, to foreign trading corporations and trading or financial corporations formed within the limits of the Commonwealth, and to external affairs (the power to implement treaties and other international agreements). The Commonwealth exercises its control over trade and commerce principally through the issue of export and import licences, which are subject to the Commonwealth's environmental regulatory regime.

The power to regulate and control the trading activities of corporations also enables the Commonwealth to do so taking account of environmental considerations. The external affairs power enables the Commonwealth Parliament to pass laws to implement any of Australia's international obligations, no matter what their subject matter may be. Since the United Nations Conference on the Human Environment in Stockholm in 1972, Australia has ratified about 43 multilateral conventions to do with the environment (Tsamenyi 1996). These conventions, including those concerned with the conservation of biological diversity, migratory species, wetlands, the world heritage and combating desertification, impose restrictions on land use and are of real consequence for Australia, not only because it has a unique natural environment and is economically reliant on primary industries such as mining, energy, agriculture and forestry (Tsamenyi 1996), but also because their implementation is dependent on a high degree of intergovernmental cooperation.

Furthermore, for constitutional and historical reasons, there is a severe imbalance between the revenue-raising and revenue-spending responsibilities of the levels of government, in the Commonwealth's favour. Revenue redistribution from the Commonwealth to the states in the form of tied grants is another enabling mechanism for the Commonwealth's participation in policy areas over which it has little or no constitutional authority.

The legal and political arrangements for forest management in Australia, involving the three spheres of government, are complex and have 'provided challenges for all parties in developing a national approach to forest management' (Forestry Technical Services 1997). As an illustration of this complexity, Table 16.2 sets out the Commonwealth and state legislative framework over different land tenures for various aspects of forest management in Victoria as they applied in 1997. There have been attempts in the past, at various constitutional conventions, to assign a more explicit role to the Commonwealth in forest management (Carron 1993) and to sort out and allocate broad policy areas to one or other level of government (Galligan & Fletcher 1993). According to Galligan and Fletcher (1993), however, the appropriate strategy is to recognize that the Australian system of governance is 'essentially concurrent' and, for complex policy areas such as natural resources, it is necessary to devise structures for coordinating the joint participation of the spheres of government. While Australia has a tradition of establishing federal institutions which emphasize cooperation and adaptation to changing circumstances (Groenewegen 1994),

determining where responsibility should lie for integrating economic and environmental values and achieving this integration across the levels of government has, in the past, proved problematical (Papadakis & Moore 1994).

16.2.3 The development of Australian forest policy

Relations between the Commonwealth and the states on issues of forest management were fairly untroubled—and pursued in a spirit of co-operation—from federation until the late 1960s. The states established administrative frameworks for managing their forests during this period. Following a Royal Commission on Forests which sat from 1897 to 1901, with 'much of (its) activity in the early years directed towards repairing damage done to the forest resource through previous poor land management practices' (Office of Auditor-General 1993), the Victorian Parliament passed the first act establishing a State Forests Department in 1907. In New South Wales, state forests were gazetted for protection from the turn of the century and the New South Wales Forestry Act of 1916 required that forests be managed on multiple-use principles that provided for the conservation of wildlife and plants, protection of soil and water and for recreation, as well as for the production of timber (Kennedy 1993; Brand 1997). The Tasmanian Forestry Act of 1920 created the Tasmanian Forestry Commission to manage state forests on a similar basis (Kennedy 1993). Development of such state legislation was uneven, and a comprehensive forestry act was not introduced in Queensland until as late as 1959 (Mercer 1995).

At the federal level, heads of the state forest services first met as a group shortly after federation and discussed issues which demanded 'collective consideration in the interests of the whole Commonwealth' (Carron 1993). In 1924 the Commonwealth appointed a Forestry Adviser and established a Forestry Bureau in Canberra to assume responsibility for professional forestry education, forestry research and advice on an Australia-wide basis (Rule 1967; Carron 1993). A national professional forestry school was also established at Canberra in 1927 (Rule 1967).

Although the state forest services made 'good

Table 16.2 Commonwealth and state legislative framework for forest management in Victoria, 1997. From: Commonwealth and Victorian Regional Forest Agreement (RFA) Steering Committee (1997).

| Legislation | Public land | | | Private land |
	National parks and reserves	State forests	Other land	
A. PLANNING				
1. Land use	Environment Conservation Council Act 1997			Planning and Environment Act 1987
2. Land management	National Parks Act 1975 Reference Areas Act 1978 Heritage Rivers Act 1992	Forest Act 1958, Code of Forest Practices 1996	Crown Land (Reserves) Act 1978 Land Act 1958	Planning and Environment Act 1987
3. Timber production	NA	Forests Act 1958 (1990 amended)	NA	Planning and Environment Act 1987 (1993 Amended)
4. Fire management	Forests Act 1958, Code of Fire Practice 1995			Country Fire Authority Act 1958
5. Threatened flora and fauna	Flora and Fauna Guarantee Act 1988 Commonwealth Endangered Species Protection Act 1992			
6. Native vegetation clearance	National Parks Act 1975 Reference Areas Act 1978 Wildlife Act 1975	Forests Act 1958, Code of Forest Practices 1996	Crown Lands (Reserves) Act 1978 Land Act 1958	Planning and Environment Act 1987
7. Woodchip export licensing	NA	Commonwealth Export Controls Act 1982	NA	Coummonwealth Export Controls Act 1982

Supporting legislation
1. State: Environment Protection Act 1970; Archaeological and Aboriginal Relics Preservation Act 1972; Land Conservation (Vehicle Control) Act 1972; Victorian Conservation Trust Act 1972; Environment Effects Act 1978; Conservation Forests and Lands Act 1987; Water Act 1989; Mineral Resources Development Act 1990; Heritage Rivers Act 1992; Catchment and Land Protection Act 1994; Extractive Industries Development Act 1995

2. Commonwealth: Environment Protection (Impact of Proposals) Act 1974; Australian Heritage Commission Act 1975; World Heritage Properties Conservation Act 1983

B. ADMINISTRATION				
1. Land use planning	Environment Conservation Council Act 1997 — Environment Conservation Council (repealing and replacing the Land Conservation Act 1970 and the Land Conservation Council)			Planning and Environment Act 1987 — Department of Infrastructure
2. Land management	Conservation, Forests and Lands Act 1987 — Department of Natural Resources and Environment			Planning and Environment Act 1987 — Department of Infrastructure

Supporting agencies
Commonwealth: Commonwealth Environment Protection (Impact of Proposals) Act 1974 — Commonwealth Environment Protection Agency; Commonwealth Australian Heritage Commission Act 1975 — Australian Heritage Commission; Commonwealth Export Control Act 1982 — Department of Primary Industries and Energy

NA, not applicable.

progress' during the 1920s and 1930s in improving the condition of the degraded forest estate and in building up productive capacity (Florence 1994), the Second World War, and subsequent postwar reconstruction, placed increased pressure on the forest resource. During the war, the Commonwealth exercised complete control over the demand and supply of wood as a material of war under a Controller of Timber working through deputy controllers in the states (Carron 1993). After the war, demand for timber and the introduction of new and more efficient harvesting technologies resulted in previously inaccessible areas of mountainous forest being harvested 'at rates in excess of sustainable yields and often in large clearfelled coupes' (Florence 1994). In the period 1955–60, native forests were harvested at rates that were more than twice the average annual cut during the Second World War. Imports of forest products in that period were four times greater than the pre-war level (Rule 1967).

The need for long-range planning on a national basis was given institutional expression in 1964, through the formation of the Australian Forestry Council, comprising appropriate Commonwealth and state ministers (Carron 1993). At its first meeting, the Council agreed that the national timber deficiency should be addressed through the rapid expansion of softwood plantations (Rule 1967). Consequently, in 1967 the Commonwealth government approved and financed the expansion of the coniferous afforestation programme through the Softwood Forestry Agreements Act (referring to its power under Section 96 of the Constitution, that is the power to grant financial assistance to the states on such terms and conditions as the Parliament thinks fit). The purpose of the first Softwood Forestry Agreement Act was to increase the rate of softwood planting in Australia to an average of 75000 acres (30 350ha) a year over the following 35 years. The act provided for a series of 5-year agreements with the states whereby the states were advanced interest-free, long-term loans to establish the plantations. Although the plantation policy aimed for reforestation of marginal and abandoned lands and catchments, much of the increase in the area of pine plantations was to be accomplished through the clearing of state forests assessed as having low timber productivity (Dargavel 1995; Mercer

1995). According to Florence (1994), 'the primacy of wood production' had gained institutional acceptance and the implicit strategy was to deplete the sawlog resource in native forest while developing 'a more efficient softwood plantation resource' as the alternative.

At about the same time, Japan became an export market for woodchips. The resulting demand for pulpwood, while welcomed by some as a means for utilizing native forests more economically (Florence 1994; Brand 1997), was deplored by others as leading to 'the large-scale use of forests for industrial purposes' (Bennison & Jones 1975). Woodchipping involved clearfelling, often in very large coupes where the visual impact was severe. Royalties were seen by some to be inadequate, regeneration strategies poor or nonexistent, and the end-product a wasteful use of the resource (Bennison & Jones 1975; Young 1996). Woodchipping became the pre-eminent forestry issue, with slogans like 'Save our Forests: Ban Woodchipping'. This only strengthened the role of the Commonwealth in forestry, because the Commonwealth was responsible for issuing the necessary export licences (Carron 1993). This, in turn, created considerable tension between the states and the Commonwealth, because the Commonwealth used this leverage to intervene in negotiations between Japanese companies and various state governments and private interests over contract prices (Fisher 1987; Carron 1993). The Commonwealth's dominance on the two issues of woodchipping and pine plantations also meant that the Commonwealth became the political focus of environmental groups. As a result of the formation of a national environment body, the Australian Conservation Foundation, in 1965–66, these groups had developed organizational coherence and influence (Carron 1993). Indeed, Dargavel argues (1995) that it was the environmental movement which set the public agenda of forest issues from the 1970s on. However, the Commonwealth's forestry initiatives also came under pressure from farmers who opposed the expansion of the softwood plantation programme through the purchase of agricultural land. The farmers were concerned about the poor cash-flow of tree crops and the effect that whole-farm tree planting would have on the provision of basic services to their communities (Shea 1998).

Changing scenes: the rise of the environmental movement and institutional restructuring

With the emergence of the environmental movement as a political force both in Australia and internationally (McCormack 1992), Australia entered into a phase of institutional restructuring, whereby attempts were made to integrate concerns about the environment into the policy-making process (Papadakis & Moore 1994). The Labour government that was elected federally in December 1972 introduced a range of environmental legislation, including the Environment Protection (Impact of Proposals) Act 1974. Under this act, the federal government could require new or expanding woodchip mills to comply with its Environmental Impact Assessment (EIA) process. It established a Committee of Inquiry into the National Estate, leading to the Australian Heritage Commission Act 1975, which provides for the establishment of a Register of the National Estate to list places of national heritage, and the National Parks and Wildlife Conservation Act 1975. At the international level it ratified the UNESCO Convention Concerning the Protection of the World Cultural and Natural Heritage (the World Heritage Convention) in 1974 and signed the Convention on International Trade in Endangered Species of Wild Fauna and Flora in 1973.

The Committee of Inquiry into the National Estate found that the programme of softwood planting was of dubious economic benefit and undertaken for the benefit of overseas interests (Committee of Inquiry into the National Estate 1974). The Committee particularly questioned the clearing of native forest for pine plantations, and the expansion of woodchipping operations. It also seriously questioned the ability of the state forest services to assure that management of the forest resource would be based on long-term biological research (Committee of Inquiry into the National Estate 1974), and challenged the view that forest policy is best left to the experts:

> Forest authorities and managers of the forests should take responsible account of the needs of the public and should ensure that the forest environment and all its values are properly husbanded for future

generations. Australians have so far entrusted the forestry profession with this task. They have not appointed them to decide the fate of the forests. (Committee of Inquiry into the National Estate 1974).

Conflict over forest areas subsequently listed on the Register was to become a feature of Commonwealth–state relations on forests over the next two decades, largely because of the community perception that listing is synonymous with the declaration of a conservation reserve (Office of Auditor-General 1993). However, listing does not directly restrict land-use options for landowners or for state and local governments; rather, it prohibits Commonwealth ministers, departments and authorities from taking actions which adversely affect places listed on the Register, unless 'there are no prudent or feasible alternatives'.

The softwood plantation programme, which was intended to be implemented over 35 years through the creation of 5-year legal regimes requiring a new act for each period, was terminated in 1976 after only two periods of planting. The third act provided for the maintenance of plantations established under the two previous programmes (Fisher 1987). The Commonwealth also commissioned its Senate Standing Committee on the Social Environment to conduct a public inquiry on the environmental impact of woodchipping in November 1974.

The historic deposing of Prime Minister Whitlam's Labour government in November 1975 saw the matter transferred to the Senate Standing Committee on Science and the Environment, which submitted its final report in May 1977. The Committee found that there were no environmental grounds for curtailing the existing woodchip industry but recommended against the issuing of licences for new projects until environmental threats could be fully researched and means devised for countering them. It also called for a detailed examination of the future use of forests 'in the light of multidisciplinary, overall land-use planning, preferably on a national basis' (Senate Standing Committee on Science and the Environment 1976). The new government announced, however, that it was not averse in principle to new woodchip projects given appropriate environmental safeguards (Carron 1993). Bitter disputes ensued throughout the 1980s.

Environmental disputes of major national consequence

Three disputes involved Tasmania. The first, known as the Tasmanian Dam case, arose from the decision of the Tasmanian Hydro-Electric Commission, an instrument of the Tasmanian government, to build a dam on the Franklin River. The area fell within the site successfully nominated for inclusion on the World Heritage List by the Commonwealth government in cooperation with the Tasmanian government. The proposed dam became a campaign issue during the 1983 federal election with the Australian Labour Party (ALP) promising, if elected, to halt construction of the dam by exercising a previously untested power in relation to environmental issues: the Commonwealth's power over external affairs. Following its election victory, the new ALP government enacted the World Heritage Properties Convention Act 1983, 'the only statute in the world enacted to ensure domestic implementation of the World Heritage Convention' (Opeskin & Rothwell 1995). The legislation, which was aimed specifically at halting construction of the dam, was challenged immediately by Tasmania on the grounds that it was beyond the Commonwealth Parliament's constitutional competence over external affairs to legislate on this matter. The High Court decided in favour of the Commonwealth and, by holding that the Commonwealth had broad powers to implement Australia's treaty obligations, brought a virtually limitless range of matters within the Commonwealth power given the large range of topics dealt with in international treaties. A further consequence, according to Economou (1992), was that what started out as a once-only commitment by the Australian Labour Party in order to attract 'green' votes in the marginal electorates of Melbourne and Sydney, developed into an 'unprecedented rise in the relevance of environmentalism to national politics'.

A few years later, in December 1986, the Commonwealth and the Tasmanian Government again met in court, over the Lemonthyme and Southern Forests. After a protracted series of intergovernmental negotiations over woodchipping and logging in forests listed in the National Estate, the Commonwealth Cabinet called for a 12-month moratorium on logging these forests while their World Heritage value was assessed. While conservationists believed that National Estate areas could be protected from domestic logging by the Commonwealth using its 'trading corporations' power to amend the Australian Heritage Commission Act, the government of the time preferred to rely on its external affairs power rather than take the 'highly politically contentious' action of extending its powers of intervention (Toyne 1994). Even so, the Tasmanian government refused to be bound by the moratorium. It continued to allow logging for sawlogs, and since this logging did not involve the production of woodchips for export, the Commonwealth was powerless despite the fact that the forests were listed on the Register of the National Estate (Toyne 1994).

The Commonwealth established an inquiry under special legislation to determine if either of the two forest areas were of World Heritage quality or contributed to the adjacent World Heritage area. Having established this special legislation, the Commonwealth then obtained an injunction against Tasmania to stop its logging operations. When Tasmania challenged the validity of the legislation, the High Court not only reaffirmed its decision in the Tasmanian Dam case on the external affairs power, but it held that the Commonwealth could provide interim protection to areas pending World Heritage identification, where there is a reasonable basis for supposing that obligations under the World Heritage convention may arise (Tsamenyi *et al.* 1989).

The inquiry, known as the Helsham Inquiry, produced a divided result 'which put the federal government in an even worse political situation than it had been before' (Tsamenyi *et al.* 1989). Two of the three commissioners ruled that less than 10% of the area warranted World Heritage nomination, while the third found the whole of the area, plus an adjoining area, worthy of World Heritage listing. The Cabinet debate on what to do with the outcome of the Helsham Inquiry lasted 14 hours and took place over several weeks

(Toyne 1994). Eventually a compromise was worked out such that 70% (Toyne 1994) to 80% (Tsamenyi *et al.* 1989) of the area was nominated for World Heritage listing in return for a compensation package to the Tasmanian government including a grant of AUS$50 million (US$29.5 million) and various agreements and concessions to do with woodchip licences, pulpwood tonnages and future World Heritage List proposals. The generosity of the package constituted, in the minds of some commentators, 'an invitation to suspect policy making' on the part of the states (Tsamenyi *et al.* 1989).

The proposal in 1989 to build a large, chlorine-bleach pulp mill in Wesley Vale, Tasmania, saw the Commonwealth government again drawn into a policy dispute despite the proposal being entirely consistent with its policy objectives for industry (Economou 1992). Intensive lobbying by the environmental movement resulted in the Commonwealth imposing more stringent environmental standards than had been applied by the state, leading to the abandonment of the project (Hay & Eckersley 1993).

Another forestry-related dispute occurred over the Daintree Rainforest in Queensland. The dispute culminated in 1988 when the Queensland government, following an unsuccessful High Court challenge, took its opposition to the Commonwealth's nomination of the area for World Heritage listing to the World Heritage Committee meetings in Paris and Brasilia and was again unsuccessful (Carron 1993). Protracted Commonwealth–state negotiations concerning the renewal of woodchip export licences in the south-eastern forests of New South Wales took place throughout the second half of the 1980s, prompting considerable agitation on the part of industry (Carron 1993).

Laying the foundations for sustainable forest management

Reformulation of the Commonwealth's constitutional rights over environmental policy was pivotal to attempts to resolve the environment versus development conflicts of the 1980s (Papadakis & Moore 1994). However, pressure on governments to incorporate environmental con-

cerns into mainstream policy was also apparent at the state level. State legislation enacted during this period included the Flora and Fauna Guarantee Act 1988 in Victoria. Codes of forest practices were introduced under specific legislation or amendments to existing forest acts and narrowly focused forestry agencies were amalgamated into broader land-management agencies (Dargavel 1995; Mercer 1995). Wilderness and rainforest conservation policies were also implemented by most states during the late 1980s and early 1990s, sometimes through agreements with the Commonwealth (Dargavel 1995). From 1986 to 1990, Victoria, Western Australia, New South Wales and Tasmania prepared timber industry strategies that attempted to achieve a balance between economic and environmental values. A key policy direction in Victoria was to reduce timber harvesting in native forests to a level that would sustain, at a regional level, all forest values in perpetuity (Office of Auditor-General 1993).

Further, two states developed strategies specifically aimed at resolution of disputes. In Western Australia, a strategy was devised to defuse the potential for conflict over National Estate areas, by conducting joint regional assessments by the Western Australian Department of Conservation and Land Management and the Australian Heritage Commission. From this assessment, an agreement or Memorandum of Understanding (MOU) on forest management plans would be produced for all forests in the region, including those listed on the National Estate register (Carron 1993; Dargavel 1995). One such agreement, for the Southern Forests in Western Australia, was reached in February 1992, although industry claimed that the process required a legislative base for it to provide resource security (Carron 1993). Joint regional assessments were also undertaken in Victoria's Central Highlands and East Gippsland regions (Dargavel 1995).

In addition, Victoria had a well-established mechanism for arriving at decisions on the use of public land. Since 1970, decisions had been based on the recommendations of the Land Conservation Council, a statutory body comprising an independent chairman and representatives from government and community organizations (Office of Auditor-General 1993). It had been established following a dispute over an area in Victoria's Little Desert, which the government had proposed to convert to freehold for farming (Christoff 1998). The Council's investigations were 'to have regard for the preservation of areas of conservation and recreation value, taking into account the social and economic implications of its recommendations' (Christoff 1998) and involved the production of a descriptive background report followed by two rounds of public consultation. Largely as a result of the Council's work, land incorporated in formal parks and reserves in Victoria increased from 1% of total land area in 1970, to approximately 15% in 1996 (Commonwealth and Victorian Regional Forest Agreement Steering Committee 1996). Further, it is argued that the Land Conservation Council's ability to offer widely supported and independent policy advice to government diminished the impact in Victoria of the intense resource conflicts of the 1980s (Christoff 1998).

Victoria had also established the Cutting Areas Review Committee in 1983 following protest actions in the Errinundra Plateau, East Gippsland. This Committee included representatives from the Conservation Council of Victoria, the Victorian Association of Forest Industries and various departmental experts. The Committee reviewed short-term operational plans in relation to the harvesting of particular coupes included in a Wood Utilization Plan. Where agreement was not unanimous, the Plan was returned to relevant regional planners, for the identification of alternative coupes. The Minister was required to make a decision if agreement could still not be reached. Of 1653 coupes examined in three logging seasons over the period 1990–92, almost a third were objected to, mostly in Tambo and East Gippsland Forest Management Areas, for which forest management plans had not at that stage been prepared. By contrast, where a forest management plan had been formulated (as was the case for the Otways Forest Management Area) the Committee approved the Wood Utilization Plan without further review. The Committee was disbanded in December 1992 to make way for a regional review and approval process (Office of Auditor-General 1993).

One outcome of the 'unhappy' and draining events of the 1980s was the determination by the Federal Government to avoid battles over specific land-use issues (Economou 1992). By the early 1990s, the preceding two decades had begun to be interpreted as a period of transition when forest policy as a wood-production strategy was reformulated in a piecemeal fashion in response to the environmental movement (Commonwealth of Australia 1992a). While the re-elections of the Hawke government in 1987 and 1990 were accredited to its policy record on the environment and to the development of 'a pragmatic relationship between the leaderships of the parliamentary Australian Labour Party and the environmental movement' (Economou 1992), the 1990 post-election period saw a significant change in the approach of the government. There was a shift away from a concentration on specific issues towards a 'focus on the processes of land-use and resource policy in general' (Economou 1992), and towards developing new institutional mechanisms for defusing conflict (Papadakis & Moore 1994). Unemployment grew to 9.0% in March 1991, placing further pressure on the government to re-orientate its policy emphases towards industry development (Economou 1992).

The first two attempts at providing policy leadership in this new regime—described as the implementation of a 'rational-comprehensive approach' by Economou (1992) and a 'bewildering stream' of 'initiatives, issues and interactions' by Carron (1993)—failed, or at least fell short of expectations.

The Resource Assessment Commission

The Resource Assessment Commission (RAC) was set up in 1989 to 'reconcile the irreconcilable' (Economou 1996); the first three issues to be investigated were the forestry industry, the coastal resource, and mining and land-use within the Kakadu National Park. The RAC was commissioned to conduct an inquiry into the 'options for the use of Australia's forest and timber resources' in November 1989.

The final report of the inquiry, which was held over 2 years and involved more than 500 submissions, 3500 pages of transcripted evidence and numerous public hearings and consultants' reports (Dargavel 1995; Mercer 1995), reiterated the 'need for long-range planning on a national basis', a theme which had been present in the forestry debate more or less since federation. Dargavel's (1995) view of the report was that it 'provided no single, coherent national vision for the future of Australia's forests' despite being urged by the Commonwealth, through the Department of Primary Industries and Energy to be more prescriptive in identifying policy options for government (Commonwealth of Australia 1992a). Economou (1996) wrote in a similar vein: 'By acknowledging that the management of timber resource policy lacked co-ordination, was riddled with confusion and conflicting goals, and was not adequately assessing environmentalist-orientated value-inputs the inquiry's final report was not really adding anything new to the forestry debate'. The report called for 'political judgements' to be made in the face of the difficulty in finding 'win–win' solutions through the application of various methodologies including 'population viability analysis, financial evaluation, benefit–cost analysis, contingent valuation, computer simulation modelling and multicriteria analysis' (Resource Assessment Commission 1992). Indeed, the Commission described, somewhat ingenuously, its own multicriteria analysis of possible national scenarios as inherently subjective, with results dependent on the priorities set (Resource Assessment Commission 1992).

The Commission sought to mediate on the perennially vexing issue of woodchipping by separating the matter of woodchip exports, policy on which it felt should be driven by commercial considerations, from forest management practices on which it provided a similar response to that of the Senate Inquiry of the 1970s. Management practices, the Commission concluded, should be modified 'if the main point of concern is that the logging practices associated with the production of woodchips from native forests are unacceptable on ecological or other grounds' (Resource Assessment Commission 1992). The Commission's views on plantations—that they should develop under normal market conditions but that they posed greater potential environmental impacts than management of native forests for wood production—and its failure to rule out the logging

of old-growth forests were seen as a rejection of environmentalists' claims (Dargavel 1995). While the softwood plantation programme of the 1960s and 1970s had been vigorously opposed by the environmental movement, the establishment of native hardwood plantations on agricultural land was promoted by conservation groups as an alternative to the logging of native forests.

The Commission possibly also fell foul of conservation groups by changing its thinking on management by the state agencies of forests for sustained yield. This followed a very strong reaction by the states to the Commission's draft report, the states having described the Commission's computer-generated model as 'simplistic', its scenarios as 'useless' and its input data as 'wrong' (Commonwealth of Australia 1992a).

The Commission's final report, while not prescriptive and gaining the support of neither industry nor conservation groups, contained significant variations on the theme of the 'need for long-range planning'. It raised the need for flexible, regionally based planning arrangements, recommending that implementation of the proposed national forest strategy 'should be based on the model of the "balanced panel of experts" operating in a regional context' and the notion of integrated and comprehensive forest management 'capable of maintaining forest ecosystems and responding to changing community attitudes while also providing security and certainty for industry' (Resource Assessment Commission 1992). These were themes that were being taken up by other forest nations (Murphy *et al.* 1993; Jones *et al.* 1995; Lamas & Fries 1995) and they were eventually given form by Australian governments when the time for applying political judgement could no longer be deferred.

Economou (1996) argues that the real significance of the RAC inquiry lay in its conduct, reporting and achievement of 'comprehensive and inclusive participation'; 'even if the inquiry had not really told the community anything not already heard about forest policy, it did succeed in bringing a more co-operative national culture to the forestry debate'. Within 4 years the RAC was dead, with the Chief Commissioner expressing his regret 'that the commission had failed to become a permanent fixture in the national land and resource-use policy-evaluation process' (Economou 1996).

Forest Conservation and Development Bill 1991

The Commonwealth's attempt to introduce resource security legislation, through the Forest Conservation and Development Bill 1991 which 'represented the most explicit statement of the re-ordering of the government's priorities in the environment-development debate' (Economou 1992) also came to nothing when it was voted down by the Senate in May 1991. The proposed legislation, aimed at guaranteeing access to forest resources for large-scale, value-adding developments such as pulp-mills, 'contained the key characteristics of a rational comprehensive approach (to future development)' (Economou 1992). It aimed to formulate long-term goals for conservation and development based on comprehensive assessments of the range of concerns—environmental, social, cultural, scientific, and those of indigenous peoples—undertaken by most of the key players in previous disputes. Environmental groups declined to participate in the process, perceiving that the room to manoeuvre between the levels of government would be significantly curtailed by such involvement (Economou 1992). Within the government, the legislation was opposed by the Treasurer, Paul Keating, who was pro-development but against the offer of protection to investors contained in the legislation, and who was also engaged in a battle for parliamentary leadership. The Resource Assessment Commission had also criticized the legislation as not comprehensive enough, given that it would apply only to developments of at least AUS$100 million (US$59 million), and would 'entrench the current practice of reactive project-by-project assessment' (Resource Assessment Commission 1992).

Environment policy was addressed, again, in Prime Minister Hawke's 1990 'New Federalism' initiative, which was primarily concerned with achieving microeconomic reform through forging a closer partnership between the levels of government. Under 'New Federalism', states were offered incentives including a review of tied grant programmes and the possibility of change

in Commonwealth–state fiscal relations to co-operate in implementing the Commonwealth's environmental policies of 'ecologically sustainable development' (Galligan & Fletcher 1993). The Commonwealth had become committed to this process following the release of the report *Our Common Future* by the World Commission on Environment and Development in 1987, which advocated an integration of ecological dimensions into national economic, trade, agricultural and energy policy frameworks (McCormack 1992).

16.2.4 The structures for sustainable forest management

The Council of Australian Governments (COAG) formed in 1992 as part of the New Federalism initiative endorsed in its first year of operation three national environmental 'strategies': the National Strategy for Ecologically Sustainable Development (ESD); the National Greenhouse Response Strategy and the National Forest Policy Statement. Heads of government also signed the Intergovernmental Agreement on the Environment. The Agreement called for a cooperative national approach to reduce the number of disputes by sharing data, by conducting joint regional assessments of heritage and other values, by accrediting state Environmental Impact Statement (EIS) procedures so that one study could serve both spheres of government, and by involving states in international matters such as the Convention on Biological Diversity and World Heritage listing.

The ESD process set up nine ESD Working Groups covering different industrial sectors, one of them forestry, 'to find agreed ways for designing a sustainable future' (Dargavel 1995). Unlike previous inquiries which considered submissions from interested groups, the working groups variously consisted of representatives of the protagonists—industries, services and unions, environmentalists and scientists. However, the Australian Conservation Foundation left the Forest Working Group after it recommended that old-growth forests could be logged subject to 'comprehensive assessment, examination of the need for protection, and planning to grow some forests to maturity and senescence' (Dargavel 1995). This approach was similar to that taken by the Resource Assessment Commission which was not prepared to express its preferences for either of the two options it presented for dealing with old-growth forests. It had argued that conservation and logging of old-growth could be accommodated within detailed and flexible planning schemes which aimed to maintain the spectrum of regeneration stages in order to conserve 'the suite of ecological processes and species diversity that characterizes the forest of a region' (Resource Assessment Commission 1992).

16.2.5 National Forest Policy Statement

The National Forest Policy Statement, signed in December 1992 by the Commonwealth and every state except Tasmania (Tasmania signed in 1995), was seen as a 'collation' of the attitudes and principles which had evolved during the various inquiries, strategies and agreements at the national level during the preceding years (Dargavel 1995). However, it also drew on the experiences of those states that had adopted models for joint regional assessments and consultative land-use decisions. While finalization of the policy statement was held over until the submission of the Resource Assessment Commission report, it is clear from the report of the 28th Meeting of the Australian Forestry Council, February 1992, that a position paper of the Standing Committee on Forestry (comprising heads of the state and territory forestry agencies) had been drawn upon in the preparation of the National Statement (Commonwealth of Australia 1992a). At that meeting, a representative of the Chairman of the Council and the Minister of Resources, Alan Griffiths, introduced the policy stating that:

> until recently governments in Australia have rarely led in policy formulation; rather they have responded to changing social perception and needs. The effects of this decision framework has meant that there have now been two decades of transition in which piece-meal reductions have been made in land allocated to wood-production. The uncertainty created had an unsettling effect on forest planning and on industrial and social stability—all the more so at a time of continuing dependence on wood supply from

the native forests. As with the danger of over-development, so there is danger of environmental over-reaction, of moving beyond that point where adequate account is taken of the wider public interest. . . .We should not underestimate the importance of this policy. It will mean for the first time, Australia will have a national policy on one of the more controversial areas of public policy—the conservation and judicious use of Australia's forests (Commonwealth of Australia 1992a).

National Reserve Criteria

The policy provides for both industry development and conservation, calling for the development of an efficient, value-adding, internationally competitive and ecologically sustainable wood products industry and the establishment of a comprehensive, adequate, representative (CAR) forest reserve system to protect forest biodiversity, old growth and wilderness (see Chapter 2, this volume, for details). The criteria for establishing a CAR reserve system in Australia were developed subsequently by the Joint ANZECC/ MCFFA National Forest Policy Implementation Sub-committee (JANIS 1997). CAR reserve targets are 15% of the 'pre-1750' extent of all forest types, 60% of all existing old-growth forest, and 90% of all existing high-quality wilderness. The national policy provides for 'ecologically sustainable forest management' by incorporating codes of practice, forest management planning, operational planning and monitoring of impacts, and contains a set of nationally agreed principles of forest practices relating to wood production to be applied to all public and private native forests. Finally, the policy provides for long-term, durable Commonwealth–state agreements to cover all aspects of ecologically sustainable management of forest resources on a regional basis. These agreements, known as Regional Forest Agreements, are to specify land-use boundaries, forest management guidelines and consultative arrangements between governments. They are to be entered into after signatory states have identified and mapped old-growth forest areas through 'comprehensive regional assessments', and planned to protect areas of high conservation value in a system of representative reserves. The states were to halt logging until the mapping exercises were complete, but the Commonwealth permitted some logging to proceed through approving export licences, a move which confirmed conservation groups' view of the policy as 'securing resource industries against on-going environmental claims' (Toyne 1994).

Other commentators expressed doubt that the policy could provide an enduring peace: it was described as 'a carefully worded but rather bland document that seeks to be all things to all people' (Mercer 1995). Dargavel conceded that 'while the proposed comprehensive regional assessments would provide a sensible basis for displaying the options for both preservation and development, and agreements might be arrived at between governments', it was doubtful that the inherent conflicts would be resolved (Dargavel 1995). Carron (1993) claimed that the policy's failure to quantify wood-production targets meant that it had not addressed 'the basis of the contentions about forest use and forest management'.

16.2.6 The Regional Forest Agreement Process

The Regional Forest Agreement (RFA) is a lengthy process, and the time for it to give effect to the broad goals of the National Forest Policy Statement has drawn understandable criticism. The first step of the process involves a state government inviting the Commonwealth to participate in the assessment of a specific region with a view to developing a Commonwealth–state forest agreement for that region. A 'scoping agreement', identifying government obligations, objectives and interests and broad forest uses, and specifying arrangements for managing the process including the completion of comprehensive regional assessments, is then formulated. The Commonwealth and state jointly identify and assess environmental and heritage values, economic opportunities and social impacts of resource-use options. Existing data, such as those collected through the joint assessments involving the Australian Heritage Commission, are used where possible. Options for the use of the forest resource are then developed with the participation of local govern-

ment, industry, unions, regional economic development organizations, conservation groups and other interested parties. The environmental, economic and social impacts of these options are then assessed and considered by governments. These impact assessments may be sufficient for future proposals for forest use that are covered by the regional forest agreement, and the proposals are then a matter for negotiation between the Commonwealth and the state.

The agreement provides details of management and use of the forests that aim to ensure that environmental obligations are met and that industry has a prescribed level of access to forest resources. The agreement must define the duration of the agreement and provisions for review.

As of June 1999, four Regional Forest Agreements (East Gippsland Victoria, Central Highlands Victoria, Tasmania and South-west Forests Western Australia) have been finalized. Others are in preparation, with New South Wales coming close to a final agreement with the Commonwealth. So far, the Commonwealth has allocated approximately AUS\$200 million (US\$118 million) for the Comprehensive Regional Assessments. The Agreements were overseen by joint state–Commonwealth steering committees and included expert advisory groups and units with members drawn from state and Commonwealth, and in some cases, private agencies, with the assistance of local government. Stakeholder groups were not represented on the steering committees, except in New South Wales. In other states, public and interest groups were consulted through surveys, workshops and meetings for the preparation of regional assessments and through public forums at the options-development stage. For the proposed RFAs in New South Wales, public participation was also secured from the outset through a Regional Forest Forum, on which interest groups were directly represented. The Forum was set up to provide a channel of communication between the Steering Committee and regional communities and to serve as a consultative focus for the Steering Committee. A similar panel has been established for the South-east Queensland RFA.

The legislative framework was provided through amendments to regulations of the Export Control Act 1992 to remove all licence requirements for the export of wood products derived from plantations that are managed in an environmentally responsible way or derived from regions covered by a regional forest agreement (Montreal Process Implementation Group 1997). The Commonwealth's Regional Forest Agreement Bill, which the Commonwealth agreed to propose as part of the Tasmanian Regional Forest Agreement, was introduced on 30 June 1998. The purpose of the Bill was to provide certainty to the termination and compensation provisions of existing and future agreements, but because amendments proposed by the Parliamentary upper house (the Senate) have not been accepted by the lower house, it has not been enacted.

Ecologically sustainable forest management

Through the RFA process, the various parties have begun to define and chart the territory of 'ecologically sustainable forest management'. The term has been defined as:

> the integration of commercial and non-commercial values of forests so that the welfare of society (both material and non-material) is improved, whilst ensuring that the values of forests, both as a resource for commercial use and for conservation, are not lost or degraded for current and future generations. Whilst forests can be assessed for a range of values in the comprehensive regional assessment process (e.g. conservation, heritage, resource, social, economic) an assessment of ecologically sustainable forest management provides the basis for the integrated and sustainable management of *all these values* [our emphasis] (Hoare 1996).

This definition or variations of it appears in glossaries attached to the documentation produced for the regional assessments. A definition for old-growth has also been produced:

> Old-growth forest is ecologically mature forest where the effects of disturbances are now negligible (JANIS 1997).

Both these definitions may be seen as examples of the imperative 'to affirmatively avoid definitions and applications that rather directly foster conflict almost by definition between resource sustainability and human needs', which Webster (1993) advocates for forest management in Canada. The definition of old-growth in particular cuts across the perceptions that it is not possible to create old-growth forests and that it is unlikely that old-growth will regenerate naturally, even over several centuries. Nevertheless, the potential for conflict remains. This definition does not preclude the use of the term 'old-growth' to be applied to any area which people want to preserve: even forest that has been logged could be regarded as old-growth as long as the effects are deemed negligible.

The National Forest Policy Statement proposes to give effect to ecologically sustainable forest management through:
- integrated planning processes and management systems;
- codes of Practice and environmental prescriptions;
- management plans incorporating sustainable yield harvesting practices;
- management of native forests outside the reserve system complementing the objectives of nature conservation reserve management (Commonwealth of Australia 1992b).

The Ecological Sustainable Forest Management (ESFM) report for East Gippsland, the first of its kind in Australia, was prepared by an Expert Advisory Group comprising research scientists from the Commonwealth Scientific and Industrial Research Organization (CSIRO) and two consultants with public service backgrounds in forest management and heritage issues (Commonwealth and Victorian Regional Forest Agreement Steering Committee 1996). The Group adopted an 'environmental management system' approach, that is an examination and assessment of the overall management system and processes covering policy and commitment, planning, implementation, information and monitoring, and review and improvement. It did not attempt to examine specific on-ground practices for their effect on ecosystems, given the lack of established performance indicators or benchmarks for ecologically sustainable forest management.

The ESFM report for East Gippsland (Commonwealth and Victorian Regional Forest Agreement Steering Committee 1996) provided the basis for a state-wide assessment of ESFM that was independently reviewed, both the assessment and the review being incorporated in the Central Highlands Comprehensive Regional Assessment report (Commonwealth and Victorian Regional Forest Agreement (RFA) Steering Committee 1997). The review found that the ESFM report provided a positive, overall assessment of Victoria's management systems, but identified a need for further improvement, particularly in the provision of staff training in ESFM planning and management, and in the level and type of research to assure ESFM. The review found that forest-related legislation and policy provided a 'comprehensive and robust framework' for the consideration, assessment and protection of forest values, although some aspects needed to be updated and made more comprehensible to the public. Land-use planning processes, the integration of reserve and off-reserve management and the development of competitive prices for publicly owned native timber were also identified as in need of improvement. These measures were neither major nor exceptional but nevertheless necessary for continuing improvement in management systems and in guarding against complacency: 'Victoria is in a good position to deal with these issues providing it remains adaptive and accepting that sustainable forest management is a goal to be pursued vigorously, not an antique to be admired' (Commonwealth and Victorian Regional Forest Agreement (RFA) Steering Committee 1997).

16.2.7 The international dimension of forest policy in Australia

As yet, there is no broad international agreement on forests (Boer 1997). Nevertheless, the development of Australian forest policy as described above fits well with those international forest policy prescriptions which do exist. The United Nations Conference on Environment and Development (UNCED) in Rio in 1992 resulted in the adoption of a non-legally binding set of forest principles and Chapter 11 of Agenda 21, the major output of the conference, was devoted to combating deforestation. Following UNCED, the United Nations established the Commission on

Sustainable Development (CSD) to oversee the implementation of Agenda 21. An Intergovernmental Forum on Forests, established under the aegis of the CSD, recommended in April 2000 that a United Nations Forum on Forests be established, with a brief to develop a legal instrument within 5 years.

Principle 3(a) of the Statement of Forest Principles, referred to above, states that 'national policies and strategies should provide a framework for increased effort, including the development and strengthening of institutions and programmes for the management, conservation and sustainable development of forests and forest lands.'

Objectives included in Chapter 11 of Agenda 21 on forests also emphasize the strengthening of forest-related national institutions to enhance the scope and effectiveness of management activities, conservation and sustainable development, and the means recommended for implementing these objectives emphasize provision of scientific means, human resource development and capacity building (Boer 1997). Boer argues that for Australia to effectively implement its international obligations, the recent 'transformation' of administrative regimes, the articulation of aspirations for the establishment of world-class environmental management systems and the development of sustainable forest management policies and strategies, need to be followed up by a comprehensive review of all forestry legislation in order to achieve nationally consistent philosophies and practices (Boer 1997). Tsamenyi (1996) argues that it is international reporting requirements which are the key to ensuring that the 'concepts and objectives underpinning international environmental conventions become a part of public sector dialogue'. He sees the 1992 Australian Intergovernmental Agreement on the Environment as a specific instrument for ensuring prompt and open communication between the different state environmental departments (Tsamenyi 1996). He also believes that the emergence of a global perspective on environmental issues has led to a gradual erosion of the traditional boundaries of state sovereignty. This is in spite of Principle 21 of the Stockholm Declaration (Tsamenyi 1996), and of the traditional perception in Australia that the states have authority and political jurisdiction over most of Australia's continental resources (as we have detailed above). Tsamenyi goes on to argue that 'international environmental imperatives provide one of the best opportunities for the creation of an Australian national identity', citing Justice Murphy's observation in the Tasmanian Dam case:

> increasingly, use of the external affairs power will . . . be . . . a regular way in which Australia will harmonize its internal order with the world order. The Constitution in its reference to external affairs and to matters arising under treaties . . . recognizes that while most Australians are residents of states as well as of the Commonwealth, they are also part of humanity. Under the Constitution, parliament has the authority to take Australia into the "one world", sharing its responsibilities as well as its cultural and natural heritage.

The Montreal Process, and Australia's response

It is our argument that implementation of Australia's National Forest Policy Statement through the Regional Forest Agreement process is a decisive step towards meeting international requirements for institution-building and provision of nationally consistent legislative bases. An element missing, however, from all frameworks for sustainable forest management, national and international, as described so far, is a set of measures or operational definitions, to 'bring to ground' the ethereal concept of sustainability (Jones *et al.* 1995). In Australia, the need for operational definitions is addressed through forest management plans and codes of practices implemented at the local level and is beginning to be addressed through the application of internationally acceptable performance measures for sustainable forest management through the 'Montreal Process' and the associated 'Santiago Declaration' (Working Group 1995). Participation in the Montreal Process could well be seen as an explicit attempt on Australia's part to harmonize its internal order with the world order and this is demonstrated by the interdependence of the regional forest agreement process and the Montreal Process.

The members of the Montreal Process are 12 countries including Australia that together contain more than 90% of the world's temperate and boreal forests. The process was initiated in 1993 by the Conference on Security and Cooperation in Europe at a conference in Montreal, where a Working Group was established to develop criteria and indicators for the conservation and management of temperate and boreal forests, which were agreed in the Santiago Declaration of 1995. Broad forest values that society seeks to sustain were agreed to by the Montreal Process, and they cover biological diversity, productive capacity, ecosystem health and vitality, soil and water resources, global carbon cycles, socioeconomic benefits, and an effective legal, institutional and economic framework. At the national level, there are 67 indicators, providing measures of temporal change in these criteria. In 1995, the Ministerial Council on Forestry, Fisheries and Aquaculture endorsed the use of the Montreal Process criteria and indicators as the basis for assessing sustainable forest management at the national level. It also approved the development of a regional framework of indicators, based on the Montreal Process, to be used at the regional level in the RFA process. This framework was released in August 1998.

Australia submitted its first 'approximation report' on meeting criteria in June 1997 (Montreal Process Implementation Group 1997). The report found that the availability and quality of information used to measure or describe indicators associated with the seven criteria was 'highly variable'. Better quality data were available for forested parts of Australia with a history of both commercial and conservation interest than for the remote and drier regions. Furthermore, while long-term management of public forest land is accommodated in institutional frameworks, there is little formal planning for the management of privately owned forests. On the positive side, the report found that:

• Australia has on average 11% of its forests and woodlands in nature conservation reserves (including 23% of closed forests and 16% of open forests), as against the 10% generally seen to be an international standard (Resource Assessment Commission 1992).

• Nationally timber yields are now below sustainable levels.

• All state and territory governments have either achieved sustainable yields or have adopted plans to work towards sustainable yields.

• Major forest health concerns caused by air pollution do not affect Australian forests to any significant extent and the available data suggest that managed forests and woodlands are a net sink of carbon.

For the biological indicators the picture is less clear. The report concluded that the data are not sufficient to allow a definitive conclusion to be reached about the sustainability of forest management in Australia as a whole, given issues relating to the applicability of indicators and the need to establish the connection between some of the indicators and sustainability. The report states that a concentrated effort will be required to develop cost-effective and sensitive monitoring systems, and to establish 'how information or trends over time should be interpreted within the context of the criteria to which they refer and how they relate to the overall assessment of sustainability' (Montreal Process Implementation Group 1997).

Finally, the report places great emphasis on the role of the RFA process in confirming sustainable forest management. Good-quality baseline data for use against a number of indicators should emerge from the comprehensive regional assessments, and the process will provide for the implementation of improved monitoring based on criteria and indicators and could result in mechanisms being developed to report against some indicators for which data are unavailable. For example, more efficient indicators for some criteria might be based on the implementation of codes of practice which themselves are based on sound research. An aspect of the assessment of such indicators would then be the extent to which the code is implemented (Montreal Process Implementation Group 1997).

On the social side the report indicates that there is much debate on how recognition of the principle of customary and traditional rights of indigenous people on publicly owned land should be implemented. The report noted that 'all levels of forest planning make provision for public participation', either through formalized stakeholder membership of advisory committees or through mandatory requirements for public consultation and participation during preparation of management plans and often also in subsequent management (Montreal Process Implementation Group 1997).

In reviewing Australia's response to the Montreal Process for the Ecological Society of Australia, Peacock *et al.* (1997) acknowledged that 'establishing an explicit basis for assessing sustainability is one of the most challenging aspects of contemporary forest management, irrespective of whether a forest is being managed primarily for consumptive uses or for biodiversity conservation'. They found that the Montreal Process provides a new model for assessing ecologically sustainable forest management in temperate and boreal forests whereby indicators can be used to test explicit assumptions about desired ecosystem condition through active adaptive management or through scenario modelling. Peacock *et al.* (1997) noted that there had been difficulties in monitoring some forest ecosystem properties (e.g. soil and water quality/productivity) and concluded that 'the detailed description of the individual components of sustainable forest management will remain an on-going challenge from both a technical and a policy perspective'.

16.3 FOREST MANAGEMENT FOR SUSTAINABILITY IN THE CENTRAL HIGHLANDS

16.3.1 The Central Highlands Regional Forest Agreement

Of the three RFAs signed to January 1999, the two Victorian agreements are relatively short and reasonably similar in content. Both contain 15 'milestones', or actions to be taken by specified dates. The key elements of the Central Highlands Regional Forest Agreement are that, in relation to resource security, Commonwealth controls on woodchip exports are removed, and industry is guaranteed supply of at least current per annum licensed volume of D+ grade sawlogs for the next 20 years. This provision is not as open-ended as it first appears. Sustainable yields are legislated and there is a legislative requirement to review sustainable sawlog yields every 5 years. Industry licences are allocated on the basis of regional sustainable yield and, at present, licensed volumes are below sustainable yield estimates pending the incorporation of information derived from improved data and modelling systems before the next legislated review in 2001. Sustainable yield in the Central Highlands is dominated by regrowth forests for which there is reliable information on growth rates and net productive area; furthermore, wood availability will increase as the resource matures. A decision has therefore been made that any correction that may be required to current forecasts of sustainable yield is likely to be offset by potential increases in sustainable yield through the long-term resource capacity of the region.

The areas reserved for conservation meet the JANIS reserve criteria and, under the agreement, an additional 116000ha of public land is to be included in the reserve system. Through the agreement, Victoria's forest management system is accredited but regional prescriptions for timber production, the Central Highlands Forest Management Plan and management plans for national and state parks, must be completed by specified dates. The agreement sets priorities for the management of vegetation communities and endangered species, and for forest research, and it provides for the development of state-wide guidelines for the management of cultural heritage values in forests, parks and reserves. In addition, the identification, protection and management of National Estate areas is agreed. The Agreement provides for 5-yearly reviews with public input to monitor implementation of the agreement and requires the Victorian Government to publish audits of compliance with the Code of Forest Practices for Timber Production. Victoria is also required to implement a package of measures to ensure appropriate management of Aboriginal heritage, including measures to ensure formal consultation with local Aboriginal communities, and to review pricing and allocation policies for government-owned forestry operations. There are also provisions for termination of the agreement and appropriate compensation arrangements.

16.3.2 Forest areas

The Central Highlands covers 1.1 millionha of which 0.63 millionha is public land (Commonwealth and Victorian Regional Forest Agreement (RFA) Steering Committee 1997; Department of Natural Resources and Environment Victoria 1998); 29% of the public land is in conservation reserves of various types, including national parks (Fig. 16.2a). State Forest covers a total of 38 960000ha, 62% of total public land.

Within State Forest (Fig. 16.2b), 256 300ha is listed as General Management Zone, of high

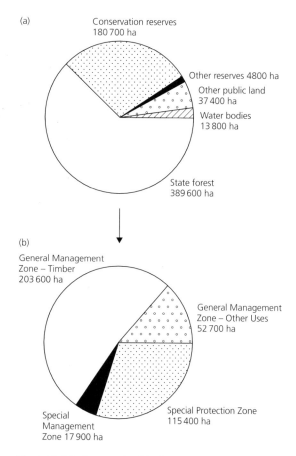

Fig. 16.2 (a) The use of public land within the Central Highlands. (b) The division of State Forest in the Central Highlands within various management zones. (Data from Commonwealth and Victorian Regional Forest Agreement (RFA) Steering Committee 1997.)

priority for timber production. Within this General Management Zone, 203 600 ha (52% of State Forest) is available (e.g. topography) and of a quality suitable for sawlog production (General Management Zone–Timber, 30% of State Forest, Fig. 16.2b); 115 400 ha is listed as Special Protection Zone which is excluded from logging and managed for conservation and protection of a range of values including retention of old-growth, rainforest and surrounding buffers, protection and management of fauna and flora, stream buffers, national estate values, etc.

The ash-type eucalypts (the wetter forests) cover 183 622 ha of public land in the Central Highlands, of which 113 800 ha is mountain ash. Overall, 86 000 ha of the ash-type eucalypts is formally protected (in conservation reserves and Special Protection Zones, Fig. 16.3) and 78 600 ha is in General Management Zones–Timber). Of the 113 800 ha of mountain ash forest, 50% is formally protected (31% in conservation reserves including national parks, the rest in Special Protection Zones) and 39% (44 390 ha) is in General Management Zones–Timber (Table 16.3).

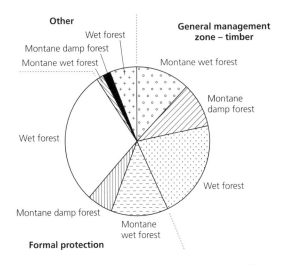

Fig. 16.3 The reservation status of mountain ash forest (= wet forest, other ecological vegetation classes are defined in Table 16.4) in the Central Highlands of Victoria. (Data from Commonwealth and Victorian Regional Forest Agreement (RFA) Steering Committee 1997.)

16.3.3 Meeting the targets for a comprehensive, adequate and representative reserve system

The comprehensive, adequate, representative (CAR) forest reserve system outlined earlier in this chapter, and in Chapter 2, aims for reserve targets of 15% of the pre-1750 (i.e. pre-European settlement of Australia) extent of all forest types, 60% of all existing old-growth forest, and 90% of all existing high-quality wilderness.

Of the wetter forest types in the Central Highlands (Table 16.4), 90% of the pre-1750 area remains. For the ash-type eucalypts (wet forest, montane damp forest and montane wet forest,

Table 16.3 The distribution of mountain ash forest in the Central Highlands within reserves and management zones. (Data from Commonwealth and Victorian Regional Forest Agreement (RFA) Steering Committee 1997.)

Category	Area (ha)	% of total
Conservation reserves	35 290	31
Old-growth	4 850	
Regrowth and other	30 440	
Special Protection Zone	21 620	19
Old-growth	200	
Regrowth and other	21 420	
Special Management Zone + other uses + unstocked	12 520	11
General Management Zone — timber	44 390	39
No old-growth forests, 97% regrowth forests (post-1939)		

Table 16.4 The reservation status of forests within ecological vegetation classes in the Central Highlands as percentages of estimated pre-1750 areas. (Data from Commonwealth and Victorian Regional Forest Agreement (RFA) Steering Committee 1997.)

Ecological vegetation class	Area (ha)			Area protected as % of pre-1750 area			
	Pre-1750	Current	Current area as % pre-1750 area	Conservation reserves	Special Protection Zones	Protected total	Private land as % of pre-1750 area
Montane wet forest (*E. nitens,* *E. denticulata*)	50 319	49 678	98.7	33.6	14.4	48.0	0
Montane damp forest (various species including *E. obliqua,* *E. delegatensis*)	20 506	20 150	98.3	7.7	22.6	30.3	0.1
Wet forest (*E. regnans*)	123 752	120 068	97.0	28.7	18.6	47.3	5.0
Damp forest (various species including *E. obliqua,* *E. cypellocarpa,* *E. regnans*)	198 726	162 307	81.7	16.6	13.9	30.5	9.3
Cool temperate rainforest	12 984	12 970	99.9	43.8	37.7	81.5	0.6
Total	406 287	365 173	89.9	22.8	16.6	39.4	6.3

Table 16.4), 97.6% of the pre-1750 area remains and 96.6% of that area is public forest. Almost 100% of the pre-1750 area of cool-temperate rainforest remains (Table 16.4). Some 37% of the total area of the Central Highlands has been cleared for urbanization, agriculture and water reservoirs. It is therefore clear that the loss of pre-1750 vegetation is greatest in the drier vegetation types; 12 out of a total of 40 ecological vegetation classes (EVCs) recognized by the Joint Commonwealth and Victorian Regional Forest Agreement (RFA) Steering Committee (1997) now fall below the 15% of pre-1750 area criterion, with the worst being native grassland, which now covers only 0.2% of the pre-1750 area. Within State Forest, at least 30% of common EVCs, 30–90% of uncommon EVCs and at least 90% of rare EVCs are protected within conservation reserves and Special Protection Zones.

The JANIS (1997) definition of old-growth forest as 'ecologically mature' has been expanded within the states to provide definitions that are practical. In Victoria, old-growth forest was defined as that which has at least 10% of the total crown cover as senescent trees in the upper canopy, less than 10% of total crown cover as regrowth, and has not been significantly disturbed (Woodgate *et al.* 1994; Department of Natural Resources and Environment Victoria 1996b; Commonwealth and Victorian Regional Forest Agreement (RFA) Steering Committee 1997). Some 84% of old-growth forest in the Central Highlands is in the drier vegetation classes, and at least 60% of these forests are protected within conservation reserves and Special Protection Zones. In the wetter forests, all stands of the relatively small areas of old-growth which survived the 1939 bushfire are in interconnected conservation reserves or Special Protection Zones.

In summary, the land-use basis for conservation and protection of the forests of the Central Highlands fully meets the defined criteria for a comprehensive, adequate and representative reserve system.

16.4 THE ECOLOGY AND SILVICULTURE OF MOUNTAIN ASH

16.4.1 A summary of the ecology of mountain ash

The mountain ash forests of Victoria have been extensively reviewed, scientifically (Ashton 1981; Squire *et al.* 1991; Ashton & Attiwill 1994; Attiwill 1994), historically (Winzenried 1986; Griffiths 1992) and in television documentaries (Attenborough 1995). Extensive bibliographies are available (e.g. King *et al.* 1994; Attiwill *et al.* 1996; Incoll *et al.* 1997). The ecology of mountain ash is presented here in summary form, and only more recent references than those cited by Ashton (1981), Ashton and Attiwill (1994) and Attiwill (1994), and references for quotes are cited.

Mountain ash is the world's tallest flowering plant, reaching a height up to 110 m. It is a fire-dependent species. There are no seedlings or advanced growth beneath the forest canopy. The forest regenerates after fire of stand-replacing magnitude and intensity. Thus, the forests include large areas of monospecific, even-aged stands, and ecotones between mountain ash and other eucalypts are generally quite narrow. Two or three age classes (and, very rarely, four) are found where less severe ground-fires have not killed the mature trees and have provided the right conditions for germination and establishment. McCarthy and Lindenmayer (1998) found that only 9% of 373 three-hectare sites in the Central Highlands included trees within more than one age class.

Mountain ash forest has many elements that can be interpreted as supporting Mutch's (1970) hypothesis: 'natural selection has favoured development of characteristics that make (fire-dependent communities) more flammable'. Or from Ashton and Attiwill (1994): 'the tolerance of the high forests to severe fire once every one, two or three centuries is due to their low resistance to it'. The leaves contain resins, waxes and volatile oils. The understorey and shrub layer are variable, but often quite dense; the litter layer is a mull, with a large amount of large, woody material. The outer bark of the trees peels off and hangs down

from the trees in long ribbons. The climate is temperate, with a generally reliable rainfall. However, given extreme climatic conditions that produce the combination of drought and strong, dry, northerly winds, the mountain ash forests with their high fuel loads are ripe to burn. Fire spreads rapidly through the litter layer and ignites the crown by lighted material being carried into the crown. Once in the crown, the fire moves rapidly, creating a 'fire storm'. Short-distance spotting—ignition 1–2 km ahead of the fire by flaming bark 'firebrands'—increases the intensity of the fire storm as individual fires converge, and long-distance spotting has been recorded up to 30 km ahead of the fire front. There are also reports of balls of fire—supposedly the ignited, volatilized oils of the eucalypts—leaping far ahead of the fire front.

Fires ignited by lightning over millions of years and by humans over thousands of years have been a force throughout the evolution of the eucalypts. The frequency of fire has increased following European settlement and there have been fires of catastrophic magnitude. In Victoria, the 'Black Thursday' fire, 6 February 1851, burnt some 76 000 km², the 'Red Tuesday' fire, 1 February 1898, burnt 2600 km²; the 'Black Sunday' fire, 14 February 1926, killed 31 people, and a total of 426 fires over the summer burned 3940 km², most of it in the mountain areas of central and south Gippsland; the 1932 bushfires killed nine people and a total of 307 fires over summer burnt 2040 km², most of it in the Central Highlands; the catastrophic bushfire of 'Black Friday', 13 January 1939, burnt 13 800 km², including most of the ash-type forests, and killed 71 people; and finally, the 'Ash Wednesday' bushfire, 16 February 1983, burnt 870 km² including 140 km² of ash forest. Fire protection and suppression to meet the infrequent, high extremes of fire danger in south-eastern Victoria is impossible, in terms of both cost and technology. As Christensen *et al.* (1989) commented after the 1988 fires in Yellowstone National Park: 'Both fire suppression and protection pivot on a paradox. The only way to eliminate wildland fire is to eliminate wildlands'.

Mountain ash trees are killed by crown fire. The bark is relatively thin and offers little protection to dormant buds, although the butt-bark of mountain ash is sufficiently thick to protect the tree from low-intensity surface fires with a return interval greater than 25 years. Mountain ash rarely coppices and does not sprout by root suckers. Regeneration therefore depends entirely on seed germination. The seeds are very small (<0.5 mm) and are quickly harvested by ants so that there is little seed stored in the soil. By contrast, the canopy carries up to 14 million seeds ha^{-1}, held in woody capsules. Fire accelerates the shedding of this stored seed. The capsules are killed by crown fire, and their valves open so that the seeds are shed within hours or weeks.

'The ecology of (mountain ash) forests (therefore).... hangs on an extraordinarily fine thread' (Ashton & Attiwill 1994)—how do the small capsules protect the tiny seeds during intense crown fire? The answer may well lie in the oil content and high flammability of the crown, creating intense heat for a very short time, short enough to preclude lethal temperatures within the capsules.

Seedlings germinate profusely on the soil burned bare by fire, at a rate up to 2.5 million ha^{-1}. They grow rapidly as a result of both lack of competition from other species and the ash bed created by intense fire (Fig. 16.4). Germination, growth and competitive development all increase with increasing fire intensity. Mortality is high, the number of trees reducing to about 4000 stems ha^{-1} at age 7 years and to 20–40 stems ha^{-1} in the mature forest aged 250 years, at which age leaf area is quite low (leaf area index of about 2).

Fig. 16.4 Regeneration of mountain ash and the understorey species silver wattle (*Acacia dealbata*, with pinnate leaves) and hazel pomaderris (*Pomaderris aspera*, with heavily veined leaves) 6 months after fire.

16.4.2 Silvicultural systems in mountain ash

Following the pioneering ecological and silvicultural studies of Ashton (1956), Gilbert (1958), Cunningham (1960), Cremer (1960), Grose (1960) and Gilbert and Cunningham (1972), clearfelling followed by high-intensity burning of the logging residues (Fig. 16.5) and broadcast sowing of seed has been extensively and successfully used to achieve regeneration of the ash-type eucalypts after logging (Fig. 16.6). Clearfelling with the retention of habitat trees will continue to be the predominant silvicultural system for the ash-type eucalypts of the Central Highlands (Department of Natural Resources and Environment Victoria 1998). However, other systems such as the retained overwood system are being trialed for fauna conservation.

The decision to continue with clearfelling was not reached lightly. Clearfelling is perceived by many to be damaging to landscape, water and habitat. Furthermore, clearfelling for sawlogs generates a large amount of wood which is not suitable for processing to sawn timber. It is either left on the ground or converted to woodchips. A perpetual argument over timber harvesting is whether or not clearfelling for sawlogs is driven by the export market for woodchips.

Fig. 16.5 A coupe in mountain ash forest immediately after clearfelling and burning.

Fig. 16.6 Regeneration of mountain ash following clearfelling and burning in 1975. The trees are 24 years old; the mean diameter at breast height (over bark) is 26 cm, and the mean dominant height is 41 m.

Following increasing environmental concerns through the 1980s, the Timber Industry Strategy for Victoria (Government of Victoria 1986) identified a need for a comprehensive programme of research, and this resulted in the Silvicultural Systems Project (SSP), which aimed to 'understand, evaluate and, where appropriate, develop alternative silvicultural systems for major production forests including mountain ash' (Squire *et al.* 1991). The results are summarized in two extensive reports (Campbell 1997a,b). In the most general terms, successful germination and survival of mountain ash can be achieved under a range of silvicultural treatments, although the results are highly variable (Campbell 1997b). However, seedling growth, vigour and density are reduced even with a small amount of retained overwood, the magnitude of reduction increasing with increasing retention of overwood (Dignan *et al.* 1998).

There are other problems with silvicultural systems that are based on retention of a proportion of overwood. The bark of mountain ash is thin and is easily damaged by the trees that are felled; the damage allows fungal infection resulting in degradation of the retained trees. The retention of overwood involves greater costs of felling, roading, extraction and site preparation per unit volume of timber harvested. And finally, the risk of injury when working in tall mountain ash forests is greatly increased where overwood is retained.

Is clearfelling an ecologically defensible procedure? We have argued (Attiwill 1994) that the effects of clearfelling, planned and managed at the landscape level within an integrated system of land use, can be accommodated within the ecological effects of natural disturbance. Despite tight definitions and the categorical statement 'Obviously, because of the removal of timber and the amount of soil disturbance, clearfelling . . . and bushfire are different sorts of disturbance' (Attiwill 1994), our argument that the effects of clearfelling can be ecologically incorporated has been quite wrongly misrepresented by statements such as 'clearfelling *mimics* natural disturbance' (e.g. Lindenmayer 1995; Ferguson 1996; Church & Richards 1998). In our view, the initial argument (Attiwill 1994) stands. We should study the ecology of natural disturbance as the basis for designing silvicultural systems and for implementing these systems so that their ecological effects are incorporated within landscapes.

16.4.3 Managing diversity and ecological processes

The Forest Management Plan for the Central Highlands lists the various measures to meet the requirements of the Victorian Fauna and Flora Guarantee Act 1988 and the Code of Forest Practices (Department of Natural Resources and Environment Victoria 1996a). The difficulty is, of course, that our knowledge of the life cycles of many species is limited. As we cannot predict the effects of management with certainty, such effects must be constantly monitored. Thus, both management and defined areas must be adaptive and flexible: 'by establishing a formal monitoring process based on the Montreal criteria and indicators, NRE (Department of Natural Resources and Environment, the state land manager) will become more effective in assessing trends in biodiversity and in progressing towards ecological sustainable forest management' (Department of Natural Resources and Environment Victoria 1998). Three examples are now used to illustrate

this fundamentally essential feedback between management and monitoring in mountain ash forest in the Central Highlands.

The management of understoreys in mountain ash forests

There is increasing evidence that past timber-harvesting operations in ash-type forests have resulted in significant change in the composition of the understorey (e.g. Mueck & Peacock 1992; Ough & Ross 1992; Hickey 1994; Murphy & Ough 1997; Ough & Murphy 1997). The causes of this change include compaction of soil where machines have travelled, and damage to plant structures such as lignotubers, rhizomes and above-ground apices from which many species resprout vegetatively (Ough & Murphy 1998). These species include a number of shrubs and woody trees, ground-ferns and the tree-ferns *Dicksonia antarctica* and *Cyathea australis*. The decrease in the numbers of tree-ferns in regrowth forests following timber harvesting is of further concern because tree-ferns provide habitat for a number of epiphytes. The abundance of epiphytic plants on tree-ferns is greatest in old-age (>200 years) forest, due to the greater size of the tree-ferns and to the more sheltered and wetter environment of the understorey (Hickey 1994; Muirhead 1998).

The composition of understorey vegetation in forests that have regenerated following timber-harvesting operations is indicative of a drier forest relative to the understorey of undisturbed forest. Since the major cause of this change is the damage to soil and vegetative structures by logging machines, measures to reduce these impacts are being evaluated (Ough & Murphy 1998). These trials are based on the concept of 'understorey islands'—areas of 0.1–0.2 ha within which timber is still harvested but logging machinery is excluded, thereby ensuring that disturbance of the soil is minimized. The results are encouraging; within the understorey islands, 96% of the shrub *Olearia argophylla* survived compared with 32% elsewhere within the coupes. Survival of the tree-fern *Dicksonia antarctica* in understorey islands was 69% compared with 15% survival elsewhere. Ough and Murphy (1998) concluded that understorey islands offer 'a simple, low cost means of retaining in-coupe biodiversity', and they recommended that these islands might occupy 10–15% of the coupe area. Furthermore, the concept of understorey islands allows planning both around structural elements (habitat trees, fallen logs, etc.) and within the landscape mosaic of reserves, wildlife corridors and river and stream buffers (Ough & Murphy 1998).

The management of animals that depend on nesting hollows

Of the large number of species of animal that depend on nesting hollows in mountain ash trees (e.g. possums, gliders, owls, cockatoos), the species that has been given most prominence in the international literature by Lindenmayer and colleagues (and many papers referenced in Incoll *et al.* 1997, e.g. Lindenmayer & Possingham 1994) is Leadbeater's possum (*Gymnolbelideus leadbeateri*).

Leadbeater's possum was discovered in the last half of the nineteenth century, well south of its present distribution. It was adopted as the State of Victoria's faunal emblem, but was so rare that, by the mid-twentieth century, it was considered to be extinct. Leadbeater's possum was rediscovered in 1961, and is now quite widely distributed through some 3500 km² of the Central Highlands. It is now classified as endangered under the Fauna and Flora Guarantee Act (Government of Victoria 1988).

The relatively rapid increase in populations of Leadbeater's possum is undoubtedly the result of the extensive bushfire of 1939 that swept through much of the ash forest in the Central Highlands. Regrowth after the fire included a rich component of fire-dependent *Acacia* spp. Nesting hollows in the old trees (many of which were killed by fire) and food (gum from *Acacia*) were both in plentiful supply. The problem now is that the principal *Acacia* spp. that provide food for Leadbeater's possum are relatively short lived (their seed remains in the soil, to germinate after the next fire), and the trees still standing after the 1939 bushfire are decaying and falling at the rate of 3.6% per year. Trees do not develop suitable

nesting hollows until they are 150–200 years old. The risk of extinction remains high (Lindenmayer & Possingham 1994); clearly, managing the forest for viable populations of Leadbeater's possum is not just a problem for those parts of the forest used for timber production but must be integrated across all uses.

Management of Leadbeater's possum within the *Forest Management Plan for the Central Highlands* (Department of Natural Resources and Environment Victoria 1998) was developed from the work of Lindenmayer mentioned at the beginning of this section, Macfarlane and Seebeck (1991) and the *Flora and Fauna Guarantee Action Statement* (Department of Conservation and Natural Resources Victoria 1995). The known range of Leadbeater's possum is divided into 21 management units of average size 8000 ha, with the aim of maintaining viable populations of Leadbeater's possum within each unit. Habitat within each management unit is then classified:

• Zone 1A habitat contains >4 living, hollow-bearing trees per hectare in patches >3 ha. Zone 1A habitat will be critical for the future management of Leadbeater's possum, and is either in reserves or within Special Protection Zones so that timber harvesting is excluded.

• Zone 1B habitat contains >4 dead or living, hollow-bearing trees per hectare in patches >10 ha, with a basal area of *Acacia* spp. >5 m² ha⁻¹; Zone 1B is prime habitat for Leadbeater's possum now, and timber harvesting is excluded from Zone 1B habitat within the General Management Zone.

• Zone 2 habitat is the remainder of the forest.

The overall aim of the management plan for Leadbeater's possum is to retain at least 600 ha of ash forest in each management unit in patches of 50–100 ha, the patches to be linked where possible with the reserve system. Where this aim cannot be achieved through the retention of Zones 1A and 1B habitat, the plan is to retain patches of 1939 regrowth forest that will begin to develop suitable nesting hollows in 50–100 years.

In summary, management for Leadbeater's possum must be pro-active and flexible. The only way to plan for the long-term survival of Leadbeater's possum is through 'timely implementation of active and adaptable management strategies, further research and close liaisons between wildlife biologists and forest managers' (Menkhorst & Lumsden 1995).

The management of mountain ash forests as water catchments

The Central Highlands extend across parts of five river basins—the Goulburn, Thomson, Latrobe, Yarra and Bunyip Rivers (Commonwealth and Victorian Regional Forest Agreement (RFA) Steering Committee 1997). Water for the city of Melbourne comes from some $1550\,km^2$ of forested catchment in the Central Highlands, about one-half of which is mountain ash forest. This one-half yields about 80% of the streamflow because of the high rainfall (Vertessy *et al.* 1996).

Streamflow from the mountain ash water catchments has been measured for decades. Streamflow from regrowth forests that regenerated after the 1939 bushfire decreased markedly relative to streamflow from old-age forest in the years before the fire. According to Kuczera's (1987) model of regional streamflow, streamflow from regrowth forest decreases over the first 20–30 years to about 50% of that from old-age forest and then increases to equal that from old-age forest at about 150 years. Mountain ash is a strongly self-thinning species, and both tree density and leaf area index decrease from relatively high in the young forest to low in the old forest. The greater streamflow in old forest than in young forests has been attributed to less interception of rainfall and to less transpiration due to smaller leaf area and smaller sapwood basal area (Dunn & Connor 1993; Jayasuriya *et al.* 1993; Vertessy *et al.* 1995).

The age structure of the mountain ash forest of the Central Highlands is now well within the range where streamflow from the catchments is increasing, and the Management Plan (Department of Natural Resources and Environment Victoria 1998) is based on that premise. It has been argued that the financial return from the forests favours water rather than timber, but the arguments for and against depend on economic assumptions that are outside the scope of this case study. Given, however, that timber harvesting in mountain ash forest is restricted to 39% of

the total area (Table 16.3), and supposing a rotation of 80 years is maintained, 0.5% of the forest area will be harvested each year. If the mean age of the remaining forest continues to increase (a difficult assumption in the face of the probability of major bushfires), the effect of timber harvesting on streamflow will be small.

The Forest Management Plan for the Central Highlands (Department of Natural Resources and Environment Victoria 1998) therefore concentrates on control of water quantity in critical catchments with limited streamflow, and on control of water quality through buffer and filter strips, limits to design and placement of logging tracks and log landings, and particularly design, placement and use of roads (e.g. Grayson *et al.* 1993).

16.5 CONCLUDING COMMENTS

The case study presented in this chapter traces the history of conflicts over the management of native forests for timber. The conflicts have been between the states (the constitutional owners and managers of public land) and environmentalists, and between the states and the Commonwealth (which has constitutional rights in certain matters such as the export of products). In recent years, the Commonwealth has taken the initiative to resolve these conflicts by establishing Regional Forest Agreements between the Commonwealth and state governments. These agreements are directed at world standards of reservation, and of implementation and documentation of ecologically sustainable forest management.

Then the Regional Forest Agreement for the Central Highlands, Victoria was outlined. Mountain ash (*Eucalyptus regnans*) is one of the dominant species of the Central Highlands, and is of major importance to the Victorian timber industry. Some aspects of the ecology and silviculture of mountain ash forest that have been at the forefront of on-going debate have been outlined. We have attempted to present some of the most critical issues underlying the Forest Management Plan for the mountain ash forests of the Central Highlands. Above all, we have aimed to emphasize flexibility rather than rigidity. A Regional Forest Agreement (and the consequent Forest

Management Plan) is based simply on the application of the best knowledge of the day to the social and political aspirations and structures of the day. All of these will change. Our knowledge of silvicultural systems (e.g. size and shape of coupes, the significance of boundaries) will increase. More will be found out about the life histories of the plants and animals we are managing. Community expectations of publicly owned forests will change. Local and overseas manufacturing processes and markets will change. Plantations will be established. All of these and other changes will have to be incorporated into forest management. Thus any plan for sustainable forest management can, like the forest itself, never be stable, harmonious and enduring.

Regional Forest Agreements are the latest approach that aims to provide solutions to the long history of intense debate on the future of Australia's native forests. Whether or not they will succeed remains to be seen. Deficiencies and potential problem areas have already been documented. They include the extent of public involvement (especially at any serious level of decision-making), the scientific methods and judgements used in the selection of reserves, and the contrast between resource security for industry on the one hand and the limited flexibility within the terms of the agreements for management agencies to adapt to new developments on the other (Dargavel 1998; Kirkpatrick 1998). After signing the Regional Forest Agreement for the South-west Forests, Western Australia, the long-running and bitter dispute over logging of old-growth karri forest intensified to the point where the State had to back down substantially from the Agreement. New protagonists have emerged, and the interpretations of 'ecologically sustainable forest management' have only widened. Dargavel's (1998) conclusion, that 'most probably (Regional Forest Agreements) will be seen as but a stage in the history of forest use, one showing both new strengths and old flaws', will undoubtedly be proven correct.

ACKNOWLEDGEMENTS

This chapter was prepared with the assistance of funding from the Australian Research Council, and from the Scientists for Sustainability Project funded by Forests and Wood Products Research and Development Corporation, Standing Committee on Forestry (Department of Agriculture, Forestry and Fisheries, Australia) and National Association of Forest Industries.

REFERENCES

Ashton, D.H. (1956) *The autecology of* Eucalyptus regnans *F. Muell.* PhD thesis. University of Melbourne, Melbourne.

Ashton, D.H. (1981) Fire in tall open-forests (wet sclerophyll forests) In: Gill, A.M., Groves, R.H. & Noble, I.R., eds. *Fire and the Australian Biota*, pp. 339–66. Australian Academy of Science, Canberra.

Ashton, D.H. & Attiwill, P.M. (1994) Tall open-forests. In: Groves, R.H., ed. *Australian Vegetation*, pp. 157–96. Cambridge University Press, Cambridge.

Attenborough, D. (1995) *The Private Life of Plants*. BBC Books, London.

Attiwill, P.M. (1994) The disturbance of forest ecosystems and the conservative management of eucalypt forests in Australia. *Forest Ecology and Management* **63**, 301–46.

Attiwill, P.M., Polglase, P.J., Weston, C.J. & Adams, M.A. (1996) Nutrient cycling in forests of south-eastern Australia. In: Attiwill, P.M. & Adams, M.A., eds. *Nutrition of Eucalypts*, pp. 191–227. CSIRO, Melbourne.

Beadle, N.C.W. & Costin, A.B. (1952) Ecological classification and nomenclature. *Proceedings of the Linnaen Society of New South Wales* **77**, 61–82.

Bennison, G.L. & Jones, R. (1975) Woodchipping in Tasmania. In: Jones, R., ed. *The Vanishing Forests? Woodchip Production and the Public Interest in Tasmania*, pp. 1–37. The Environmental Law Reform Group, University of Tasmania, Hobart.

Boer, B. (1997) Developments in international law relating to forests. *Environmental and Planning Law Journal* **14**, 378–84.

Boland, D.J., Brooker, M.I.H., Chippendale, G.M. *et al.* (1992) *Forest Trees of Australia*, 4th edn. CSIRO Publications, Melbourne.

Brand, D.G. (1997) Forest management in New South Wales, Australia. *Forestry Chronicle* **73**, 578–85.

Campbell, R. (1997a) *Evaluation and Development of Sustainable Silvicultural Systems for Multiple-purpose Management of Mountain Ash Forests. Discussion Paper. Value Adding and Silvicultural Systems Program*, VSP Technical Report 28. Centre for Forest Tree Technology, Forests Service, Department of Natural Resources and Environment, Victoria, Melbourne.

Campbell, R. (1997b) *Evaluation and Development*

of Sustainable Silvicultural Systems for Multiple-purpose Management of Mountain Ash Forests. SSP Scientific Knowledge Base. Value Adding and Silvicultural Systems Program, VSP Technical Report 27. Centre for Forest Tree Technology, Forests Service, Department of Natural Resources and Environment, Victoria, Melbourne.

Carron, L.T. (1993) Changing nature of federal-state relations in forestry. In: Dargavel, J. & Feary, S., eds. *Australia's Ever-Changing Forests II, Proceedings of the Second National Conference on Australian Forest History*, pp. 207–40. Australian National University, Canberra.

Christensen, N.L., Agee, J.K., Brussard, P.F. *et al.* (1989) Interpreting the Yellowstone fires of 1988. *Bioscience* **39**, 678–85.

Christoff, P. (1998) Degreening government in the garden state: environment policy under the Kennett government 1992–97. *Environmental and Planning Law Journal* **15**, 10–32.

Church, R. & Richards, M.T. (1998) Native forest harvesting in Australia: an ecological economics approach to the question. *Australian Forestry* **61**, 137–40.

Committee of Inquiry into the National Estate (1974) *The National Estate: Findings and Recommendations*. Committee of Inquiry into the National Estate, Canberra.

Commonwealth and Victorian Regional Forest Agreement Steering Committee (1996) *Comprehensive Regional Assessment East Gippsland: Ecological Sustainable Management Report*. Commonwealth and Victorian Regional Forest Agreement Steering Committee, Canberra.

Commonwealth and Victorian Regional Forest Agreement (RFA) Steering Committee (1997) *Central Highlands Comprehensive Regional Assessment Report*. Commonwealth Forests Taskforce, Department of the Prime Minister and Cabinet, Canberra.

Commonwealth of Australia (1992a) *The Australian Forestry Council Twenty-Eighth Meeting: Summary of Resolutions and Recommendations*. Department of Primary Industries and Energy, Canberra.

Commonwealth of Australia (1992b) *National Forest Policy Statement*. Australian Government Printing Service, Canberra.

Cremer, K.W. (1960) Problems of eucalypt regeneration in the Florentine Valley. *APPITA* (Journal of the Australian and New Zealand Pulp and Paper Industry Technical Association) **14**, 71–7.

Cunningham, T.M. (1960) *The Natural Regeneration of Eucalyptus Regnans*. Bulletin 1. School of Forestry, University of Melbourne, Melbourne.

Dargavel, J. (1995) *Fashioning Australia's Forests*. Oxford University Press, Melbourne.

Dargavel, J. (1998) Politics, policy and process in the forests. *Australian Journal of Environmental Management* **5**, 25–30.

Department of Conservation and Natural Resources Victoria (1991) *Forests of Victoria: 1:1 000 000 Mapsheet*. Department of Conservation and Natural Resources, Victoria, Melbourne, Victoria.

Department of Conservation and Natural Resources Victoria (1995) *Flora and Fauna Guarantee Action Statement no. 62: Leadbeater's Possum*. Department of Conservation and Natural Resources Victoria, Melbourne.

Department of Natural Resources and Environment Victoria (1996a) *Code of Forest Practices for Timber Production. Review no. 2*. Department of Natural Resources and Environment, Victoria, Melbourne.

Department of Natural Resources and Environment Victoria (1996b) *Study of Old-growth Forest in Victoria's Central Highlands*. Forest Service Technical Reports 96–3. Department of Natural Resources and Environment, Victoria, Melbourne.

Department of Natural Resources and Environment Victoria (1998) *Forest Management Plan for the Central Highlands*. Department of Natural Resources and Environment, East Melbourne, Victoria.

Dignan, P., King, M., Savenah, A. & Walters, M. (1998) The regeneration of *Eucalyptus regnans* F. Muell. under retained overwood: seedling growth and density. *Forest Ecology and Management* **102**, 1–7.

Dunn, G.M. & Connor, D.J. (1993) An analysis of sap flow in mountain ash (*Eucalyptus regnans*) forests of different age. *Tree Physiology* **13**, 321–36.

Economou, N. (1992) Resources security legislation and national environment policy: new objectives, old dynamics. *Current Affairs Bulletin* **68**, 17–26.

Economou, N. (1996) Australian environmental policy making in transition: the rise and fall of the Resource Assessment Commission. *Australian Journal of Public Administration* **55**, 12–22.

Ferguson, I.S. (1996) *Sustainable Forest Management*. Oxford University Press, Melbourne.

Fisher, D.E. (1987) *Natural Resources Law in Australia*. The Law Book Company, Sydney.

Florence, R.G. (1994) Forestry and the environment in Australia. *Forestry Chronicle* **70**, 36–7.

Forestry Commission of Tasmania (1994) *State of the Forests*. Forestry Commission of Tasmania, Hobart.

Forestry Technical Services (1997) *Review of sustainable forest management certification and labelling*. Report to the Standing Committee on Forestry. Forestry Technical Services, Acton.

Galligan, B. & Fletcher, C. (1993) *New Federalism, Intergovernmental Relations and. Environment Policy*. Federalism Research Centre, Australian National University, Canberra.

Gilbert, J.M. (1958) *Eucalypt–rainforest relationships and the regeneration of the eucalypts*. PhD thesis, University of Tasmania, Hobart.

Gilbert, J.M. & Cunningham, T.M. (1972) Regeneration of harvested forests. *APPITA* (Journal of the Australian and New Zealand Pulp and Paper Indnstry Tecnical Association) **26**, 43–5.

Government of Victoria (1986) *Victoria Timber Industry Strategy*. Government Printer, Melbourne.

Government of Victoria (1988) *Flora and Fauna Guarantee Act no. 47*. Government Printer, Melbourne.

Grayson, R.B., Haydon, S.R., Jayasuriya, M.D.A. & Finlayson, B.L. (1993) Water quality in mountain ash forests—separating the impacts of roads from those of logging operations. *Journal of Hydrology* **150**, 459–80.

Griffiths, T. (1992) *Secrets of the Forest. Discovering History in Melbourne's Ash Range*. Allen & Unwin, St Leonards, New South Wales.

Groenewegen, P. (1994) The political economy of federalism since 1970. In: Bell, S. & Head, B., eds. *State, Economy and Public Policy in Australia*, pp. 169–93. Oxford University Press, Melbourne.

Grose, R.J. (1960) Effective seed supply for the natural regeneration of *Eucalyptus delegatensis* RT Baker, syn *Eucalyptus gigantea* Hook f. APPITA (Journal of the Australian and New Zealand Pulp and Paper Industry Technical Association) **13**, 141–8.

Hay, P.R. & Eckersley, R. (1993) Green politics: lessons from Tasmania's Labor–Green accord 1989–91. *Current Affairs Bulletin* **69**, 10–15.

Hickey, J.E. (1994) A floristic comparison of vascular species in Tasmanian oldgrowth forest with regeneration resulting from logging and wildfire. *Australian Journal of Botany* **42**, 383–404.

Hoare, J. (1996) *Assessing Sustainable Forest Management in Australia*. Department of Primary Industries and Energy, Bureau of Resource Sciences, Canberra.

Incoll, R.D., Lindenmayer, D.B., Gill, A.M., McCarthy, M.A. & Mullen, I.C. (1997) *Montane Ash and Associated Cool Temperate Rainforest Environments in South-east Australia; a Bibliography*. Working Paper 1997/3. Centre for Resource and Environmental Studies, Australian National University, Canberra.

JANIS (1997) *Nationally Agreed Criteria for the Establishment of a Comprehensive, Adequate and Representative Reserve System for Forests in Australia*. Joint ANZECC/MCFFA National Forest Policy Statement Implementation Committee, Canberra.

Jayasuriya, M.D.A., Dunn, G., Benyon, R. & O'Shaughnessy, P.J. (1993) Some factors affecting water yield from mountain ash (*Eucalyptus regnans*) dominated forests in south-east Australia. *Journal of Hydrology* **150**, 345–67.

Jones, J.R., Martin, R. & Bartlett, E.T. (1995) Ecosystem management in the United States—the Forest Services response to social conflict. *Society and Natural Resources* **8**, 161–8.

Kennedy, R.W. (1993) Recent developments in forest policy in Australia. *Forestry Chronicle* **69**, 40–5.

King, M., Vale, K. & Dignan, P. (1994) *Ecology, Silviculture and Management of Victoria's* Eucalyptus regnans *Dominated Ecosystems: A Bibliography*.

Value-adding and Silvicultural Systems Project (VSP) Technical Report 24. Department of Conservation and Natural Resources, Melbourne.

Kirkpatrick, J.B. (1998) Nature conservation and the Regional Forest Agreement Process. *Australian Journal of Environmental Management* **5**, 31–7.

Kuczera, G. (1987) Prediction of water yield reductions following a bushfire in ash-mixed species eucalypt forest. *Journal of Hydrology* **94**, 215–36.

Lamas, T. & Fries, C. (1995) Emergence of a biodiversity concept in Swedish forest policy. *Water, Air and Soil Pollution* **82**, 57–66.

Lindenmayer, D.B. (1995) Forest disturbance, forest wildlife conservation and the conservative basis for forest management in the mountain ash forests of Victoria—comment. *Forest Ecology and Management* **74**, 223–31.

Lindenmayer, D.B. & Possingham, H.P. (1994) *The Risk of Extinction. Ranking Management Options for Leadbeater's Possum using Population Viability Analysis*. Centre for Resource and Environmental Studies, The Australian National University, Canberra.

McCarthy, M.A. & Lindenmayer, D.B. (1998) Multi-aged mountain ash forest, wildlife conservation and timber harvesting. *Forest Ecology and Management* **104**, 43–56.

McCormack, J. (1992) *The Global Environmental Movement*, paperback edn. Belhaven Press, London.

Macfarlane, M.A. & Seebeck, J.H. (1991) *Draft Management Strategies for the Conservation of Leadbeater's Possum* Gymnobelideus leadbeateri, *in Victoria*. Technical Report Series 111. Arthur Rylah Institute for Environmental Research, Department of Conservation and Environment, Melbourne.

Menkhorst, P.W. & Lumsden, L.F. (1995) *Gymnobelideus leadbeateri*. In: Menkhorst, P.W., ed. *Mammals of Victoria*, pp. 104–5. Oxford University Press, Melbourne.

Mercer, D. (1995) *A Question of Balance: Natural Resources Conflict Issues in Australia*, 2nd edn. Federation Press, Leichhardt, New South Wales.

Montreal Process Implementation Group (1997) *Australia's First Approximation Report for the Montreal Process*. MIG Secretariat, Forests Branch, Department of Primary Industries and Energy, Canberra.

Mueck, S. & Peacock, R.J. (1992) *Impacts of Intensive Timber Harvesting on the Forests of East Gippsland, Victoria*. VSP Technical Report 15. Department of Conservation and Natural Resources, Melbourne.

Muirhead, N. (1998) BSc (Hons) thesis, *Substrates for epiphytes in the tall open forests of the Central Highlands, Victoria*. Botany Department, University of Melbourne, Melbourne.

Murphy, A. & Ough, K. (1997) Regenerative strategies of understorey flora following logging in the Central Highlands. *Australian Forestry* **60**, 90–8.

Murphy, P.J., Rousseau, A. & Stewart, D. (1993) Sustainable forests: a Canadian commitment, National Forest Strategy and Canada Forest Accord process and results. *Forestry Chronicle* **69**, 278–84.

Mutch, R.W. (1970) Wildland fires and ecosystems—a hypothesis. *Ecology* **51**, 1046–51.

Office of Auditor-General (1993) *Special Report No 22 on the Timber Industry Strategy.* Auditor-General's Office, Melbourne, Victoria.

Opeskin, B. & Rothwell, D. (1995) The impact of treaties on Australian federalism. *Case Western Research Journal on International Law* **27**, 1–59.

Ough, K. & Murphy, A. (1997) The effect of clearfell logging on tree-ferns in Victorian Wet forest. *Australian Forestry* **59**, 178–88.

Ough, K. & Murphy, A. (1998) *Understorey Islands: A Method of Protecting Understorey Flora During Clearfelling Operations.* Value Adding and Silvicultural Systems Project, VSP Internal Report 29. Department of Natural Resources and Environment, Melbourne.

Ough, K. & Ross, J. (1992) *Floristics, Fire and Clearfelling in Wet Forests of the Central Highlands, Victoria.* VSP Technical Report 11. Department of Conservation and Environment, East Melbourne.

Papadakis, E. & Moore, A. (1994) Environment, economy and the state. In: Bell, S. & Head, B., eds. *State, Economy and Public Policy in Australia*, pp. 334–51. Oxford University Press, Melbourne.

Peacock, R.J., Williams, J.E. & Franklin, J.F. (1997) Disturbance ecology of forested ecosystems: implications for sustainable management. In: Klomp, N. & Lunt, I., eds. *Frontiers in Ecology: Building the Links*, pp. 67–78. Elsevier Science, Oxford.

Resource Assessment Commission Australia (1992) *Forest and Timber Inquiry, Final Report.* Australian Government Printing Service, Canberra.

Rule, A. (1967) *Forests of Australia.* Angus and Robertson, Sydney.

Senate Standing Committee on Science and the Environment (1976) *The Impact on the Australian Environment of the Current Woodchip Industry Program: Interim Report.* Senate Standing Committee on Science and the Environment, Canberra.

Shea, S. (1998) *Western Australia's Development and Future Prospects for Tree Crop Industries.* Department of Conservation and Land Management, Perth.

Specht, R.L. (1970) Vegetation. In: Leeper, G.W., ed. *The Australian Environment*, pp. 44–67. CSIRO and Melbourne University Press, Melbourne.

Squire, R.O., Campbell, R.G., Wareing, K.J. & Featherston, G.R. (1991) The mountain ash forests of Victoria: ecology, silviculture and management for wood production. In: McKinnell, F.H., Hopkins, E.R.

& Fox, J.E.D., eds. *Forest Management in Australia*, pp. 38–57. Surrey Beatty & Sons, Chipping Norton, New South Wales.

Toyne, P. (1994) *The Reluctant Nation: Environmental Law and Politics in Australia.* ABC Books, Sydney.

Tsamenyi, M. (1996) The domestic consequences of international environmental imperatives. In: Alston, P. & Chaim, M., eds. *Treaty-making and Australia: Globalisation Versus Sovereignty?*, pp. 163–73. Federation Press in association with the Centre for International and Public Law, Faculty of Law, Australian National University, Leichhardt, New South Wales.

Tsamenyi, M., Bedding, J. & Wall, L. (1989) Determining the World Heritage Values of the Lemonthyme and Southern Forests: lessons from the Helsham Inquiry. *Environmental and Planning Law Journal* **6**, 79–93.

United Nations Conference on Environment and Development (1992) *Agenda 21.* United Nations Conference on Environment and Development, UNO, New York.

Vertessy, R.A., Benyon, R.G., O'Sullivan, S.K.O. & Gribben, P.R. (1995) Relationships between stem diameter, sapwood area, leaf area and transpiration in a young mountain ash forest. *Tree Physiology* **15**, 559–67.

Vertessy, R.A., Hatton, T.J., Benyon, R.G. & Dawes, W.R. (1996) Long-term growth and water balance predictions for a mountain ash (*Eucalyptus regnans*) forest catchment subject to clearfelling and regeneration. *Tree Physiology* **16**, 221–32.

Webster, H. (1993) Some thoughts on sustainable development as a concept, and as applied to forests. *Forestry Chronicle* **69**, 531–3.

Wilkinson, G. (1994) *Native Forest Silviculture.* Technical Bulletin 5. Forestry Commission of Tasmania, Hobart.

Winzenried, A.P. (1986) *Britannia Creek. Wood Distilling in the Warburton District.* APW Productions, Melbourne.

Woodgate, P.W., Peel, W.D., Ritman, K.T. *et al.* (1994) *A Study of the Old-growth Forests of East Gippsland.* Department of Conservation and Natural Resources, Victoria, Melbourne.

Working Group (1995) *Criteria and Indicators for the Conservation and Sustainable Management of Temperate and Boreal Forests ('The Montreal Process').* Canadian Forest Service, Natural Resources Canada, Hull, Quebec.

World Commission of Environment and Development (1987) *Our Common Future.* Oxford University, Oxford.

Young, A. (1996) *Environmental Change in Australia Since 1788.* Oxford University Press, Melbourne.

17: Sustainable Management of Malaysian Rain Forest

SIMMATHIRI APPANAH

17.1 INTRODUCTION

The attention that management of Malaysian rain forests has received appears disproportionate to the size of the forests. In fact, much bigger parcels of forests of similar stature exist in Indonesia, and existed in the Philippines. Moreover, Malaysia is not the largest producer of tropical wood products. That credit belongs to Indonesia. Nevertheless, there exist a few sound reasons for the heightened interest in Malaysian rain forests. Among the many forest types in Malaysia, the important ones are the dipterocarp forests. These forests have been described in superlative terms as the climax forests, the oldest forests, the most diverse in terms of species richness, and so on (Richards 1952; Ashton 1969; Whitmore 1975). Thus, they attract a lot of attention from biologists. The dipterocarp forests of the aseasonal South-East Asian countries also have greater timber volume per hectare than the vast tracts of forests in tropical South America and Africa. This is due to the dominance of the emergent storey dipterocarp family. At least 60% of the trees in the emergent layer may be members of this timber family. Above all, the attention given to Malaysian forestry originates from the history of its development, which is unique in tropical forest management, and is known as a rich mixture of science and enterprise.

Malaysian forestry was the birthplace of many exciting developments in tropical forestry following the introduction of scientific forest management during the British colonial era. These developments were followed with interest by foresters from the rest of the tropics, and some of the silvicultural systems became standards for sustainable forest management practices. Malaysia is a good example of silvicultural development in the humid tropics, as forestry in its truest sense has been practised here for probably longer than anywhere else. Furthermore, the rate of development has been rapid in the country, and the large local market for timber and for export brought about some remarkable changes. Efforts towards the sustainable management of Malaysian rain forests require special attention, and the progress made in Malaysia may contain useful guidance for other tropical countries that still have old-growth forest.

Malaysia consists geographically of three regions: Peninsular Malaysia, and the two northern states of the Borneo Island, Sabah and Sarawak. There are some differences in the evolution of forest management and application of practices in the three regions. However, the developments began in the peninsula and subsequently were applied in the other two states. For a case study, this chapter will be confined mainly to a discussion of the events and developments as they occurred in Peninsular Malaysia.

17.2 GEOGRAPHY, GEOLOGY AND CLIMATE

Peninsular Malaysia extends from the Kra Peninsula in the north (latitude 6°45') to just above Singapore island in the south (latitude 1°10'). Its longitude ranges from 100°40' to 104°30' E. Its extreme length is 780 km and its greatest width 320 km. Its total area is 131 598 km². The country is generally hilly to mountainous with about 40% of the land area rising above the 150-m contour and 23% above the 300-m contour. The mountains occur as a series of parallel ranges running north–south, the largest of which extends over three-quarters of the country and acts as the backbone that divides the peninsula into east and west coastal areas.

Peninsular Malaysia is usually described as a land mass built around an intrusive core of solid granite, and its uneven roof reaches the surface to form the main mountain ranges. Sedimentary rocks older than the granite, of Palaeozoic and Mesozoic age, occur along the flanks of the mountain ranges. Recent, Quaternary and Tertiary sediments have accumulated since the emplacement of the granite and are confined mainly to the coastal areas and the main river basins and valleys of the interior.

Peninsular Malaysia falls in the equatorial climate zone. The region enjoys a uniform temperature but not the high temperatures experienced in continental Asia. Both the annual and diurnal ranges of temperatures are very small. The mean annual temperature is around 27°C, and the mean daily minimum and maximum are 22°C and 32°C, respectively. Rainfall in the peninsula is from: (i) cumuliform cloud development at the boundary of two different currents of air; (ii) orographic rains from the south-west and north-east monsoons; and (iii) local tropical showers in the lowlands. As a result, the rainfall is evenly distributed throughout the year, and the region does not experience droughts on a regular basis. The annual rainfall is high, varying between 2000 and 3000 mm over the whole peninsula. With a particularly high rainfall, even during the intermonsoonal months, the effect on vegetation and silvicultural practice is insignificant compared with subtropical regions.

17.3 DESCRIPTION OF FORESTS

The Malaysian flora is among the richest in the world and attracts a lot of attention. The most valuable and characteristic flora of the region are in the family Dipterocarpaceae. Although not present in all the forest types, this family makes up a high proportion of the upper storey of the forests, and is the dominant timber group. The vegetation in the peninsula is grouped into six main classes (Wyatt-Smith 1963): Lowland Evergreen Rain Forests, Lower Montane Forests,

Upper Montane Forests, Swamp and Low-lying Forests, Regenerated Forests and a few Miscellaneous types that are deficient in soil moisture. The important groups in terms of commercial timber production are the Lowland Evergreen and the Swamp and Low-lying forests. Within the first group are the Lowland Dipterocarp Forests, White Meranti-Gerutu Forests, Heath Forests, Hill Dipterocarp Forests and the Upper Dipterocarp Forests. In the Swamp category, the timber-producing forests include the Mangrove, Peat Swamp and the Freshwater Swamp Forests. Of the groups above, the most important timber forests used to be the Lowland Dipterocarp Forests. These were mostly converted to cash crop plantations in the 1970s and 1980s. Currently, the Hill Dipterocarp Forests and, to a lesser extent, the Peat Swamp Forests are the main timber producers. The Dipterocarp Forests typically have three storeys of trees, comprising thousands of species, often with over 200 tree species per hectare, as well as shrubs, herbs and woody climbers. The emergent storey can reach about 45 m high and contain as many as four or five trees per hectare; girths in excess of 3 m may be found.

17.4 DEVELOPMENTS IN EXPLOITATION AND UTILIZATION

The historical sequence of forest exploitation and utilization in Peninsular Malaysia has taken the usual historical path from selective harvesting of a few highly durable species in the beginning (1900–45) to that of more complete utilization of a broad range of species today. It is known that as early as the eighth century Chinese traders came to Peninsular Malaysia for crystalline camphor (obtained from *Dryobalanops aromatica*) (Burkhill 1935), and later for the highly durable cengal timber (*Neobalanocarpus heimii*).

Modern forest management began in the early 1900s, when the Malayan colonial officers became alarmed at the reckless cutting of *Palaquium gutta* for gutta percha used for coating marine cables and the highly durable species of *N. heimii* and *Intsia palembanica* for railway sleepers. These species were becoming scarce from accessible forests. As a result, forest reserves were developed and demarcated in several Malayan states, and harvesting of timbers and fuelwoods became controlled (Hill 1900). A variety of silvicultural systems were also progressively introduced to keep pace with the rapid changes in the demand for timber.

Through the 1920s, the demand for naturally durable timbers, mostly for railway sleepers, remained strong. Research extended the range of species considered suitable for this use, and *Vatica* sp. (resak) and those species of the genera *Hopea* (giam) and *Shorea* (balau) with heavy hardwood timber were also utilized. Other common but less durable light hardwood species, especially those belonging to *Shorea*, were beginning to be exploited in a small way. In the 1930s, medium-powered sawmills replaced the hand sawyer in many districts. The demand for timber increased as a result, and this was met by the less durable timber species which were previously considered as weed species and were either used for fuelwood or killed by girdling to open the canopy for favoured species.

After the Second World War (1945–50), demand for timber of all categories increased sharply, locally and worldwide. Timber production increased, and more species were utilized as sawmill production improved. Extraction methods like winch lorry and the construction of forest roads were introduced to reduce traditional hauling by buffalo. As a consequence, it became economically necessary to extract all utilizable trees in a single operation. Besides sawmills and more modern extraction methods, increased botanical and wood technological knowledge of the species, improved knowledge of timber qualities, especially modern methods of wood preservation, and general education of the public to the potential use of the variety of timber species all contributed to the utilization of a wide range of species. This trend toward wider species utilization and increased demand for timber is shown in Table 17.1. In 1925 only 19 species were utilized, whereas in the 1990s, the number had reached over 400 of the estimated 890 species that can reach a harvestable size of 45 cm d.b.h. (diameter at breast height). Currently, more and more of the unutilized species are marketed as 'lower valued' mixed hardwoods. This again has been possible because of changes in the processing industry. There has been a big shift in the type of timbers being processed: the ratio of heavy hardwoods (high density, naturally durable timbers) to light hardwoods (low to medium density, less durable timbers) has shifted over time—in the 1920s it was 1:2, and at present it is 1:10 (Table 17.2).

This trend continued into the early 1970s when the downstream industry began to expand, and overseas demand for tropical timbers increased rapidly. About this time, large tracts of the Lowland Dipterocarp Forests were converted to cash crop plantations such as rubber and oilpalm. These conversions provided bountiful amounts of timber, and only the highly valuable species were extracted. The wood-processing industry was flooded with an abundance of cheap timber, and it grew unchecked. It was still primarily a sawmilling industry, although some plywood and moulding mills were also constructed. Following the conversion of the timber-rich Lowland Dipterocarp Forests, thereafter the forest estate was mainly limited to Hill and other less accessible Dipterocarp Forests, which are less well stocked in timbers. Before the end of the 1970s, there was apprehension that the huge and growing downstream timber industry would face severe timber shortages (Chong 1979). To avoid such a shortfall, fast-growing exotic timber plantations, mainly with *Acacia mangium*, were established (Yong 1984).

At the same time the downstream industry began to mature, from rudimentary sawmilling to more sophisticated industries like plywood, veneer, moulding and furniture manufacturing. With the almost unbridled growth of such industries, timber shortages are now occurring. While the furniture industry has successfully turned to rubberwood (*Hevea brasiliensis*) for their raw material, the acacia plantations have not met with expectations of yielding much sawn timber (Hashim *et al.* 1990). Under the circumstances, improvization is going on. The variety of timber species being used is high—at present almost any tree that can reach the minimum size of 30 cm d.b.h. is potentially usable. The industry is also working on reducing wastage, and other sources of fibre are being included in the manufacturing of wood products. Despite the efforts, the natural forests are, at present, unable to meet the needs of the industry. Today timber is being imported to meet the production deficit from Malaysian forests.

Species	1925	1950	1960	1980	1990
HHW					
Cynometra spp.				----	----
Cotylelobium spp.				----	----
Dialium spp.			----	----	----
Fagraea fragrans					----
Hopea spp.		----	----	----	----
Intsia palembanica		----	----	----	----
Madhuca utilis					----
Mesua ferrea		----	----	----	----
Neobalanocarpus heimii		----	----	----	----
Palaquium spp.			----	----	----
Parinari spp.		----	----	----	----
Shorea spp.		----	----	----	----
Streblus elongatus		----	----	----	----
Vatica spp.		----	----	----	----
MHW and LHW					
Agathis dammara				----	----
Alstonia spp.				----	----
Amoora spp.			----	----	----
Anisoptera spp.		----	----	----	----
Annonaceae spp.					----
Aquilaria malaccensis			----	----	----
Artocarpus spp.				----	----
Burseraceae spp.		----	----	----	----
Calophyllum spp.		----	----	----	----
Campnosperma spp.		----	----	----	----
Cananga odorata			----	----	----
Canarium spp.					----
Cinnamomum parthenoxylon					----
Coelostegia griffithii			----	----	----
Cratoxylon arborescens		----	----	----	----
Dacrydium spp.			----	----	----
Dacryodes spp.			----	----	----
Dialium spp.			----	----	----
Dillenia spp.		----	----	----	----
Dipterocarpus spp.		----	----	----	----
Dryobalanops spp.		----	----	----	----
Durio spp.		----	----	----	----
Dyera costulata		----	----	----	----
Endospermum malaccense			----	----	----
Eugenia spp.		----	----	----	----
Ganua motleyana			----	----	----
Gluta spp.				----	----
Gonystylus spp.				----	----
Gordonia concentricicatrix			----	----	----
Heritiera javanica				----	----
Hopea spp.		----	----	----	----
Koompassia malaccensis					----
K. excelsa		----	----	----	----
Kostermansia malayana			----	----	----
Lauraceae spp.				----	----

Table 17.1 Addition to timber species harvested in Peninsular Malaysia over the decades. (Source: Annual Reports of Forest Department.)

Species	1925	1950	1960	1980	1990
Litsea grandis			-----	-----	-----
Lophopetalum spp.		-----	-----	-----	-----
Mangifera spp.			-----	-----	-----
Myristicaceae spp.			-----	-----	-----
Neesia altissima			-----	-----	-----
Ochanostachys amentacea		-----	-----	-----	-----
Palaquium spp.		-----	-----	-----	-----
Parashorea spp.					-----
Paratocarpus triandra			-----	-----	-----
Parishia spp.			-----	-----	-----
Parkia spp.				-----	-----
Payena spp.			-----	-----	-----
Pentace spp.				-----	-----
Pentaspadon spp.				-----	-----
Pithecellobium splendens				-----	-----
Podocarpus spp.					-----
Pometia spp.				-----	-----
Pouteria malaccensis			-----	-----	-----
Rhizophora spp.		-----	-----	-----	-----
Sandoricum koetjape			-----	-----	-----
Santiria spp.		-----	-----	-----	-----
Sapotaceae spp.					-----
Scaphium spp.				-----	-----
Schima noronhae		-----	-----	-----	-----
Scorodocarpus borneensis		-----	-----	-----	-----
Shorea spp.		-----	-----	-----	-----
Sindora spp.		-----	-----	-----	-----
Strombosia javanica		-----	-----	-----	-----
Swintonia spp.					-----
Terminalia spp.		-----	-----	-----	-----
Tetramerista glabra				-----	-----
Toona serrata		-----	-----	-----	-----
Triomma malaccensis			-----	-----	-----
Vatica spp.				-----	-----
Xylopia spp.			-----	-----	-----

Table 17.1 *continued*   HHW, heavy hardwoods; LHW, light hardwoods; MHW, mixed hardwoods.

17.5 SILVICULTURAL DEVELOPMENTS

17.5.1 Improvement fellings

At the beginning of the twentieth century, when it was realized that the desirable species of gutta-percha (*Palaquium gutta*), cengal (*Neobalanocarpus heimii*) and merbau (*Intsia palembanica*) would disappear from accessible areas, an attempt was made to enhance the growth of the individual trees and secure their populations through silvicultural interventions. A form of the shelterwood system was introduced to favour the commercially valuable species (Hodgson 1932). Poles and immature Class I (naturally durable, heavy hardwoods) trees were released by improvement fellings, which included the removal of all the non-commercial species that were shading out commercially important species. Regeneration of cengal and merbau was often insufficient, and

Table 17.2 Production of heavy and light hardwood in Peninsular Malaysia over the decades (timber from plantations are excluded). (Source: Annual Reports of Forest Department.)

Year	Heavy hardwoods (m³)	Light hardwoods (m³)	Ratio
1925	50 348	102 004	1 : 2.0
1935	79 657	200 805	1 : 2.5
1939	177 615	523 001	1 : 2.9
1950	134 718	903 654	1 : 6.7
1955	121 889	1 445 790	1 : 11.9
1960	173 021	2 076 796	1 : 12.0
1970	486 501	6 058 523	1 : 12.5
1980	892 038	9 561,366	1 : 10.7
1990	948 546	11 870 829	1 : 12.5
1996	708 798	7 425 900	1 : 10.6

Table 17.3 Sequence of operations in the Regeneration Improvement Fellings.

Year	Operation
N − 1	Unmarked and unwanted species felled for poles
N	First marked seeding felling of unwanted species
N + 1 to 3	First cleaning
N + 2 to 4	Second marked seeding felling of unwanted species
N + 4 to 6	Marked final felling of economic species
N + 5 to 7	Cleaning after final felling

N = year of harvest.

these two species were planted along lines. Such enrichment planting proved expensive and was unsuccessful (Strugnell 1938).

Even in the 1920s, the demand was still for highly durable timbers. With research, additional species of *Vatica* (resak), *Hopea* (giam) and *Shorea* (balau) with heavy hardwood timbers could also be utilized. By the 1920s, improvement fellings were considered ineffective in enhancing the growth of the commercial species, but instead were found effective in the establishment of young regeneration (Strugnell 1936). Thereafter, such prescriptions were called regeneration improvement fellings, and a series of them were usually undertaken to establish the young regeneration before the removal of the economic crop. The fellings were undertaken over a 5–8-year period, and the sequence of operations is shown in Table 17.3.

No forests were opened for logging without these felling and cleaning operations. Doubts began about the usefulness of the cleanings, undertaken even when there was no regeneration on the ground.

17.5.2 Malayan Uniform System

During the Second World War, cleaning and felling operations were completely abandoned. Postwar inspections clearly indicated that the regeneration of light hardwood *Shorea* species, the dominant commercial species for harvesting by then, had benefited from the uncontrolled loggings without any of the prerequisite fellings and cleanings (Walton 1948). The regeneration present on the ground at the time of felling had survived undamaged and had grown rapidly to form the dominant crop. This indicated that prefelling operations were not required in forests where young regeneration of timber species was abundant. A single operation of harvest felling would be all that was needed to release the young regeneration. There were other non-silvicultural factors that also influenced the development of such a silvicultural system. Increase in demand for timber, introduction of sawmills, and development in wood preservation techniques all allowed more species to be harvested. Also, mechanical harvesting techniques and the cost of making roads and their high cost of maintenance under tropical conditions made it economically necessary to harvest as much timber as possible within one area in a single felling operation. Both factors, ecological and economic, led to the birth of the renowned silvicultural system called the Malayan Uniform System in 1948 (Walton 1948; Wyatt-Smith 1963). All the silvicultural operations from pretreatment fellings and cleanings were abandoned for one of single felling and post-treatment. The Malayan Uniform System is therefore a system for converting the virgin tropical Lowland Dipterocarp Forest, a rich, complex, multispecies and multi-aged forest, into a more or less even-aged forest with a greater proportion of the commer-cial species, by a single felling release of selected natural regeneration, aided by systematic girdling of unwanted species (Wyatt-Smith 1963).

In the Malayan Uniform System all harvest-able species down to 45 cm d.b.h. are removed in one single operation. This is followed by poison girdling of defective relics and non-commercial species down to a minimum diameter of 15 cm and 5 cm, respectively. The operations would release the selected natural regeneration, mainly the light-demanding light hardwood dipterocarp species. About 5–7 years following felling, a linear strip sampling is carried out to verify the presence of sufficient regeneration on the ground, and to determine if additional silvicultural treatments are required.

Several factors were required for successful implementation of the Malayan Uniform System. These include: (i) an adequate and well-distributed stocking of seedlings of economic species at the time of logging — this is determined by milliacre (2 × 2 m) linear sampling of the area. If stocking is inadequate, logging is deferred until after a seed year; (ii) complete removal of the canopy through poison girdling of all defective trees and all unwanted species down to 5 cm d.b.h.; (iii) maintenance of sufficient new canopy to minimize the regrowth of climbers; and (iv) linear samplings at regular intervals to ensure the seedlings are regenerating well. The sequence of operations originally formulated is given in Table 17.4 (Wyatt-Smith 1963):

The Malayan Uniform System was successfully applied in the Lowland Dipterocarp Forests. Nevertheless, some concern was expressed. A drastic opening of the canopy was considered more favourable for pioneer growth. In that context, the poison girdling of species down to 5 cm d.b.h. was regarded to be too drastic even for forests where the light hardwood *Shorea* species dominate. Next, for a variety of reasons, including security problems from communist insurgency, there were difficulties in delaying logging until regeneration was adequately established. Further-more, the mandatory prefelling sampling could not be carried out, and regeneration could only be

Table 17.4 Sequence of operations in the Malayan Uniform Systems.

Year	Operation
N − 1½	Linear sampling (2 × 2 m) of regeneration, and enumeration of merchantable trees
N to N + 1	Exploitation, followed by poison girdling down to 5 cm d.b.h.
N + 3 to N + 5	Linear sampling (5 × 5 m) of new crop, followed by cleaning, climber cutting and poison girdling as required
N + 10	Linear sampling (10 × 10 m) of new crop, followed by treatment as required or passed as regenerated
N + 10, N + 40 etc.	Sampling and thinning as required

N = year of harvest.

attended to after the logging. It must also be rec-ognized that the Malayan Uniform System was devised for forests rich in the regeneration of the light hardwood dipterocarp species, and forests currently harvested do not have a rich representa-tion of such species. In such areas, modifications of the system would be required. With these con-siderations, slight modifications were introduced in the mid 1960s (Table 17.5) (Wyatt-Smith 1963).

Other implementation difficulties were also present during this time. These included incom-plete utilization for lack of a market, steep and hilly terrain, or prolonged interrupted felling due to security reasons. Peninsular Malaysia was then undergoing the Emergency Period as a result of activity of communist guerillas who disrupted normal civilian activities. Many prefelling inven-tories were not undertaken, nor were postfelling tendings adequately met. As a consequence, the Malayan Uniform System could only be imple-mented in its full form on a small number of sites.

17.5.3 Modified Malayan Uniform System

The necessity of delaying logging until adequate seedling regeneration was established was re-garded as a constraint. In the late 1960s, interest in enrichment planting took off with enthusiasm. With this new 'tool', it was thought that the lack

Table 17.5 Revised sequence of operations in the Malayan Uniform System.

Year	Operation
N – 3 to N – 7	Canopy opening at time of good seedfall, in areas where seed regeneration and seed bearers are scarce
N – 3 to N – 5	Treatment of areas heavily infested with bamboo
N – 3	Treatment of areas with dense *Eugeissona*
N – 2	Climber cutting in areas of heavy vine infestation
N to N – $1^1/_2$	Milliacre sampling and enumeration of utilizable stems
N	Exploitation of all merchantable trees ($\geqslant$45 cm d.b.h.). Poison girdling all unwanted stems down to 15 cm d.b.h. in heavily opened patches, and down to 5 cm d.b.h. in shaded patches. Retention of any potentially valuable stem below 45 cm d.b.h.
N + 2	Commence tending in areas designated as heavy hardwoods
N + 3 to N + 5	5 m^2 sampling. Treatment indicated by diagnostic sampling, e.g. climber cutting, eradication of weed tree spp., etc.
N + 8 to N + 10	10 m^2 sampling. Such treatment as indicated by diagnostic sampling

d.b.h., diameter at breast height; N = year of harvest.

of regeneration need not be a limitation to logging, as it could be remedied with intensive line plantings (Ismail Ali 1966). This stance necessitated modifications to the Malayan Uniform System, and the revision was labelled the 'Modified Malayan Uniform System'. This decision to abandon the seedling requirements was a major departure in silvicultural thinking: it disregarded an important asset of dipterocarp forests—their potential to regenerate very heavily at supra-annual intervals. Instead, intervention enrichment planting was to make up for poor natural regeneration.

Enrichment planting turned out to be unreliable. Often, the plantings failed, and the task is very uneconomical. Technically, artificial regeneration with indigenous species is feasible (Appanah & Weinland 1993). However, there are several difficulties to be surmounted. First is the availability of seedlings. Dipterocarp fruiting is not a regular event, and the seeds are highly recalcitrant. Therefore the nursery work must be highly organized to capture irregular fruiting and produce adequate seed. Next, the canopy of the forest must be kept open for quite some time, at least 8–10 years for the seedlings to establish and enter the growth phase. Unless the gaps are big enough, the canopy will close over after a year or so, and the plants will not grow. If inappropriate species are used, the plantings often fail miserably too. In the end the Modified Malayan Uniform System turned out to be a failure. The occasional successes were not by design, as in some situations natural regeneration happened to be good, and the forest recovered.

17.5.4 Selective Management System

In the mid 1970s Malaysia went through an intensive development programme to diversify its agriculture. Huge tracts of timber-rich Lowland Dipterocarp Forests were converted into cash crop plantations. As a consequence, forestry was relegated to less accessible forests, undulating terrain, poor soil areas and the hills, areas which are less suitable for agriculture. These sites are also quite poor in timber stocking. Following the loss of the lowland forests, most production forest is now mainly limited to the Hill Dipterocarp Forests.

These Hill Forests occur at elevations ranging from 300 to 600 m above sea level and, in the main, are poorer in timber stocking Lowland Dipterocarp Forests except for the ridge tops where rich stands of seraya (*Shorea curtisii*) dominate. Distribution of the young regeneration is patchy. The suitability of using the Malayan Uniform System under such conditions was reviewed (Burgess 1972). Other constraints included the danger of erosion on steep slopes and the incidence of bertam (*Eugeissona triste*) and other secondary growth, both of which do not favour drastic opening of the canopy. Furthermore, the poison girdling of species down to 5 cm d.b.h. in the Malayan Uniform System was considered too drastic. Moreover, many of the species that were originally girdled became economically valuable in subsequent years. These and the difficult terrain made foresters conclude that the Malayan Uniform System cannot be applied to the hill forests. While there were oversimplifications in what constituted the Hill Dipterocarp Forests, the general decision was to switch to other management systems.

When the forest estate was switched to the hills in the late 1970s, a polycyclic system was tested (Griffin & Caprata 1977; Rasid 1998). It was derived from the bicyclic cutting of dipterocarp forests first introduced into the Philippines and later in Indonesia (Appanah & Weinland 1990). The Philippine system resembled a selection system, and the Indonesian cutting method was more of a selective felling based on diameter limits. The system introduced for Malaysian forests was similar to that tested in Indonesia, a selective felling. Called the Selective Management System, it relies on the advanced regeneration of the light hardwood species to form the next cut, unlike with seedling regeneration for the next rotation in the Malayan Uniform System. In the Selective Management System, harvests are twice, first at half the rotation period, but with less timber extracted at each harvest (Thang 1987).

The selective fellings of the Selective Management System are carried out on the basis of the prefelling stocking inventory. Depending on the stocking results, the felling diameter limit is set (Griffin & Caprata 1977; Thang 1987). The lowest cutting limit for dipterocarps is set at 50 cm d.b.h., and 5 cm less d.b.h. for non-dipterocarps. If the stocking is higher, then the diameter cutting limits are raised proportionately. All cuttings must maintain a minimum residual stand left behind for the next cut. In the event the minimum number of residuals cannot be achieved after meeting the economically har-

vestable portion, then the stand is managed on the Modifed Malayan Uniform System whereby all harvestable material above 45 cm d.b.h. is removed and the stands are put on rotations of 50 years. A prefelling inventory is carried out to estimate the stocking so the appropriate minimum cutting limits can be set.

The derivation of the number of residuals to be left behind in the Selective Management System is complicated. In Philippine and Indonesian practices, intermediate residuals were those in the 30–45 cm d.b.h. class. In the Selective Management System, where such intermediate classes cannot be found in abundance in the hill forests, even smaller classes of 15–30 cm d.b.h. were included. Three such individuals are considered equivalent to one individual in the 30–45 cm d.b.h. class. The equivalent classes are given in Table 17.6 (Thang 1987).

In this system, a tree of >45 cm d.b.h. is equivalent to two trees in the 30–45 cm d.b.h. class, and the small tree of 15–30 cm d.b.h. equivalent to one of the intermediate residual class of 30–45 cm d.b.h. The obligatory 32 residual trees which must be saved for the next cut are calculated on this basis. The minimum number of trees from the various diameter classes for retention are based on the overall stock results of the National Forest Inventory of 1973 for unexploited/undisturbed forests, considered to be the average stocking capacity of the natural forests. These stock data were increased between 30% and 50% for trees over 2 cm d.b.h. to take into consideration natural mortality, windthrow and growth stagnation of some of the residuals (Griffin & Caprata 1977).

Table 17.6 Equivalent classes used to modify the Selective Management System developed in Indonesia for use in Malayan dipterocarp forests.

Class	d.b.h. (cm)	Trees equivalent	Minimum no. of trees ha^{-1}
Merchantable	>45	2	25
Ingrowth	30–45	1	32
Small trees	15–30	$^1/_3$	96

d.b.h., diameter at breast height.

Table 17.7 Sequence of operations in the Selective Management System.

Year	Operations
N – 2 to N – 1	Prefelling forest inventory using systematic line plots and determination of cutting regimes
N – 1 to N	Climber cutting to reduce damage during logging. Tree marking incorporating directional felling. No marking of residuals for retention
N	Felling of all trees as prescribed
N + 2 to N + 5	Postfelling inventory using systematic line plots to determine residual stocking and appropriate silvicultural treatments

N = year of harvest.

The cutting cycles are fixed at 25–30 years (Thang 1987), based on a tentative gross volume increment of commercial species at $2.20\,m^3\,ha^{-1}$ $year^{-1}$. The sequence of operations is given in Table 17.7.

The Selective Management System was considered to have attributes that are preferable over the Malayan Uniform System, and became the choice for Peninsular Malaysia. In addition, the Selective Management System was preferred as it was supposed to confer four advantages.

1 There is flexibility to manage a very complex forest with very variable conditions.

2 It is rationally based on the inherent characteristics of the forest

3 It allows management to consider the socioeconomic conditions of the area.

4 It allows an economic cut, sustainability of the forest, and minimum investments for forest development (Thang 1987).

Unlike the Malayan Uniform System which relies on seedlings and saplings for the next rotation, the Selective Management System depends totally on the intermediate residual stock for the next cut. The second cut has been determined to be at around 30–35 years. The cutting cycles were set on the basis of inventory sample plots and experimental cutting and silvicultural treatment plots. Table 17.8 shows the annual growth and mortality rates of trees over 30 cm d.b.h. that were adopted (UNDP/FAO 1978).

Values of 2.0–$2.5\ m^3\ ha^{-1}\ year^{-1}$ in commercial growth volume would support removal of about $40\ m^3\ ha^{-1}\ year^{-1}$ at 30-year cutting cycles, which is close to the current average production level for virgin hill forest stands. However, this would

Tabel 17.8 Growth and mortality of dipterocarp forests (UNDP/FAO 1978).

Diameter growth (cm $year^{-1}$)	
All marketable species	0.80
Dark and light red meranti	1.05
Medium and heavy hardwoods	0.75
Light non-meranti species	0.80
Non-marketable species	0.75
Gross volume growth ($m^3\ ha^{-1}\ year^{-1}$)	
All marketable species	2.20
All species	2.75
Annual mortality %	
All marketable species	0.9
Annual in-growth %	
All marketable species (>30 cm d.b.h.)	0.6

obviously depend on the accuracy or applicability of the growth rates and limitation of logging damage to the intermediate residual stands of not more than 30%. Felling damage to the remaining intermediate-sized trees of 30 cm d.b.h. and above was estimated to be around 30%, and wastage due to bucking and breakage was about 8% of the gross timber volume (Griffin & Caprata 1977). However, there is much concern over the growth rates, which in reality are far below that estimated (Tang 1976; Wyatt-Smith 1988). Equally disturbing is the high rate of damage to the residuals and soils during harvesting, which may thwart all efforts to achieve sustainability in the future (Appanah & Weinland 1990).

Both the Selective Management System and the Modified Malayan Uniform System have now been in practice for almost two decades. Based on the original assumption, it should be possible to return to the logged stands in another decade. It has been suggested that the minimum economic cut based on log prices, estimated cost of logging and government charges is currently between 40 and $50\,m^3\,ha^{-1}$. The success of the management regimes would be to achieve such cuts in the future while maintaining the ecological integrity of the forests.

17.6 SUSTAINABLE MANAGEMENT

While the concept of sustainable forest management has received much attention in this decade, it was practised in spirit throughout the early history of forestry practices in Malaysia (Barnard 1950). Truthfully, while conceptually we have shifted from sustainable production to management of ecosystems, much of it is still rhetoric. There was much more care for the environment in the past, although the language for such effort was not developed. Several developments and practices indicate a conscious effort towards sustainable forest management, without the 'fanfare' attached to the term. Some of the practices that can be identified are described below.

17.6.1 Forest reserves

The concept of forest reserves was introduced, and forested areas were identified and gazetted for the specific end-use as forests at the very beginning of the twentieth century. This ensured that large tracts of land were not converted to other forms of land use except within the context of forestry practices. As a result, 50% of the country is still under forest cover. Harvesting inside these forest reserves is regulated.

17.6.2 National Forestry Acts and State Forest Enactments

There have been several national forestry acts which guaranteed permanency of the forests in the country, and at least half the land area is to be retained under forests in perpetuity. These forest enactments provided the guidelines for management and protection of forests within the states. While their implementation may have encountered difficulties, the laws for forest protection

have been clearly defined from the beginning of the last century.

17.6.3 Forest dwellers

Besides the creation of forest reserves, enactments were also introduced to protect forest dwellers. Laws were passed to allow them to collect forest produce like timber for their own needs, as well as the permission to harvest minor forest produce (non-timber forest products (NTFPs)) such as rattan, bamboo, fruits and resin for purposes of commerce. Some forested areas were also set aside for their permanent dwelling and cultivation.

17.6.4 Minor forest produce

The enormous publicity given in recent times to the potential minor forest products (non-timber forest products (NTFPs)) offer as an alternative to timber harvesting may not have fully appreciated the past practices in Malaysia. Historically, minor forest produce represented a significant portion of the forest value, and timber was proportionately smaller (Table 17.9). In the 1950s, the contribution of the two was in the ratio 1:1500, but in the 1980s, the ratio was 1:3500. (Similar figures for the 1990s were not available for comparisons.) It indicates that the value obtained from timber production has almost tripled while that from minor forest produce has remained static. The growth in timber production occurred mainly in the 1970s, but this has not resulted in any displacement for the forest dwellers who are given full opportunities to collect minor forest produce, enshrining one major criterion in sustainable forest management in the country since the beginning of modern forestry.

17.6.5 Shelterwood systems

Shelterwood systems of harvest and forest regeneration introduced to Peninsular Malaysia in about the 1920s emphasized timber production. This was the era when no more than a handful of species was exploited, so the overall impact on the forest's structure and species composition from logging and silvicultural operations was minimal. By the time sawmilling became mecha-

Table 17.9 Minor forest produce (NTFPs) from Peninsular Malaysia: 1957 and 1981 (when data were available are compared).

Produce	1957	1981
Timber	2 208 244 m³ = $363 891 000	10 226 261 m³ = $1 022 626 000 (estimated)
Minor forest produce		
Rattan	$74 769	$179 374
Bamboo	$26 193	$101 232
Damar	$10 528	$616
Jelutong latex	$33 911	$938
Gutta-percha	$192	–
Nipah	$84 805	$7 311
Wood Oil	$122	$703
Total for minor forest produce	$230 520	$290 174

In addition, quarrying for stone earned $134 869 in 1957, and $1 397 540 in 1981.

nized, and a larger number of timber species could be exploited, more trees were harvested from an area, but the impact was still minimal. With the introduction of the Malayan Uniform System, the impact on species composition would have been greater by virtue of the system. The Malayan Uniform System aimed to raise a more uniform crop of timber trees. To accomplish this, heavy girdling of trees of poor form and non-merchantable species was required. This would generally reduce the biodiversity of the forest in favour of commercial species in the stand. In addition, other practices such as cutting lianas were introduced—the lianas sometimes held the crowns of many trees together and this may result in many trees coming down together when one of them is felled. Liana cutting can contribute to reduced biodiversity, but the resulting reduction in damage to the residuals may offset the loss from cutting lianas. On the positive side, under the Malayan Uniform System, ameliorative measures such as establishing virgin jungle reserves within every forest compartment were introduced to ensure that species diversity is maintained, and the system is reversible.

17.6.6 Selective fellings

When the timber-rich lowland forests were converted to agriculture, forestry was restricted to the hill forests. Here the Malayan Uniform System was deemed unsuitable, and selective fellings were introduced. These fellings have many drawbacks as far as sustainable practices go, but the major flaw is not the selective felling itself, but the extraction method employed. Tractor-crawler systems are used, and these disturb large areas of forests causing much damage to the soil and resultant problems to waterways. Besides the logging damage to the valuable residuals during felling, extraction using tractors further injures the standing trees. Damage can be so severe that the second and subsequent harvests are threatened. Nevertheless, unlike the Malayan Uniform System, selective fellings are not likely to have much effect on the biodiversity of the forests. However, if the forests lose their timber production capacity after two harvests, pressure to convert them to other forms of land use would mount.

17.6.7 Enrichment plantings

Such planting trials were carried out from the very beginning, from about the 1910s, mainly to ensure that the forests are well stocked with young seedlings of the preferred species. Wyatt-Smith (1963) concluded that the plantings were costly, and failures too frequent, and the preferred method remains management and improvement of the natural regeneration. However, subsequent trials have shown promise (Tang & Wadley 1976), and failures can be reduced with better understanding of the silvics of the trees (Appanah &

Weinland 1993). One major cause for failure (observed all over the Asian region) appears to be the lack of canopy openings for enough years for the plants to become established and reach the canopy. However, frequent tendings for keeping the canopy open would raise the cost of the undertaking. Therefore, enrichment planting as an option to ensure the regeneration of desired species needs further improvement. Trials are under way to plant larger saplings in bigger holes, and for mechanizing the operations, to reduce the need for repeated follow-up tendings, greater survival and lower costs.

17.7 NEW DEVELOPMENTS IN SUSTAINABLE FOREST MANAGEMENT

17.7.1 Reduced impact logging

Regardless of the silvicultural system employed, logging damage both to the residual stand and soils will inevitably jeopardize sustainable management efforts. Reduced impact logging techniques are now being promoted (Marsh *et al.* 1996). The techniques involve limiting the tractor from free ranging in the forest. Instead, the vehicle is limited to movement along predetermined skid roads and the main road. Likewise, felling damage is minimized through directional felling. Researchers are also investigating the possibility of using mobile skyline systems and long-range cable logging using fibreglass cones attached to the front of the timber and hauling the log to the main road without the vehicle going to the log site. Overhead cable winch systems have also been tested successfully. Further improvements to this involve using mobile cable crane systems. Such a system is in the design stages now. A helicopter hauling system was tried in Sarawak, though without much success. The cost of harvesting was high, and the operations were extremely dangerous.

17.7.2 Malaysian criteria and indicators

The International Tropical Timber Organization Year 2000 Objectives, to which Malaysia is a signatory, requires that all timber should originate from sustainably managed forests by the year

2000. A more holistic management approach is required which pays attention to several issues that include resource security, continuity of production, planning, socioeconomic benefits, environmental impacts, and conservation of flora and fauna. To achieve sustainable management as expressed in the Year 2000 Objectives, a set of Malaysian Criteria and Indicators have been drafted. They are currently being revised. These Malaysian criteria and indicators have been developed on the basis of current understanding of the forests in terms of growth, regeneration, conservation of biodiversity, minimizing wastage, introduction of reduced impact logging techniques, and protection of riparian zones, wildlife and soils (Appanah & Thang 1997; Thang 1997). The Malaysian criteria and indicators also make provision for the welfare of the indigenous people dependent on the forest resources. All the guiding principles promulgated at international conventions and meetings are translated into a set of criteria, with relevant indicators which can be measured as proof that the implementation is undertaken correctly. Internal assessment systems are being developed in order to train the forest managers in implementing all the management rules now being strongly recommended (Appanah *et al.* 1999). The certification of forest units for implementing sustainable forest management in Malaysia would be coordinated by a new agency called the National Timber Certification Council of Malaysia. External assessments may become mandatory in the future for the timber to be certified as originating from sustainably managed forests. This could become a requirement for timber to be sold in the international markets.

17.7.3 Model forests

Trials were carried out to implement sustainable forest management in model schemes. One has been successfully completed in a 56 000-ha area of the Deramakot Forest Reserve in Sabah (Anjin & Kleine 1997). The management of this forest reserve has been certified as sustainable by an independent assessor. Similar models are now being tested in Jengai Forest Reserve in Terengganu, Peninsular Malaysia (Rasid 1998). The plan is now to expand such models throughout the country.

17.7.4 New research initiatives

A major thrust in research has been initiated in order to fulfil the requirements of the Malaysian Criteria and Indicators. Upon development of the criteria and indicators, a number of gaps were identified which lacked quantitative measurable standards for Malaysian conditions. For instance, issues of biodiversity conservation and ecosystem integrity and areas needed for maintenance of wildlife were not considered in depth in past management procedures. Research has now begun on what are the minimum number of species and what are the critical species and ecological linkages that are required to maintain ecosystem integrity. Results from these research findings will be used to modify management procedures. There are also uncertainties in the standards that have been already adopted. For example, a volume increment of $2.5\,m^3\,ha^{-1}\,y^{-1}$ for Hill Dipterocarp Forests has been used to calculate the cutting cycle and the annual allowable cut for the forests throughout Peninsular Malaysia. Enormous local variation in growth exists, and for some locales, this level of growth cannot be achieved. Hence new local growth figures are being developed to ensure such forests are sustainably cut.

17.7.5 Plantations

It is already recognized that following enforcement of all the management rules, log production may decline appreciably in Malaysia. Industrial milling capacity will far exceed the future production estimates. Besides stop-gap measures like importing timber, high-yield timber plantations will be needed to raise the production of timber over much smaller parcels of land. New initiatives are being introduced in this direction. Several attempts were made to establish timber plantations in Malaysia, but most of them were unsuccessful (Appanah & Weinland 1993). For example, in the early 1950s, a major study was undertaken to test the performance of Australian eucalypts in Malaysia. The eucalypts did not succeed except in the highlands. Later, in the same decade, tropical pines (*Pinus merkusii* and *Pinus caribaea* var. *hondurensis*) were tested for the production of pulp. But the planned paper mill was shelved, and the plantings, although promising, had a similar fate. In the late 1970s another major investment was undertaken to plant fast-growing tropical hardwoods (*Gmelina arborea, Eucalyptus* spp., *Acacia mangium*) for general utility timber. Almost 200 000 ha of *A. mangium* were planted. Unfortunately, the species had problems of heartrot, loose knots and poor form. As a consequence, this species is now considered only suitable for pulp. Much has been learned since then about plantations, and there is enough research knowledge to approach plantation establishment with confidence. A new directive for developing plantations has been announced, but it will be undertaken by the private sector, viz. the industries that need to secure their resource supply. Plantations of both short-rotation industrial wood and long-rotation high-quality timber are planned. Both exotics and indigenous species are being proposed, with the emphasis on the latter. Species such as *Acacia* (*mangium* × *auriculiformis*) hybrid and *Maesopsis eminii* are being planned for industrial wood plantations. The high-quality timber species being tested are *Tectona grandis, Azadirachta excelsa, Hevea brasiliensis, Hopea odorata, Shorea macrophylla* and several others. Smallholders have already started plantations to replace idle lands and non-profitable cash crops with timber species. There are plans for bigger plantations by the bigger agencies.

17.8 CONCLUSIONS

The concept of sustainable management in forestry has evolved considerably, from one of narrow application for sustainable wood production to that of holistic management of the whole environment. Although the present concept of sustainable forest management was not overtly expressed in the early years of forest management, one cannot help but realize that it was constantly upheld when forest management issues were deliberated in Peninsular Malaysia. Thereafter, harvesting practices and sawmilling became more mechanized, making it necessary to harvest large areas for economic reasons. The cognoscenti in forestry have always maintained that silviculture should not be dictated by nor be subservient to forest management. But, as the case with Peninsular Malaysia reveals, when dealing with natural vegetation in the tropics, such a dictum has been difficult to maintain (Appanah 1998). By a fortunate coincidence, the Malayan Uniform System had all the qualifications of an economically viable silvicultural system. Unfortunately, forests where the Malayan Uniform System could be implemented are mostly gone. The remaining forests devoted to timber production cannot be entirely managed using this system. Selective fellings, if gauged correctly, remain an economically viable system more or less. While it is possible to find several weaknesses with selective fellings, they provide an opportunity for foresters to maintain most of the vegetation in the stand, and most of the original biodiversity. Until better management systems can be introduced, selective fellings will be used. Application of selective fellings will be severely limited if the residual stand and the soils are heavily damaged as a result of using heavy machinery. If such damage is avoided, the management of these forests can be sustained.

REFERENCES

Anjin, H. & Kleine, M. (1997) *Sustainable forest management: the Deramakot Model.* Theme Paper Presented at the Fourth Conference on Forestry and Forest Products Research, Forest Research Institute Malaysia, Kepong, Malaysia.

Appanah, S. (1998) Management of natural forests. In: Appanah, S. & Turnbull, J.M., eds. *A Review of Dipterocarps: Taxonomy, Ecology and Silviculture,* pp. 133–50. CIFOR Publications, Indonesia.

Appanah, S. & Thang, H.C. (1997) International initiatives on forest management certification. In: Appanah, S., Samsudin, M., Thang, H.C. & Ismail, P., eds. *Forest Management Workshop. Proceedings,* pp. 5–28. Forest Research Institute Malaysia, Kepong, Malaysia.

Appanah, S. & Weinland, G. (1990) Will the management systems for hill dipterocarp forests stand up? *Journal of Tropical Forest Science* 3, 140–58.

Appanah, S. & Weinland, G. (1993) *Planting Quality Timber Trees in Peninsular Malaysia—A Review.* Malayan Forest Records no. 38. Forest Research Institute Malaysia, Kepong, Malaysia.

Appanah, S., Ismail, H. & Kleine, M. (1999) *Internal Assessment Procedures for Sustainable Forest Management in Malaysia.* Consultancy report to the 'Forest Certification Project'. Deutsche Gesellschaft fuer Technische Zusammenarbeit (GTZ), Forest Department Peninsular Malaysia, Kuala Lumpur.

Ashton, P.S. (1969) Speciation among tropical forest trees: some deductions in the light of recent evidence. *Biological Journal of Linnean Society* 1, 155–96.

Barnard, R.C. (1950) The elements of Malayan silviculture, 1950. *Malayan Forester* 13, 112–19.

Burgess, P.F. (1972) Studies on the regeneration of the hill dipterocarp forests of the Malay peninsula: the phenology of dipterocarps. *Malaysian Forester* 35, 103–23.

Burkhill, I.H. (1935) *Dictionary of Economic Products of the Malay Peninsula,* 2 Vols. Crown Agents for the Colonies, London.

Chong, P.W. (1979) The growing domestic demand for timber and its influence on forest management. *Malaysian Forester* 42, 378–89.

Griffin, M. & Caprata, M. (1977) *Determination of cutting regimes under the Selective Management System.* Paper presented at ASEAN Seminar on Tropical Rain Forest Management, 7–10 November 1977, Kuantan, Malaysia.

Hashim, M.N., Maziah, Z. & Sheikh, A.A. (1990) The incidence of heartrot in *Acacia mangium* Willd. plantations: a preliminary observation. In: Appanah, S., Ng, F.S.P. & Roslan, I., eds. *Proceedings of the Conference on Malaysian Forestry and Forest Products Research,* pp. 54–9. Forest Research Institute Malaysia, Kepong, Malaysia.

Hill, H.C. (1900) *The Presence of Forest Conservancy in the Straits Settlements with Suggestions for Future Management.* Government Printer, Singapore.

Hodgson, D.H. (1932) The elements of Malayan silviculture. *Malayan Forester* 1, 85–91.

Ismail Ali (1966) A critical review of Malayan silviculture in the light of changing demand and form of timber utilization. *Malayan Forester* 29, 228–33.

Marsh, C.W., Tay, J., Pinard, M.A., Putz, F.E. & Sullivan, T.E. (1996) Reduced impact logging: a pilot project in Sabah. In: Sculte, A. & Schone, D., eds. *Dipterocarp Forest Ecosytems: Towards Sustainable Management,* pp. 293–307. World Scientific, Singapore.

Rasid, M.I. (1998) The development towards sustainable forest management in the state of Terengganu. In: Chin, T.Y., Krezdorn, R. & Yong, T.K., eds. *Proceedings of the Workshop on the Malaysian–German Sustainable Forest Management and Conservation Project in Peninsular Malaysia, 11–12, February, 1998. Kepong, Selangor, Malaysia,* Forest Research Institute Malaysia, Kepong, Malaysia. pp. 107–20.

Richards, P. (1952) *The Tropical Rain Forest. An Ecolog-*

valuable from a conservation perspective (see Fig. 18.4), being largely free of human influence beyond an annual burning as a fire protection measure.

Silvicultural treatment of pine stands is simple and productive. In each rotation trees have been planted at 2.7–3.0 m spacing (1100–1400 stems ha^{-1}), low pruned at 5 years for access and fire protection, and clearfelled at ages 15–16 years. At harvest stem wood yield averages 250–350 m^3 ha^{-1}. There is no thinning and, until recently, no fertilizing of plantations. Harvested sites are usually replanted within a few weeks and certainly by the next wet season. The average mean annual increment (underbark stem volume to 7 cm top diameter) of the whole forest is about 19 m^3 ha^{-1}.

18.1.2 The Usutu pulp mill

In 1962 a Kraft pulp mill was constructed beside the Great Usutu River and virtually all forest production has been for this one end-use. The mill has operated almost continuously for the 36 years since construction, though its capacity has more than doubled and currently exceeds 200 000 tonnes of dried pulp per annum. This requires a pulpwood furnish somewhat greater than one million tons, all of which is supplied from the company's own forest plantations in Swaziland.

The Usutu Forest represents a simple but intensive form of plantation forestry practised on mediocre soils in a moderately favourable climate. Thus it is legitimate to ask how sustainable the pine plantations might be over successive rotations? If declining yields are to characterize plantation forestry—and hence be inherently unsustainable—then the pine plantations at Usutu are a *prima facie* candidate for such problems. It is also legitimate to ask what impacts has such large-scale afforestation had on environmental features and what steps can be taken to promote such features as part of sustainable management. Again the scale, simplicity of purpose and long-term commitment make Usutu a significant case study. In this case study we address the related question: can pulpwood crops be grown continuously without harm to the sites on which they are grown or without harm to the wider environment?

18.2 SUSTAINABILITY OF WOOD PRODUCTION

18.2.1 Assessment methods

First rotation

When research into assessing productivity began in 1968, about 40% of the first rotation (1R) crops had already been felled without a useful record of yield data. Thus, the first step was to recover sufficient 1R data by means of internodal stem analysis of sample trees. A network of 92 plots was established covering a range of stands 11–14 years old. This original network was confined to *P. patula* stands, as it was by far the predominant species. These 92 1R plots were matched with a similar number of young 5- and 6-year-old second rotation (2R) *P. patula* crops, planted in 1963 and 1964, of comparable site characteristics based on a site assessment survey. All the long-term productivity plots were established and first measured between October 1969 and June 1970.

Second rotation

The 92 second rotation plots of the original plot-pairs were remeasured in both 1973 and 1977 to capture growth data at mid-rotation (9 and 10 years) and near the end of rotation (13 and 14 years). In 1977, the location of each (2R) plot was surveyed accurately to permit re-establishment in the third rotation, but without any permanent on-site marking to avoid all risk of exceptional or favourable treatment. Thus, second rotation productivity was measured at three stages during the life of the crop.

In 1973, 38 additional second rotation plots were established in the p67 *P. patula* age class, i.e. planted in 1967, to broaden the spread of age classes sampled. Most were only measured at two stages: ages 5–6 years and 9–10 years.

Third rotation

In 1986, as a result of a preliminary survey in 1984, long-term productivity plots began to be systematically relocated in the third (3R) rotation on exactly the same sites as in the previous (2R) crop. The new plot centre was usually within

5 m of the previous one. Measurements followed identical procedures and at the same three stages of early, mid and late rotation. Owing to the company's felling operations age classes were more variable and more frequent assessment visits were required to capture growth data at the optimum time for accurate comparisons. Also some 3R plots were no longer *P. patula* but had been replanted to *P. elliottii* or *P. taeda* for the third rotation.

At the time of writing (1999) almost all available third rotation plots have received a final late rotation assessment.

Plot establishment and measurement

An account of the methodology of plot establishment and yield assessment was published by Evans (1975). Throughout the research the same procedures and measurement conventions have been followed to ensure consistency: only an outline of them is given here. A further contribution to accuracy is that all assessments made over the 30+ years since 1968 have been carried out by the author and, since 1973, all with help from the same Swazi research forester, Mr Milton Nkambhule.

The long-term productivity plots are rectangular with 8 rows by 12 rows of trees covering about 0.072 ha. Plot area is measured accurately so that stand density can be determined. All trees are measured for diameter at breast height (d.b.h.) to the nearest millimetre. Using the d.b.h. distribution three trees are selected: the median tree and the upper and lower percentile trees ± 30% of the median diameter. These are measured for total height and a local volume function, of the generalized form $V = D^2H$, is used to calculate under-bark volume. (The selection of the median and upper and lower percentile trees for height and volume measurement was adopted in 1969 as a simple, quick and practical field procedure when working on site. For consistency of comparison across rotations it has continued to be used.) Parameters of stocking and basal area per hectare are also calculated.

This case study is mostly confined to reporting the comparison between second (2R) and third (3R) rotations. This is the most accurate comparison because it is based on precisely relocated plots compared sequentially and measured in the same way and at the exact reference ages. First rotation data come from stem analysis of trees and from plots which were on matched but not the same sites, and are thus less directly comparable.

18.2.2 Productivity in successive rotations

Comparison between first and second rotation Pinus patula

The results comparing the first two rotations were reported in 1978 (Evans 1978). In summary, the second rotation initially grew more vigorously, owing to largely weed-free conditions, compared with the first crop which had been planted into dense grassveld. This improvement did not persist and by the end of the rotation, overall growth was slightly poorer than the first rotation, but not significantly so. However, this overall comparison masked considerable variation. In one area of the forest, in part of management Block A, almost all second rotation plots grew poorer and showed a statistically significant growth decline. Over the rest of the forest, more than 85% of the area, there was no difference at all between the rotations with some plots better and some poorer. The reason remained unexplained, until the work of Morris (1986) showed that all the poor plots were on soils derived from gabbro lithology which is a very poor phosphate source. Most of the forest is on granite geology and soils possess a somewhat better phosphate status. Fertilizer experiments later confirmed this hypothesis: response to phosphate occurred in Block A but not elsewhere.

Productivity of third rotation Pinus patula

Research on *P. patula* productivity has been the principal investigation concerning narrow-sense sustainability. By the end of 1997 a total of 43 third rotation plots of *P. patula* had been measured at 13/14 years of age. The data presented incorporate changes to volume estimation for second and third rotation *P. patula* described in Evans and Boswell (1998) and Evans (1999b). Plots are grouped in the two broad categories following Morris's work on the impact of underlying geology (Morris 1986): the plots on granitic and gneiss soils—most of Usutu Forest (Table 18.1), and those on Usushwana complex (gabbro-dominated) soils which occupy about 13% of the forest (Table 18.2). Figures 18.1 and 18.2 show average height growth of plots over all three rotations grouped into these two underlying geology types.

Table 18.1 shows that third rotation mean height is highly significantly superior to second rotation—almost 5% greater—but that volume per hectare has not changed significantly. If the effects of poorer third rotation stocking are compensated for (corrected volume (see definition in Table 18.1)), third rotation volume appears better though the increase is not quite statistically significant. Correlation coefficients (R) show a reasonable amount of variation accounted for (about 40%), indicating that fertile second rotation sites are also proving fertile in the third rotation (as one would hope!).

Table 18.2 shows that there is no difference between second and third rotation growth in the parts of the forest on the phosphate-poor gabbro-derived soils though, as elsewhere, third rotation stocking is inferior. There is no correlation

Table 18.1 Comparison of second and third rotation *Pinus patula* on granite- and gneiss-derived (non-Usushwana) soils at 13/14 years of age (means of 32 plots). (From Evans 1999.)

Rotation	Stocking (SPH)	Mean height (m)	Mean d.b.h. (cm)	Mean tree vol. (m³)	Uncorrected vol. (m³ ha⁻¹)	Corrected vol. (m³ ha⁻¹)*
Second	1381	17.43	20.15	0.2171	292.2	293.6
Third	1267	18.28	20.99	0.2334	291.1	304.9
Change†		+0.85	+0.84			+11.3
		(+4.9%)	(+4.2%)			(+3.8%)
t		3.43				1.32
Probability and significance		$P = 0.002$				$P = 0.197$
		**				NS
R		0.642				0.585

* Corrected vol. is volume per hectare corrected for differences in stocking between rotations.
† Figures in parentheses are % change.
** Highly significant; NS, not significant; SPH, stems per hectare.

Table 18.2 Comparison of second and third rotation *Pinus patula* on Usushwana geology (gabbro-dominated) soils at 13/14 years of age (means of 11 plots). (From Evans 1999)

Rotation	Stocking (SPH)	Mean height (m)	Mean d.b.h. (cm)	Mean tree vol. (m³)	Uncorrected vol. (m³ ha⁻¹)	Corrected vol. (m³ ha⁻¹)*
Second	1213	16.71	20.01	0.2063	244.7	244.0
Third	1097	16.79	21.68	0.2272	241.1	255.3
Change†		+0.08	+1.67			+11.3
		(+0.05)	(+8.3%)			(+4.6%)
t		0.14				0.73
Probability and significance		$P = 0.890$				$P = 0.480$
		NS				NS
R		−0.043				−0.243

*† See notes to Table 18.1.

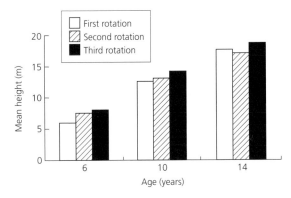

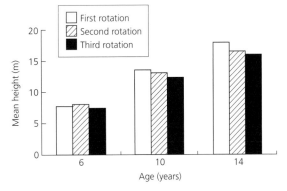

Fig.18.1 Mean height development, recorded at age 6, 10 and 14 years, for three rotations of *Pinus patula* on granite-derived soils in the Usutu Forest, Swaziland (most of forest).

Fig. 18.2 Mean height development, recorded at age 6, 10 and 14 years, for three rotations of *Pinus patula* on gabbro-derived soils in the Usutu Forest, Swaziland (small part of forest).

between sites judged as good in height growth in one rotation with sites judged as good in the next.

Assessments of the younger p67 second rotation age class series in the third rotation (section 18.2.1) has confirmed the small but general improvement of third rotation productivity.

18.2.3 Discussion of sustainability

Sustainability of Pinus patula *productivity over three rotations*

The data confirm earlier reports (Evans 1996; 1999b; Evans & Boswell 1998) that the third rotation is growing at least as well as the second throughout the forest. Only a very few plots still remain for a final assessment, so that this overall conclusion is unlikely to change in the future. Third rotation height growth is consistently superior to second but actual volume per hectare is almost identical owing to somewhat poorer third rotation stocking in most stands. If adjustment is made for this (corrected volume), using relationships described in Evans (1976), third rotation volumes are mostly greater than second rotation ones, though the improvement is not statistically significant. Taking all plots together, and using corrected data, almost two-thirds of third rotation plots have a greater volume per hectare than second rotation plots.

Of particular interest is that plots on the Usushwana complex (gabbro-dominated) soils (Table 18.2) have not shown further yield decline from second to third rotation. There was a significant drop between first and second rotations—see Evans (1996) and Fig. 18.2—which appears to have been arrested in the third rotation for no very clear reason. It is also interesting to note that mean yield from these plots at age 13/14 years is nevertheless poorer at about 250 m^3 ha^{-1} whereas for the bulk of the forest it is around 300 m^3 ha^{-1} (Tables 18.1 and 18.2).

Productivity comparison data in context

The productivity data reported here for third rotation pine come from stands that were subjected to the very severe drought that affected southern Africa in 1991 and 1992 and yet show at least comparable growth to the second rotation (Morris 1993a). Moreover, none of the plots measured in this research has yet benefited from fertilizer application. Genetically the second rotation came mostly from unselected seed purchased from South Africa, though in the early years of the rotation special collections were made from a particularly favoured stand near Tzaneen (Morris 1987). From 1986 third rotation seed was a mix of orchard quality material and unselected purchased from Mondi; it is not known which plots in the long-term productivity network benefited from the improved seed. However, the orchard grade material was from selections made for saw-timber rather than pulpwood where yield is not necessarily the most important trait. Neverthe-less, it is reasonable to conclude that third rota-tion *P. patula* has benefited from some genetic improvement which may explain why it has grown as well as it has despite the severe drought. However, this argument is weak since tree breed-ing generally improves net primary productivity which is unlikely to be fully realized if one or more nutrients is deficient.

On several plots the third rotation was planted to a different species, most notably changing from *P. patula* to *P. taeda*. This change appears not to have affected productivity greatly, the details of which are reported in Evans (1999b). *P. taeda* height growth was generally poorer but diameter greater. Owing to less persistent branches and thicker bark the advantage of switching species will not be in production but in savings in low pruning and other silvicultural operations carried out for fire protection.

Thus, these data provide evidence for main-tained yield, including sites on which species have been changed, and provide confirmation that in the Usutu Forest there is no threat to narrow-sense sustainability of wood yield over three rotations of plantation forests.

Several long-term productivity plots have already been felled and replanted with the *fourth rotation* and the first assessments of these is expected in about 2002. These plots have bene-fited from genetically improved seed—initially of Zimbabwean seed orchard origin and later plots from Usutu's own selections and breeding pro-gramme. Results from trials suggest that a 15% improvement in growth rate should be forthcom-ing. Also all fourth rotation sites in Block A will be fertilized with phosphate.

18.2.4 Conclusions concerning biological productivity

1 Throughout the Usutu forest the third rotation of *P. patula* is growing at least as well as the second rotation and probably a little better. Narrow-sense sustainability of wood production is being maintained.

2 In view of (a) the intense management carried out at Usutu and (b) the fact that third rotation stands experienced an extremely severe drought in the early 1990s, the results support the practice of plantation silviculture.

3 If existing plantation practices are not causing site decline, then future yields in the fourth rota-tion can be expected to be better owing to the use of genetically improved seed. Clear evidence of the size of the benefit should be forthcoming in first productivity assessments due in 2002.

18.3 ENVIRONMENTAL MANAGEMENT

18.3.1 Introduction

Although the Usutu Forest has been at the forefront of sustained yield experiments for many years, as outlined above, only recently has the concept of sustainable forest management changed from the narrow-sense definition to a broader, holistic management of the environ-ment. This has resulted in steps being taken to develop an environmental management programme which conforms to the ISO 14001 requirements and takes cognizance of corporate policy, pertinent national and regional environ-mental issues and the sociopolitical requirements of a developing African country. The complex rural social structure, the existence of conflicting and often outdated legislation and the absence of baseline information has provided challenges when attempting to apply the international prin-ciples outlined in ISO 14001 to the local level.

18.3.2 Background

Assessing the conservation status of the natural resource base, understanding the extent of the impacts of the operation on the natural and social environment, and developing strategies to assist with implementation and to monitor progress with actions taken were recognized as being important components of the environmental programme and key requirements for developing an ISO 14001 environmental management sys-tem. The increasing importance of forest certifi-cation on international markets underlined the need for Usutu to develop an environmental management system which would be recognized internationally.

In the southern African subregion, major issues influencing sustainable commercial forestry management have been identified as being the impact of afforestation on available water resources and soil fertility, the possible loss of biodiversity through the alteration of species-rich habitats within plantations and the impact that large-scale development has on social issues (Kruger & Everard 1997; Olbrich *et al.* 1997). Whilst all issues are of regional significance, the importance value attached to each issue is influenced by differences in legislation, social history and the economies of each country.

The key components of Usutu's environmental programme, namely environmental resource assessment, the identification of impacts of forestry on the environment and developing strategies to monitor progress with actions, will be discussed with reference to the above factors.

18.3.3 Key components of environmental management programme

Environmental resource assessment

In order to ensure that Usutu's natural assets are managed in the most appropriate manner, it was first necessary to ascertain what was present on the property, identify the conservation significance of unplanted areas and prioritize sites of exceptional conservation importance. These sites include areas that are aesthetically pleasing such as waterfalls, wetlands, and prominent river zones, vantage points delivering scenic vistas and tracts of natural forest and grassland forming firebreaks, some of which contain rare or unusual species. An evaluation of the conservation status of the forest was subsequently carried out at the start of the programme (Masson 1994a), identifying the extent of different land-use types on the property, rating, on a subjective scale, the conservation significance of unplanted areas and assigning priority status to sites of exceptional conservation importance. Establishing a preliminary database of environmental information enabled guidelines to be drawn up, which identi-

fied the recommended management action to be taken for the feature of note.

Social issues

Understanding land tenure issues, the social fabric of the surrounding communities and identifying avenues for communication is important for maintaining good relations with neighbours and for appreciating social problems which may impact on the forest. Ownership of land in Swaziland is broadly differentiated into three main categories: private title deed land, Swazi Nation Land which is divided into chiefdoms, and Crown Land belonging to the State. A smaller, fourth category is land held in trust by the King of Swaziland and it is in this category that the Usutu Forest belongs, being part of the Usutu Royal Trust.

Communities bordering Usutu are governed by local chiefs through whom dialogue and disputes are settled. Subsistence cultivation and cattle grazing are practised widely in rural areas where income generation is low. The perception of Usutu being a provider of employment has created dilemmas for management who must provide for a large workforce, supported by labour unions, despite rising costs of production and competition from technologically advanced companies in other parts of the world.

Water availability, grazing requirements and the provision of employment were identified as important social issues to neighbouring communities which could also, directly or indirectly, affect the sustainable nature of Usutu's management.

18.3.4 Environmental impacts

Hydrology

Understanding the impact that the Usutu Forest has on the hydrology of the region was identified as an important area of investigation, due to the direct dependency of communities living outside of the forest on water sources originating on Company property and possible increase in water demand by downstream development.

The Usutu River Basin in which the forest is

situated is the largest of five river basins in Swaziland and extends beyond the western boundary into neighbouring South Africa (Fig. 18.3).

Differences in legislation pertaining to afforestation in the two countries plays an important role in determining the extent of afforestation in the subregion and hence the impact that forestry has on water resources. Although both countries require that permits be issued prior to afforestation, the criteria used for assessing afforestation permit applications differ. In South Africa, the evaluation of permits is based on water conservation principles (Van der Zel 1995). In Swaziland, agricultural productivity of the land is considered to be the deciding factor when authorizing afforestation development. In the absence of legislation directed at water conservation, it is the responsibility of the landowner to ensure that water resources downstream are not excessively affected by forestry activities.

Possible implications of afforestation on water catchment yield have been the subject of considerable research attention in South Africa over several decades. A number of studies have shown that in the southern African subregion, pine plantations will reduce stream flow relative to indigenous vegetation (Nanni 1970; Bosch & Hewlett 1982) and that pine plantations contribute to the lowering of the water table in specific situations (Versveld & van Wilgen 1986; Versveld 1993). It is now known that a number of factors affect water availability, and these include the proportion of catchment area afforested, age of trees within the given area, the underlying geology of the area and soil depth and climatic variation determining low flows and water table level.

The considerable size of the Usutu Forest relative to the country (it covers roughly 4% of Swaziland) and the lack of information relating to water use in Swaziland resulted in the commissioning of a study to determine the possible implications of Usutu's afforestation on the hydrology of the region. In the absence of base flow data, simulated rainfall models, age class distributions and underlying geology were used to determine the effect of Usutu's afforestation on runoff. Results from this forest indicated a reduction of roughly 100 million m^3 $year^{-1}$ or a steady abstraction rate of 3.17 m^3 s^{-1}, if all water was drawn from the Great Usutu River (Scott & Olbrich 1996). According to Murdoch (personal communication) this abstraction rate would serve 3000–4000 ha of irrigated crop land. Although this represents a significant reduction in water yield (there are about 30 000 ha of irrigated crop land downstream from the Usutu Forest), some reduction can be justified by the important role the forestry industry plays in the economy of Swaziland. Furthermore, the contribution that plantations make in maintaining hydrological properties of soils and hence the functioning of the catchment should be viewed in a positive light when compared with alternative land uses (Scott & Olbrich 1996). The negligible effect that forestry practices have on stormflow contrasts sharply with that of many other activities such as the injudicious grazing of grassland and land-use practices which lead to compaction, erosion and denudation of humic top soil, and result in a decline in water quality and reduced flood control characteristics of catchments (Lake 1993).

Recommendations for improving water availability included, in the short term, keeping riparian zones free of woody weeds and planted pine and, in the longer term, to plan for the creation of a normal age-class distribution as opposed to a single age class within a catchment area.

Recognizing the importance of water to neighbours and downstream users, and taking note of recommendations made from the report, strategies for improving water catchment management were formulated. First, all principal water courses on Usutu's property were identified and their status (either planted through or possessing natural vegetation) within compartments noted. Catchments and subcatchments of principal waterways were demarcated, end-users were identified and strategies were suggested for improving the management of waterways and thus restoring aquatic habitats and streambank functions. These included removal of alien vegetation from the vicinity of the riparian zone, and the redefining of new planted stand boundaries, following felling, some distance away (appoximately 20 m) from principal drainage lines.

An estimate of river length management requirements was calculated (Masson 1994b, 1997), followed by a more detailed examination

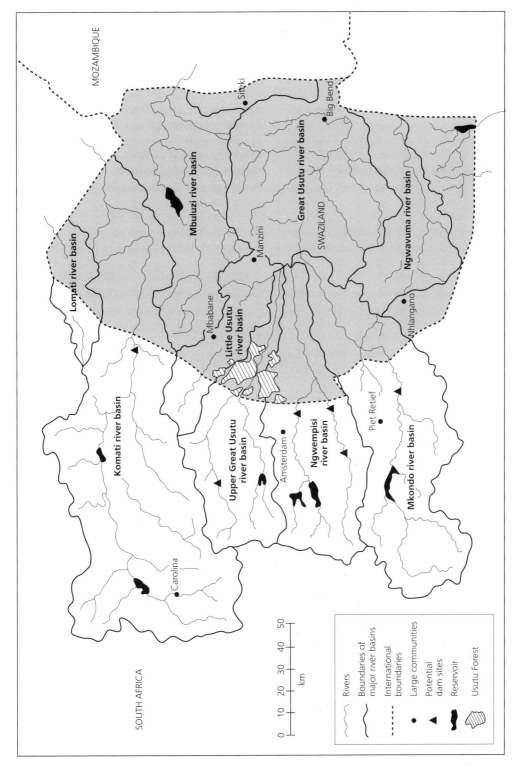

Fig. 18.3 Location of the Usutu Forest in relation to the major river basins of Swaziland.

of practical requirements for creating riparian buffer zone management units (P.H. Masson, unpublished results). The objective of this exercise was to provide a strategic overview of remedial actions to be taken throughout the plantation. Conservation areas or riparian buffer zones inside operational stands scheduled for clearfelling or weeding are attended to, along with silvicultural operations, enabling progress with conservation management actions to be measured against a strategic management plan.

Fire management

Grassveld fires are common on the highveld in winter, traditionally being used by communities as a means of promoting the flush of new growth for grazing with the onset of spring rains. Under controlled conditions, fires do not pose a threat to the plantation and can contribute positively to a reduction in fire hazard (de Ronde 1997). However, uncontrolled fires can spread through the plantation, causing serious losses in timber production and contributing to environmental degradation through soil erosion and soil modification.

Recently, an increasing number of wildfires has caused substantial losses in planted area. In 1998, this amounted to 2880 ha and in 1999 to approximately 3200 ha. The causes of fires are unknown, but may be attributed to accidental ignitions on Swazi Nation Land which quickly rage out of control, honey-hunting forays of young boys, or arson resulting from disputes related to unemployment or a lack of grazing land.

The practice of not burning after clearfelling, coupled with a slow litter decomposition rate at high altitudes (Morris 1993b, 1995), has resulted in excessive fuel load accumulating in some parts of the Forest. An assessment of the fire problem was carried out, taking into account regional fuel and fire behaviour modelling, fire hazard ratings at the regional and plantation level and existing fire protection measures on Usutu. Creating multipurpose buffer zones which prevent the spread of fires and, at the same time, function as a conservation management area was identified as a possible solution to reducing losses due to fire and promoting appropriate management of unplanted areas. The proposed location of these multipurpose buffer zones is along principal river courses and firebreaks following ridge tops, hence combining the dual functions of fire protection and conservation area within a single unplanted zone. The size of buffer zones will vary according to location and strategic importance. Smaller, internal buffer zones follow paths of riparian zones and existing grassland firebreaks, whereas the largest of the fire protection buffer zones were positioned along the axis of greatest fire risk and incorporated both unplanted and planted stands. Within planted stands, a change from a total fire exclusion policy to one of selective, prescribed burning of forest floor litter was proposed as a means of managing the high fuel loads and reducing fire risk (de Ronde 1997; de Ronde & Masson 1998).

A reassessment of existing firebreaks was carried out and new buffer zones were identified, some of which coincided with the riverine management programme. Main buffer zones identified as principal lines of defence, and internal buffer zones, the secondary lines of defence, are considerably wider than existing firebreaks (500–1000 m) but are not necessarily all unplanted area. Replacement of *P. patula* with the more fire-resistant *P. elliotii* was recommended alongside the buffer zones, allowing controlled burning to take place underneath the canopy, and minimizing the loss of production area. According to Evans (1999b) the impact on plantation productivity of this species change will probably be negligible. Temporary belts and existing firebreaks which are not part of the planned fire protection system would be maintained and managed until the buffer zones were established. After this, their conservation value would be evaluated, and if not of high value they could be planted, thus recouping some of the area which may have been lost when establishing the main buffer zones.

Biodiversity

Despite the recognition of the importance of maintaining biological diversity in the southern African subregion, few studies have been undertaken to determine the requirements for doing this. In 1989, the first comprehensive review of biotic diversity was undertaken (Huntley 1989), and in 1992 added impetus was given to the concept through the Rio Convention on Biological Diversity. Swaziland became a signatory to the Convention in 1994, and, in so doing, pledged its support to promoting the conservation and maintenance of biodiversity in the country. The process of developing a biodiversity strategy and action plan has recently been completed, identifying information gaps, priority actions to be taken and the projected time frame for implementation (Goverment of Swaziland 1999).

Commercial afforestation has been identified as a significant land-use activity in Swaziland, covering over 6% of the country's area (Scharfetter 1987). Usutu's area of approximately 70 000 ha equates to 4% of Swaziland's total area and almost 14% of the Highveld region. As custodian to a significant percentage of Swaziland's area, Usutu can play an important role in maintaining biodiversity through appropriate management of unplanted areas within the plantation (Masson 1996a).

Management activities and interventions that are appropriate for both timber production and biodiversity conservation are poorly understood and are the subject of debate between conservationists and plantation managers. A recent review of the impacts of afforestation on native biota in South Africa (Armstrong & van Hensbergen 1996) listed what is currently known on the impact of afforestation on indigenous habitats (Table 18.3) and stressed the need for further research on impacts and measures to reduce them. This call was endorsed by nature conservation authorities (Armstrong *et al.* 1998).

In the absence of proven recommendations for management, nature conservation authorities have preferred to follow the 'precautionary principle', which is usually interpreted as a 'minimal disturbance approach', making recommendations which are often perceived by forest management to be unrealistic to implement. The need for a pragmatic approach to sustainable forest management was recognized by Kruger and Everard (1997) and also applies to biodiversity conservation and management. In this regard, an attempt was made to identify practical management prescriptions, promoting environmentally acceptable forest management (Anonymous 1995),

and this document has been widely distributed throughout the southern African forestry industry.

On Usutu, the need for research to develop a clearer understanding of the impacts of forestry operations on the environment was motivated by the lack of baseline information for all biota, and the presence of what appeared to be floristically rich firebreaks (Masson 1996b). Firebreaks serve as 'conservation corridors' forming a network on unplanted habitat and provide an indication of the type of vegetation which would have existed in the area prior to afforestation (Fig. 18.4).

A checklist of woody flora has been compiled and an investigation of firebreak floral diversity in relation to environmental factors is in progress. Preliminary indications are that species diversity is high on certain firebreaks on Usutu. This may be a result of reduced grazing activities within the plantation in contrast to grazing pressure on Swazi Nation Land. It may also be related to environmental features such as soil type and the presence of habitats such as rocky ridges on Usutu, which have not been substantially altered through human intervention.

18.4 CERTIFICATION OF SUSTAINABILITY

Various types of forest certification have developed in response to market requirements, the most widely known being that of the Forest Stewardship Council (FSC). In the southern African subregion, a number of organizations have subscribed to FSC certification (Chihambakwe *et al.* 1997). While acknowledging that certification can produce better forestry management, it has a number of associated costs which developing countries can ill afford (Chihambakwe *et al.* 1997). The cost of managing in accordance with certification standards and maintaining the necessary monitoring requirements may be prohibitive if not in place already prior to seeking certification. Furthermore, employing overseas certification agencies to evaluate and assess performance adds to the cost of certification which, in the absence of a willing consumer market to absorb the cost, must rest on the shoulders of the producer.

Table 18.3 Impacts of afforestation. (Compiled from references in Armstrong & van Hensbergen 1996.)

Focus area	Research topic	References
Physical environment		
Soils	Soil fertility	Morris 1984; Musto 1991; 1992
	Soil acidification	Good *et al.* 1993
	Water-holding capacity	Bosch & Hewlett 1982
	Soil decomposition	Watts 1951; Musto 1992
	Soil compaction	Musto 1992
Nutrients	Nutrient sinks	De Ronde 1993
	Nutrient removal	Morris 1992
Water use	Streamflow	Nanni 1970; Boschd & Hewlett 1982; Versveld & van Wilgen 1986; Versveld 1993
	Water table	Midgeley *et al.* 1994; Musto 1991
Biotic environment		
Species diversity	Comparison of plant species richness in the planted and unplanted environment	Cowling *et al.* 1976; Bigalke 1980; Richardson & van Wilgen 1986; Richardson *et al.* 1989
	Species regeneration under pine plantations	Geldenhuys *et al.* 1986
Species dynamics	Faunal community dynamics	Allen-Rowlandson 1986; Versveld & van Wilgen 1986
	Invertebrate populations	Watts 1951; Samways & Moore 1991
	Bird populations	Winterbottom 1968, 1972; Allen *et al.* 1995
	Mammal populations	Schutz *et al.* 1978; Willan 1984, 1992; Droomer 1985
Plantation management		
Pest control measures	Non-target species	Wirminghaus & Schroder 1994
Plantation design	Windbreaks and shading on invertebrate populations	Samways & Moore 1991
Fire management	Fire exclusion	Bond 1980; Tainton & Mentis 1984; Breytenbach 1986; Manders 1989; Porter *et al.* 1990; van Wilgen & Forsyth 1992
	Frequency of burning	Everson & Tainton 1984
	Timing of burning	Kruger 1984; Tainton & Mentis 1984

Where markets do not require certified products, the motivation for pursuing certification lies in improving forest management through the development of an environmental management system such as ISO 14001.

Usutu's decision to pursue ISO 14001 as opposed to FSC was determined by the relatively unsophisticated eastern market requirements to which most of Usutu's pulp is sold, and the realization of the compounded benefits to be derived for effective management, by having a structured management system in place. The importance of operating within the realms of the law, planning adequately for operational implementation and being able to monitor progress with management actions has been recognized through the development of various initiatives on Usutu.

18.4.1 Planning

The importance of planning environmental actions at both the strategic and operational level became apparent when attempting to identify the scope of environmental activities needing atten-

Fig. 18.4 Unplanted firebreak in the Usutu Forest, showing a rock field and scattering of indigenous trees, that contributes both to forest fire protection and conservation of biodiversity. © J. Evans.

tion on Usutu and the most cost-effective means of implementing actions. The phased silvicultural management plan for water resource conservation and the cooperative management plan combining water catchment management requirements with fire protection requirements have been discussed. Planning for harvesting and silvicultural activities over the long-term (5-year period) and in the short term (1 year) is the function of the planning department, with shorter-term implementation planning conducted by forest staff in-field.

18.4.2 Sustainable management at the operational level

Measuring the quality of silvicultural operations on a regular basis enables a high standard of workmanship to be maintained, enables records of operations undertaken to be kept, and allows for the detection of any potential problems arising on the plantation. To this effect, a silvicultural quality reporting system has been developed which also includes the reporting of environmental issues and activities and the identification of where non-compliance occurs and needs to be rectified.

Another form of measurement which is in use on SAPPI's (South Afican Pulp and Paper Industries) plantations is the environmental audit. Although only recently introduced to Usutu, the environmental audit has served as an important tool for creating awareness of environmental issues and broadening perspectives on environmental management at the operational level. Included in the environmental audit are, among other things, an evaluation of actions taken to minimize the impacts of silvicultural and harvesting operations, an appraisal of efforts made to create environmental awareness within the workforce and in neighbouring communities, and the recognition by the forester of important unplanted areas and their management.

18.5 CONCLUSIONS

Research in the Usutu Forest in Swaziland shows that silvicultural practices are sustainable in the narrow-sense of maintained biological productivity as rotation has followed rotation of intensively grown pine. However, it is also clear that this narrow-sense sustainability neglects the wider issues, particularly the contribution of a large landowner to important environmental and ecological parameters of national importance. The measures adopted in forest management since the early 1990s will ensure this wider or 'broad-sense' sustainability is equally thoroughly addressed.

For both aspects, they can only be demonstrated if long-term trials, field measurements and assessments are made and maintained for posterity.

ACKNOWLEDGEMENTS

The research summarized here has been supported by several organizations but principally the Usutu Pulp Company of Swaziland, now a part of SAPPI Forests Pty. Ltd and the UK Department for International Development (DFID). Particular thanks go to Dr A.R. Morris and Mr Arnulf Kanzler of Usutu Research and to their always supportive staff, especially Milton Nkambhule. Successive forest managers have been generally supportive of the long-term nature of both the productivity research and progress to address the wider environmental imperatives.

REFERENCES

Allan, D.G., Harrison, J.A. & Navarro, R.A. (1995) *The Impacts of Commercial Afforestation on Bird Populations in the Eastern Cape Province of South Africa—Insights from Bird Atlas Data.* Avian Demography Unit Research Report 11. University of Cape Town.

Allan-Rowlandson, T.S. (1986) *An autecological study of bushbuck and common duiker in relation to forest management.* PhD thesis, University of Natal.

Anonymous (1995) *Guidelines for Environmental Management in Forestry Plantations in South Africa.* Environmental Committee. Forest Owners Association of South Afican, Pietermaritzburg.

Armstrong, A.J. & van Hensbergen, H.J. (1996) The impacts of afforestation with pines on assemblages of native biota in South Africa. *South African Forestry Journal* **175**, 35–42.

Armstrong, A.J., Benn, G., Bowland, A.E. *et al.* (1998) Plantation forestry in South Africa and its impact on biodiversity. *South African Forestry Journal* **182**, 59–65.

Bigalke, R.C. (1980) Plantation forests as wildlife habitats in Southern Africa. In: *Proceedings of the Joint Symposium on Plantation Forests as Wildlife Habitats and Problems of Damage IUFRO Athens*, pp. 5–11. Interational Union of Forest Research Organisations, Vienna.

Bond, W. (1980) Fire and senescent fynbos in the Swartberg, Southern Cape. *South African Forestry Journal* **114**, 68–71.

Bosch, J.M. & Hewlett, J.D. (1982) A review of catchment experiments to determine the effect of vegetation changes on water yield and evapotranspiration. *Journal of Hydrology* **55**, 3–23.

Breytenbach, G.J. (1986) Impacts of alien organisms on terrestrial communities with emphasis on communities of the south-western Cape. In: Macdonald, I.A.W., Kruger, F.J. & Ferrar, A.A., eds. *The Ecology and Management of Biological Invasions in Southern Africa*, pp. 229–38. Oxford University Press, Cape Town.

Chihambakwe, M., Mupudzi, R. & Mushove, P.T. (1997) Forestry certification: a developing world viewpoint. *Commonwealth Forestry Review* **76**, 191–3.

Cowling, R.M., Moll, E.J. & Campbell, B.M. (1976) The ecological status of the understorey communities of pine forests on Table Mountain. *South African Forestry Journal* **99**, 13–23.

Droomer, E.A.P. (1985) Volume and value loss owing to samango monkey damage in pine stands in the northern Transvaal. *South African Forestry Journal* **134**, 47–51.

Evans, J. (1975) Two rotations of *Pinus patula* in the Usutu Forest, Swaziland. *Commonwealth Forestry Review* **53**, 57–62.

Evans, J. (1976) The influence of spacing in a pulpwood plantation. *South African Forestry Journal* **96**, 23–6.

Evans, J. (1978) A further report of second rotation productivity in the Usutu Forest, Swaziland—results of the 1977 assessment. *Commonwealth Forestry Review* **57**, 253–62.

Evans, J. (1992) *Plantation Forestry in the Tropics*, 2nd edn. Clarendon Press, Oxford.

Evans, J. (1996) The sustainability of wood production from plantations: evidence over three successive rotations in the Usutu Forest, Swaziland. *Commonwealth Forestry Review* **75**, 234–9.

Evans, J. (1997) The sustainability of wood production in plantation forestry. In: *Proceedings of the XI World Forestry Congress, Antalya October 1997*, Vol. 3, pp. 35–41. UN Food and Agriculture Organization, Rome.

Evans, J. (1999a) Sustainability of plantation forestry: impact of species change and successive rotations of pine in Usutu Forest, Swaziland. *Southern Africa Forestry Journal* **184**, 63–70.

Evans, J. (1999b) *Sustainability of Forest Plantations: The Evidence.* Issues paper, Department for International Development (DFID), London 64pp.

Evans, J. & Boswell, R.C. (1998) Research on sustainability of plantation forestry: volume estimation of *Pinus patula* trees in two different rotations. *Commonwealth Forestry Review* **77**, 113–18.

Everson, C.S. & Tainton, N.M. (1984) The effect of thirty years of burning in the highland sourveld of Natal. *Journal of the Grassland Society of South Africa* **1**, 15–20.

Geldenhuys, C.J., Le Roux, P.J. & Cooper, K.H. (1986) Alien invasion in indigenous evergreen forest. In:

Macdonald, I.A.W., Kruger, F.J. & Ferrar, A.A., eds. *The Ecology and Management of Biological Invasions in Southern Africa*, pp. 119–31. Oxford University Press, Cape Town.

Good, J.E.G., Lawson, G.J. & Stevens, P. (1993) *Natural Environment. Study No 8. Shell/WWF Tree Plantation Review.* SIPC/WWF, Godalming, UK.

Government of Swaziland (1999) *Biodiversity Strategy and Action Plan for Swaziland* (unpublished).

Huntley, B.J. (1989) *Biotic Diversity in Southern Africa: Concepts and Conservation.* Oxford University Press, Cape Town.

Kanowski, P.J. (1997) Afforestation and plantation forestry: plantation forestry in the 21st Century. In: *Proceedings of the XI World Forestry Congress, Antalya, October 1997*, Vol. 3, pp. 23–34. UN Food and Agriculture Organization, Rome.

Kruger, F.J. (1984) Effects of fire on vegetation structure and dynamics. In: Booysen, P. de V. & Tainton, N.M., eds. *Ecological Effects of Fire in South African Ecosystems*, pp. 219–43. Springer Verlag, Berlin.

Kruger, F.J. & Everard, D.A. (1997) The sustainable management of the industrial plantation forests of South Africa: policy development and implementation. *South African Forestry Journal* **179**, 39–44.

Manders, P.T. (1989) Experimental management of a *Pinus pinaster* plantation for the conservation of *Diastella buekii*. *South African Journal of Botany* **55**, 314–20.

Masson, P.H. (1994a) *Conservation Status Review: Plantations.* Usutu Pulp Company Forest Research Document 14/94. Internal Report (unpublished).

Masson, P.H. (1994b) *Riverine Conservation Requirements in the Section South of the Usutu River.* Usutu Pulp Company Forest Research Document 9/94. Internal Report (unpublished).

Masson, P.H. (1996a) *The Opportunities and Constraints to the Usutu Pulp Company's Involvement in Promoting the Conservation and Sustainable Management of Swaziland's Biological Diversity.* Usutu Pulp Company Forest Research Document 19/96. Internal Report (unpublished).

Masson, P.H. (1996b) *Usutu's Firebreaks: Floristic Diversity and Veld Condition: Research Proposal and Interim Report.* Usutu Pulp Company Forest Research Document 11/96. Internal Report (unpublished).

Masson, P.H. (1997) *River Lengths North of the Usutu River: An Assessment of Conservation Management Requirements.* Usutu Pulp Company Forest Research Document 6/97. Internal Report (unpublished).

Morris, A.R. (1984) A comparison of soil nutrient levels under grassland and two rotations of *Pinus patula* in the Usutu Forest—Swaziland. In: *Proceedings of the IUFRO Symposium on Site and Productivity of Fast Growing Plantations.* South African Forestry Research Institute, Pretoria, pp. 881–92.

Morris, A.R. (1986) *Soil fertility and long-term productivity of* Pinus patula *in the Usutu forest, Swaziland.* PhD thesis, University of Reading, UK.

Morris, A.R. (1987) A review of Pinus patula seed sources in the Usutu Forest, 1950–86. Usutu Pulp Company Forest Research Document 8/87, Swaziland (unpublished).

Morris, A.R. (1992) Dry matter and nutrients in the biomass of an age series of *Pinus patula* plantations in the Usutu Forest, Swaziland. *South African Forestry Journal* **163**, 5–11.

Morris, A.R. (1993a) *Observations of the Impact of the 1991/92 Drought on the Usutu Forest.* Usutu Pulp Company Forest Research document 6/93.

Morris, A.R. (1993b) Forest floor accumulation under *Pinus patula* in the Usutu Forest, Swaziland. *Commonwealth Forestry Review* **72**, 144–117.

Morris, A.R. (1995) Forest floor accumulation, nutrition and productivity of *Pinus patula* in the Usutu Forest, Swaziland. *Plant and Soil* **1668**, 271–8.

Musto, J.W. (1991) Impacts of plantation forestry: a review of existing knowledge and current research projects. In: *ICFR* (Institute for Commercial Forestry Research, Pietermaritzburg). *Annual Research Report*, pp. 208–15.

Musto, J.W. (1992) Impacts of plantation forestry on various soil types. In: *ICFR Annual Research Report*, pp. 38–51.

Nanni, U.W. (1970) Trees, water and perspective. *South African Forestry Journal* **75**, 9–17.

Olbrich, K., Christie, S.I., Evans, J., Everard, D., Olbrich, B. & Scholes, R.J. (1997) Factors influencing the long-term sustainability of the South African Forest Industry. *South African Forestry Journal* **178**, 53–8.

Pandey, D. (1995) *Forest Resources Assessment 1990. Tropical Forest Plantation Resources.* FAO Forestry Paper 128. FAO, Rome.

Porter, R.N. (1990) Future afforestation and the potential impacts on nature conservation in Natal. In: Erskine, J.M., ed. *The Physical, Social and Economic Impacts of Large Scale Afforestation in Natal Kwazulu. Proceedings of Forestry Impacts Workshop*, Institute of Natural Resources, University of Natal, Pietermaritzburg, pp. 29–46.

Richardson, D.M. & van Wilgen, B.W. (1986) Effects of thirty five years of afforestation with *Pinus radiata* on the composition of mesic fynbos near Stellenbosch. *South African Journal of Botany* **52**, 309–15.

Richardson, D.M., MacDonald, I.A.W. & Forsyth, G.C. (1989) Reductions in plant species richness under stands of alien trees and shrubs in the fynbos biome. *South African Forestry Journal* **149**, 1–8.

de Ronde, C. (1997) *Fuel Modelling and Management Plan for Section II of the Usutu Forest.* Internal Consultancy Report for the Forest Manager. Usutu Pulp Company, Swaziland.

de Ronde, C. & Masson, P.H. (1998) *Integrating fire management with riparian zone and conservation management programmes on the Swaziland Highveld.* Paper presented at the Third International Conference on Forest Fire Research, Luso–Coimbra, Portugal.

Samways, M.J. & Moore, S.D. (1991) Influence of exotic conifer patches on grasshopper assemblages in a grassland matrix at a recreational resort, Natal, South Africa. *Biological Conservation* **57**, 117–37.

Scharfetter, H. (1987) Timber resources and needs in southern Africa. *South African Journal of Science* **83**, 256–9.

Schutz, C.J., Kunneke, C. & Chedzey, J. (1978) Towards an effective buck repellant. *South African Forestry Journal* **104**, 46–8.

Scott, D.F. (1978) *A study of the bird and mammal life of three age classes of* Pinus radiata *stands, Jonkershoek plantation.* Honours thesis, Department of Nature Conservation, University of Stellenbosch.

Scott, D.F. & Olbrich, W. (1996) *Hydrological Implications of Afforestation and Forest Management on the Usutu Forest.* Internal Consultancy Report for the Forest Manager. Usutu Pulp Company, Swaziland.

Tainton, N.M. & Mentis, M.T. (1984) Fire in Grassland. In: Booysen, P. de V. & Tainton, N.M., eds. *Ecological Effects of Fire in South African Ecosystems*, pp. 115–47. Springer-Verlag, Berlin.

Van Wilgen, B.W. & Forsyth, G.C. (1992) Regeneration strategies in fynbos plants and their influence on the stability of communities boundaries after fire. In: van Wilgen, B.W., Richardson, D.M., Kruger, F.J & van Hensbergen, H.J., eds. *Fire in South African Mountain Fynbos. Ecosystem, Community and Species Response at Swartboskloof*, pp. 54–80. Springer-Verlag, Berlin.

Versveld, D.B. (1993) The forestry industry and management for water conservation. In: van der Sijde, H.A., ed. *South African Forestry Handbook*, pp. 657–74. Southern African Institute of Forestry, Pretoria.

Versveld, D.B. & Van Wilgen, B.W. (1986) Impact of woody aliens on ecosystem properties. In: MacDonald, I.A.W., Kruger, F.J. & Ferrar, A.A., eds. *The Ecology and Management of Biological Invasions in Southern Africa*, pp. 239–46. Oxford University Press, Cape Town.

Watts, J.C.D. (1951) *Some comparative studies of the fauna in soils developed under natural forest, pine and blue gum.* MSc thesis, Rhodes University, Grahamstown.

Willan, K. (1984) Rodent damage in South Africa: a review. *South African Forestry Journal* **18**, 1–7.

Willan, K. (1992) *Problem Rodents and their Control.* University of Natal, Durban.

Winterbottom, J.M. (1968) A check-list of the land and freshwater birds of the western Cape Province. *Annals of the South African Museum* **53**, 1–276.

Winterbottom, J.M. (1972) Ecological distribution of birds in Southern Africa. *Monographs of the Percy Fitzpatrick Institute, University of Cape Town* **1**, 1–82.

Wirminghaus, J.O. & Schroder, J. (1994) Coumatetralyl efficiency trial against rodents in a young pine plantation at Linwood, Natal Midlands. *South African Forestry Journal* **169**, 21–4.

Synthesis and Conclusions

When we began *The Forests Handbook* project in 1997 we invited contributors 'to synthesize scientific principles and practical knowledge about the world's forests to demonstrate and provide foundations which underlie sustainable management. Put simply, how can an understanding of forests and forest processes lead to their better management and, ultimately, people's better stewardship of this immensely important resource? We went on to say that the message of Volume 1 should be 'this is what we know about forests and how we find out new things', and of Volume 2 'this is how we apply scientific knowledge to address issues and problems that keep cropping up in forestry.'

Now at the conclusion of the project it is fair to ask: what has been learnt from such a wide-ranging review of forest science and its application in practice collated in the two volumes that constitute the *The Forests Handbook*? How far have we travelled on the journey we began, while always recognizing that in the very nature of the exercise one rarely reaches a destination, but one does, to press the metaphor, pass important places on the way? It is in this sense that I attempt a synthesis of Volumes 1 and 2 to draw out some conclusions and pointers to help us in sustainable management of forests (SFM). In some cases I relate them to particular chapters (Volume:Chapter). I have tried to avoid anodyne or absurdly obvious remarks while being careful to restate well-known points where these have emerged with renewed force or greater clarity. However, the points are not a summary but rather emerging themes. This same synthesis occurs in both Volumes 1 and 2.

Of course, in one sense, in a book with 45 authors writing 33 chapters, agreed definitive conclusions cannot really be assembled. Each author presents cogently his or her analysis. But it is worthwhile drawing out the themes that emerge to inform sustainable forest management and to present them together to help forward our thinking and ultimately our striving towards this worthwhile goal. As Peter Attiwill and Jane Fewings (2:16) quote from the Australian state of

Victoria's Regional Forest Assessment: '. . . sustainable forest management is a goal to be pursued vigorously, not an antique to be admired.'

THE RECORD OF HISTORY

1 Knowing what has happened to forests in the past and why (1:1, 2:1, 2:3 and 2:14) teaches crucial lessons, one of which is that a forest that is used, and the benefits of which are enjoyed, is a forest that survives (2:14).

2 The record of history is particularly important in understanding the evolution of policy towards forests and their management, and the way it has developed (2:2, 2:11 and 2:16). A fundamental underpinning of informed SFM is to know why things are done, as well as how.

3 SFM needs a policy framework—administrative, environmental, production, social—to function successfully (1:14, 2:1–2, 2:11 and 2:16). SFM won't just happen.

4 Recent history shows rapid migration to urban areas, particularly in developing countries, and that worldwide the majority of people will soon live in massive conurbations. While this is only touched on indirectly (1:13 and 2:12) the globalization of trade and fair sharing of resources increases the SFM imperative since most people will have no direct control and often very little say in what happens to forests.

ROLE OF FOREST SCIENCE

5 Knowledge about forests (1:2–3), their functioning as biological systems and their interaction with the environment (1:4–8, 1:10–12) is the only sound basis for informed and potentially sustainable management. It is obvious that scientific and related research is largely responsible for building this knowledge base.

6 Knowledge informs understanding of *impacts* and *consequences* of forest operations and processes, which is fundamental to SFM. However, general knowledge must be applied to the particular in terms of sites, forest and species types and circumstances (2:3 and Volume 2 case studies).

7 Great advance has been achieved in the last 30 years, not least in our understanding of forest ecological processes and interaction between forest and the wider environment (1:3–5, 1:7–8, 1:10–12, 2:5, 2:7). It has become clear that many processes need to be understood at the landscape scale (1:5–6, 1:8–10, 1:12), i.e. at a scale which has received relatively little attention compared with the molecular, cellular, tree or stand scale.

8 The capture of scientific knowledge with the long timespans of forestry may not be cheap, will often involve complex investigations of processes including large-scale field studies (1:7–12) and crucially requires some role for long-term experimentation in forest science (2:5, 2:13, 2:17–18).

9 Modelling is a powerful tool in forest science to help predict consequences of actions and to raise questions about different options, such as their sensitivity (1: 6–10, 2:7–9). It can assist but not substitute for 8 above. High data quality to calibrate models is crucial for informed decision taking – see examples in 1:10 and 2:7.

10 Decision support can only be developed on the basis of detailed relevant knowledge, e.g. 'understanding disease epidemiology is vital' for successful disease management (2:9). A crucial component is maintaining a critical mass of expertise as the whole of *The Forests Handbook* shows!

SILVICULTURE AND MANAGEMENT

11 Understanding of forest ecology informs silvicultural actions, for example (a) to allow improvements in biodiversity, whether in plantation restructuring to deliver multiple benefits (2:15), maintaining or rebuilding natural populations (2:16) or more generally (1:4–5, 2:4), or (b) to optimise the protective role of forests (2:3).

12 Holistic and integrated approaches are central to SFM, in pest management (2:8), in soil husbandry (2:7) and particularly of organic matter (1:7), and more generally in planning and executing forest operations. For example, divorce of logging from regeneration will almost always lead to unsustainable actions (1:7, 2:5, 2:6–7, 2:15 and 2:18), hence 13.

13 Sustained yield rests on two assumptions (2:5) which separate forest management from forest exploitation: (a) commitment to successful and productive regeneration; and (b) harvesting balance must not exceed growth increment.

14 In the vast majority of forests sustainability is largely about understanding how forests function and working with that knowledge, but in the massively altered and unnatural urban environment, it is more to do with understanding how the built environment impacts on urban forest, and especially street (shade) trees, and to modify management accordingly (1:13, 2:12).

PEOPLE AS STAKEHOLDERS

15 Involvement of stakeholders (people's participation) is today's received wisdom (1:14, 2:10), but clearly in the sense of stewardship or custodians 'communities of protection' in Saxena's terminology (2:11) it greatly aids adoption of SFM in many instances, since it confers both benefits and responsibilities.

16 Non-timber forest products (NTFPs) often oil the mechanism to develop successful SFM in less developed countries (1:15, 2:11, 2:17).

17 Stakeholder involvement is also essential for achieving SFM policies in developed countries (1:14, 2:2 and 2:16).

These very general points and principles are what, for me as a committed Christian, I see as going to the heart of stewardship in terms of humanity's responsibility towards the world's resources. We are to be careful and caring custodians and not exploiters of them or of people who benefit from them. Today we find that concern for such resources, and for forests and forest peoples especially, is becoming enshrined in new institutions and processes of which 'certification' is increasingly the guarantor of SFM, albeit perhaps not yet in a wholly satisfactory form.

The world's forests are more than of utilitarian value. SFM acknowledges and focuses attention on this fact. And we are not alone. Uniquely in the creation account (Genesis 2:9, NIV translation) is added precisely this extra dimension when the Lord God declares of trees (and of nothing else He made): '. . . trees that *are pleasing to the eye* and good for food.' We would humbly agree.

Julian Evans

Index

Note: This is primarily a subject and topic index. In the text there are numerous references to tree species, both by vernacular and scientific names as appropriate, and also reference to most countries of the world. Not all such references to species and countries have been indexed. The reader is referred to the relevant topic, e.g. boreal forest, savannahs, tropical forest, etc., or regions, e.g. Africa, North America, Oceania, etc., to help access information sought. Where appropriate, cross-references are made, and readers will often find reference to tree genera under a particular topic.

Page numbers in **bold** indicate tables; those in *italics*, illustrations